Y0-BSD-719

ORGANIC TRACE ANALYSIS

ELLIS HORWOOD SERIES IN ANALYTICAL CHEMISTRY

Series Editors: Dr. R. A. CHALMERS and Dr. MARY MASSON
University of Aberdeen

APPLICATION OF ION SELECTIVE MEMBRANE ELECTRODES IN ORGANIC ANALYSIS
G. E. BAIULESCU and V. V. COŞOFREŢ
EDUCATION AND TEACHING IN ANALYTICAL CHEMISTRY
G. E. BAIULESCU, C. PATROESCU and R. A. CHALMERS
HANDBOOK OF PRACTICAL ORGANIC MICROANALYSIS
S. BANCE
FOUNDATIONS OF CHEMICAL ANALYSIS
O. BUDEVSKY
INORGANIC REACTION CHEMISTRY Volume 1: Systematic Chemical Separation
D. T. BURNS, A. TOWNSEND and A. G. CATCHPOLE
INORGANIC REACTION CHEMISTRY Reactions of the Elements and their Compounds Volume 2, Part A: Alkali Metals to Nitrogen and Volume 2, Part B: Osmium to Zirconium
D. T. BURNS, A. TOWNSHEND and A. H. CARTER
AUTOMATIC CHEMICAL ANALYSIS
J. K. FOREMAN and P. B. STOCKWELL
FUNDAMENTALS OF ELECTROCHEMICAL ANALYSIS
Z. GALUS
LABORATORY HANDBOOK OF PAPER AND THIN LAYER CHROMATOGRAPHY
J. GASPARIČ and J. CHURÁČEK
HANDBOOK OF ANALYTICAL CONTROL OF IRON AND STEEL PRODUCTION
T. S. HARRISON
ORGANIC ANALYSIS USING ATOMIC ABSORPTION SPECTROMETRY
SAAD S. M. HASSAN
HANDBOOK OF ORGANIC REAGENTS IN INORGANIC CHEMISTRY
Z. HOLZBECHER, L. DIVIŠ, M. KRÁL, L. ŠŮCHA and F. VLÁČIL
GENERAL HANDBOOK OF ON-LINE PROCESS ANALYSERS
DAVID HUSKINS
QUALITY MEASURING INSTRUMENTS IN ON-LINE PROCESS ANALYSIS
DAVID HUSKINS
ANALYTICAL APPLICATIONS OF COMPLEX EQUILIBRIA
J. INCZÉDY
PARTICLE SIZE ANALYSIS
Z. K. JELÍNEK
OPERATIONAL AMPLIFIERS IN CHEMICAL INSTRUMENTATION
R. KALVODA
METHODS OF PROTEIN ANALYSIS
Edited by I. KERESE
ATLAS OF METAL-LIGAND EQUILIBRIA IN AQUEOUS SOLUTION
J. KRAGTEN
GRADIENT LIQUID CHROMATOGRAPHY
C. LITEANU and S. GOCAN
STATISTICAL THEORY AND METHODOLOGY OF TRACE ANALYSIS
C. LITEANU and I. RÎCĂ
SPECTROPHOTOMETRIC DETERMINATION OF ELEMENTS
Z. MARCZENKO
LABORATORY HANDBOOK OF CHROMATOGRAPHIC AND ALLIED METHODS
O. MIKES
SEPARATION AND PRECONCENTRATION METHODS IN INORGANIC TRACE ANALYSIS
J. MINCZEWSKI, J. CHWASTOWSKA and R. DYBCZYŃSKI
HANDBOOK OF ANALYSIS OF ORGANIC SOLVENTS
V. SEDIVEČ and J. FLEK
ELECTROCHEMICAL STRIPPING ANALYSIS
F. VYDRA, K. ŠTULÍK and B. JULÁKOVÁ
ANALYSIS WITH ION-SELECTIVE ELECTRODES
J. VESELÝ, D. WEISS and K. ŠTULÍK

ORGANIC TRACE ANALYSIS

KLAUS BEYERMANN
Institute of Inorganic and Analytical Chemistry
Mainz University

Translation editor:
R. A. Chalmers
Department of Chemistry
University of Aberdeen

ELLIS HORWOOD LIMITED
Publishers · Chichester

Halsted Press: a division of
JOHN WILEY & SONS
New York · Chichester · Brisbane · Toronto

First published in 1984 by

ELLIS HORWOOD LIMITED
Market Cross House, Cooper Street, Chichester, West Sussex, PO19 1EB, England

The publisher's colophon is reproduced from James Gillison's drawing of the ancient Market Cross, Chichester.

Distributors:

Australia, New Zealand, South-east Asia:
Jacaranda-Wiley Ltd., Jacaranda Press,
JOHN WILEY & SONS INC.,
G.P.O. Box 859, Brisbane, Queensland 40001, Australia

Canada:
JOHN WILEY & SONS CANADA LIMITED
22 Worcester Road, Rexdale, Ontario, Canada.

Europe, Africa:
JOHN WILEY & SONS LIMITED
Baffins Lane, Chichester, West Sussex, England.

North and South America and the rest of the world:
Halsted Press: a division of
JOHN WILEY & SONS
605 Third Avenue, New York, N.Y. 10016, U.S.A.

Translated from the German: *Organische Spurenanalyse*
Published by Georg Thieme Verlag, Stuttgart, FDR

British Library Cataloguing in Publication Data
Beyermann, Klaus
Organic trace analysis. –
(Ellis Horwood series in analytical chemistry)
1. Chemistry, Organic 2. Trace elements – Analysis
I. Title II. Organische Spurenanalyse. *English*
547.3 QD271

Library of Congress Card No. 84-6717

ISBN 0-85312-638-0 (Ellis Horwood Limited)
ISBN 0-470-20077-4 (Halsted Press)

Typeset in Press Roman by Ellis Horwood Limited.
Printed in Great Britain by Unwin Brothers of Woking.

Table of Contents

Preface to the English Edition 11

Preface to the German Edition 13

Chapter 1 – Introduction 15

Chapter 2 – General aspects of organic trace analysis

2.1 Definitions 18
2.1.1 'Trace' 18
2.1.2 Organic compound 19
2.2 Evaluation of the importance and difficulties of organic trace analysis, from a statistical literature survey 20
2.2.1 Statistical survey 20
2.2.2 The importance of organic trace analysis 20
2.3 Difficulties in organic trace analysis 23
2.3.1 The number of compounds 23
2.3.2 The necessity for specific methods with sufficient sensitivity 23
2.3.3 Similarity of organic compounds 25
2.3.4 The instability of organic trace constituents 25
2.3.5 Search for organic traces with unknown chemical identity 25
2.3.6 Contamination control in organic trace analysis 28
2.4 The aim of organic trace analysis 28
2.4.1 The purpose of any analysis 28
2.4.2 Analytical function 29
2.4.3 Sensitivity *S* 31
2.4.4 Types of error 31
2.4.5 Remarks on nomenclature 33
2.4.6 Measures of imprecision (dispersion) 33
2.4.7 Standard deviation 35
2.4.8 Confidence interval, tolerance interval 38

2.4.9 Hypothesis testing 40
2.4.10 Limit of detection, limit of determination 42
2.4.11 Regression analysis, correlation 46
2.4.12 Analysis of variance 51
2.5 Methods for calibration and checking the results of organic trace analysis 51
2.5.1 Hierarchy of analytical methods 51
2.5.2 Reference and calibration materials 52
2.5.3 Methods of validating the results of organic trace analysis 56
2.5.4 Systematic control of the entire procedure 60
2.5.5 Use of different and independent methods 61
2.5.6 Combination of control methods 61
2.6 Interlaboratory control and collaborative studies 61
2.7 Good laboratory practice (GLP) 62
2.7.1 Introduction 62
2.7.2 Current government implementation of GLP 64
2.7.3 Definitions, terminology 64
2.7.4 Quality assurance programme 68
2.7.5 Rooms and facilities 68
2.7.6 Laboratory for organic trace work 69
2.7.7 Equipment and reagents 70
2.7.8 Standard operating procedures 71
2.7.9 Study plan 71
2.7.10 Conduct of the study and recording of data 76
2.7.11 Final reporting of study results 80
2.7.12 Archives, storage and retrieval of data, length of storage 80
2.7.13 Quality control 82
2.7.14 Safety regulations 84
2.7.15 Disposal of used reagents and solvents 85
References 85

Chapter 3 – Sampling in organic trace analysis
3.1 General aspects 91
3.2 Sampling of special matrices 92
3.2.1 Sampling of organic trace constituents in air 92
3.2.2 Sampling of water 98
3.2.3 Stripping of volatile components from water 102
3.2.4 Sampling in clinical chemistry, biochemistry and for pharmaceutical analysis 102
3.2.5 Techniques using Extrelut®, Sep-Pak® cartridges and similar materials for sampling/enrichment of traces 104
3.2.6 Head-space analysis 105
References 107

Chapter 4 – Treatment of samples before analysis
4.1 Storage 115
4.2 Influence of impurities during storage and sample preparation 118
4.3 Preparation of samples for analysis 123
4.4 Extraction from solids 124
4.5 Deproteination 126
References 126

Chapter 5 – Separation methods and concentration steps
5.1 General considerations 131
5.2 Evaporation of solvents and distillation techniques 136
5.2.1 Evaporation of solvents from solutions 136
5.2.2 Drying of the solution before evaporation 141
5.2.3 Steam distillation and co-distillation 141
5.2.4 Sublimation 143
5.3 Liquid-liquid distribution 143
5.3.1 General aspects 143
5.3.2 Equipment and techniques 145
5.3.3 Special treatment after partitioning 147
5.3.4 Mechanization of liquid-liquid distribution 147
5.3.5 Multistep liquid-liquid distribution processes 149
5.4 Separation by chromatography 149
5.4.1 General aspects 149
5.4.2 Types of liquid-solid chromatography 149
5.4.3 Theoretical background of chromatographic separations 150
5.4.4. Principles of column operation 151
5.4.5 Adsorbent materials used as stationary phases in liquid chromatography 152
5.4.6 Filling of columns 152
5.4.7 Application of the solute mixture 153
5.4.8 Detection and collection of column effluents 153
5.4.9 Combination of columns, column switching 154
5.4.10 Problems in liquid chromatography 154
5.4.11 Gel chromatography 155
5.4.12 High-pressure liquid chromatography (hplc) 155
5.4.13 Thin-layer chromatography (tlc) 165
5.4.14 Ion-exchange chromatography 169
5.4.15 Scheme for selection of a liquid chromatographic method 169
5.5 Gas chromatography 170
5.5.1 General aspects 170
5.5.2 Types of columns 170
5.5.3 Stationary phases in gas chromatography 172
5.5.4 Special forms of column operation 173

5.5.5 Bleeding of septa and columns . . . 174
5.5.6 Injection techniques in gas chromatography . . . 174
5.5.7 On-column derivative-formation techniques . . . 174
5.5.8 Pyrolysis . . . 176
5.5.9 Fraction collection . . . 176
5.5.10 Carbon skeleton technique . . . 177
5.6 Zone melting . . . 180
5.7 Precipitation and co-precipitation . . . 181
5.8 Separation with membranes . . . 182
5.9 Diasolysis . . . 183
5.10 'Adsubble' methods and 'solvent sublation' techniques . . . 183
5.11 Electrophoresis . . . 184
5.12 Plasma electrophoresis (gaseous electrophoresis) . . . 185
5.13 Checking the efficiency of separation steps by radiochemical methods . . . 185
5.14 Treatment of sample materials to alter their separation characteristics . . . 186
5.14.1 General considerations . . . 186
5.14.2 Continuous and discontinuous techniques for forming derivatives . . . 187
5.14.3 Examples of the application of derivative formation . . . 189
5.15 Sample transfer . . . 196
References . . . 198

Chapter 6 – Methods for the determination or detection of compounds
6.1 General aspects . . . 221
6.2 General remarks on spectroscopic methods . . . 222
6.3 Spectrometry with ultraviolet and visible light . . . 223
6.4 Infrared absorption . . . 224
6.4.1 General aspects . . . 224
6.4.2 Developments and improvements in ir instrumentation . . . 224
6.4.3 Sample preparation . . . 225
6.4.4 Micro specular reflectance . . . 226
6.4.5 Influence of temperature in ir spectrometry of traces . . . 227
6.4.6 Matrix isolation spectroscopy . . . 227
6.4.7 Coupling of ir with gc . . . 230
6.4.8 Combination of ir with liquid chromatography . . . 232
6.4.9 Application of ir methods in organic trace analysis . . . 232
6.4.10 Advantages of ir methods . . . 233
6.5 Raman spectroscopy . . . 234
6.6 Fluorescence and phosphorescence . . . 235
6.7 Optoacoustic spectroscopy . . . 236

6.8 Mass spectrometry . . . 238
6.8.1 General aspects . . . 238
6.8.2 Principles of mass spectrometry . . . 239
6.8.3 Further developments in ms instrumental design . . . 242
6.8.4 Combination of ms with other methods . . . 245
6.8.5 Combination of mass spectrometry with gas chromatography . . . 245
6.8.6 Coupling of liquid chromatography with mass spectrometry . . . 247
6.8.7 Application of mass spectrometry to the analysis of waters, effluents etc. . . . 247
6.8.8 Application of mass spectrometry to determination of organic traces in air, tobacco smoke and particulate matter from air . . . 248
6.8.9 Application of mass spectrometry in life sciences . . . 248
6.8.10 Applications of mass spectrometry in forensic chemistry . . . 249
6.8.11 Application of mass spectrometry in pharmaceutical trace analysis . . . 251
6.8.12 Application of mass spectrometry to the analysis of samples of ecological interest to man and of food samples . . . 253
6.8.13 Some comments on mass spectrometry . . . 256
6.9 Detectors used in combination with chromatography . . . 257
6.9.1 Detectors for liquid chromatography . . . 257
6.9.2 Detector systems for gas chromatography . . . 258
6.10 Electrochemical methods . . . 270
6.10.1 Potentiometric methods . . . 270
6.10.2 Coulometric methods . . . 271
6.10.3 Voltammetry . . . 272
6.11 Chemical detection or determination . . . 277
6.12 Enzymatic reactions . . . 278
6.13 Immunological reactions . . . 280
6.14 Protein binding . . . 286
6.15 Biological methods . . . 286
References . . . 290

Chapter 7 – Special topics in organic trace analysis
7.1 Topics related to groups of substances . . . 317
7.2 Determination of trace compounds without separation . . . 318
7.3 Methods to improve the limit of detection of organic trace analytical methods . . . 320
7.4 Some ways to increase the specificity of methods . . . 327
7.5 Screening methods in organic trace analysis . . . 330
7.6 Local analysis and analysis of surfaces . . . 332
7.7 Telemetry in trace analysis . . . 334

7.8 Application of trace analysis to recognition of diseases by analysis profiles of constituents of body fluids 337
7.9 The future 338
References 338

Index 345

Preface to the English Edition

Only two years after the appearance of the first German edition of *Organische Spurenanalyse* the English version will be published, which amply indicates the increasing significance of organic trace analysis. Furthermore, its growing importance can be demonstrated by the fact that the literature of the additional years examined (1980-1982) contributes about 40% of all the references in the English version.

The author is greatly indebted to Dr. R. A. Chalmers of the University of Aberdeen who turned the teutonic-English manuscript into the present form. His extremely careful and accurate work of editing improved this book greatly. His help is highly appreciated.

Mainz, March 1984 Klaus Beyermann

Preface to the German Edition

Organic trace analysis is steadily growing in importance. This scientific discipline is in an expansive phase and is in the process of becoming to some extent an independent art. The absence of an introductory comprehensive review of organic trace analysis in the form of a monograph is all the more regrettable. The present publication, which selectively incorporates the literature on this subject up to the beginning of 1980, represents an attempt to close this gap.

Dr. Gorbach, Hoechst AG, participated in the writing of Chapter 1.* His contribution is an essential one from the viewpoint of the practical trace analyst. I should like to thank him for this valuable contribution and also for going through and discussing the rest of the book.

Mainz, February 1982 Klaus Beyermann

*Chapter 2 in the English edition.

1

Introduction

Organic trace analysis is a rapidly growing field of analytical chemistry with many practical applications and implications, and many interdisciplinary connections. New developments in pharmaceutical products, discoveries of new toxic substances, the appearance of new drugs amongst addicts, all necessitate the creation of analytical methods of sufficient sensitivity and specificity. Control of environmental samples results in co-operation of the trace analyst with colleagues from biology, microbiology, economics and many other fields.

Legislation has many effects on organic trace analysis. After a regulation has been made, the analytical investigations necessary to enforce it have to be made. This may require much organizational ability from the analyst. Ideally, the analytical procedure should be developed before the government action is taken. This needs some 'political sensitivity' of the analyst as to which methods are needed or are to be improved. Thus, legislation forces analytical chemists to react to legislative acts concerning chemistry.

On the other hand, legislation must react to modern developments in analytical organic trace chemistry. The elaboration of better tools or methods, e.g. the electron-capture detector [1], has greatly stimulated current discussions on ecology. Investigation of the worldwide distribution of halogenated pesticides, identification of their origin, and their replacement by new and less persistent pesticides, would never have taken place without the refined method of gas chromatography with the electron-capture detector.

Many more such multifaceted interactions between organic trace analysis and other areas of science, politics, sociology, economy etc. can be traced. Ecological problems can become driving forces for the development and improvement of trace analytical procedures.

In spite of its great significance, organic trace analysis is still in the early stage of development. So far, it has not received the attention that has been paid to inorganic trace analysis. Even the word 'trace' itself seems to be largely reserved for inorganic analysis. It is used about four times as often in titles of publications on inorganic trace constituents as in those for organic traces, although the number of publications in the two fields is about the same.

Another deficiency becomes apparent from a survey of the literature of this area. Only a few reviews have been published and these only lately [2-4]. Selected topics have recently been described in textbooks [5, 6] covering the important area of chromatography in organic trace analysis. An international symposium has been held, which was dedicated entirely to 'Trace Organic Analysis, a New Frontier in Analytical Chemistry' (10-13 April, 1978, Gaithersburg, Md, USA), and its proceedings have been published [7].

It is the aim of this book to demonstrate some of the interest that organic trace analysis finds and some of its uses. At the same time, the great difficulties of organic trace analysis will be dealt with. There are very interesting proposals made in the literature, for practical laboratory work to solve some of the problems in this field. It is the author's intention to sketch these developments and their principles. This will be done by reviewing the literature, mostly the more recent work. Since about 1000 publications on organic trace analysis now appear each year, it is obviously necessary to select from these the more important and descriptive examples. A complete catalogue of the literature would defeat the object of this book, since the trivial would tend to be undistinguishable from the important. The intention is to give rapid primary access to the literature for the majority of problems by quoting a few selected examples.

Besides, within the limits of an introductory book it is impossible to give full details of the procedures. In order to give a broad survey, the data are mostly collected into tabular form. The sorting of the published information into these tables was sometimes a problem, however, as it was intended to quote a publication once only. Thus, a paper might be referred to in a table describing a certain detection method, and information about the separation involved get lost. The arrangement is mostly substance-orientated, though, and the reader will find all the information, distributed over several tables and chapters, by looking up the index entry for the substance of interest.

The number of entries in the tables gives a fair indication of the current interest in themes, substances, methods and problems.

The results of organic trace analysis are of interest to many persons who are not trained analysts, as already pointed out. Science administrators, biologists, politicians, forensic experts etc. might possibly be interested in some of the methods, problems, limitations and high-lights of organic trace analysis. Thus, it is hoped that the rather simple style of the treatment will be appreciated by such readers as well as by the professional analyst.

REFERENCES

[1] J. E. Lovelock and S. R. Lipsky, *J. Am. Chem. Soc.*, **82**, 431 (1960).
[2] K. Beyermann, *Angew. Chem., Intern. Ed.*, **13**, 224 (1974).
[3] K. Beyermann, *Pure Appl. Chem.*, **50**, 87 (1978).
[4] H. S. Hertz, W. E. May, S. A. Wise and N. S. Chesler, *Anal. Chem.*, **50**, 429A (1978).

[5] J. F. Lawrence (ed.), *Trace Analysis,* Vol. 1 (*HPLC*), Academic Press, New York (1981).
[6] J. F. Lawrence, *Organic Trace Analysis by Liquid Chromatography,* Academic Press, New York (1981).
[7] H. S. Hertz and S. N. Chesler (eds.), *Trace Organic Analysis: A New Frontier in Analytical Chemistry,* NBS, Washington (1979).

2

General aspects of organic trace analysis

2.1 DEFINITIONS

2.1.1 'Trace'

A trace is a constituent which is found in a minor concentration in another material, called the 'matrix'. There is no general agreement about the concentration level at which use of the term 'trace constituent' becomes justified. About 30 years ago a constituent at a concentration of 0.1% was considered to be present in trace amounts. With increasing sensitivity of analytical methods the limit of detectability has shifted towards much lower concentrations. In addition, new analytical problems have created a demand for the analyst to deal with considerably lower concentrations than 0.1%. Today (and in this book), ppm concentrations (1 part of trace in 1 million parts of matrix = 1 part per million = 10^{-4}%) are generally regarded as trace concentrations, but determinations are very often needed in the 10^{-7}% region (1 ng per g of matrix) and even lower. The latter concentration level is often called the ppb (parts per billion) region, but the use of the word 'billion' is inherently ambiguous, since it means 10^9 in America, but 10^{12} in Europe. The use of ppt for parts per trillion (see Table 2.1) is even worse, since not only is the trillion differently defined (10^{18} in Europe), but ppt was formerly frequently used as an abbreviation for parts per thousand (in discussion of relative error). There is now a general move to unify

Table 2.1 – Abbreviations and units often used in organic trace analysis.

ppm part per million	$1:10^6$	10^{-4}%	μg/g	mg/kg
ppb part per billion	$1:10^9$	10^{-7}%	ng/g	μg/kg
ppt part per trillion	$1:10^{12}$	10^{-10}%	pg/g	ng/kg
ppqud part per quadrillion	$1:10^{15}$	10^{-13}%	fg/g	pg/kg

μg microgram (10^{-6}g); ng nanogram (10^{-9}g); pg picogram (10^{-12}g); fg femtogram (10^{-15}g); ag attogram (10^{-18}g)

the definition of billion, trillion etc. in the American sense, but to give better agreement with SI nomenclature and avoid ambiguity completely, the weight/weight, volume/volume or weight/volume scales should always be used; use of ppm etc. is often made further ambiguous by failure to indicate the units used, especially in the case of gas mixtures.

It has been suggested [1] that trace concentrations should be expressed in one of the three forms:

a pp 10^x or
a pT x (semi-logarithmic) or
pTr (completely logarithmic)

For example, the concentration of a trace Tr at the level 0.005 ppm (= 5 ppb) could be written as 5 pp10^9 or 5 pT 9 or pTr 8.10. Many scientists tend to use the first form [2]. Unfortunately, many data in the literature are not clear enough to permit such a conversion. Sometimes, for practical reasons, concentrations are given in rather unconventional units, e.g. aflatoxin contents are found quoted in μg/egg and concentrations of polycyclic aromatic hydrocarbons (PAHs) are sometimes quoted as ng/cigarette.

Trace methods must not be confused with micro methods. In the latter, very small weights of matter are handled, e.g. mg-amounts or less in micro methods; μg-amounts are analysed in ultramicro methods, but usually the concentration of the species to be determined is fairly high. Micro methods are extremely useful in trace analysis, of course, but obviously require a preliminary isolation and concentration (preconcentration) of the trace species from the matrix. Micro methods also have the advantage of decreasing the consumption of reagents, problems of reagent waste disposal, and health hazards.

2.1.2 Organic compound

This book considers determination of organic compounds at the trace concentration level. The matrix may be organic (plants, meat, tissue, milk, foodstuffs, faeces, blood, industrial organic samples, polymers etc.) or inorganic (e.g. water, air, minerals) or of mixed composition (e.g. aqueous solutions of organic compounds, such as wine, beer, urine, plasma).

The trace constituent may be purely organic or of mixed organic/inorganic composition. Examples of the latter are (a) metal organic compounds with covalent organic/inorganic bonds (e.g. alkylated mercury compounds); (b) inorganic constituents chelated or otherwise complexed with organic ligands; (c) inorganic compounds more loosely bound to organic macromolecules, such as proteins, DNA.

This book will not deal with most of these mixed species, nor will the analytical chemistry of traces of polymer substances be considered, as these require very special methods of separation and detection. Such substances, besides the synthetic polymers from industry, are the bipolymers, such as DNA, RNA, proteins etc. Although they play a most important role in biochemistry they will not be dealt with here, partly on account of the specific biochemical methods which are used for their determination at trace levels. Analogously, their precursors – amino-acids, nucleosides etc, – will be mentioned only briefly.

2.2 EVALUATION OF THE IMPORTANCE AND DIFFICULTIES OF ORGANIC TRACE ANALYSIS, FROM A STATISTICAL LITERATURE SURVEY

2.2.1 Statistical survey

During recent years *Analytical Abstracts* has summarized about 10000 analytical papers per year. Table 2.2 indicates that about 60% of these deal with organic substances, while only about 25% are concerned with inorganic materials.

Table 2.2 – Distribution of analytical publications in different areas (calculated from *Analytical Abstracts*, 1978 and 1979).

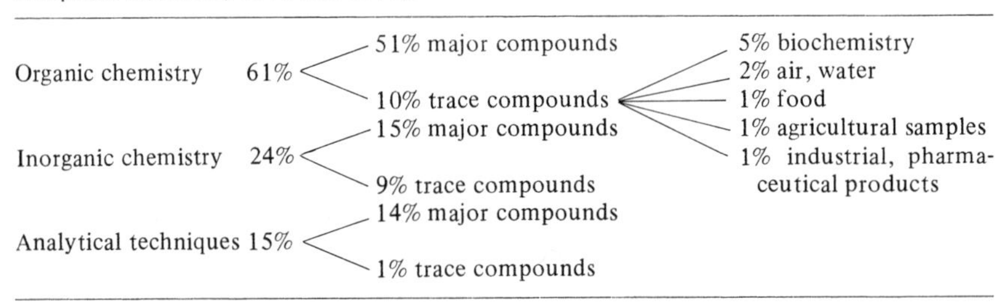

The proportion of papers on trace analysis is smaller in organic chemistry than in inorganic chemistry, however, only about a sixth of the papers on organic analysis being concerned with traces, in comparison with a third for inorganic analysis.

2.2.2 The importance of organic trace analysis

Table 2.3 lists the fields of application of organic and inorganic trace analysis in order to compare their significance (the data are calculated from *Analytical Abstracts* for 1978 and 1979).

Table 2.3 – Different fields of application of organic and inorganic trace analysis.

Field of application	Percentage of papers, dealing with organic traces	inorganic traces
biochemistry	5	1
analysis of air, water	2	2
analysis of food	1	1
samples from chemical or pharmaceutical industry	1	4
agricultural samples	1	1
total	10	9

Organic trace analysis plays an important role in the life sciences and ecology. About 5% of analytical publications deal with the determination of organic traces in food samples, specimens from agriculture, air or water sources. The trace analysis of organic compounds directly relevant to mankind obviously contributes to the development of those fields that are currently of public interest, such as protection of the environment or purity of food, and also to biochemistry, clinical chemistry and medicine. In this connection Hertz *et al.* [3] may be quoted: "Until recently the major emphasis in trace analysis has been in the determination of inorganic substances. However, we are now coming to realize that many of our most pressing problems require competence in organic trace analysis. These analyses are needed to protect our health and our environment and to ensure the nutritional value of our food. Recent US Federal Legislation recognizing these needs included the Federal Water Pollution Control Act (1972), Federal Environmental Pesticide Control Act (1972), Safe Drinking Water Act (1974), Toxic Substances Control Act (1976) and others. The enforcement of such legislation is ultimately based on the ability of analysts to accurately identify and quantify trace levels of organic substances in diverse matrices."

Analogous legislative acts have since been passed in many other countries (Table 2.4). Public agencies try to control the realization of these regulations. This requires man-power, know-how and a lot of co-operation in many respects. It is one of the tasks of scientists to provide the information needed to introduce objectivity into any discussion about environmental toxocology, its economic implications etc.

Table 2.4 – Examples of legislative acts, regulations or regulatory agency proposals, which greatly rely on organic trace analysis or which influence analytical developments.

EPA guidelines for registration of pesticides	(USA)
Pflanzenschutzgesetz	(FRG)
(Umwelt-)Chemikaliengesetz	(FRG)
WHO – standards in PAHs in water	(WHO)
Deutsche Trinkwasserverordnung	(FRG)
European Economic Community directive on pesticide content of fruits and vegetables	(EEC)
Aflatoxin-Verordnung	(FRG)
Toxic substances control act (TOSCA)	(USA)
Good clinical practices	(USA)
Bioequivalence/bioavailability	(USA)

For further regulations – mainly from the USA – see Freeburg [4]

To typify such regulations it may be mentioned that the 'Aflatoxin-Verordnung' (30.11.1976, BGBl.1, p. 3313) sets a permissible total concentration of 10 μg/kg for aflatoxins B_1, B_2, G_1 and G_2, the limit for aflatoxin B_1 being 5 μg/kg.

The fields of interest and application of organic trace analysis can be divided into two major categories, according to whether the traces are produced by man or are naturally occurring. They can be further classified according to the origin of the samples (environmental samples, industrial samples, samples of pharmacological or forensic interest, samples of biological or medical interest).

The category of man-made traces often has legal aspects. Practically all environmental applications are included in it, as they deal with alteration of the environment as a consequence of man's activity. Such changes are becoming more and more controlled by human society, through legislation, though sometimes it seems that the control is dictated by subjective feelings rather than objective assessment.

An area of growing importance is the control of organic traces in technical products. Quality control of many products of chemical industry and almost all substances from the pharmaceutical industry is becoming more and more necessary, again partly as a consequence of legal requirements. Often enough, industry is forced to introduce its own control of intermediates and final products. Colourless products must in fact be free from any colour. Solvents and many other reagents must meet certain purity standards. For use in the pharmaceutical industry many compounds must be highly purified and tested for impurities. For example, 2,4,5-T – a frequently used herbicide – is required to contain less than 0.2 ppm of TCDD [5].

The interdependence of organic trace analysis and legal regulations is apparent in forensic and crime-investigation applications. As the application of a pharmacologically active substance also has its legal side, again the interaction is obvious.

There is no general legislative influence, however, for the second category of naturally occurring traces, such as those which give biochemical, biological or psychological effects, (e.g. hormones, flavours, antibiotics, drugs from plants, poisons from flowers or animals, body constituents with diagnostic value for clinical chemistry), although food must sometimes be analysed for such traces (which would come into this category, if they are not man-made). Aflatoxins, mycotoxins and other biogenic contaminants are typical examples.

A greatly expanding field of organic trace analysis is research on drug metabolism. Any new drug developed must be described – according to national and international conventions and regulations – with respect to uptake, excretion and biochemical or metabolic alteration in the body. To obtain these data many analyses are done in which concentrations as low as a few ng per ml of plasma or urine must be determined. Further, the kinetics must be examined at such levels. Consequently, the most reliable, sensitive, rapid, simple and cheap methods are to be applied. Much analytical research is therefore devoted to comparison of methods in terms of these chemical and economic parameters.

2.3 DIFFICULTIES IN ORGANIC TRACE ANALYSIS

2.3.1 The number of compounds

About 10^5 new organic compounds are synthesized each year [6]. This leads increasingly towards new problems of recognition, detection, identification and quantification. Even if the analyst sets a limit on the number of possible compounds to be taken into consideration in a particular analysis, there will still be many groups of substances possibly present.

Consequently, there is practically no scheme available that is comparable to the separation schemes used in inorganic trace analysis, which take into consideration a limited number of inorganic ions (or compounds). Since the number of organic compounds is several orders of magnitude greater than that of inorganic compounds such a systematic scheme which would take all organic substances into account cannot be expected. At best, only group separations can be hoped for.

The vast number of organic substances has the consequence that methods for the determination of organic traces are usually developed only *after* some alarming or calamitous event, e.g. the discovery of the carcinogenic properties of *N*-nitrosamines in cigarettes [7], the toxic properties of aflatoxins [8] or the dramatic contamination of a large area after a factory accident (Seveso) [9]. One or two years after the publication of such a triggering event articles on the necessary methodology begin to appear in analytical journals or there is a great increase in interest with a resulting increase in numbers of publications (see Fig. 2.1).

Similarly, analytical chemists had to react to the challenge of the street drug problem, such as the increasing availability of diamorphine (heroin). A similar picture could be drawn for the cannabis alkaloids [10] which became of interest in the 1960s, for LSD etc. Many other examples serve to show that the organic trace analyst has to react to a problem which is suddenly given to him and which is unpredictable. With the vast number of compounds known, no advance preparation for the solution of the problem can be made. This is in contrast to inorganic analysis where the number of elements is limited, the tradition is older, the analytical possibilities are better defined, and so more procedures are developed and kept in stock in the literature. The necessity for the organic trace analyst to be more flexible in approach adds to the difficulties of the field, but on the other hand makes it also interesting and challenging.

2.3.2 The necessity for specific methods with sufficient sensitivity

Another problem in organic trace analysis is the lack of methods that are both specific for a given compound and sufficiently sensitive. Recently developed methods and equipment have begun to close this gap. Often, such new techniques are very expensive, however, and must be used by analysts with a high degree of special training.

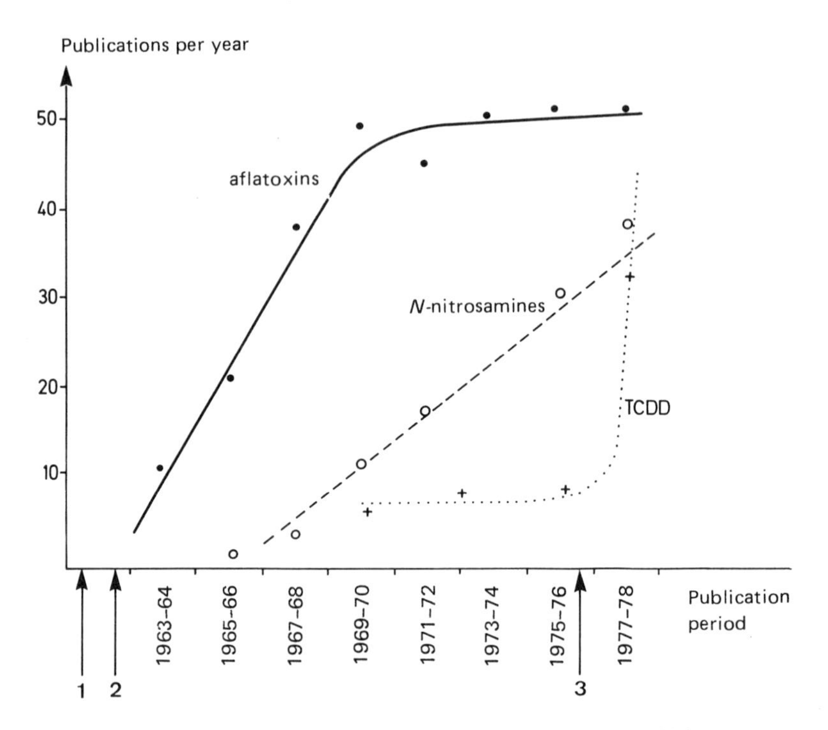

Fig. 2.1 – Number of organic trace analysis publications per year which appear after some triggering event. The curves show that after some event the interest of the organic trace analyst is aroused. The intensity of the response to the problem – here expressed in terms of publications per year – depends on the significance of the problem and the availability of the methods. A problem of very acute interest (e.g. Seveso, No. 3 in the figure) results in rapid response. The answer to the *N*-nitrosamine problem (1962, No. 2) was rather slow. The reaction to the first publication on aflatoxin toxicity (1961, No. 1) was more prompt and drastic. After a certain period a saturation level is observed, which indicates that enough publications have been offered for solution of the problem [for the analysis of tetrachlorodibenzodioxins (TCDD) this level has not been reached yet!], international congresses have been held, collaborative studies have been started and standard methods have been agreed upon; except for minor details the problem is considered to be of less interest than that of new tasks.

To solve the very complex problems resulting from modern technical innovations it is often necessary to apply many methods from the arsenal of instrumental analysis. One example [11] may serve to illustrate this.

Diesel-engined vehicles are becoming increasingly popular, but this also increases the possibility of pollution from this source. In the USA, Congress has directed the EPA (Environmental Protection Agency) to study diesel particulate emissions and their health effects (The Clean Air Act, August 1977). The findings of these studies formed the basis of regulations concerning diesel particulate emissions for the model year 1982. Consequently, investigators have intensified their efforts to characterize diesel exhaust emissions [10–14]. "The complexity of the samples and the need for as complete characterization as possible has led to the use of highly sophisticated analytical systems, such as gc–ms (gas chromatography-mass spectrometry), gc–fid (gas chromatography with flame ionization detection), hplc (high performance liquid chromatography), gc–ft–ir (gas chromatography with Fourier transform infrared analysis). Bioassay has also been used for highly mutagenic fractions. The use of complementary analytical techniques, e.g. investigation by gc–ms, gc–fid and gc–ft–ir is the only means of reducing uncertainties in the identification. The gc–ms technique can easily give molecular weights and some information about structure, but is often of less use in the identification of functional groups, which can be done rather easily, on the other hand, by gc–ft–ir. The latter method, however, cannot distinguish between members of homologous series" [15]. The example clearly indicates the need to apply the best and most informative methods, although they might be expensive. The cost has to be related to the importance of the object of the investigation, which is often a question of pollution control, with great economic implications.

Although instrumental methods of organic analysis have made great progress during recent years it is still practically impossible to investigate matrices for organic traces by direct methods which require very little or no preseparation. In about 96% of all cases reported in the literature at least one separation step is necessary.

2.3.3 Similarity of organic compounds

Very often compounds of great similarity are to be separated and determined in organic trace analysis. The problem of isomers, for instance, is almost entirely unknown to the inorganic trace analyst, but is quite common in organic trace analysis (see Table 2.5). Different isomers very often exhibit different pharmacological or toxicological activity. For a correct evaluation of these properties the analyst not only has to determine the concentration of the compound(s) in question but also the ratio of isomers.

2.3.4 The instability of organic trace constituents

Many organic compounds exhibit instability towards hydrolysis, oxidation and microbiological attack. Some compounds are light-sensitive. The consequent losses in organic trace analysis are difficult to control. At least, they are always to be kept in mind as possible sources of errors.

2.3.5 Search for organic traces with unknown chemical identity

Occasionally, the chemical structure of a compound is not known, although its effect is known. This substance (or group of substances) has to be isolated and determined by procedures which are developed during the analysis itself.

Table 2.5 – A few examples of separation or determination of isomers at trace level

Isomer	Matrix	Conc. or amount	Method	Reference
22 tetrachlorodibenzo-*p*-dioxin isomers	soil, fish		gc–ms	16
	fish	25 ng/kg	gc–ms	17
	environ. samples	sub-ng	hplc–gc–ms	18
10 hexachlorodibenzo-*p*-dioxin isomers			gc–ms	19
3 tetrachlorodibenzofuran isomers	environ. samples		gc–ms	20
$\alpha, \beta, \gamma, \delta, \epsilon$-isomers of hexachlorocyclohexane	vegetables	2–4 ppb	gc	21
	water	200 pg/μl	gc	22
	soil	1–10 ppb	gc	23
nitrophenol isomers	water	0.2 ng/μl	gc–ms	24
ethylbenzene, xylene isomers	air	ppb	gc–ms	25
polyheterocyclic compounds with different positions for the hetero-atom	coal products	500 ng	Fourier–ir	26
cis-trans isomers of abscisic acid		1 ng	hplc	27
cis-trans isomers of retinoic acid	rats	500 ng	hplc	28
Pseudoephedrine enantionmers	plasma	60 ng/ml	ria*	29
pyrethroid insecticide enantiomers		0.1 μg	hplc	30
propranolol enantiomers	plasma	36 ng/ml	hplc	31, 32
carprofen enantiomers	blood, urine	0.6 μg/ml	hplc	33
sulphoxides, amines, amino-acids, hydroxy-acids, thiols, enantiomers			hplc	34
permethrin enantiomers	soil	> 0.1 ng	gc	35
methadone enantiomers		1 ng	ria	36
carboxylic acids		0.1 ng	hplc	37

*ria = Radioimmunoassay.

As the number of organic substances is practically unlimited, the organic trace analyst must take great care when he decides which substance to exclude from a search of the compound(s) of interest.

The possibilities in such problems can be summarized by a scheme:

The identity of an organic trace compound is

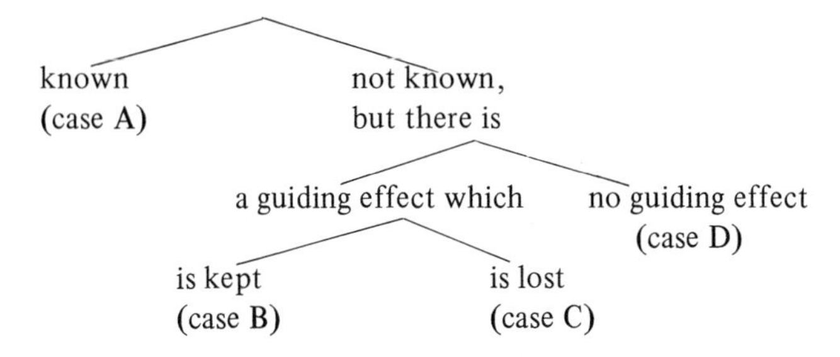

Case A. The identity of the trace compound is known. Examples are assays for additives to food, such as antioxidants, vitamins, flavours, antibiotics or for pharmacologically active compounds present in rather low concentrations in drug preparations, or for pesticide residues in agriculture. There are many examples of organic trace analysis which belong to this important category.

Case B. The identity of the trace compound is not known, but a specific effect can be monitored and is used to guide the analyst. The effect can be chemical, biological or physical in nature. An example is the first identification of physiologically active substances – drugs, poisons, hallucinogens etc. – in biological material. The isolation of hormones or other compounds with biochemical effects belongs to this category. In ecology, the isolation of unidentified odorous substances in water, air, food etc. is another example. Generally, after the isolation the structure or chemical identity of the trace substance is determined. Case B is less often found in the literature than examples of category A. Problems from category B are more tedious to solve than those from group A.

Case C. A compound of unknown identity gives a defined effect which is lost during the analysis. For example, polymeric plastics sometimes contain spots of opaque material, which reduces the transparency (and consequently the economic value of the material). During the analysis the opaque material is dissolved and analysed. In such an analysis much knowledge about the sample origin, sample fate and substances giving similar behaviour is necessary. An elective search for some suspected substance is to be made by the analyst. Obviously, he will solve these rather rare but puzzling cases only with a lot of know-how and general chemical knowledge and intuition.

Case D. This is the most complicated and is illustrated by the determination of a trace substance which has undergone chemical transformation in the matrix, e.g. the biotransformation of pesticides in soil, plants and animals. The metabolism of drugs in animals or man is another example. The chemical identity of the resulting trace(s) is not known, although often some speculation on their nature is possible. The nature of substances giving analogous behaviour may be known and the pattern of metabolism is limited and very often repeated by nature [38]. It is rather difficult, however, to do such investigations without the aid of isotopically labelled tracers (usually radioactively labelled). Examples of category D are more frequent than those of categories B and C.

This general description indicates the strategy of the organic trace analyst. He always uses such information about the trace compound as he can in order to reduce the number of possibilities to a practicable level for investigation. This requires experience and responsibility and can by no means be done automatically. In inorganic trace analysis the application of a systematic analytical scheme is useful to a much greater extent, although it can always be shortened by a clever analyst on the basis of his knowledge of the sample and of inorganic chemistry in general. In organic trace analysis almost any given problem will require its own individual solution, which again is in great contrast to inorganic trace analysis.

2.3.6 Contamination control in organic trace analysis

So far, because of the relatively insensitive methods used, organic trace analysis has not been as much handicapped as inorganic trace anlysis by contamination problems. With the greater sensitivity of modern methods, however, the organic trace analyst now has to take background contamination into consideration. Any trace analysis at the pg/g-level necessitates such control [39]. This is especially important since several compounds of interest to the trace analyst, e.g. PAHs, pesticides, polychlorinated biphenyls (PCBs), are distributed worldwide in the air. Trace analysis for these compounds, which are of great ecological significance, demands good contamination control during the analysis. The reported reduction of the PCB content in the North Atlantic is probably just a consequence of improved analytical techniques and less a result of an increase in the ecological conscience of mankind [39].

2.4 THE AIM OF ORGANIC TRACE ANALYSIS*

2.4.1 The purpose of any analysis

As a rule the analytical result is required as a basis for decisions. Common questions which lead to the necessity of decision-making are: "Does the concentration found exceed a maximum tolerated level?" or: "Does the stated content certify a guaranteed level?" or: "Can the observed signal be clearly distinguished from noise", etc.

*This section has been written by Dr. S. Gorbach, Farbwerke Hoechst A.G, to whom the author wishes to express his thanks for this valuable contribution, which adds the expperience of an organic trace analyst working in industry.

It is obvious that stringent safety margins should be set in the analytical control of a pharmaceutical preparation when small variations in the content of the active ingredient would be extremely hazardous for the patient; i.e. the risk of accidental release of a faulty product should be minimized. The same is true for the emission of a hazardous substance from a production plant, etc.

The risk of a wrong decision can only be evaluated in connection with the magnitude of the damage or unwanted event which will or may be caused [40]. It is necessary to take into consideration the profits and losses which may arise from erroneous decisions, including the cost of the test procedures (which is dependent on the nature and magnitude of the necessary assay).

Numerous decisions are made in our daily life without a data-base which would allow the asistance of mathematical statistics in the decision-making process. Normally, we judge our risk by intuition on the basis of experience.

Our decisions are influenced by two strategies inherent in man. The first is the strategy of the optimistic explorer and the second the strategy of the more pessimistic critic. The first would be prepared to take a higher risk of erroneously rejecting the true hypothesis (H_0, the null hypothesis), in favour of a more promising but wrong alternative hypothesis. This erroneous rejection of the true null hypothesis H_0 is called an error of the first kind (error I). Such an over-optimistic explorer would try to minimize the risk of erroneously rejecting a true alternative and accepting without justification an incorrect null hypothesis. The erroneous acceptance of an incorrect null hypothesis is known as an error of the second kind (error II). The critic would react in the opposite direction. He would try to minimize the risk of error I and would accept a higher risk for error II.

In mathematical statistics the null hypothesis H_0 is the hypothesis that no significant difference has arisen and is the negation of an event or a difference a scientist believes to have been detected. The occurrence of a significant event or difference is called the research hypothesis and is denoted as H_1.

The rationalization of this aspect requires an objective technique (test) for accepting or rejecting a hypothesis. Details of such techniques are described in the literature [41,42]. Here we shall give only a few facts which offer some help for decision-making processes in the daily routine of an organic trace analyst.

The mechanism of a statistical test is first to define a null hypothesis, and to reject it as soon as a result occurs which would be highly improbable in the case of a valid null hypothesis.

Before the procedures of hypothesis testing are discussed some properties of the analytical result will be given.

2.4.2 Analytical function

In analytical chemistry the measured signal y is related to the content c of analyte in the sample by the analytical function f

$$c = f(y) \tag{2.1}$$

This function must be definite and unequivocal. It is obtained from the calibration function by inversion

$$y = g(c) \tag{2.2}$$

g being derived from measurement of calibration standards.

Generally, the analytical function f applies only over certain intervals. For example, with a function with a maximum at c_m, the signal will at first increase with increasing concentration, and for concentrations greater than c_m the signal will decrease. Such a function with a maximum is obviously not unequivocal. (This case is found with fluorescence measurements.) There is one definite relation between signal y and concentration c at $c < c_m$ and another at $c > c_m$. The analyst has to decide which interval of concentration he is dealing with. (This can be done by addition of the analyte and determination of the resulting signal.)

Given an interval with an unequivocal, reasonable calibration function, the following more severe conditions hold for the existence of the analytical function f:

(1) f must be describable by a function which can be differentiated, and
(2) the derivative must not be zero.

For practical reasons we try to find a working interval in which the calibration function exhibits linear dependence of the signal y on concentration c:

$$y = bc + d \tag{2.3}$$

If this is impossible, we try to convert the function into a linear function. Often the function is exponential, e.g.

$$y = kc^n \tag{2.4}$$

and can be linearized in logarithmic form:

$$\log y = \log k + n \log c \ , \tag{2.5}$$

with $\log k$ as an intercept and n as the slope.

Here y is the dependent variable, which is mainly influenced by random errors. On the other hand, the values used for the independent variable may also show some fluctuation. Various steps involved in the preparation of the calibration standards (e.g. weighing) will not be free from random errors, and this will result in uncertainty in the abscissa values.

This becomes very obvious when a functional connection between two independent analytical methods is to be investigated. The reliability of one new analytical method may be estimated by comparison with values which have been obtained by an independent method. Good agreement [i.e. $f_1(y_1) = f_2(y_2)$] would result in a plot of c_1 against c_2 being linear with a slope 1. In this case the values of the measurements are independent from each other and the scatter should be about equal in both directions. The better the agreement the smaller the scatter, and the closer the slope approaches unity.

2.4.3 Sensitivity S

It is generally agreed that the sensitivity S of a method is defined by the first derivative of the calibration function:

$$S = g'(c) = \frac{dy}{dc} \tag{2.6}$$

The sensitivity is constant if c and y are linearly related. The stipulation that the calibration function must have a non-zero derivative is significant with respect to measurement strategies. A method with zero or very small sensitivity is of no use to the analyst. The sensitivity has limited significance for the description of the quality of a method, however, since S can be increased quite simply, e.g. by electronic amplification.

It must be pointed out, however, that in the literature other definitions of sensitivity are often used. In particular, the limit of detection is very often called (erroneously) the sensitivity.

2.4.4 Types of error

Random errors

When a number of samples, the analyte contents of which are precisely known, are analysed the measurements will be scattered around the calibration line, as indicated in Fig. 2.2. Replicate analysis of one particular sample would show that the measurements (y) also vary in a certain range.

A more thorough study of the variability of y would show that the frequency distribution of y is bell-shaped, the values in the middle of the range having higher frequency than those at its limits, and the true mean value μ can be expected, with a probability of 68%, to be included in the range $\bar{y}_i \pm s_y$ (s_y = standard deviation of the measured value; $\bar{y}$ = arithmetic mean). Each point in Fig. 2.2 stands for the mean value $\bar{y}_i$ of the corresponding population of n measurements of y_i for a particular sample $[\bar{y}_i = (\Sigma y_i)/n]$.

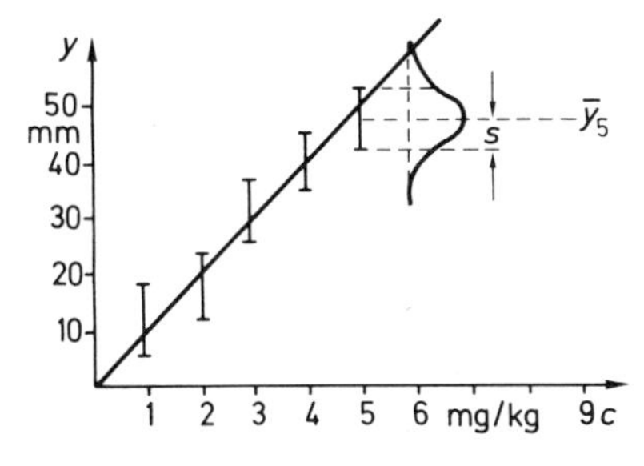

Fig. 2.2 – Random errors within a series of measurements.

Constant systematic error

A constant systematic error would cause a parallel shift of the analytical function along the ordinate and the function would become $y = a + bc$ (see Fig. 2.3). This error leads to a constant contribution to the signal y, and its relative effect decreases with increasing magnitude of the signal. This type of error can be either positive or negative.

A well known positive constant error is a (constant) blank value. Sometimes, however, a constant amount of an organic trace constituent is held irreversibly by the sample matrix under the conditions of the separation procedure used. The resulting loss would cause a negative shift of the analytical function, and a signal would be expected only when the content of the trace constituent exceeded the fixed loss.

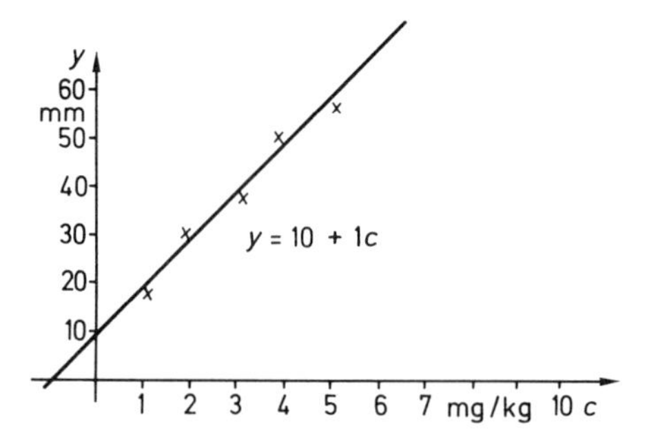

Fig. 2.3 – Constant systematic error.

Proportional systematic error

When the error is a constant fraction of the analytical result it is referred to as a proportional systematic error. It would cause a change in the slope b of the analytical function (see Fig. 2.4). A typical error of this kind could be caused by using a calibration standard supposed to be 100% pure, but in reality only 90% pure. Recovery experiments, e.g. in pesticide residue analysis (see Table 5.2, p. 132), often show a constant recovery, e.g. a mean of 80%, which indicates a proportional error of 20% in the average.

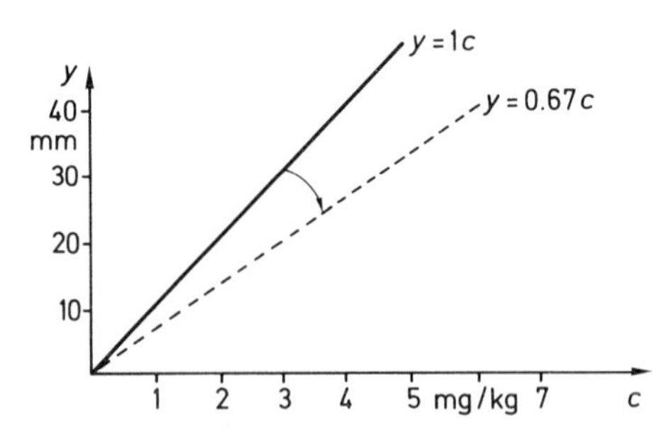

Fig. 2.4 – Proportional error.

In practical analysis all three types of error are found, often enough simultaneously. It must also be taken into account that random errors, as well as systematic errors, can be proportional to the measured value. Generally speaking, in organic trace analysis the systematic errors are the most important and of greatest influence.

2.4.5 Remarks on nomenclature

There has been (and still is) a lot of confusion about nomenclature in this field. Clinical chemistry has greatly contributed to clarification of the situation [43, 44] and some of the recommendations of clinical chemists can well be used in organic trace analysis (see Table 2.6).

Table 2.6 – Definitions of precision, accuracy, specificity and sensitivity [43, 44] (by permission of the copyright holders, Walter de Aruyter, Berlin).

criterion	definition
precision	agreement between replicate measurements within a set; it has no numerical value
imprecision	variation of the results in a set of replicate measurements; the design of the set (within-day, between-day, between laboratories) should be specified
accuracy	agreement between the best estimate of a quantity and its true value; it has no numerical value
inaccuracy	numerical difference between the mean of a set of replicate measurements and the true value
specificity	the ability of an analytical method to determine solely the component(s) it purports to measure
sensitivity	the slope of the calibration curve
determination limit	the lowest analytical result which will be sufficiently precise and accurate to yield a satisfactory quantitative estimate of the true value.

2.4.6 Measures of imprecision (dispersion)

The range w

The simplest measure of imprecision is the range w

$$w = y_{max} - y_{min} \tag{2.7}$$

w being the difference between the highest and lowest results in a set. Obviously, the range is very sensitive to outliers. As with increasing replication of a measurement the probability of outliers increases, the estimation of imprecision by the range is restricted to small numbers of observations. When only a small number of results can be obtained and no further information on the kind of distribution of the results is available, it is recommended that the range should be stated rather than a standard deviation, which requires a normally distributed set of data.

Gutermann's measure of imprecision

A slightly more powerful estimate of the imprecision has been proposed [45]. When the kind of distribution of the results is unknown, the range w of n results is used to obtain an estimate of the standard deviation (s_G):

$$s_G = \frac{w}{2}\sqrt{\frac{n}{n-1}} \qquad (2.8)$$

With increasing n, the square root term approaches 1 and s_G is practically equal to half the range ($s_G \sim w/2$).

Interdecile range

The interdecile range is a measure for scattering which – in contrast to the range w – is scarcely affected by outliers. Furthermore, it does not require any knowledge of the type of distribution. Consequently, interdeciles can be calculated in all cases, without regard to the distribution pattern. In ecology and toxicology calculations based on interdeciles are often applied for the evaluation of data from 'real world' investigations. Very often the data resulting from the measuring process itself are normally distributed, whereas the results obtained by application of the same measuring system to normally distributed populations are *not* normally distributed. Often enough, non-normally distributed results are encountered. A prediction of the distribution pattern is thus difficult to make. It requires many measurements and a statistical analysis. Interdecile calculations are much simpler to make and they give some valuable information, especially when only a few measurements have been made.

To calculate the interdecile range, the values of the n analytical results are arranged in ascending order of magnitude and this ranking order is divided into 10 equal parts, the dividing lines being called the deciles. Thus the sixth decile is the point on the measurement scale below which six tenths of the distribution of results will be found. Thus for the sixth decile we count along the set of results until we reach the $0.6n$th value (obtained by interpolation in the set). The interdecile range I_i is defined as

$$I_i = D_i - D_j \qquad (2.9)$$

where D_i and D_j are the hypothetical results corresponding to the ith and jth deciles.

Two useful properties of this scheme should be mentioned.

(1) The median corresponds to the fifth decile, since equal numbers of observations occur above and below the median. The median is more useful than the arithmetic mean for the evaluation of data (e.g. in toxicology).

(2) The ninth decile may be defined as an upper limiting value which covers 90% of all the values. It is also the 90th percentile and is of significance for the consideration of tolerance limits.

Example
Twelve values (R1–R12) have been obtained:

R	1	2	3	4	5	6	7	8	9	10	11	12
Value	0.53	0.63	0.65	0.71	0.74	0.77	0.79	0.83	0.86	0.90	0.98	1.06

Since there are twelve results, the median must lie half-way between the sixth and seventh values, so the fifth decile is

$$(0.77 + 0.79)/2 = 0.78$$

For an odd number of results the median would be the central value in the set.

The 90th percentile (9th decile) corresponds to the 0.9nth value, i.e. $R_{10.8}$, which by interpolation between R_{10} and R_{11} has the value

$$0.90 + \frac{(0.98 - 0.90) \times 8}{10} = 0.96$$

2.4.7 Standard deviation

The best estimate of the imprecision of a set of n results is the estimated standard deviation s, the distribution being assumed to be normal.

$$s = \sqrt{\frac{\Sigma(y_i - y)^2}{n - 1}} \tag{2.10}$$

A form which is more convenient for computation is

$$s = \left[\frac{\Sigma y_i^2 - (\Sigma y_i)^2/n}{n - 1}\right]^{1/2} \tag{2.11}$$

The computation can be done easily by use of even a small programmable desk computer.

It should always be checked that the frequency distribution of a given set of data can be approximated by a normal distribution. To do this the results are converted into a cumulative frequency distribution and the cumulative relative frequencies are plotted against the upper class boundaries on arithmetic probability paper. If a straight line results, a normal distribution can be assumed.

Example
Thirteen parallel determinations have yielded the results

0.55; 0.62; 0.65; 0.71; 0.74; 0.77; 0.79; 0.83; 0.86; 0.88; 0.90; 0.98; 1.06 mg/kg

These results are grouped into classes of equal range, e.g. for a range of 0.1, the result 0.55 falls into the first class, 0.62 and 0.65 into the second, etc. (Table 2.7).

Table 2.7 – Test for normal distribution

Class (mg/kg)	Frequency of results in a class		Cumulative frequency (%)
	absolute	relative (%)	
0.5 – 0.6	1	8	8
0.6 – 0.7	2	15	23
0.7 – 0.8	4	31	54
0.8 – 0.9	3	23	77
0.9 – 1.0	2	15	92
1.0 – 1.1	1	8	100

The plot of these values on arithmetic probability graph paper is shown in Fig. 2.5 and indicates a normal distribution.

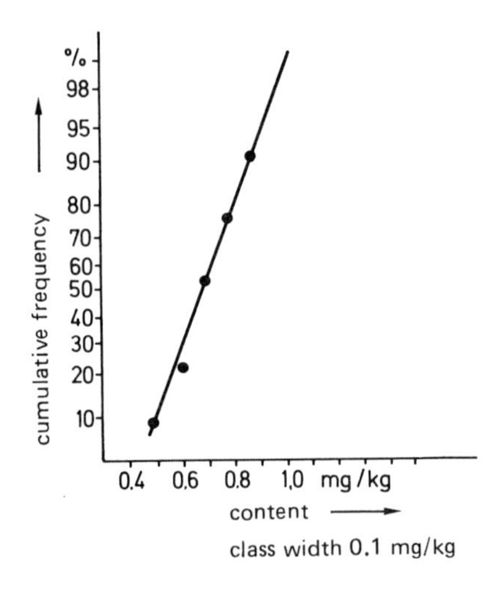

Fig. 2.5 – Graphical test of normal distribution.

Measurements which are not symmetrically distributed on both sides of the arithmetic mean are said to be skewed. This is found rather frequently for sets of measurements close to the limits 0% or 100% and in those cases where the random error of a signal is almost equal in magnitude to the signal itself. Therefore, the frequency distribution of results from trace analysis usually gives an oblique, skewed distribution curve (Fig. 2.6).

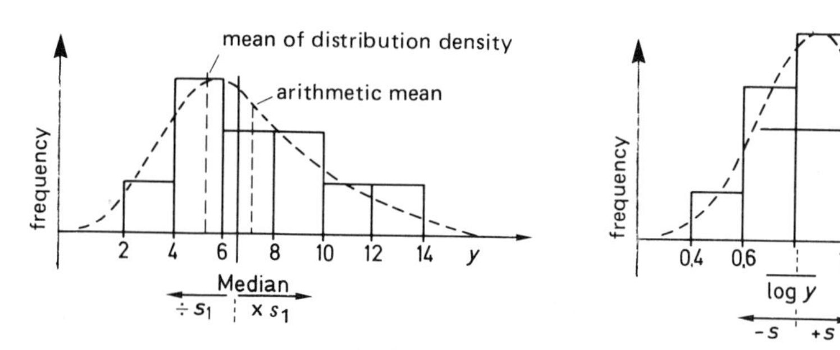

Fig. 2.6 – Log-normal distribution.

As can be seen from Fig. 2.6 the parameters of this type of distribution curve are quite different from those of the normal distribution curve. The median still denotes the value on the abscissa which divides the distribution into two equal parts, each containing 50% of the values, but the standard deviation is replaced by the deviation factor s_1. Multiplication of the median by this factor gives the upper boundary of the 90% probability range, and division of the median by it gives the lower boundary.

Skewed distributions can often be approximated by a log-normal distribution. For this, the logarithms of the measurement values are used in computation of the deviation factor (see the following example).

Example

The measurements y_s of the analytical signal are given in the first column of Table 2.8. In the second column their logarithms and in the third column the squares of the logarithms are tabulated. The deviation factor is then calculated by applying Eq. (2.11).

Table 2.8 – Values in a log-normal distribution

y_s	$\log y_s$	$(\log y_s)^2$
2.5	0.3979	0.15836
4.1	0.6127	0.37550
4.8	0.6812	0.46409
5.6	0.7482	0.55979
6.7	0.8261	0.68240
7.5	0.8751	0.76573
8.4	0.9243	0.85429
9.3	0.9685	0.93796
10.0	1.0000	1.00000
13.0	1.1139	1.24087

The standard deviation of the log-normal distribution curve is given by

$$s_{\log y} = \sqrt{\frac{7.0390 - 8.1479^2/10}{10 - 1}} = \sqrt{0.04446} = 0.2109$$

The deviation factor becomes

$$s_1 = \text{antilog}\ s_{\log y} = \text{antilog}\ 0.2109 = 1.26$$

and the median is given by

$$\text{median} = \text{antilog}\left(\frac{8.1479}{10}\right)$$
$$= 6.53\ .$$

To find the 90% probability range the median is multiplied by s_1 ($6.53 \times 1.65 = 10.8$) and divided by s_1 ($6.53 \div 1.65 = 4.0$)

2.4.8 Confidence interval, tolerance interval

Even when systematic errors are totally excluded, random errors are unavoidable. Consequently, it is always uncertain whether an observed result corresponds to the true content of the sample. Repeated analysis and inspection of the results will demonstrate that these accumulate around a mean value. The frequency distribution will be bell-shaped if the values are approximately normally distributed.

With normal distribution the probability function $f(x)$ gives a symmetrical bell-shaped curve for the frequency distribution. The variable x can vary from $-\infty$ to $+\infty$ and the probability has a maximum at the mean value μ. The probability of finding values of x far from μ decreases with $|x - \mu|$. The two parameters σ and μ determine the normal distribution completely. The mean value μ determines the distribution on the abscissa and the standard deviation σ determines the shape of the curve. A large value of σ results in a broad and flat peak, while a small value of σ leads towards a narrow and sharp peak.

Obviously, the mean value of the analytical results gives the best basis for an estimation of the true value, but because it is very unlikely that the mean will *be* the true value, it is much better to calculate two values bracketing the true value and to state the probability that the true value lies between them. These values are called 'confidence limits'. The true value lies between them in $> 100\ (1 - 2\alpha)\%$ of all cases (or in other words; it lies outside them in $\leqslant 200\alpha\%$ of cases), where $(1 - \alpha)$ is the desired probability (or degree of confidence).

It is also possible to state only an upper or a lower confidence limit, and these are called one-sided confidence limits. They include in $\geqslant 100\ (1 - \alpha)\%$ of all cases the true value or do not include it in $\leqslant 100\alpha\%$ of all cases. In science confidence limits of 0.95, more rarely of 0.01, are usual. That means that 2α (two-sided) or α (one-sided) takes the value 0.05 and 0.01 respectively. The confidence intervals for the true value μ can be stated as follows (with prob = probability for):

one-sided upper prob $(-\infty < \mu < \bar{x} + ks) = 1 - \alpha$ (2.12)

lower prob $(\bar{x} - ks < \mu < \infty) = 1 - \alpha$ (2.13)

two-sided prob $(\bar{x} - ks < \mu < \bar{x} + ks) = 1 - 2\alpha$ (2.14)

As a rule, only very rarely will enough values be available in trace analysis for σ to be used for the calculation of the confidence interval, and the estimated standard deviation s is used according to Eq. (2.11), and k in Eq. (2.12) or (2.13) has to be replaced by $t_{(n-1),\alpha}/\sqrt{n}$ (one-sided) or $t_{(n-1),2\alpha}/\sqrt{n}$ (two-sided). Values for t as a function of the number of degrees of freedom $r = n-1$ and of α can be taken from tables [46]. Most 'scientific' pocket calculators have a built-in program for t-calculation.

Analytical results are very often to be judged with respect to limiting values, e.g. in toxicology, ecology, chemical production etc. In such cases it is required that a certain percentage of the total number of values should be below (or above, as the case may be) the limiting tolerance value (one-sided) or between two tolerance limits (two-sided). These percentiles relate to the total population of results. It remains to say how they can be estimated. The difference between such a percentile and its estimate is demonstrated by the following example.

In this example, a percentile describes a certain percentage of all values found for pesticide residues on the individual apples in a lot collected from a treated orchard. The 95th percentile would denote an upper limiting residue value such that 95% of all individual residue values in the lot would be below that figure. Generally, we cannot determine the residue on all the individual apples in the lot in order to find out which residue value is equal to the 95 percentile. Instead, we try to estimate the 95 percentile from random samples and describe the confidence interval for the estimated 95 percentile.

There is no general procedure by which the percentile and its confidence limits can be estimated. It largely depends on knowledge of the type of distribution of the relevant population. If the distribution is normal the 95 percentile becomes equal to the one-sided tolerance limit for 95% safety. Generally, the relevant tolerance limits are:

one-sided	upper limit	prob $(-\infty < x < \bar{x} + ks) = 1 - \alpha$	(2.15)
	lower limit	prob $(\bar{x} - ks < x < \infty) = 1 - \alpha$	(2.16)
two-sided		prob $(\bar{x} - ks < x < \bar{x} + ks) = 1 - 2\alpha$	(2.17)

In these equations s is estimated from a sample with $n < 100$ and k is taken from tables [47]. A confidence interval for the percentile can be stated. With a log-normal distribution a transformation is necessary (see page 37), after which the procedure is as given above.

With an unknown type of distribution only an estimation of the percentile is possible, as demonstrated on page 39. The n values of the sample population are arranged in increasing order. The first integer (m_u) higher than $0.95n$ and the first one (m_l) lower than $0.95n$ are determined. The estimated value for the 95 percentile (two-sided) is then found between the values of the m_uth and m_lth of the arranged data. Interpolation gives an estimated value for the percentile. Its confidence interval can also be determined. It is necessary to perform a certain minimum number of measurements to be able to state a 95 percentile. For $n = 5$ m_u is practically equal to 5, since $0.95 \times 5 = 4.75$, and the estimated value is thus rather close to the highest value. The estimate becomes rather sensitive to accidental high values.

Enough samples should be taken to guarantee a statistical resolution of one measurement value; e.g. the 95 percentile results in a resolution of 5%, and 1 value is 5% of the total number if $n = 20$. An example for the mode of setting tolerances in residue analyses is given by Weinmann and Nolting [48].

2.4.9 Hypothesis testing

Analytical results are needed to test hypotheses, as stated at the beginning of Section 2.4. A question often encountered asks whether an unknown entity corresponds to a known one or a hypothetical one. The question could be, for example, "Has a plant breeding experiment resulted in a new species of apples which has a higher content of vitamin C than a standard comparison species?" Testing in this case is done by analysis of a number of samples for vitamin C content. It is investigated whether $\mu_{\text{standard}} - \mu_{\text{new cultivar}} \neq 0$. If a difference can be proved by means of a statistical test with a significant probability, the null hypothesis H_0 ($\mu_{\text{standard}} - \mu_{\text{new cultivar}} = 0$) is rejected and the alternative hypothesis ("there is a difference") is accepted. The probability of making a type I error is α (one-sided) or 2α (two-sided). In the case of the one-sided test only one limit (upper or lower) is checked. An example is the specification of a sample in which the concentration of a trace is not permitted to exceed a given value. In this case the lower limit is of no significance as all values below the upper limit are permitted.

The power of a test is the probability $(1 - \beta)$ of detecting a difference if one actually exists (see Fig. 2.7). The power of a test is usually increased by taking a larger number of samples. Furthermore, the power is the greater, the more is known about the populations which are compared. One-sided tests are more powerful than two-sided ones!

Important guidelines for decision making are as follows.

(1) If the intention is to prove that two populations are in agreement and if the test does not find a significant difference, the two populations should be taken as the same until further experiments disprove this verdict.

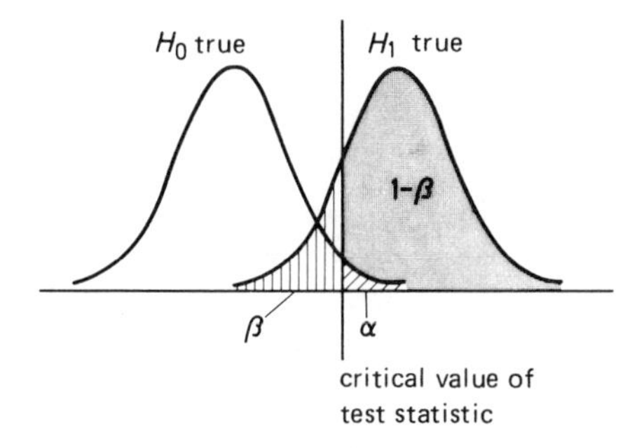

Fig. 2.7 – Power of a test.

(2) If the intention is to prove by the experimental design that there is a difference between two populations and the corresponding test does not demonstrate a difference, then the conclusion should be: "For the samples taken no difference can be proved on the basis of the preselected probability of significance".

When only a small number of samples can be taken (as happens often enough in chemistry and biochemistry) any conclusion on the type of distribution of the population is quite uncertain. In these cases mere guessing should be avoided, e.g. postulation of a normal distribution when there is none, and a more powerful test should be applied. The probability of encountering an error of the first or second kind will vary considerably in such circumstances, and it is strongly recommended to make a conservative decision and apply tests which do not require a defined form of distribution. Table 2.9 presents a selection of seven tests and an indication of how much knowledge about the distribution is needed when such a test is used.

Table 2.9 – Some tests for deciding whether $\mu_0 - \mu_1 = 0$

Test	Conditions with respect to population
Student (t-test)	normal distribution
Lord	,, ,,
Mid-range (Walsh)	,, ,,
Walsh	symmetrical distribution
Zeichen	none
Maximum (Walter)	,,
Wilcoxon	,,

Besides the tests given in Table 2.9 (after Documenta Geigy) the F-test should be mentioned, which is used for testing the homogeneity of variances. A description of the χ^2-test and many others can be found in the literature [41, 42, 46, 49, 50].

2.4.10 Limit of detection, limit of determination

In organic trace analysis it is important to have a clear understanding of the meaning of the terms limit of detection and limit of determination. In the analytical literature there is a plethora of mathematical expressions and a widely ranging terminology. For example, a procedure is often said to be very sensitive when what is meant is that the limit of detection is low. Other authors use 'limit of detection' to refer to the minimum amount which can be determined with a preselected relative standard deviation, e.g. 10%.

Surveys of the different definitions have been published [41, 51, 52].

Random errors are unavoidable and therefore limits for detection and determination have to be accepted because it makes no sense to claim to determine quantities which cannot be distinguished from random errors. A solution of the problem has been presented [41, 51] and provides the basis for the following presentation.

For convenience, the symbols used are defined as follows:

blank

$\bar{y}_B$ limiting mean

y_B observed value

σ_B standard deviation

s_B estimated standard deviation

gross signal

$$\bar{y}_u = \bar{y}_{(S+B)} \tag{2.18}$$

limiting mean

$$y_u = (y_S + y_B) \tag{2.19}$$

observed individual value caused by the analyte in the sample plus blank

σ_{S+B}

standard deviation

net signal

$$\bar{y}_S = \bar{y}_{S+B} - \bar{y}_B \tag{2.20}$$

limiting mean

$$y_S = (y_S + y_B) - y_B \tag{2.21}$$

value derived from a pair of observations

$$\sigma_S = (\sigma^2_{(S+B)} + \sigma^2_B)^{1/2} \tag{2.22}$$

$$= (\sigma^2_u + \sigma^2_B)^{1/2} \tag{2.23}$$

standard deviation

In the discussion of the problem of detection we have to distinguish between two fundamental aspects:

(1) a net signal is observed and we must decide whether a real signal has been detected

$y_S > 0$ (*a posteriori* decision)

(2) a complete specified analytical procedure is given and we must estimate the minimum mean true net signal $\bar{y}_S$ which will yield an observed net signal y_S, sufficiently large to be detected (*a priori* estimate of the detection capability).

The decision in (1) is subject to two possible errors:

(*a*) it may be decided that the substance is present when it is not (error of the first kind, α).

(*b*) it may be decided that the substance is not present, although in fact it is (error of the second kind, β).

The decision may be based on a critical level L_C, which is established by means of the maximum acceptable value for the error of the first kind α, and the standard deviation, σ_0, of the net signal y_S when the limiting mean of the net signal is zero ($\bar{y}_S = 0$); (the presence of a systematic error is excluded at this stage of discussion). The mathematical expression for the decision limit is

$$L_c = k_\alpha \sigma_0 \tag{2.24}$$

where k_α is the abscissa of the standardized normal distribution corresponding to the probability level $1-\alpha$ (see Fig. 2.8). A signal y_S larger than the decision limit L_c is interpreted as "component detected". The fraction $(1-\alpha)$ corresponds to the correct decision "not detected" while the fraction α corresponds to the wrong decision "component present" when in reality it is not.

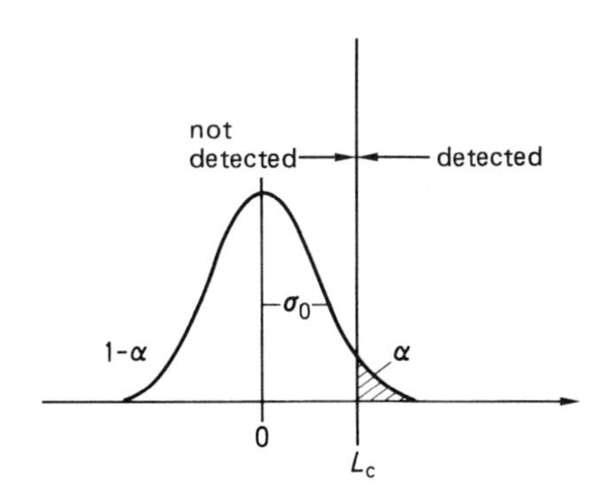

Fig. 2.8 – Critical level.

In organic trace analysis, in most cases the standard deviation σ of a net signal y_S is not known, and the signal and its standard deviation have to be estimated by means of replication. So y_S is replaced by $\bar{y}_S$ and σ by $s/\sqrt{n}$. The confidence level may be calculated from $t_{1-\gamma/2}\ s/\sqrt{n}$. Here s represents the standard deviation estimated from n observations and $t_{1-\gamma/2}$ represents the critical value of the Student's t-distribution corresponding to $r = n - 1$ degrees of freedom and the chosen probability level $1 - \gamma/2$. Values for t are given in statistical tables [46]. Table 2.10 summarizes the decision scheme.

Table 2.10 – Decision scheme for detection

Observation	Decision	Confidence interval	
$\bar{y} > L_c$	detected	$\bar{y}_S \pm t_{1-\gamma/2} \cdot \frac{s}{\sqrt{n}}$ symmetrical	(2.25)
$\bar{y}_S < L_c$	not detected	$\bar{y}_S + t_{1-\gamma/2} \cdot \frac{s}{\sqrt{n}}$ one sided	(2.26)

It is obvious that the decision level is only of limited use as a detection limit in the sense outlined above. When $L_c = L_D$ the limiting mean $\bar{y}_S$ for the measurement of the unknown sample y_S would coincide with the decision level L_c. Accordingly, 50% of the measurements could be expected to yield the decision 'component not present' (not detected) when in reality the component *is* present (error of the second kind, $\beta = 0.5$, see Fig. 2.9). This, of course, would be a very unreliable detection.

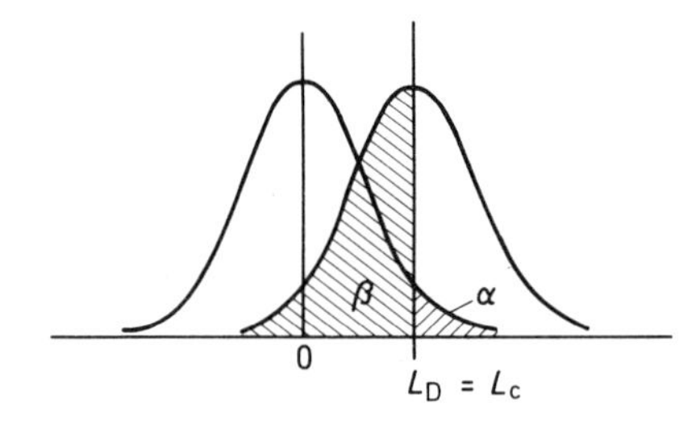

Fig. 2.9 – Coincidence of the critical level and the detection limit.

Therefore an *a priori* detection limit L_D is established by specifying L_c, and an acceptable level of β for errors of the second kind, and the standard deviation σ_D which characterizes the net signal when its true limiting mean, $\bar{y}_S$ is equal to L_D. An 'acceptable level' for β might be taken as 0.05 or 0.01.

The detection limit is expressed as $L_D = L_c + K_\beta \sigma_D$, where k_β is the abscissa of the standardized normal distribution corresponding to the probability level $1 - \beta$. The fraction $1 - \beta$ corresponds to the correct statement 'component detected', the fraction α to the incorrect statement 'component detected' when it is not. Similarly β corresponds to the wrong statement 'component not detected' when in reality it is present (see Fig. 2.10).

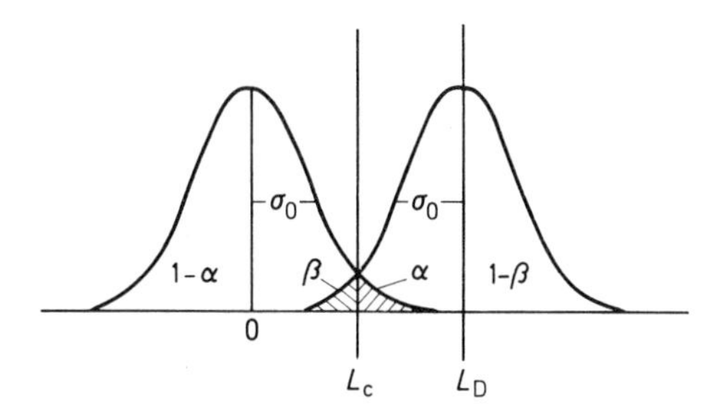

Fig. 2.10 – Sufficiently great difference between detection limit and critical level.

For quantitative analysis it is necessary to have a result which is as close as possible to the true value. Furthermore, the standard deviation must be only a small fraction of the true value. Adams *et al.* [53] have defined the 'minimum working concentration', as the lowest value for which the relative standard deviation does not exceed 10%. A similar definition is used for trace analysis in pesticide residue analysis [54]. The determination limit so defined is

$$L_Q = k_Q \sigma_Q \tag{2.27}$$

where L_Q is the true value of the net signal having the standard deviation σ_Q and $1/k_Q$ is the requisite relative standard deviation.

In summary it can be said that L_c is used to test an experimental result, whereas L_D and L_Q refer to the capabilities of measurement processes to detect and determine substances. Thus three principal analytical regions are defined [51] (Table 2.11).

Table 2.11 – The three principal analytical regions

Region		
I	II	III
0-L_D unreliable region	L_D-L_Q detection qualitative analysis	L_Q determination quantitative analysis

The risks α and β may be assigned values that are not necessarily equal, but in routine work they are usually considered to be equal and accepted to be 0.05. In that case $k_\alpha = k_\beta = k$, and

$$L_c = k\,\sigma_0 \tag{2.28}$$

If σ is approximately constant, $\sigma_0 \approx \sigma_D$ and

$$L_D = L_c + k\,\sigma_D = 2L_c \tag{2.29}$$

The detection limit is twice the decision level. Further, the standard deviation of the net signal, σ, is given by

$$\sigma = \sqrt{\sigma_{B+S}^2 + \sigma_B^2} \quad (2.30)$$

and if the standard deviation for an observation is approximately independent of the signal level, then $\sigma_{B+S} \approx \sigma_B$, and $\sigma = \sigma_B\sqrt{2}$. Substitution of σ for σ_D in Eq. (2.29) gives

$$L_D = 2k\,\sigma_B\sqrt{2} \quad (2.31)$$

If we take $k_Q = 10$, then we obtain

$$L_Q = k_Q\,\sigma_Q \approx 10\,\sigma \quad (2.32)$$

These assumptions result in the working expressions for L_C, L_D and L_Q given in Table 2.12.

Table 2.12 – Working expressions for L_C, L_D and L_Q (adapted by permission from L. A. Currie, *Anal. Chem.*, 1968, **40,** 586. Copyright 1968 American Chemical Society).

	L_C	L_D	L_Q
paired observations	$2.33\sigma_B$	$4.65\sigma_B$	$14.1\sigma_B$
'well known' blank	$1.64\sigma_B$	$3.29\sigma_B$	$10.0\sigma_B$

The values in the first row are derived from pairs of observations of samples and blanks, where the blanks have the same matrix as the sample but do not contain the analyte. The values in the second row differ from those in the first by a factor of $1/\sqrt{2}$, on the curious assumption that a long history of observations of the blank in some way makes the standard deviation of the net signal equal to σ_{B+S} [51].

In organic trace analysis, especially of food, authentic control samples are not always available. In this case the estimation of the detection limit is difficult. Proposals have been made for dealing with such cases [55, 56].

2.4.11 Regression analysis, correlation

It is part of the daily routine for trace analysts to evaluate analytical results on the basis of linear regression lines. Calibration lines, adsorption isotherms, kinetic rates, and degradation and elimination rates are some examples.

In all cases the functional interdependence or a correlation between two specific characters is evaluated. The correlation can be linear or non-linear. Whenever possible the analyst tries to deal with linear correlations and it is common practice to linearize non-linear correlations, e.g. by plotting the data on a logarithmic/linear scale or a logarithmic/logarithmic scale. Non-linear curves are discussed by Schwartz [57].

With linear correlations it is necessary to distinguish between two different situations, viz. functional (causal) and stochastic (conjectural) relationships.

A stoichiometric analytical reaction is an example of a strongly functional relation. The signal y depends on the content c. It is, in a way, predetermined.

Two examples of stochastic correlations may be given. (a) The loss of a pesticide from lettuce may be correlated with the time of observation. In many such instances a linear correlation is shown when the data are plotted semi-logarithmically. (b) When two analytical methods are compared the measurements made on a sample by method I are correlated with those obtained by method II. There is no functional relationship between the two and the measurements are independent of each other. A good correlation and a slope of unity indicates that the analyte can be determined equally well by both methods. In both examples, the data are first examined to see whether a good correlation exists between the two variables. If this stochastic correlation is established it is described by a functional correlation (regression).

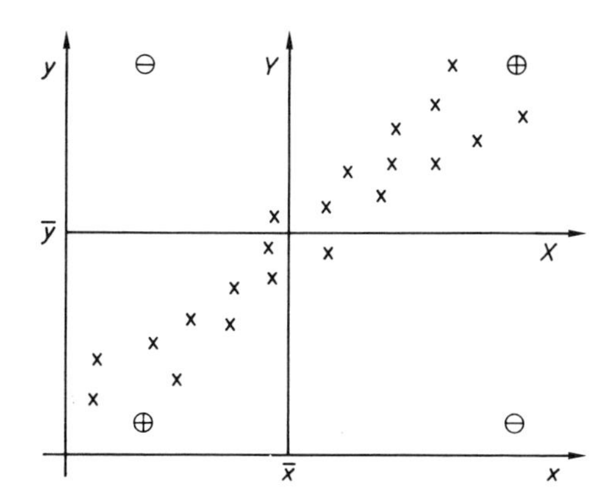

Fig. 2.11 – Geometrical illustration of correlation. Y, X co-ordinates of centre of gravity. —, + Negative or positive quadrants in which x_i, y_i are negative or positive.

To test for correlation, first the mean value $\bar{x}$ is calculated from all the individual values x_i. Correspondingly, $\bar{y}$ is calculated from all individual y_i. Then $\bar{y}$ and $\bar{x}$ give the centre of gravity of the points x_i, y_i plotted in Cartesian coordinates (see Fig. 2.11). Next, for each of the n values the difference from the mean value is calculated:

$$Y_i = y_i - \bar{y} \text{ and } X_i = x_i - \bar{x}$$

The coefficient of correlation r is then defined by

$$r = \frac{\Sigma X_i Y_i}{\sqrt{\Sigma X_i^2 \, \Sigma Y_i^2}} \qquad (2.33)$$

or

$$r = \frac{\overline{X}\,\overline{Y}}{s^{(x)}\, s^{(y)}} \tag{2.34}$$

in which $s^{(x)}$ and $s^{(y)}$ are the empirical standard deviations, determined separately for the x and y values.

The denominator of Eqn. (2.33) is always positive because of the use of X^2 and Y^2. The numerator, however, can be either positive or negative, so r is also positive or negative, depending on the relative positions of the points. If r approximates to or becomes zero, no correlation between x and y exists. Complete correlation is expressed by $|r| = 1$, and in this case a formal connection is established but not necessarily a *causal* connection.

Regression is the description of a stochastic connection by means of a functional connection. Thus, the regression line must not be interpreted as a strong interdependence. It shows only the strength of the stochastic relation between the two variables [58].

When we have only pairs of data x_i. y_i the simplest relation is the equation of the straight line:

$$y_i = a + bx_i \tag{2.35}$$

The intercept with the ordinate is a and the slope is b (also known as the regression coefficient). If b is positive the measurement y increases with increasing x. Because it is not easy to estimate the coefficients a and b graphically from a cloud of points, the estimation is accomplished mathematically by the least-squares method, which searches for that particular regression line which, out of all possible alternative lines, gives the least value for the sum of the squares of the differences between the observed values, y_i, and the corresponding values on the line, $\overline{y}_i$:

$$\Sigma\, (\overline{y}_i - y_i)^2 \longrightarrow \text{minimum}$$

In the example of the pesticide on lettuce [(a) on p. 47)], the time of sampling is predetermined and thus (as a rule) does not possess a significant error. In this case x is compared with y, a practically error-free variable. Example (b) on p. 47 is different. The values of c_I obtained by method I exhibit random fluctuation, as do those of (c_{II}) obtained by method II [$c_I = f(y_I)$; $c_{II} = f(y_{II})$]. Thus two regression lines are possible, with the slopes $b_{c_I c_{II}}$ and $b_{c_{II} c_I}$ (see Fig. 2.12). The angle between these two lines can be a measure of the correlation coefficient r, which is the geometrical mean of the two slopes.

$$r = \sqrt{b_{c_I c_{II}}\, b_{c_{II} c_I}} \tag{2.36}$$

An exact correlation exists if the angle between the lines is zero, and only one regression line is obtained.

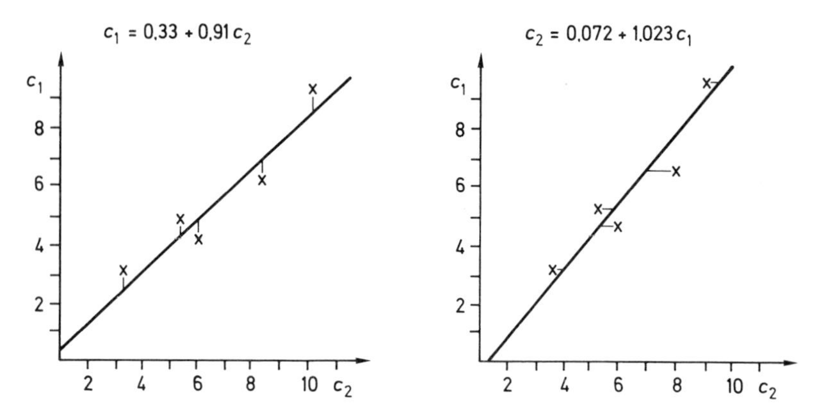

Fig. 2.12 – Demonstration of regression lines. Regression of c_1 on c_2 and of c_2 on c_1.

So far, regression has been discussed on the assumption that one value of y corresponds to one value of x. A cloud of individual points results. In practice, however, (see p. 31) after repeated measurements of y for constant x, scattered results are obtained. If the confidence interval of each mean value is introduced (displayed as a vertical line, called an error bar, see Fig. 2.13) description of the cloud of points is improved. The regression line calculated on the basis of these mean values enables a better judgement of the quality of the measurements. For example, if the regression line does not pass through all the individual confidence limits the original assumption of a linear regression may be incorrect. Another mathematical function can then be sought that will give a better fit to the set of measured data, the method of least squares again being used for the purpose. Figure 2.13 shows the confidence intervals increasing as x decreases, giving the form of a funnel for the confidence limit bands; this is generally the case in trace analysis.

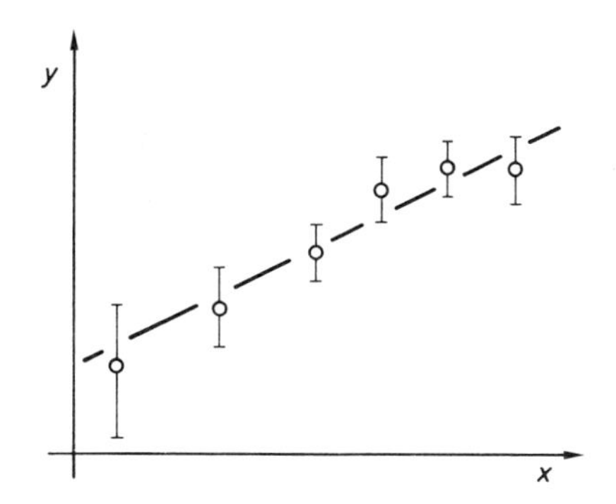

Fig. 2.13 – Confidence intervals and regression line for x-values without error and y-values with error.

Details of the computation of the best fitting line and the correlation coefficient are given in the literature [42, 49]. These operations can be carried out quite simply with scientific pocket computers which are programmed for the purpose. For example, the HP-65 can be used with programs written on magnetic cards or chips. The subsequent application of the programs STAT 1-05 A and STAT 1-22A allows calculation of a, b and r for the linear regression $y = a + bx$ from a set of data (y_i, x_i), and also the standard deviations of the regression coefficient and intercept.

An example will demonstrate the procedure. The dissipation of a plant protection agent residue on a treated plant is measured as a function of time, the following data being obtained (x_i = time, $y_i = c_i$ = concentration of residue on plant):

x_i	0	1	3	5	7	14	days
c_i	90	62	36	21	8.5	1.3	mg/kg

In such experiments it is reasonable to expect that the data fit a regression line

$$\log y = a + bx$$

The logarithmic transformation of the y_i values gives:

c_i	90	62	36	21	8.5	1.3	mg/kg
$\log c_i$	1.954	1.792	1.556	1.322	0.929	0.114	

Using the programs already mentioned gives the following values:

$$\begin{aligned} a &= 1.938 \\ b &= -0.1321 \\ r^2 &= 0.9955 \\ s_{c,x} &= 0.0508 \\ s_a &= 0.0305 \\ s_b &= 0.0045 \end{aligned}$$

The coefficient of correlation $r^2 = 0.996$ indicates almost perfect correlation. The equation for the regression is:

$$\log c = 1.938 - 0.1321\,x$$

According to this formula the concentration of the residue 10 days after the application of the plant protection agent is given by

$$\begin{aligned} \log c &= 1.938 - 0.1321 \times 10 \\ &= 0.617 \\ c &= 4.1 \end{aligned}$$

At the tenth day a residue of 4.1 mg/kg can be expected.

2.4.12 Analysis of variance

Analytical procedures are complex. Each step contributes to the systematic and random errors. This becomes even more noticeable when interlaboratory comparisons are to be made, e.g. in collaborative studies.

In such cases it is necessary to detect sources and magnitudes of variance between samples, between replicates, between work on different days within one laboratory, between several laboratories etc. To detect statistical deviations a system of analysis of variance (ANOVA) has been introduced. It comprises the breakdown of the total precision into its components. Once these components are known, the one with greatest influence on the total precision (and which hence should be optimized first) can be identified.

Of course, in collaborative testing ANOVA will not detect a systematic error inherent in the analytical method if all the participants use the same method. That can only be done when different analytical methods are used and the ANOVA technique is applied. Reviews on the application of ANOVA are given in the literature [59, 60]. ANOVA is used successfully by the Deutscher Ausschuss für Pestizid Analytik, a panel of the Collaborative International Pesticide Analytical Committee.

2.5 METHODS FOR CALIBRATION AND CHECKING THE RESULTS OF ORGANIC TRACE ANALYSIS

2.5.1 Hierarchy of analytical methods

The methods available to the analyst can be classified in a hierarchy [43, 44] according to the quality of the results (Table 2.13).

Table 2.13 –Hierarchy of analytical methods and analytical results [by permission from stamm, D. (1982). A New Concept for Quality Control of Clinical Laboratory Investigations in the Light of Clinical Requirements and Based on Reference Method Values. *J. Clin. Chem. Clin. Biochem.*, **20**, 817–824, Table 1].

Method		Result
		"true value"
Definitive method		definitive value
Reference method		reference method value
Class A:	tested with definitive method	
Class B:	not tested with definitive method, but highly purified, defined standard available, reliability of method assured	
Class C:	no homogeneous standards of known composition available, testing with definitive method not possible	
Routine method		assigned values
Class A:	systematic error known (selected method = ausgewählte Methode)	
Class B:	systematic error not known	

Further definitions of methods are given by the recommendations of the Expert Panel on Nomenclature and Principles of Quality Control of the International Federation of Clinical Chemists [61] (see Table 2.14).

Table 2.14 – Description of methods [by permission, from Büttner, J., Borth, R., Boutwell, J. H. and Broughton, P. M. G. (1975), International Federation of Clinical Chemistry, Committee on Standards. *J. Clin. Chem., Clin. Biochem.*, **13**, 523–531, Table p. 530].

Term	Concept	Value obtained for calibration or control material
definitive method	no known source of inaccuracy	definitive value = best known approximation to 'true value'
reference method	after exhaustive testing; inaccuracy $= 0 \pm \Delta$, Δ is negligible as compared to between-laboratory-imprecision	reference value (stated or certified)
method with known bias	known bias as determined	assigned value (stated or certified)
method with indeterminate bias	bias not known	assigned value (stated or certified)

A prerequisite for the evaluation and comparison of routine methods is the existence of reference methods [43], which must be highly specific and not exhibit any matrix effects. The description of such fundamental methods should include the principle of the method; obtaining and handling of specimens; description of materials and procedures; reliability of the procedures with a description of their precision and accuracy; determination of the limit of detection; practicability of the method; statement of the reference ranges.

Reference methods are usually complicated and tedious. For routine purposes simpler methods are developed, which may be based on the same principle as the reference method.

To develop a reference method it is necessary either to compare the result with a defined value obtained by a 'definitive method' (class A reference methods) or to use a value which is defined for or given by a standard material (class B reference method). Standard materials consequently play an important role. Reference methods in clinical chemistry and also in other areas of general and international importance are built up by worldwide collaboration [62].

2.5.2 Reference and calibration materials

A reference material is defined by the International Organization for Standardization (ISO): "A material or substance, one or more properties of which are sufficiently well established for the calibration of an apparatus or for the verification of a measurement method. A reference material makes possible the transfer of the value of a measured quantity between one place and another".

A 'certified reference material' is a reference material accompanied by a certificate stating the property values concerned, issued by an organization, public or private, which is generally accepted as technically competent [63].

A 'standard reference material' (SRM) is defined by the U.S. National Bureau of Standards as a material to which three methods of measurement have been applied to ensure the accuracy of the results: (1) a definitive method; (2) a reference method; (3) an independent method. For organic trace analysis. however, there is very often a lack of adequate methods. "Thus, we are in the stage of development of appropriate SRMs for trace level organics in natural matrices" [64]. Probably, there will be only a limited number of matrices, since it is difficult to guarantee sufficient stability. Several surveys have been published [63, 65-68]. Several institutions supply reference materials, but mainly those of interest only to the inorganic trace analyst [69].

A calibration standard is very often required in laboratory experiments. It is prepared by the investigator himself, and, unlike reference materials, is not intended to be transferred for use in other laboratories. In organic trace analysis the preparation of such a calibration standard is usually the first step of a whole sequence of actions. Accordingly, all errors made at that critical initial phase are introduced into the final analytical scheme. There are several ways to prepare calibration standards.

In many cases a concentrated solution of the compound of interest can be made, and suitably diluted. This procedure becomes problematic, however, if the compound is to be dissolved in trace concentrations in a solvent in which it is only sparingly soluble, e.g. most halogenated pesticides in water. Very long shaking or stirring of the mixture of solvent and solid, followed by filtration is necessary to obtain saturated solutions. An alternative is to impregnate silica gel with the material to be dissolved, by evaporation of a solution of it (e.g. in hexane) on the gel; water is then passed through a column of the gel and more rapidly becomes saturated owing to the large surface area of the gel. This method has been proposed for preparation of an aqueous PAHs standard reference material [64] and of a standard reference material for PCBs [70]. For pesticide standard solutions the solvent should have low volatility (e.g. 1,2,4-trimethyl-pentane or toluene), the volume of the solution should be greater than 100 ml, the solution should be stored in a refrigerator, and the container should be sealed by a screw-cap with a precision-made seal [71]. Insecticide standards containing accurately measured amounts (1 mg) in hexane, toluene or acetone have been prepared. Aliquots were sealed into glass ampoules and stored at $-20°$ and $+20°C$. After two years the mean concentrations of all 6 compounds studied were within 0.9-3.6% of the original concentrations [72]. An SRM for antiepilepsy drugs in serum has been prepared from pooled human serum with diphenylhydantoin, phenobarbital and primidone added in known concentrations [73].

Occasionally, solids containing a standard concentration of a trace compound are needed. The preparation of such standards is another problematic procedure. On the other hand, such standards have advantages, as they keep better than solutions [74]. To prepare such standards a solution containing the matrix and trace constituent can be evaporated to give a dry residue, and this is then homogenized. Another possibility is the addition of the trace constituent (in solution or as an already prediluted mixture with a solid carrier), drying, and homogenization by grinding. There is a danger of hydrolysis and oxidation, however, and this becomes the more pronounced the longer the homogenization procedure takes and the more the surface area of the solid material is increased by grinding. It is always necessary to monitor the entire procedure, e.g. by means of radioactive tracers. An atmosphere of dry nitrogen can be useful to prevent oxidation. The incorporation of traces into the interior of tissue cells without disturbance of the histological structure is possible only by application of the labelled trace compound to the growing organism, and monitoring of the uptake.

Gases with two level concentrations of organic constituents are required for calibrating mass spectrometers, gas chromatographs and sampling devices. The stability of diluted gas mixtures can be critical. It depends on the concentration of the trace species, and the purity and concentration of the other components of the mixture. The walls of most 1-litre sample containers have an adsorption capacity of about 50 μg [75]. Filling such a container with air containing a trace constituent at the ppm-level (~ 1 μg) could theoretically result in complete loss by adsorption in the walls. There is a dynamic equilibrium, however, between species already present as adsorbates (e.g. water, other air components and impurities) and the trace compound of interest. The importance of thorough flushing of the sample container is evident. Such losses of trace components to the walls of containers such as Teflon bags [76], other plastics [77] or steel containers [78] have been described, and take place within a matter of hours. The rather large coefficients of variation in interlaboratory quality-control experiments (see Table 2.15). are attributed to such effects.

Table 2.15 – Recovery experiments with trace gas mixtures

Substance in synthetic gas mixture	Concentration prepared (ppt w/v)	Concentration found in collaborative studies (ppt w/v)	Relative standard deviation (%)
$CFCl_3$	150	140	16
CF_2Cl_2	250	253	43
CCl_4	150	141	68
CH_3CCl_3	104	88	34

On the other hand, a 10 ppm mixture of vinyl chloride with air is stable [79]. Traces of CH_3Br in air are slowly degraded when stored in stainless-steel containers; CH_3I disappears more rapidly [80].

Methods for the preparation of standard gas mixtures include single- or multi-step gas-blending procedures, using exponential dilution equipment [81-84]. Other methods use the gas-permeation tubes, saturation of carrier gases or controlled diffusion [85-91]. Mixtures containing the vapours of even explosive materials, such as 2,4,6-trinitrotoluene, 2,4-dinitrotoluene and ethylene glycol dinitrate, can be generated at the 0.5-ppt w/v level [92].

Occasionally, the gaseous trace is produced from a precursor. Thus, formaldehyde is formed through thermal degradation of polyoxymethylene. The formaldehyde formed diffuses through a silicone-rubber membrane and is diluted and carried away by a stream of dry air. A precolumn for *in situ* generation of acrolein, acrylonitrile and vinyl chloride for gas chromatography has been proposed, in which a precision of better than 5% is achieved at the 5-ng level [93,94].

Methods of preparation of standard gas mixtures with very low concentrations of the gas of interest have been surveyed [95], and include

(1) drawing a stream of pure gas through a container of the standard substance in liquid form [96-99];
(2) passing a defined diffusional flux of the standard substance into a stream of gas [100-102];
(3) permeation of the standard substance through a membrane into a stream of gas [103, 104];
(4) continuously dosing the standard substance into a stream of gas [105];
(5) combination of a permeation tube and exponential dilution cell [106];
(6) saturation of a stream of gas by passage through a dilute solution of the volatile standard substance in a non-volatile solvent [107];
(7) saturation of a current of gas by passage over a solid support impregnated with the component to be volatilized [95, 108].

Finally, standardization based on an entirely different principle should be mentioned [109]. The electron-capture detector (ecd) gives nearly 100% ionization of certain halogen-containing compounds in gas chromatography, and this has prompted investigations of its use as a gas-phase coulometer. The number of electrons lost, as measured from ecd peak areas, is reported to be nearly equal to the number of sample molecules passed through the detector for compounds such as CCl_4, $CFCl_3$ and CF_2Br_2. Such gas-phase coulometry should provide a very useful method for calibration, in view of the considerable difficulties incurred in preparing reliable calibration standards at environmental concentrations. The method has since been improved [110]. It seems to have limitations, though, as correlation between electron capture detector response and chemical structure has been reported for hexachlorocyclohexane isomers [111].

2.5.3 Methods of validating the results of organic trace analysis

Many of the calibration systems – described in the preceding section – are made up without matrix constituents. As these may play an important role, it is necessary to evaluate their influence. If, as is generally the case, a reference material with known trace and matrix constitution is not available, there are two possibilities for dealing with the problem. Either synthetic samples can be used or known amounts of the trace species can be added to matrices which are free from the trace component (or contain accurately known concentrations of it). The second method can be called internal standardization.

If synthetic samples are used, a mixture of all the matrix components is made up, with addition of a known amount of the trace. This synthetic sample is analysed by the method under survey. It is often a problem, though, to obtain matrices in pure enough form and free from the trace constituent in question. How can cheese or meat be simulated, for example? Many such matrices are difficult to obtain in a native form in which the proteins are not denatured and in which trace constituents are included and fixed as in the original sample. Homogeneous distribution of traces is difficult to ensure and sometimes not even desirable, as the original matrix may also contain the trace in inhomogeneous distribution. The method of 'synthetic samples' is only useful with homogeneous systems, i.e. solutions and gases.

The other approach, the use of internal standards, adopts three different strategies (see Table 2.16).

Table 2.16 – Different types of internal standards

Method A	Method B	Method C
A known amount of the substance to be determined is added to a second sample ('spiking'). The spiked and unspiked samples are compared.	The substance to be determined is added, but with some of its atoms isotopically different ('isotope dilution')	Addition of a known amount of a substance which behaves rather similarly to the analyte but is not chemically identical with it

In method A the sample is 'fortified' or 'spiked' with a known amount of analyte. The untreated and the fortified samples are analysed by the same procedure. Provided the method works properly the quantity added will be fully recovered. The method allows correction to be made for losses if the recovery is incomplete. The precision is not very high, as the concentration of the unknown is calculated from the difference of two measurements, each of which is subject to error. The method is applicable with ease only if quantitative response throughout the entire procedure is guaranteed. Often enough, the substance to be determined in trace concentrations cannot be procured in sufficiently pure or native form. This is typical of real unknowns of biogenetic origin. Compensation of errors is a further complication, as a trace constituent

Table 2.17 – Examples of the use of isotopically labelled internal standards (D = ^{2}H; T = ^{3}H)

Compound determined	Matrix	Concn. or amount	Label atom	Separation/determination	C.o.v.* (%)	Reference
5-hydroxyindolylacetic acid homovanillic acid etc.	cerebrospinal fluid		D	partition/gc–ms		113
prostaglandin $F_{2\alpha}$	urine	5 ng/ml	D	partition/tlc/gc–ms	13	114
melatonine	plasma	1–100 pg/ml	D	gc–ms		115
etorphine	urine	2 ng/ml	D	gc–ms	5	116
methadone	biol. fluids	100 ng/ml	D	gc–ms		117
metanephrine	urine	10 ng/ml	D	gc–ms	2.5	118
pethidine	plasma	25 ng/ml	D	gc–ms		119
tryptamine, serotonin	rat brain	pg	D	chromatography/gc–ms		120
metoprolol + metabolites	plasma, urine	>0.3 ng/ml	D	partition/gc–ms	<10	121
maprotiline + metabolite	plasma	>2 ng/ml	D	partition/gc–ms	5	122
norepinephrine	blood	25–1600 pg/ml	D	chromatography/gc–ms		123
levorphanol	plasma	>1 ng/ml	D	partition/gc–ms	0.3–8.2	124
4-hydroxy-3-methoxy-phenylacetic acid	plasma	>1 ng/ml	D	ion-exchange/gc–ms		125
N-methylhistamine	plasma, urine	>5 pg	D	partition/gc–ms		126
digoxin	plasma	6 ng/ml	T	partition/hplc/ria†	4	127
LSD,tetrahydrocannabinol	biol. fluids	1 ng/ml	T	tlc/gc–ms or ria†		128
prostaglandins	urine	ng	T	partition/hplc/gc–ms		129
amino-acids	urine	>1 ng	^{13}C	ion-exchange/gc–ms		130
oestrogens	rat brain	4 pmole	^{14}C	tlc/gc		131
retinoic acid	blood	>2 ng/ml	^{14}C	partition/gc–ms		132
steroids	gonadal tissue		^{14}C	gc		133
DDT, chlorinated aromatic pollutants	fat, milk		^{14}C	chromatography		134
benzo(*a*)pyrene	petroleum substitutes	>10 ng/g	^{14}C	partition/chromatography/gc or hplc	10	135
indol-3-ylacetic acid	plant (maize)	>10 ng	^{14}C	partition/chromatography/gc		136
indol-3-ylacetic acid	Pinus sylvestris	>50 pg	^{14}C	hplc/gc–ms		137
pentachlorophenol	water	200 ng	^{18}O	partition/gc–ms	8	138
prostacyclin + metabolites	urine	1–2 ng	D,T	gc–ms		139
timolol	plasma	>0.5 ng/ml	^{13}C,D	partition/gc–ms		140
paeonol + metabolites	urine	>30 pg	^{13}C,D	tlc/gc–ms	5	141
bile acids	serum	10 pmole	T,^{14}C	chromatography	3	142
amezinium	body fluids	>1 ng/ml	T,^{14}C	chromatography		143
clonazepam	plasma	0.1 ng/ml	^{15}N,^{18}O	gc–ms		144

*C.o.v. = coefficient of variation. †ria = radioimmunoassay.

may be partially lost during the procedure while another constituent takes its place, if the method has poor selectivity. These complicated interactions operate to different extents at different concentrations.

Method B is used mainly to monitor losses during separation steps. Its applicability depends on availability of the labelled compound. Homogenization of the sample with the added standard is a critical step which is often performed only after dissolution of the sample. The decomposition step is then not subject to the control scheme. Depending on the type of isotope, the labelled compounds added are measured either by radiochemical methods (^{3}H, ^{14}C etc.) or by mass spectrometry (^{2}H, ^{13}C, ^{15}N, ^{18}O etc.). Occasionally, when the degree of labelling is excessive – as with $C_{16}D_{34}$ or $C_{32}D_{66}$ – the isotopic effect becomes so apparent that the gas-chromatographic behaviour is influenced. The labelled internal standard is eluted differently [112]. Examples of the application of isotopically labelled internal standards are given in Table 2.17.

The coefficients of variation given in Table 2.17 indicate a precision that is acceptable for the levels quoted. The recovery can be calculated from the results obtained by use of internal standards. In some cases double labelling is used, to improve the specificity or for the second label to be used as a means of determination (see also p. 186). Clonazepam [144] is an example. Many other applications of stable isotopes have been reported [145]. Simple methods for the preparation of highly enriched deuterium-labelled compounds (e.g. steroids) are known [146].

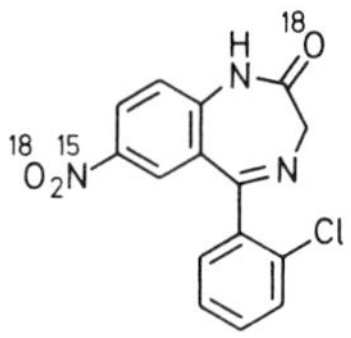

^{18}O ^{15}N-Clonazepam

Method C is used very often in the analysis of therapeutically applied drugs. A few selected examples are given in Table 2.18. Further examples can be found in the procedures for determination of members of this group of compounds by hplc (Table 5.7, p. 157) and gc-ms (Tables 6.8 and 6.9).

The success of this method depends on the similarity of behaviour of the trace and internal standard, which must be as close as possible. On the other hand, it requires very specific procedures for the separate determination of the two closely related substances. The added standard serves an additional purpose, as it becomes absorbed on the walls of containers, on adsorbents etc., similarly to the trace. The recovery of the trace is therefore improved by the resulting flushing action of the internal standard [167], but the recovery of the standard is correspondingly lowered. Some frequently used pairs of analyte and added

Table 2.18 – Examples of addition of internal standard

Compound determined	Matrix	Concentration or amount	Internal standard	Analysis	Recovery/precision	Reference
N-nitrosodimethylamine	fish meal	0.1 ppm	*N*-nitrosodipropylamine	distiln. gc–ms	recovery 75%	147
polyamines	urine		1,6-hexanediamine	partn./gel chrom./fluoresc.		148
phenylethylamine	urine	1–6 ng/ml	tolylethylamine	partn./gc	recovery 75–85%	149
nalidixic acid	plasma		flufenamic acid	partn./gc		150
nicotine	plasma	30 ng/ml	propylnornicotine	partn./gc	c.o.v. 9%	151
morphine, codeine		100 pg	ethylmorphine	gc	c.o.v. 2–5%	152
zearalenone	corn	0.1 ppm	zearalanone	partn./tlc/gc		153
chrysene, benzpyrene	air	0.1–50 ng	benzanthrone	ms	c.o.v. 2–7%	154
progesterone	blood	1 μg/100 ml	testosterone	partn./tlc/gc		155
theophylline	blood	2 μg/ml	cyheptamid	partn./gc	recovery 90–110%	156
	plasma	10–20 μg/ml	hydroxyethyltheophylline	partn./hplc	c.o.v. 4%	157
neostigmine	plasma	5 ng/ml	pyrostigmin-HBr	partn./gc		158
acebutolol	plasma	0.3 μg/ml	ethyl analogue	partn./hplc	c.o.v. 2–6%	159
chlorodesmethyldiazepam	blood	1 ng/ml	pinazepam	partn./gc–ms	recovery 92 ± 2%	160
trimethoprim	blood	20–200 ng/ml	methyl analogue	partn./hplc		161
sulphinpyrazone	plasma	10 μg/ml	phenylbutazone	partn./gc	c.o.v. 2%	162
theophylline	plasma	10–20 μg/ml	8-chlorotheophylline	polarography		163
atenolol	plasma	2 ng/ml	metoprolol	partn./hplc	c.o.v. 5%	164
phenacetin	plasma	0.1 μg/ml	*p*-bromoacetanilide	partn./gc	recovery 98%	165
nitrazepam	plasma	10–40 ng/ml	*N*-desmethyldiazepam	partn./gc	c.o.v. 13%	166

standard are given in a review [168] on sample preparation for clinical trace organic analysis (see Table 2.19, in which the differentiating group is ringed).

Table 2.19 – Examples of internal standards with slightly different structure from the analyte (after [168]).

Trace	Suggested internal standard
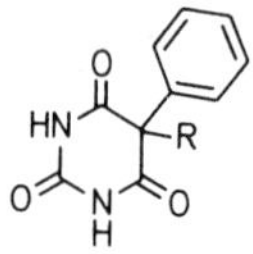 phenobarbital	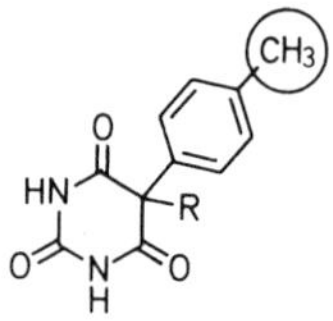 *p*-methyl analogue
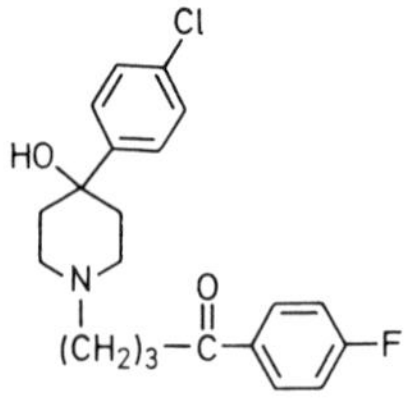 haloperidol	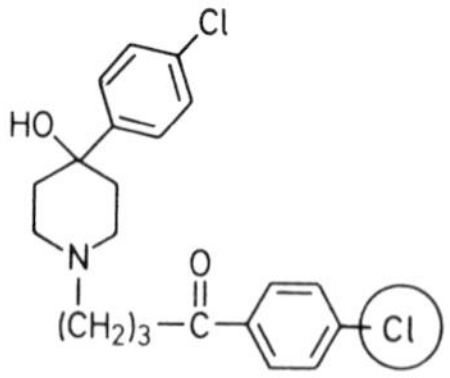 chloro analogue
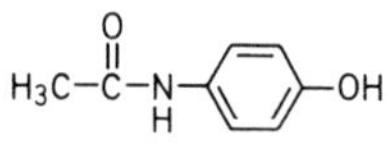 acetominophen	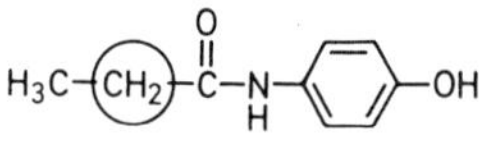 *N*-propionyl analogue
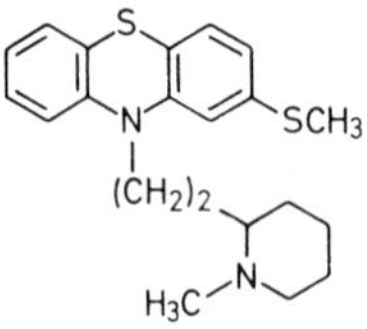 dothiepin	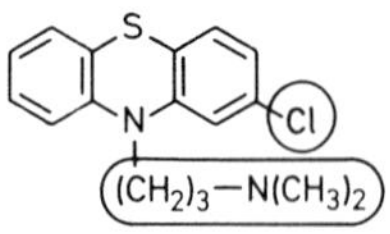 amitryptiline

2.5.4 Systematic control of the entire procedure

Some information on the performance of the analytical system can be gained by systematically checking the entire procedure step by step. Therefore, it is split up into many small consecutive steps. The determination is checked first. Afterwards the step before the determination is investigated and so on, until finally the whole scheme has been checked by the analyst. This approach is very time-consuming. It requires a lot of analytical know-how and scientific training to foresee possible implications and to check for these. Collaborative studies can be useful to economize on time investment. There is a trend towards using just a few procedures which have been checked very carefully (e.g. Normverfahren.)

2.5.5 Use of different and independent methods

There is a very good strategy for checking the analytical result, especially with respect to specificity and systematic errors (see p. 82), namely the use of at least two entirely different procedures and independent techniques for the determination. Often enough, however, in organic trace analysis not even one reliable method exists. Nevertheless with the progress in organic trace analysis the literature shows a more and more pronounced trend towards publication of comparative values obtained by two or more independent methods. (See, e.g. comparison of the results of immunoassays and other methods, Table 6.25).

2.5.6 Combination of control methods

The best procedure can be deduced from the following (translated) quotation [66]. "While reproducibility of results can easily be determined by means of measurements, the determination of the reliability is often more difficult and requires more work such as:

- analysis by as many different methods, analysts and instruments as possible; when the agreement is good, the results are assumed to be accurate;
- control analysis of standardized material, which must be as similar as possible to the material to be analysed; agreement between certified and observed values is a direct measure of accuracy of the particular type of analysis.
- participation in interlaboratory comparisons; samples used in such an intercomparison should be, as far as possible, similar in composition and concentration to the samples to be analysed on a routine basis; the agreement of results from a particular laboratory with the most probable value obtained from a statistical evaluation of all the results is a measure of the accuracy for the type of analysis under investigation."

2.6 INTERLABORATORY CONTROL AND COLLABORATIVE STUDIES

When reference methods are in the development stage, before recommendation they have to be validated. Besides comparison of the results with those obtained by independent methods, the practicability in daily routine and performance in the hands of different technicians in different laboratories has to be checked, as well. This is best done by a collaborative analysis, which is therefore a pre-requirement before a method can be approved as certified. In such an analysis a number of laboratories are asked to analyse a set of samples by the same procedure, each performing a predetermined number of assays.

In the U.S.A. a study has been made in which analyses of three degrees of difficulty had to be made by several laboratories. The results were compared. In the first step, solutions containing PAHs at the μg/g level were distributed and were to be analysed by hplc. The interlaboratory relative standard deviation was 2%, while results from all the laboratories combined resulted in an interlaboratory relative standard deviation of 11%. An analogous experiment comprising the analysis for phenols in water resulted in an interlaboratory relative standard deviation of more than 20%. The third stage of assessment was through interlaboratory comparison on 'real-life' samples. The National Bureau of Standards distributed oyster tissue homogenates and asked for an analysis for lindane and dieldrin. The results showed a relative standard deviation of 200% (i.e. the distribution of the results was strongly skewed). "In the analysis of such samples it is currently impossible to assess absolute accuracy, since extractions are often incomplete and since the matrix cannot be destroyed to release the analytes totally as in inorganic trace analysis" [3]. In a more recent experiment, mussel tissue homogenates, containing traces of hydrocarbons, were analysed by 8 laboratories. Values showed an interlaboratory relative standard deviation of 40%, "owing to sample inhomogeneity, storage instability over 9 months and analytical uncertainty" [169]. A summary of results from other collaborative studies is given in Table 2.20.

An International Mycotoxin Check Sample Programme has been set up [186, 187], and a symposium on collaborative interlaboratory studies has been held [188].

2.7 GOOD LABORATORY PRACTICE (GLP)

2.7.1 Introduction

Under the heading 'Good Laboratory Practice' we propose to describe how a laboratory should work, be organized, and how it should operate to produce valid data. GLP does not describe particular analytical methods or parts of analytical procedures. As GLP comprises all functions of the laboratory which influence the quality of the analytical work, the laboratory design and space are just as important as the equipment, the performance of the work, the laboratory management, the technical staff, the reporting, the archives etc.

In some countries legislation requires chemical manufacturers to perform laboratory tests and submit the results to a government authority for assessment of the potential hazard to human health and the environment. For this purpose an unambiguous data-base is required. Consequently, government authorities stress the implementation of the principles of good laboratory practice to ensure and promote the generation of the necessary quality of test data.

Table 2.20 – Results of some collaborative studies in organic trace analysis

Substance	Matrix	Specified or most probable value	Method of determination	Number of laboratories participating	Results	Reference
ethylenethiourea	potatoes, milk	0.1 μg/g	gc-flame phot.	8	85–97% recovery	170
hexachlorobenzene	fish, butter	0.06 μg/g	gc–ecd	8	98 ± 7% recovery	171
histamine	tuna	200 μg/g	fluorim., colorim.		99 ± 6% recovery	172
histamine	plasma		enzymat., fluorimetry	12	26% c.o.v.	173
tri-iodothyronine	serum	200 ng/g	ria	30		174
pesticides		0.02–2 μg/g	gc		30% c.o.v.	175
pesticides	water	8–150 ng/l.	gc–ecd			176
hydrocarbons	marine sediments	9–500 μg/kg	gc	8	25% c.o.v.	177
hydrocarbons	marine sediments	5–300 ng/g	gc–ms	13	8–42% c.o.v.	178
PCBs	milk		gc–ecd	10		179
vitamin D	oil		hplc or gc	9		180
methyl benzoquate	animal feed	5–15 mg/kg	chromat./fluor. or photometry		102–106% recovery 5–10% c.o.v.	181
aflatoxins		10 ng/g			70% recovery 30–40% c.o.v.	182
arprinocid	feed	50–70 μg/g	hplc	14	97–104% recovery	183
53 pollutants	environm. samples		gc–ms	5	24% c.o.v.	184
N-nitrosodimethylamine	malt		partn./gc	7		185

Unfortunately, this creates a good deal of bureaucracy which sometimes hampers sound laboratory work as much as it tends to improve it.

It is a typical feature of GLP that its application makes data clearer for internal or external control. GLP dictates how and when data are to be documented so that reconstruction of the process of data acquisition is always possible. In principle, all studies for the evaluation of toxicological risks for man and the environment should be made according to the rules of GLP. Such studies almost always include organic trace analyses. It is therefore necessary to describe the principles of GLP rather fully in this book.

2.7.2 Current government implementation of GLP

A number of countries are active in implementation of GLP, most notably the United States (see Table 2.21). The procedures proposed in the U.S.A. have been published in the *Federal Register* (1977–1979). [For further details see *Anal. Chem.*, **51**, 1207A (1979)].

Both the FDA (Food and Drug Administration) and EPA (Environmental Protection Agency) have collaborated very closely in the development of GLP regulations. The FDA enforces these by a compliance programme which includes data audits and routine inspections of test laboratories. The FDA has an interagency agreement with the EPA whereby it inspects test laboratories for the EPA. The FDA has also entered into bilateral agreements with Canada and Sweden. These conventions aim at mutual recognition of national GLP regulations and of national inspections. Interest in negotiating such bilateral agreement has been expressed to the FDA by Japan, The Netherlands, Switzerland and the United Kingdom. The EPA favours a multilateral agreement rather than several bilateral agreements. Therefore, the statements of international bodies are of interest.

In the European Economic Community the 6th Amendment of the 1967 Directive (79/831/EEC) refers only once to GLP (in the preambles to Annexes VII and VIII): "The bodies carrying out the tests shall comply with the principles of good current laboratory practice".

Within the OECD a number of member countries have recognized the need for international harmonization of GLP and prepared an OECD document (ENV/CHEM/MC/79:5). Therefore, the principles of the OECD guidelines and those published by Telling [189], together with the author's* experiences, will be taken as the basis for the following sections.

2.7.3 Definitions, terminology

Test substance means a defined chemical substance or mixture which is under investigation. (*Reference*) *control substance* means any defined chemical substance or mixture other than the test substance, used to provide a basis for

*Again, Dr. Gorbach of Farbwerke Hoechst is to be thanked, for preparation of Section 2.7 of this book.

Table 2.21 – US federal regulatory agency proposals affecting analytical chemistry

Title	Scientific Area	Primary Analytical Involvement	Agency
Good Clinical Practices (GCP)	Human Efficacy/Safety Studies	Test material characterization; human fluid/ tissue analysis	FDA (Drugs)
Bioequivalence/ Bioavailability	Biopharmaceutical Studies	Body fluids/tissues analysis; methods development (pharmacokinetic)	FDA (Drugs)
Federal Insecticide, Fungicide & Rodenticide Act (FIFRA) – Registration Guidelines	Environmental/Toxicology/ Efficacy Studies	Test material characterization; residue analysis	EPA (Pesticides)
Toxic Substances Control Act (TSCA) – Sect. 4, Chronic Testing Standards and GLP's	Human and Environmental Safety	Method development (trace levels)	EPA (Chemicals)
Toxic Substances Control Act(TSCA) – Sect. 4 Acute & Subchronic Toxicity, Mutagenic & Teratogenic Effects, & Metabolism Studies	Human and Environmental Safety	Method development (trace levels)	EPA (Chemicals)
Cancer Policy (Various Federal Agencies)	Human and Environmental Safety	Regulation of hazardous chemicals. Affect facility, design, exposure to and how, or if, chemicals can be used	OSHA CPSC EPA HEW
IRLG – Report: Guidelines for Identifying Carcinogens and Estimation of Risks	Human and Environmental Safety	Test material characterization; residue analysis	IRLG (Reps. from FDA, OSHA, CPSC, EPA)
Sensitivity of Method – Chemical Compounds in Food-producing Animals – Carcinogenic Residues	Human and Environmental Safety	Analytical methods development	FDA

comparison with the effects of a test substance. The control substance is often termed 'reference substance', which must not be mistaken for 'reference standard'.

Impurity means a chemical substance which is unintentionally present in small quantities in another chemical substance.

Batch means a specific quantity or lot of a test or reference substance of uniform character which will be confirmed by appropriate test methods.

The following terms concern the test facility. *Test facility* means the individuals, premises, and operational units that are used for the development of data. The test facility is directed either by a board (or committee) or by an individual. *Study director* means the individual responsible for the overall conduct of the study. *Quality assurance programme* means an internal control system designed to ensure the quality and integrity of the study, in compliance with the principles of GLP.

The following terms concern the study. *Test system* means any animal, plant, microbiological, sub-cellular, chemical or physical system used in a study. *Study plan* means the description of the entire scope of the study, test guidelines, the methods used. *Protocol* means a detailed description of the test and the directions which are to be followed. *Raw data* means all original records and/or documentation, including verified copies thereof, which are the result of the original observations and activities in a study which are necessary for the reconstruction and evaluation of the final report. *Specimen* means any material derived from a test system for examination, analysis or storage. *Standard operating procedure* means procedures which describe in detail how to perform certain routine laboratory tests or activities normally not specified in study plans or test guidelines.

The following definitions refer to characterization of substances. Standard operating procedures should be established for the *receipt* of test and reference substances. Records to be kept include substance characterization, date of receipt, and quantities received. *Handling and storage* procedures should be identified for each study in order that stability is ensured. Contamination or mixture of the substances is to be precluded. Storage containers should carry complete identification information. Any expiry date and specific storage instructions should be included. The proper *disposal* of test, reference and other waste substances by procedures that avoid adverse effects on human health and the environment should be established.

With respect to characterization of test and reference substances the following recommendations are made. Each test and reference substance should be appropriately identified [e.g. code number, Chemical Abstracts number (CAS), name]. For each study the identity, including batch number, purity, composition, concentrations, or other characteristics to define appropriately each batch of the test or reference substances should be known. The stability of the test and reference substances under conditions of storage should be described for all studies. A sample for analytical purposes from each batch of test substance should be retained in the case of studies lasting longer than four weeks.

When it is necessary to mix, dilute, suspend or dissolve a test or reference substance with a vehicle for administration to a test system, standard operating procedures should be established, to

(1) test the homogeneity,
(2) test the stability of the test and reference substance in the vehicle;
(3) test periodically, when appropriate, the concentration of the test and reference substance in the vehicle.

If it is known that the test or reference substance or the vehicle, or the combination thereof, has a limited shelf-life, the date of preparation and an expiry date should be shown on the container(s).

All reagents and solutions which are used in a study should be labelled to indicate source, identity, concentration or titre, and stability information as appropriate (e.g. date of preparation, an expiry date, storage conditions, and any special instructions.) Standard operating procedures should be available for the laboratory preparation of reagents and solutions.

The OECD-guidelines fix the duties of the chief management of the test facility very strictly. Test-facility management should provide for, but not be limited to, the following functions:

(1) ensuring that qualified personnel, facilities, equipment, and materials are available;
(2) maintaining a record of the qualifications, training, experience and job description for each professional and technical worker;
(3) ensuring that personnel clearly understand the functions they are to perform and, where necessary, providing training;
(4) ensuring that there is a quality assurance programme;
(5) maintaining a master schedule of all studies conducted at the test facility;
(6) ensuring for each study that a sufficient number of persons is available to form an appropriate interdisciplinary team for the timely and proper conduct of the study;
(7) designating an individual with the appropriate qualifications, training and experience as Study Director for each study and before the study is initiated.

The study director carries the responsibility for the entire carrying out of the study, its evaluation and the recording. He is responsible for adherence to the study plan and following of the protocol, including all accepted alterations of the plans. The study director should sign the final report, and ensure the validity of the data and compliance with the principles of GLP. The study director should be a scientist or other person with adequate training and practical experience.

2.7.4 Quality assurance programme

A quality assurance programme is necessary. A test facility should have procedures for verifying that the facilities, equipment, personnel, methods, and practices, including test substance characterization, procedures, records, reports and archives, conform with the principles of GLP, and for ensuring the reliability and integrity of data obtained from a study. The quality assurance procedures should be carried out by an individual or individuals designated by management and not directly involved in the conduct of the study. They should report their findings directly to management and to the study director.

The quality assurance programme should be defined in writing and provide for, but not be limited to, the following functions:

(1) periodical inspection and audit of the test facility and its equipment to ensure compliance with the principles of GLP;
(2) keeping copies of standard operation procedures and study plans in use at the test facility;
(3) promptly reporting to management and to the study director any deviations from the study plan and standard operating procedures;
(4) reviewing final reports to ensure that each report accurately describes the methods, procedures and observations, and that the reported results accurately reflect the raw data of the study;
(5) preparation and signature of a statement to be included with the final report, which specifies the dates when inspections were made and findings were reported to management and to the study director.

The quality assurance programme outlined above is specifically concerned with animals, but may also be applied in organic trace analysis concerned with environmental hazard evaluation.

2.7.5 Rooms and facilities

The test facility should be of suitable size, construction and location to meet the requirements of the study and minimize disturbances that would interfere with the validity of the study. The design of the test facility should provide a degree of separation of the different functions adequate enough to ensure the proper conduct of the study. It is necessary to prevent contamination, so there should be separate areas for

(1) receipt and storage of the test and reference substances;
(2) mixing the substances with a vehicle, e.g. feed;
(3) storage of substances and of substances in vehicles.

Suitable facilities should be available for diagnosis, treatment and control of the test system, and to minimize vermin infestations, odours, disease hazards, and environmental contamination. There should be storage areas for supplies and equipment. These should be separate from areas housing the test systems and should be adequately protected against infestations and contamination. Refrigeration should be provided for perishable commodities.

Separate laboratory space should be provided as needed for the performance of routine or specialized procedures, e.g. control areas for work with radioactively labelled substances. Separate space should be available for cleaning, maintaining and, if necessary, sterilizing equipment and supplies. Space should be provided for archives, to which only authorized personnel are allowed access.

2.7.6 Laboratory for organic trace work

These very general guidelines to GLP need supplementation by more specific comments relevant to organic trace analysis.

One of the basic definitions of trace analysis is that the term 'trace' means a component present in a concentration below 10^{-3}%; in practice the concentration may be as low as 10^{-10}% or even lower. It follows that a trace analytical laboratory starts an analysis with a very large sample ('bucket' chemistry) and ends with an ultramicro technique. Hence, as a rule, at least four different working areas should be provided in an organic trace laboratory.

The first area is for sample preparation; here is done all the work which requires macerators, mincers, shredders, blenders, mixers, etc. for comminution and homogenization of samples. In this area the subsample is prepared from the bulk, and sent to the second laboratory area.

In the second area all extraction and clean-up procedures are performed. This main working area should be treated as a solvent laboratory and all electrical equipment in it should be spark-free. The area should be well ventilated. It is important to have enough fume-cupboards and fume hoods available. Small movable fume hoods made from plastic are advantageous; they are ventilated through tubes that are connected to the main ventilation system. This saves working time because volume reduction by evaporation can be done in the working area by use of a controlled jet of compressed clean air or ventilators.

Mobile laboratory furniture is very useful; it can be adapted more easily to alterations and improvements. At a few fixed locations in the laboratory, supplies of water, electricity and – if available – vacuum and auxiliary gases such as nitrogen and compressed air should be accessible, preferably piped either from the floor or overhead from the ceiling. Movable tables with ceramic or highly resistant plastic or stainless steel tops are grouped around the service supplies. Steel frames fixed to the tables are convenient for carrying shelves or fixing stirrers and other equipment. Cupboards mounted on castors or rollers may be grouped in U or L shape to provide additional working space. The L arrangement is preferable for micro-scale work, which is best done seated and with all required tools at hand. The design and organization of such a working area has been described [190].

Gas cylinders should be kept out of the laboratory, and the gases piped into the work area.

Only small volumes of solvents should be kept on the working bench and the bulk solvents should be stored separately under the conditions prescribed by the safety regulations. Special safety screens and protective glasses and clothing are essential when working with hazardous reagents, e.g. methylating reagents and those which might cause cancer. Fire extinguishers, fire blankets and a spillage treatment kit are obligatory.

A third area should be available for delicate and sensitive equipment, such as analytical balances, and electronic and optical equipment. This room should be kept clear from corrosive reagents and any chemical operations whatsoever, with the possible exception of preparation of derivatives for gas chromatography (e.g. silylation).

A fourth area should be made available for sample storage either in a deep-freezer (−20°C) or at ambient temperature, but the reference standards and other chemicals should not be stored here.

2.7.7 Equipment and reagents

Equipment used for generation, measurement, or assessment of experimental data, and for controlling environmental factors relevant to the study (e.g. temperature, humidity, light) should be suitably located and of adequate design and capacity.

Equipment used in a study should be inspected, cleaned, and maintained and/or standardized according to standard operating procedures. Records of such procedures should be kept.

Chemicals used as reagents are a special problem in trace analysis because trace impurities may interfere in the analysis and may also be the source of blank values. This is not necessarily a handicap in trace analysis, but it can become very troublesome when the blank values have a large variance or become large in comparison with the sample signal.

The reagent specifications are therefore of utmost interest, and must take into account the specific use of the reagents. For example, those impurities which give a blank in spectrophotometry may be of no significance in gas chromatography, and vice versa. Quality control of the reagents has to take this into consideration, and sometimes more than three differently specified batches of a particular solvent or reagent have to be kept in storage to be at hand for the analysis. Major errors and even the loss of a complete set of determinations can be caused by using the wrong batch of solvent or reagent for an analysis.

To reduce the cost of the quality control it is advisable to buy a relatively large quantity of a batch of the required solvent or reagent that satisfies the specification (or to clean up a slightly inferior batch), but the cost must be balanced against that of the storage facilities needed.

In routine work it is economical to install distillation apparatus dedicated to specific solvents.

Reference substances require special attention. It is advisable to store these in the dark and at −20°C. It is useful to divide reference materials (especially solutions) into small portions (0.5–10 g) so that not much more than the amount actually needed is taken out of stock, since any surplus is discarded, and is never returned to the cold storage area. This is done because repeated opening of a cold sample container results in condensation of water in the interior of the bottle. Access of air can also be detrimental. Further, it should be strictly forbidden to handle undiluted solutions of reference materials within the trace analysis area, and all glassware that has been in contact with high concentrations of the analytes must be kept outside the trace analysis area.

Storage of solvents in glass bottles should be avoided for organic trace analysis. Aluminium containers are preferable.

2.7.8 Standard operating procedures

The test facility should have written ‘Standard Operating Procedures’, approved by the management, and adequate to ensure the quality and integrity of the data generated in the course of the study.

Each laboratory should have the relevant standard operating procedures readily available. Textbooks, journals and manuals may be used as supplements to these standard procedures. A historical file of all standard operating procedures and their revisions, including dates of such changes, should be kept.

Standard operating procedures should be available for, but not limited to, the following routine laboratory activities:

(1) receipt, identification, characterization, handling, formulation and storage of substances;
(2) testing the stability of substances;
(3) testing the homogeneity and concentration of substances in carrier matrices;
(4) collection, characterization and handling of samples;
(5) use of equipment;
(6) maintenance, cleaning, calibration and/or standardization;
(7) documentation, evaluation, reporting of data;
(8) preparation of reports;
(9) storage and retrieval of data and reports.

2.7.9 Study plan

The result of an analysis is always part of a complex investigation system and the basis for a decision-making process (see p. 40).

This is best illustrated by an example. For the toxicological evaluation of a new chemical it is quite common that small amounts (mg/kg) of the substance are mixed with animal feed. In this connection, there are two questions for the analyst.

(1) Does the spiking result in a homogeneous distribution?
(2) Does degradation of the chemical take place during storage?

The results are required by the toxicologist to decide whether or not the mixing procedure needs to be optimized and how long the food-mixture can be stored before it must be discarded and a fresh batch prepared. This example demonstrates that it is necessary to come to an *a priori* agreement with the toxicologist on the required quality of the analytical results, i.e. decision levels have to be stated at which the following conclusions are drawn:

(1) substance is stable in the food mixture or not;
(2) substance is distributed homogeneously or not.

Both the problem *per se* and its solution may be rather complex, and the course of the analysis has to be planned carefully and presented synoptically to avoid errors and confusion as far as possible.

For economic reasons it is necessary to prepare the strategy of problem solution well ahead. Only in this way is it possible to prevent mistakes and to delegate the major part of the analytical work to the co-operating technicians. A GLP inspection has to be foreseen, as well! The analytical scheme in organic trace analysis is usually rather complicated and widely branched. Again, it is necessary to plan the procedure ahead in order to run adequate checks on the whole experiment.

The study plan should be in written form before the work is started. It should contain at least the following information:

(1) a descriptive title;
(2) a brief statement revealing the nature and purpose of the study and the required quality of results (decision limits, etc.);
(3) identification of the test and reference substance by name (IUPAC), Chemical Abstracts reference number (CAS), and/or code number;
(4) name and function of the staff who developed and approved the study plan;
(5) name and address of the sponsor and the test facility, including responsible management officials;
(6) the name of the study director;
(7) the proposed starting and completion dates;
(8) the justification for selection of the test system;
(9) where appropriate, characterization of the system such as the organisms, species, strains, source of supply, number, body weight range, sex, age, and other pertinent information with respect to the test system;
(10) the procedure for identification of the test system during the study, when necessary;
(11) detailed information on the experimental design, including a description of the chronological procedure of the study, all methods, materials, conditions, type of frequency of analysis, measurements, observations and examinations to be performed.

(12) where appropriate, the rate of administration and the reason for its choice;
(13) where appropriate, the dose levels and/or concentration(s), frequency, duration, and method of administration of the test and reference substance;
(14) description of the scheme of randomization and the statistical method;
(15) the records to be kept.

As an aid for planning a study, flow sheets are very useful, especially when the study tends to be widely branched. Simple symbols for use in (organic trace) analysis have been developed [191] (see Figs. 2.14 and 2.15).

Symbol	Meaning
□	procedure, analytical step
▭	sample, subsample, fraction
→	flow of material
|—	two-phase system g = gas phase i = liquid phase s = solid phase
→	mobile phase
/	chemical rection
//═	auxiliary materials
⯃	measurement
○ ●	nodular points
◇	alternative

Fig. 2.14 – Symbols for designation of analytical steps in flow sheets.

Symbol	Meaning	Example of application	
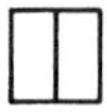	physical separation of existing phases	 l s	filtration
	distribution without auxiliary component	 g / l	head-space analysis
	distribution (without auxiliary component) with mobile phase	 g / l	distillation
	distribution with auxiliary component	 l / l	liquid–liquid distribution
	distribution (with auxiliary component) with mobile phase	 g / l	gas–liquid chromatography
	chemical reaction without auxiliary component		pyrolysis
	chemical reaction with auxiliary component		derivative formation
 A	measurement of A	 pH	pH-value
 (a)	measurement of a quantity with the unit a	 (g)	weighing
 A(B)	measurement of a dependence A (B)	 UV	ultraviolet spectrum

Fig. 2.15 – Examples of application of the symbols given in Fig. 2.14.

A schematic diagram can be drawn by use of these symbols (see Fig. 2.16). A few simple rules suffice for versatile application of this system. As an example, part of a widely branched procedure for processing a plant sample in a metabolite study is presented here both as a flow sheet (Fig. 2.16) and in written form.

(1) After harvest a portion of fresh sample was combusted immediately (A 1), and another was combusted after drying at 100°C (A 2), for the determination of radioactivity.

(2) A further portion of the plant sample was rinsed with methylene chloride. The rinsings were collected and concentrated (A 3). The radioactivity of this solution was measured and a thin-layer chromatography assay done.

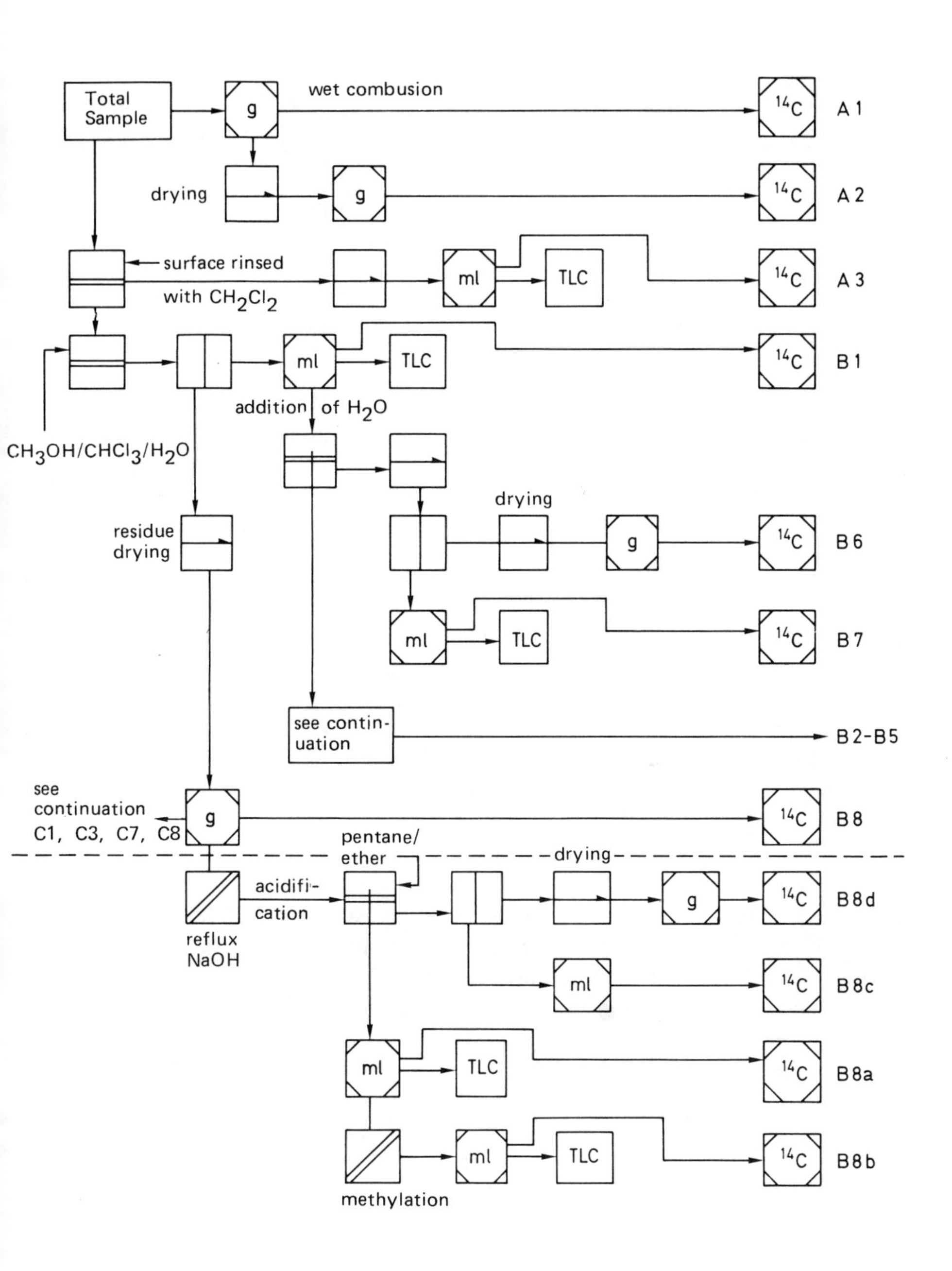

Fig. 2.16 – Example of a flow sheet for a work-up procedure.

(3) The washed plants were homogenized and then extracted with three portions (70, 40, 40 ml) of chloroform/methanol/water mixture (1:2:1 v/v). The liquid phases (B 1) were separated from the solid phases (B 8) on a sintered-glass filter funnel. Fraction B 1 was examined by thin-layer chromatography and the radioactivity of aliquots of fractions B 1 and B 8 was measured.
(4) The addition of water to fraction B 1 caused separation into a water/methanol phase and a chloroform/methanol phase.
(5) After concentration and subsequent filtration of the chloroform/methanol phase the radioactivity was measured in a fraction of the residue (B 6) and aliquot of the filtrate (B 7). The latter was also examined by thin-layer chromatography.

The end of this description is indicated by the broken line on the flow sheet.

Such flow sheets have proved to be an excellent guide for the technicians at the bench. They may be fixed at the back of the laboratory bench in plain sight. A tedious search for the next step in the written procedure is thus avoided.

2.7.10 Conduct of the study and recording of data

The study should be conducted in accordance with the study plan. All data generated and procedures followed except those generated as direct computer input, should be recorded directly, promptly, accurately and legibly by the individual entering the data, and should be signed or initialled, and dated. Any change in the raw data should be made in such a way that the previous entry is not obscured. If such an alteration is necessary the reason for the change must be indicated and the change identified by date and the signature of the individual making the change.

Data generated as direct computer input should be identified at the time of data input by the individual(s) responsible for direct entries. Corrections should be identified separately by the reason for change, and should be dated.

Practical aids have been developed to facilitate the daily recording of data both for complicated analytical procedures as illustrated in Fig. 2.17 and Table 2.22 and for normal routine analysis.

Figure 2.17 shows a page of the flow sheet for data-recording, which illustrates the mechanism of the recording. The record shows part of the data from the example given in Fig. 2.16. On its left-hand side it follows exactly, but in more detail, the original study-plan flow sheet. The measurements are recorded on the right-hand side. The designations of functions are the same in both flow sheets, and thus provide a perfect guide through the procedure. The recording flow sheet is a convenient means of obtaining a complete data record in the right order of sequence and time, and meets the requirements of GLP. The necessary raw data additional to the recorded measurements in the data-recording sheet are tabulated in a separate data sheet, as illustrated in Table 2.22. To complete the record

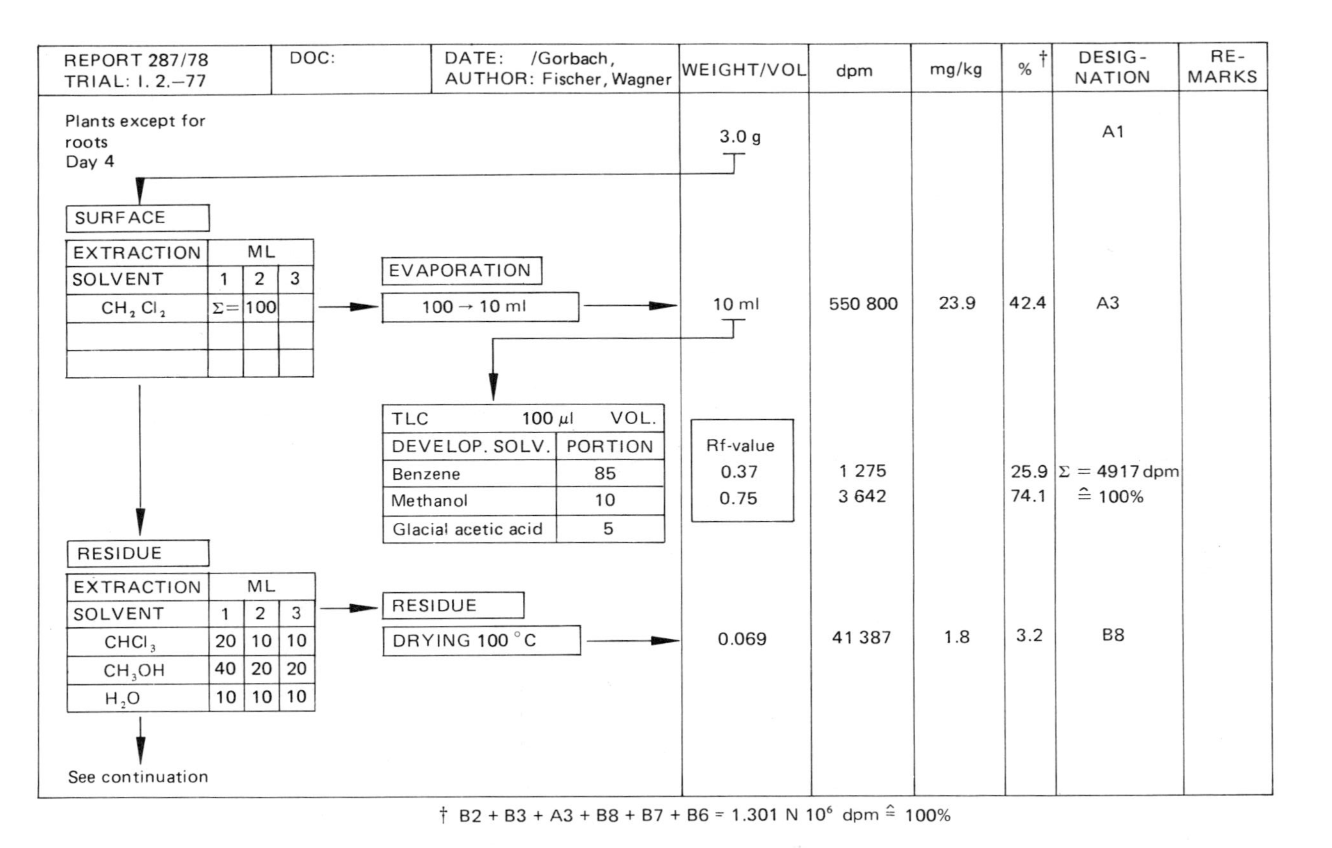

Fig. 2.17 – Example of a study-plan flowsheet which permits registration of all raw data. Flowsheet on the left-hand side, raw data entries on the right-hand side.

Table 2.22 – Example for raw data sheet

Raw Data:
Report: 287/78
Trial: I. 2. –77 Day 4

Doc.:

Date: 13.6.1977
Operator: Fischer, Wagner

SAMPLE				RADIOACTIVITY			RESIDUE CONTENT†
Designation	Amount fresh	Amount dry	Aliquotation sequence [ml] *	dpm in last vol./weight of aliquot sequence	dpm total	dpm per g of fresh material	[mg/kg] (fresh material)
A1 Plant, except roots	3.0 g						
A3 ,,			100 ml → 10 ml/0.025	$1377 = \bar{x}$; n = 3	550800	183600	23,g
B8 ,,		0.069 g	0.069 g	41387	41387	13796	1.8
B2 ,,		0.038 g	0.032 g	2203	2203	734	0.1
B3 ,,			→ 10 ml/0.025 ml	$883 = \bar{x}$; n = 3	353067	117689	15.3
B6 ,,		0.031 g	0.031 g	2411	2411	803	0.1
B7 ,,			70 ml → 10/0.025 ml	$879 = \bar{x}$; n = 3	351600	117200	15.3
A2 Roots	0.18 g	0.023 g	0.023 g	1744	1744	9689 fresh	1.3

*Sequence of volume: first volume, first aliquot etc.
†Calculated on the basis of a molecular weight of 341.

the print-out of scintillation counters, integrators etc., and the recorder graphs of gc, gc–ms and other equipment are added to the data-record sheet file. The designations in the master study-plan flow sheet make it easy to relate each data sheet to the corresponding step(s) in the procedure. The master study-plan flow sheet, the data recording sheets, and the attached data sheets, print-outs and records are all collected in a separate raw-data file which can be attached to the study report.

A similar procedure is followed with routine trace analysis. The design of the initial study-plan need not be repeated here, but an example computer print-out will demonstrate the mode of data-processing and evaluation of results. In industry, samples, e.g. in analysis of pesticide residues, are always analysed in parallel with control samples and spiked samples, so it is useful to record all measurements made in the course of the analysis (weight, volumes, peak heights etc.) in a fixed order. Even a small programmable table computer can be used to

Table 2.23 – Example of a sheet for recording raw data from a residue determination using final determination by gas chromatography.

No.	W g	V_1 ml	T_1 ml	V_2 ml	T_2 ml	V_3 ml	T_3 μl	F_S cm^2	F_T cm^2	M_T ng	added ppm	found %
1	50	2	1	1	1	1	1	0.4	1.0	0.3052		
2	50	2	1	1	1	1	1	0.3	1.0	0.3052		
3	50	2	1	1	1	1	1	0.2	1.0	0.3052		
4	50	2	1	1	1	1	12	1.4	1.4	0.6104	0.001	41
5	50	2	1	1	1	1	2	5.8	6.6	3.052	0.05	
6	50	2	1	1	1	1	1	8.2	12.2	3.052	0.1	
7	50	2	1	1	1	1	1	1.0	0.8	0.3052		
8	50	2	1	1	1	1	1	0.8	1.0	0.3052		
9	50	2	1	1	1	1	1	1.2	1.0	0.3052		
10	50	2	1	1	1	1	1	0.5	0.9	0.3052		
11	50	2	1	1	1	1	1	0.8	1.2	0.3052		

V = measured volume, T = partial volume, F_S = peak area of signal from sample, F_T = peak area of signal from standard, M_T = amount of reference test material used for calibration.

do the necessary calculations. The work is simplified by the fact that all the figures in the form are entered into the computer as they appear from left to right, line by line, in the data form (see Table 2.23).

The printer of the table computer prints the whole data sheet, including all the raw data, in a legible and instructive form as shown in Table 2.24. The two forms are a sufficient raw-data base for the report. It should be remembered that computers often print out a number of insignificant figures, and that the operator must chop the printed result to a realistic number of digits.

2.7.11 Final reporting of study results

The report should be made in a standard format and include at least the following information.

(1) Name of the study director and names of other responsible personnel involved in the study.
(2) A complete description of the course of the study, according to the study-plan, as well as all information and data laid down in the study-plan, and the changes in, or revision of, the study-plan.
(3) If appropriate, a statement concerning the characterization and stability of the test and reference substance under storage and test conditions.
(4) A presentation of the results, including transformations and calculations performed on the data, statistical methods employed and a summary of the data.
(5) An evaluation or interpretation of the result and, where appropriate, conclusions drawn.
(6) The location where all samples, specimens, raw data and the final report are to be stored.

The scientists who participated in the study must sign the part of the report for which they are responsible. The final report is to be signed by the study director. Any corrections and additions to a final report must be in the form of an amendment, which should clearly specify the reasons for correction, additions, etc., and be signed by the study director and each principal scientist from each discipline involved, if necessary. The statements of personnel responsible for the quality-assurance programme are added to the final report.

2.7.12 Archives, storage and retrieval of data, length of storage

All study plans, raw data, final reports, specimens and samples, are stored in archives. Orderly storage and swift retrieval are essential. At least one person must be named as responsible for the organization and maintenance of the archives.

The following material should be retained for the period specified by the appropriate authorities.

(1) The study plan, raw data, samples, specimens and the final report of each study.

Table 2.24 – Print-out of the results calculated from the raw data given in Table 2.23 (n.d. = not detectable).

Analytical raw data
Plan no. 1,2,3,4-502 and 503/79

Agent: phenylurea herbicide
Crop: wheat grain

Control sample	No.	W g	V_1 ml	T_1 ml	V_2 ml	T_2 ml	V_3 ml	T_3 µl	F_S cm²	F_T cm²	M_T ng	Y_B ppm		
	1	50	2.00	1.00	1.00	1.00	1.00	1.00	0.40	1.00	0.30	0.00488		
	2	50	2.00	1.00	1.00	1.00	1.00	1.00	0.30	1.00	0.30	0.00366		
	3	50	2.00	1.00	1.00	1.00	1.00	1.00	0.20	1.00	0.30	0.00244		
						1*B (ppm) = 0.0037								
Recovery experiment		W g	V_1 ml	T_1 ml	V_2 ml	T_2 ml	V_3 ml	T_3 µl	F_S cm²	F_T cm²	M_T ng	Y_U ppm	AD ppm	R %
	4	50	2.00	1.00	1.00	1.00	1.00	2.00	1.40	1.40	0.61	0,01220	0.01000	85.0
	5	50	2.00	1.00	1.00	1.00	1.00	2.00	5.80	6.60	3.05	0.05364	0.05000	99.8
	6	50	2.00	1.00	1.00	1.00	1.00	1.00	8.20	12.20	3.05	0.08205	0.10000	78.3
				2*R (%) = 87.7			LC (ppm) = 0.01							
Sample		W g	V_1 ml	T_1 ml	V_2 ml	T_2 ml	V_3 ml	T_3 µl	F_S cm²	F_T cm²	M_T ng	Y_U ppm	$Y_S^{cor.}$ ppm	
1–502–1 day 111	7	50	2.00	1.00	1.00	1.00	1.00	1.00	1.00	0.80	0.30	0.01526	0.0132	0.01
B–502–1 day 91	8	50	2.00	1.00	1.00	1.00	1.00	1.00	0.80	1.00	0.30	0.00976	0.0069	n.d.
M–502–1 day 83	9	50	2.00	1.00	1.00	1.00	1.00	1.00	1.20	1.00	0.30	0.01464	0.0125	0.01
B–502–1 day 99	10	50	2.00	1.00	1.00	1.00	1.00	1.00	0.50	0.90	0.30	0.00673	0.0035	n.d.
HH–502–1 day 88	11	50	2.00	1.00	1.00	1.00	1.00	1.00	0.80	1.20	0.30	0.00813	0.0051	n.d.

(2) Records of all inspections and audits performed under the Quality Assurance Programme.
(3) Summary of qualifications, training, experience and job descriptions of the personnel.
(4) Records and reports of the maintenance and calibration of equipment.
(5) The chronological file of Standard Operating Procedures.
(6) A record of the fate of the substances and reference materials used, after termination of the study.

Samples and specimens should be retained only as long as their condition permits evaluation.

If a test facility or an archival facility goes out of business the archives should be transferred to the archives of the sponsor(s) of the study. If the sponsor closes his business, he is obliged to offer the archives to competent and appropriate (e.g. governmental) administrative offices.

2.7.13 Quality control

The performance of the procedures used is constantly checked by quality control. Such methods have been introduced on a very broad basis by clinical chemists [192–195]. The objectives of quality control are [196]:

(1) monitoring of random errors = precision control;
(2) monitoring of systematic errors = accuracy control;
(3) monitoring of the effect of the matrix on precision, accuracy and specificity;
(4) monitoring of trends within a series, immediate recognition of deviations.

A basic programme – modelled on the Guidelines of the Medical Society of West Germany – has been presented [44], which contains:

(1) internal (intralaboratory) quality control:
    (a) precision control at the most frequent decision limit, by analysing samples of the same control specimen in every analytical run;
    (b) accuracy control over the whole relevant range of measurement by analysing an accuracy-control specimen in every fourth analytical run; the control specimen is selected from a number of different control specimens kept on hand;
(2) external (interlaboratory) control in the form of short-term interlaboratory surveys, with at least two control specimens of different concentrations.

Suggestions have been made [197] that "10 to 20% of an analyst's time should be devoted to quality control work. This proportion may seem rather large, and may be difficult to achieve while initially establishing an analytical quality control scheme in a laboratory. Nevertheless, such effort is considered a reasonable goal though initiation of a control scheme should not be prevented if only a smaller amount of effort is available initially. In general, a little control is better than none. Obviously, it is not possible to obtain direct estimates of precision and bias for every sample that laboratories analyse. Hence, they cannot prove the accuracy of every result. Rather, quality control provides valuable confirmatory evidence of the adequacy of all the various means adopted by the

Table 2.25 – Sequence of activities for building up an analytical quality control system. (from [197] by permission of the copyright holders, the Royal Society of Chemistry).

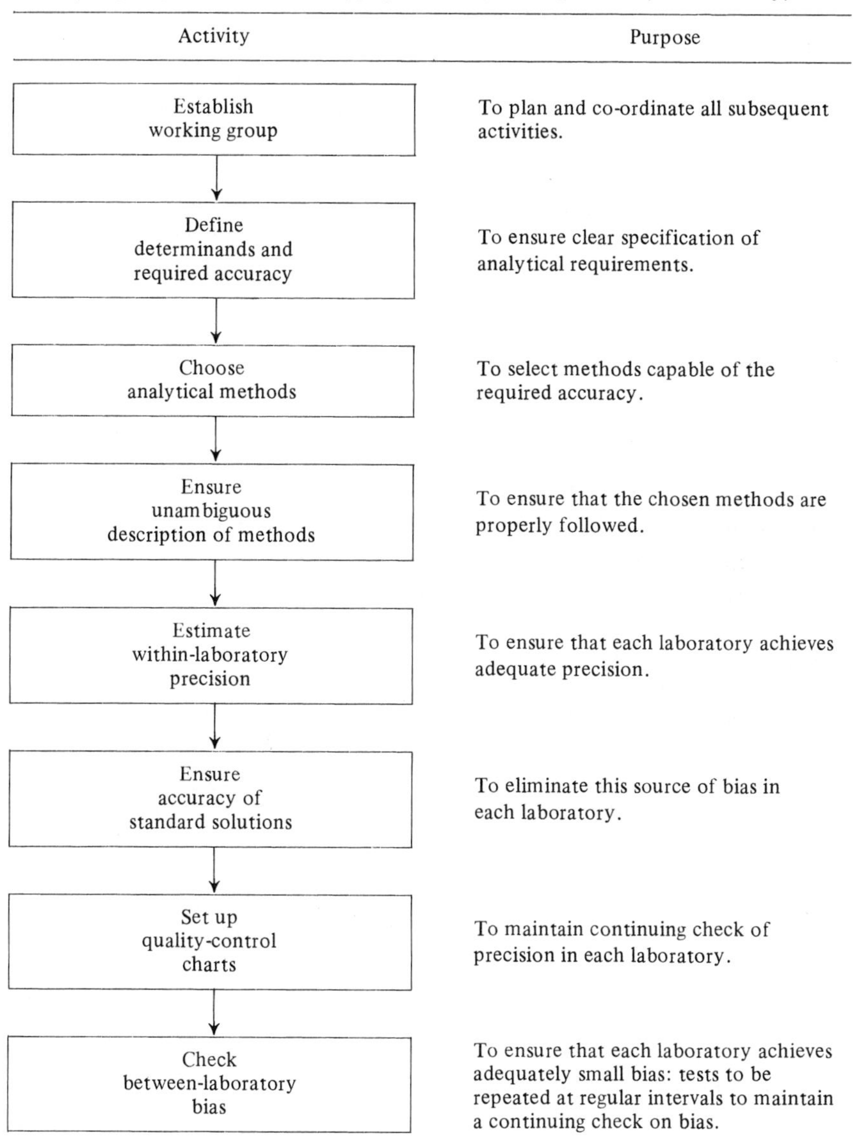

Activity	Purpose
Establish working group	To plan and co-ordinate all subsequent activities.
Define determinands and required accuracy	To ensure clear specification of analytical requirements.
Choose analytical methods	To select methods capable of the required accuracy.
Ensure unambiguous description of methods	To ensure that the chosen methods are properly followed.
Estimate within-laboratory precision	To ensure that each laboratory achieves adequate precision.
Ensure accuracy of standard solutions	To eliminate this source of bias in each laboratory.
Set up quality-control charts	To maintain continuing check of precision in each laboratory.
Check between-laboratory bias	To ensure that each laboratory achieves adequately small bias: tests to be repeated at regular intervals to maintain a continuing check on bias.

analyst to ensure the required accuracy. In other words, the continuing and critical assessment by the analyst of all that he does remains, as always, of prime importance."

A systematic approach for achieving comparable analytical results from a number of laboratories has been suggested [197]. The sequence of activities for analytical quality control is shown in Table 2.25.

2.7.14 Safety regulations

Special safety precautions and facilities are necessary with hazardous substances, e.g. methylating agents, benzene, chloroform and others. Reference may be made to MAK (Maximale Arbeitsplatz Konzentrationen = maximum allowable workplace concentrations) values and safety regulations [198, 199]. Summaries of methods for handling highly toxic materials have been given [200, 201]. Work with such substances must be done under hoods. Protective safety goggles and gloves must be worn. Adequate shoes (flat-heeled, closed and comfortable) and clothes (laboratory coats) increase safety. Fire extinguishers must be available. In laboratories where solvents are used extensively a 'dry'-type extinguisher is best. A fire-extinguishing cover or blanket, eye-wash bottles, and a set of reagents for treatment of spillage are as necessary as an alarm plan and indication of emergency exits. People working in the solvent area ought to participate twice a year in fire-extinguishing exercises. Accident-prevention information meetings should be held regularly, during which information about the reagents used and their hazards can be imparted. Participation should be mandatory and a record of attendance and the topics treated should be kept.

Solvents are categorized into different groups. Group A contains solvents which do not mix with water. They form a special hazard if they can be ignited easily (see Table 2.26). If possible, solvents of class A-I should not be used in routine analysis. The risk involved in their use becomes even more pronounced in countries with a hot climate.

Table 2.26 – Categories of inflammable solvents which do not mix with water

Class	Ignition temperature (°C)
I	21
II	21–55
III	55–100

Toxic chemicals should be labelled clearly and informatively, including the toxicity class, e.g. according to the Swiss toxicity laws (of March 1969 and February 1978) (see also Table 2.27).

Table 2.27 – Toxicity classes

Toxicity class	LD_{50}* (mg/kg)
1	<5
2	5-50
3	50-500
4	500-5000
5	50000-15000

*LD_{50}: dose at which 50% mortality of test animals (rats) is observed.

Benzene – in spite of its high toxicity being known – is still used as a solvent in liquid-liquid extraction, although it can be replaced in almost all cases by less toxic solvents. Chloroform – which has a toxicity threshold limit value of 5 ppm in workroom air – can be replaced for many purposes by 1,1,1-trichloroethane, which has a threshold value of 350 ppm [202].

It should be emphasized that the fundamental principles of hygiene should be observed. Smoking, eating and drinking should be restricted to safe areas. Routine medical checks should be made if the work involves hazardous substances. Special regulations [203] are to be followed in work with radioactive material.

2.7.15 Disposal of used reagents and solvents

In organic trace analysis, re-use of solvents is not usually possible. Consequently, their orderly disposal has to be organized. Solvents should be collected in metal containers, separated into group A (immiscible with water) and group B (miscible). When full, the containers are sent only to enterprises which are licensed for handling such types of special waste. Receipts of delivery of reagents to these firms should be stored. Any further regulations laid down by national legislation (see e.g. [199]) must be strictly observed. In this context the advantages of micro methods are again obvious – the smaller the volumes of solvents used, the less the problems of waste disposal [190].

REFERENCES

[1] O. G. Koch, *Anal. Chim. Acta,* **82,** 19 (1976).
[2] O. G. Koch, *Anal. Chim. Acta,* **107,** 373 (1979).
[3] H. S. Hertz, W. E. May, S. A. Wise and S. N. Chesler, *Anal. Chem.,* **50,** 429A (1978).
[4] F. E. Freeberg, *Anal. Chem.,* **51,** 1206A (1979).
[5] H. Vogel and R. D. Weeren, *Z. Anal. Chem.,* **280,** 9 (1976).
[6] L. A. Gubov and M. E. Elyashberg, *Zh. Analit. Khim.,* **32,** 2025 (1977).
[7] H. Druckrey and R. Preussmann, *Naturwiss.,* **49,** 498 (1962).
[8] K. Sargeant, A. Sheridan and J. O'Kelly, *Nature,* **192,** 1096 (1961).
[9] W. Forth, *Deut. Ärzteblatt,* **74,** 2617 (1977).
[10] A. M. Binder, *Deut. Ärzteblatt,* **78,** 117 (1981).

[11] J. Huisingh, R. Bradow, R. Jungers, L. Claxton, R. Zweidinger, S. Tejada, J. Bumgarner, F. Duffield, M. Waters, V. Simmon, C. Hara, C. Rodriguez and L. Snow, Application of bioassay to characterization of diesel particle emissions, in short-term bioassays in the fractionation and analysis of complex environmental mixtures, EPA-600/9-78-027, Sept. 1978.
[12] J. N. Braddock and R. L. Bradow, Emission patterns of diesel powered passenger cars, SAE Paper 750682, Houston, Texas, June 1975.
[13] C. T. Hare, K. J. Springee and R. L. Bradow, Fuel and additive effects on diesel particulate-development and demonstration of methodology, SAE Paper 760130, Detroit, Mich., February 1976.
[14] C. F. Rodriguez, Characterization of organic constituents of diesel exhaust particulate, Draft Final Report, EPA Contract 68-02-1777, Task III, June 1978.
[15] M. D. Erickson, D. L. Newton, E. D. Pellizzari, K. B. Tower and D. Dropkin, *J. Chromatog. Sci.*, **17**, 449 (1979).
[16] H. R. Buser and C. Rappe, *Anal. Chem.*, **52**, 2257 (1980).
[17] R. K. Mitchum, W. A. Korfmacher, G. F. Moler and D. L. Stalling, *Anal. Chem.*, **54**, 719 (1982).
[18] P. W. O'Keefe, R. Smith, C. Meyer, D. Hilker, K. Aldous and B. Jelus-Tyror, *J. Chromatog.*, **242**, 305 (1982).
[19] L. L. Lamparski, L. L. and Nestrick, T. J., *Chemosphere*, **10**, 3 (1981).
[20] W. A. Korfmacher and R. K. Mitchum, *J. High Resolut. Chromatog. Chromatog. Commun.*, **4**, 294 (1981).
[21] S. Waliszewski and G. A. Szymczynski, *Z. Anal. Chem.*, **311**, 127 (1982).
[22] S. Waliszewski and M. Rzepczynski, *Z. Anal. Chem.*, **304**, 143 (1980).
[23] V. G. Tsukerman, *Khim. Sel. Khoz.*, **18**, 51 (1980).
[24] K. Matsunaga, F. Shibata, N. Saito, N. Torigashi, K. Kenmochi and Y. Ogino, *Okayama-ken Kankyo Hoken Senta Nempo*, **3**, 192 (1979).
[25] K. Imamura and T. Fujii, *Bunseki Kagaku*, **28**, 549 (1979).
[26] A. A. Garrison, G. Mamantov and E. L. Wehry, *Appl. Spectrosc.*, **36**, 348 (1982).
[27] G. Görlitz and I. Halász, *Talanta*, **26**, 773 (1979).
[28] P. R. Sundaresan and P. V. Bhat, *J. Lipid Res.*, **23**, 448 (1982).
[29] J. W. A. Findlay, J. T. Warren, J. A. Hill and R. M. Welch, *J. Pharm. Sci.*, **70**, 624 (1981).
[30] M. Jiang and D. M. Soderland, *J. Chromatog.*, **248**, 143 (1982).
[31] J. Hermansson, *Acta Pharm. Suec.*, **19**, 11 (1982).
[32] J. Hermansson and C. von Bahr, *J. Chromatog.*, **221**, *Biomed. Appl.*, **10**, 109 (1980).
[33] J. K. Stolteborg, C. V. Puglisi, F. Rubio and F. M. Vane, *J. Pharm. Sci.*, **70**, 1207 (1981).
[34] W. H. Pirkle and D. W. House, *J. Org. Chem.*, **44**, 1957 (1979).
[35] R. A. Chapman and C. R. Harris, *J. Chromatog.*, **174**, 369 (1979).
[36] K. L. McGillard, J. E. Wilson, G. D. Olsen and F. Bartos, *Proc. West Pharmacol. Soc.*, **22**, 463 (1979).
[37] J. Goto, N. Goto, A. Hikichi, T. Nishimaki and T. Nambara, *Anal. Chim. Acta*, **120**, 187 (1980).
[38] E. Kutter, *Arzneimittelentwicklung*, Thieme, Stuttgart (1978).
[39] H. Ballschmiter, *Nachr. Chem. Techn. Lab.*, **27**, 542 (1979).
[40] H. Wald, *Statistical Decision Functions*, Springer, New York, (1950).
[41] D. L. Massart, A. Dijkstra and L. Kaufmann, *Evaluation and Optimization of Laboratory Methods and Analytical Procedures*, p. 146. Elsevier, Amsterdam (1978).
[42] L. Sachs, *Statische Auswertungsmethoden*, Springer Verlag, Berlin (1969).
[43] D. Stamm, *J. Clin. Chem. Clin. Biochem.*, **17**, 283 (1979).
[44] D. Stamm, *J. Clin. Chem. Clin. Biochem.*, **20**, 817 (1982).
[45] H. E. Gutermann, *Technometrics*, **4**, 134 (1962).
[46] *Dokumenta Geigy*, Table, pp. 32–35, Thieme, Stuttgart (1975).
[47] *Dokumenta Geigy*, p. 14, column 2. Thieme, Stuttgart (1975).
[48] W. D. Weinmann and H. G. Nolting, *Nachrichtenbl. Deut. Pflanzenschutz.*, **33**, 137 (1981).

[49] Graf/Henning/Stange, *Formeln und Tabellen der mathematischen Statistik*, Springer-Verlag, Heidelberg (1966).
[50] DIN 53 804, *Statistische Auswertungen messbarer Merkmale*, Sept. 1981, Beuth Verlag, Berlin.
[51] L. A. Currie, *Anal. Chem.*, **40**, 586 (1968).
[52] J. D. Winefordner and J. L. Ward, *Anal. Lett.*, **13**, 1293 (1980).
[53] C. P. B. Adams, W. O. Passmore and D. E. Campbell, Paper No. 14, Symposium on Trace Characterisation – Chemical and Physical, National Bureau of Standards (Oct. 1966).
[54] DFG, *Methodensammlung zur Rückstandsanalytik von Pflanzenschutzmitteln*, part 5, XI, Verlag Chemie, Weinheim (1979).
[55] A. Montag, *Z. Anal. Chem.*, **312**, 96 (1982).
[56] V. Luckow, *Z. Anal. Chem.*, **303**, 23 (1980).
[57] L. M. Schwartz, *Anal. Chem.*, **48**, 2287 (1976).
[58] M. Stockhausen, *Mathematische Behandlung naturwissenschaftlicher Probleme*, Part 1, Steinkopf Verlag, Darmstadt (1979).
[59] R. F. Hirsch, *Anal. Chem.*, **49**, 691A (1977).
[60] H. Heinen and G. Ortner, Trauner Verlag, in the press (1984).
[61] J. Büttner, R. Borth, J. H. Boutwell and P. M. G. Broughton, *Z. Klin. Chem. Klin. Biochem.*, **13**, 523 (1975).
[62] H. Büttner, *Z. Anal. Chem.*, **284**, 119 (1977).
[63] J. P. Cali, *Z. Anal. Chem.*, **297**, 1 (1979).
[64] E. W. May, J. M. Brown, S. N. Chesler, F. Guenther, L. R. Hilpert, H. S. Hertz and S. A. Wise, in *Trace Organic Analysis*, pp. 219–224. NBS, Washington (1979).
[65] J. P. Cali, *Anal. Chem.*, **48**, 802A (1976).
[66] R. J. A. Neider, *Z. Anal. Chem.*, **297**, 4 (1979).
[67] R. Griepink, *Anal. Proc.*, **19**, 405 (1982).
[68] B. F. Schmitt (ed.), *Production and Use of Reference Materials*, Bundesanstalt für Materialprüfung, Berlin (1980).
[69] O. G. Koch, *Pure Appl. Chem.*, **50**, 1531 (1978).
[70] J. B. Addison and M. E. Nearing, *Intern. J. Environ. Anal. Chem.*, **11**, 9 (1982).
[71] D. W. Hodgson and R. R. Watts, *J. Assoc. Off. Anal. Chem.*, **65**, 89 (1982).
[72] D. L. Suett, G. A. Wheatley and C. E. Padbury, *Analyst*, **104**, 1176 (1979).
[73] D. J. Reeder, D. P. Enagonio, R. G. Christensen and R. Schaffer, in *Trace Organic Analysis*, pp. 447–453. NBS, Washington (1979).
[74] H. Bargnous, D. Pepin, J. L. Chabard, F. Veldrine, J. Petit and J. A. Berger, *Analusis*, **5**, 170 (1977).
[75] R. E. Kaiser, *J. Chromatog. Sci.*, **12**, 36 (1974).
[76] Anon., *J. Chromatog. Sci.*, **17**, 300 (1979).
[77] T. G. Kieckbusch and C. J. King, *J. Chromatog. Sci.*, **17**, 273 (1979).
[78] R. A. Rasmussen, *Atmos. Environ.*, **12**, 2505 (1978).
[79] L. Grupinski, B. Föller, W. Staub and A. Wenk, *Wasser Luft Betrieb*, **21**, 184 (1977).
[80] D. E. Harsch, *Atmos. Environ.*, **14**, 1105 (1980).
[81] H. Nozoye, *Anal. Chem.*, **50**, 1727 (1978).
[82] J. J. Ritter and N. K. Adams, *Anal. Chem.*, **48**, 612 (1976).
[83] Anon, *Methods of Determination of Hazardous Substances*, MDHS 3, (Occup. Med. Hyg. Lab., Edgware Road, London N.W.2), (1981).
[84] J. Namiesmik, M. Bownik and E. Koztowski, *Analusis*, **10**, 145 (1982).
[85] D. D. Dollberg and A. W. Verstuyft (eds.), *Analytical Techniques in Occupational Health Chemistry*, American Chem. Soc., Washington (1980).
[86] R. P. Menichelli, *Am. Ind. Hyg. Assoc. J.*, **43**, 286 (1982).
[87] V. Kovar and L. Klusacek, *Chem. Listy*, **75**, 1295 (1981).
[88] Anon, *Methods for Determination of Hazardous Substances*, MDHS 4, (Occup. Med. Hyg. Lab., Edgware Road, London, N.W.2), (1981).
[89] N. Kashihira, *Taiki Osen Gakkaishi*, **15**, 275 (1980).
[90] J. Godin and C. Boudene, *Anal. Chim. Acta*, **96**, 221 (1978).
[91] A. Teckentrup and D. Klockow, *Anal. Chem.*, **50**, 1728 (1978).

[92] P. A. Pella, *Anal. Chem.*, **48**, 1632 (1976).
[93] D. J. Freed and A. M. Mujsce, *Anal. Chem.*, **49**, 139 (1977).
[94] K. L. Geisling, R. R. Miksch and S. M. Rappaport, *Anal. Chem.*, **54**, 140 (1982).
[95] J. Vejrosta and J. Novák, *J. Chromatog.*, **175**, 261 (1979).
[96] J. Novák, V. Vašák and J. Janák, *Anal. Chem.*, **37**, 660 (1965).
[97] W. H. King and G. D. Dupré, *Anal. Chem.*, **41**, 1936 (1969).
[98] R. S. Braman, in R. L. Grob (ed., *Chromatographic Analysis of the Environment*, pp. 82–83. Dekker, New York (1975).
[99] S. T. Kozlov, M. I. Povolotskaya and G. D. Tantsyrev, *Zh. Analit. Khim.*, **30**, 197 (1975).
[100] A. Raymund and G. Guichon, *J. Chromatog. Sci.*, **13**, 173 (1975).
[101] J. E. Lovelock, *Anal. Chem.*, **33**, 162 (1961).
[102] M. Novotny, M. L. Lee and K. D. Bartle, *Chromatographia*, **7**, 333 (1974).
[103] B. E. Saltzman and C. A. Clemons, *Anal. Chem.*, **38**, 800 (1966).
[104] A. E. O'Keefe and G. C. Ortman, *Anal. Chem.*, **38**, 760 (1966).
[105] R. S. Braman and E. S. Gordon, *IEEE Trans. Instrum. Meas.*, **IM-14**, 11, (1965).
[106] F. Bruner, C. Canulli and M. Possanzini, *Anal. Chem.*, **45**, 1790 (1973).
[107] I. A. Fowlis and R. P. W. Scott, *J. Chromatog.*, **11**, 1 (1963).
[108] D. M. Meddle and A. F. Smith, *Analyst*, **106**, 1082 (1981).
[109] E. P. Grimsrud and S. H. Kim, *Anal. Chem.*, **51**, 537 (1979).
[110] E. P. Grimsrud and S. W. Warden, *Anal. Chem.*, **52**, 1842 (1980).
[111] Y. Hattori, Y. Kuge and S. Asada, *Bull. Chem. Soc. Japan*, **53**, 1435 (1980).
[112] B. S. Middleditch and B. Basile, *Anal. Lett.*, **9**, 1031 (1976).
[113] K. F. Faull, P. J. Anderson, D. J. Barchas and P. A. Berger, *J. Chromatog.*, **163**, 337 (1979).
[114] B. Sjöquist, E. Oliw, L. Lunden and E. Anggard, *J. Chromatog.*, **163**, 1 (1979).
[115] S. P. Markey and A. J. Lewy, in *Trace Organic Analysis*, pp. 399–404. NBS, Washington (1979).
[116] S. P. Jindal, T. Lutz and P. Vestergaard, *Anal. Chem.*, **51**, 269 (1979).
[117] D. L. Hachey, M. J. Kreek and D. H. Mattson, *J. Pharm. Sci.*, **66**, 1579 (1977).
[118] D. Robertson, E. C. Heath, F. C. Falkner, R. E. Hill, G. M. Brilis and J. T. Watson *Biomed. Mass Spectrom.*, **5**, 704 (1978).
[119] C. Lindberg, M. Berg, L. O. Boreus, P. Hartvig, K. E. Karlson, L. Palmer and A. M. Thorblad, *Biochem. Mass Spectrom.*, **5**, 541 (1978).
[120] F. Artigas and E. Gelpi, *Anal. Biochem.*, **92**, 233 (1979).
[121] M. Ervik, K. J. Hoffmann and K. Kylberg-Hanssen, *Biomed. Mass Spectrom.*, **8**, 322 (1981).
[122] S. P. Jindal, T. Lutz and P. Vestergard, *J. Pharm. Sci.*, **69**, 684 (1980).
[123] J. I. Yoshida, K. Yoshino, T. Matsunaga, S. Higa, T. Suzuki, A. Hayashi and Y. Yamamura, *Biomed. Mass Spectrom.*, **7**, 396 (1980).
[124] B. H. Min, W. A. Garland and J. Pao, *J. Chromatog.*, **231**, *Biomed. Appl.*, **20**, 194 (1982).
[125] J. Girault, M. A. Lefebvre, J. B. Fourtillan, P. Courtois and J. Gombert, *Ann. Pharm. Franc.*, **38**, 439 (1980).
[126] J. J. Keyzer, B. G. Wolthers, F. A. J. Muskiet, H. F. Kaufman and A. Groen, *Clin. Chim. Acta*, **113**, 165 (1981).
[127] H. A. Nelson, S. V. Lucas and T. P. Gibson, *J. Chromatog.*, **163**, 169 (1979).
[128] A. C. Moffat, *Proc. Anal. Div. Chem. Soc.*, **15**, 237 (1978).
[129] A. R. Whorton, K. Carr, M. Smigel, L. Walker, K. Ellis and J. A. Oates, *J. Chromatog.*, **163**, 64 (1979).
[130] P. J. Finlayson, R. K. Christopher and A. M. Duffield, *Biomed. Mass Spectrom.*, **7**, 450 (1980).
[131] H. Weber, M. Hoeller and H. Breuer, *J. Chromatog.*, **235**, 523 (1982).
[132] T. C. Chiang, *J. Chromatog.*, **182**, *Biomed. Appl.*, **8**, 335 (1980).
[133] M. J. Kessler, *J. Liq. Chromatog.*, **5**, 313 (1982).
[134] B. Egestad, T. Curstedt and J. Sjovall, *Anal. Lett.*, **15**, 293 (1982).

[135] B. A. Tomkins, H. Kubota, W. H. Griest, J. E. Caton, B. R. Clark and M. R. Guerin, *Anal. Chem.*, **52**, 1331 (1980).
[136] J. D. Cohen and A. Schulze, *Anal. Biochem.*, **112**, 249 (1981).
[137] G. Sandberg, B. Anderson and A. Dunberg, *J. Chromatog.*, **205**, 125 (1981).
[138] L. L. Ingram, G. D. McGinnis and S. V. Parikh, *Anal. Chem.*, **51**, 1077 (1979).
[139] P. Falardeau, J. A. Oates and A. R. Brash, *Anal. Biochem.*, **115**, 359 (1981).
[140] J. R. Carlin, R. W. Walker, R. O. Davies, R. T. Ferguson and W. J. A. Vandenheuvel, *J. Pharm. Sci.*, **69**, 1111 (1980).
[141] K. Mimura and S. Baba, *Chem. Pharm. Bull.*, **29**, 2043 (1981).
[142] J. F. Pageaux, B. Duperray, M. Dubois and H. Pacheco, *J. Lipid Res.*, **22**, 725 (1981).
[143] M. Traut, H. Morgenthaler and E. Brode, *Arzneim. Forsch.*, **31(II)**, 1589 (1981).
[144] W. A. Garland and B. H. Min, *J. Chromatog.*, **172**, 279 (1979).
[145] H. L. Schmidt, H. Foerstel and K. Heinzinger (eds.), *Stable Isotopes, Proc. 4th Intern. Conference, March 1981*, Elsevier, Amsterdam (1982).
[146] L. Dehennin, A. Reiffsteck and R. Scholler, *Biomed. Mass Spectrom.*, **7**, 493 (1980).
[147] J. U. Skaare and H. K. Dahle, *J. Chromatog.*, **111**, 426 (1975).
[148] M. Kai, T. Ogata, K. Haraguchi and Y. Okura, *J. Chromatog.*, **163**, 151 (1979).
[149] K. Blau, I. M. Claxton, G. Ismahan and M. Sandler, *J. Chromatog.*, **163**, 135 (1979).
[150] H. Riseboom, R. H. A. Sorel, H. Lingeman and R. Bouwman, *J. Chromatog.*, **163**, 92 (1979).
[151] P. Hartvig, N. O. Ahnfelt, M. Hammaerlund and J. Vessman, *J. Chromatog.*, **173**, 127 (1979).
[152] A. S. Christophersen and K. E. Rasmussen, *J. Chromatog.*, **168**, 216 (1979).
[153] D. R. Thouvenot and R. F. Morfin, *J. Chromatog.*, **170**, 165 (1979).
[154] K. Kuwata, *Bunseki Kagaku*, **27**, 650 (1978).
[155] T. Fehér, K. G. Fehér, and L. Brodogi, *J. Chromatog.*, **111**, 125 (1975).
[156] M. Sheehan and P. Haythorn, *J. Chromatog.*, **117**, 393 (1976).
[157] P. J. Naish, M. Cooke and R. E. Chambers, *J. Chromatog.*, **163**, 363 (1979).
[158] K. Chan, N. E. Williams, J. D. Baty and T. N. Calvey, *J. Chromatog.*, **120**, 349 (1976).
[159] T. W. Guentert, G. M. Wientjes, R. A. Upton, D. L. Comb and S. Riegelman, *J. Chromatog.*, **163**, 373 (1979).
[160] J. Lanzoni, L. Airoldi, F. Narcucci and E. Mussini, *J. Chromatog.*, **168**, 260 (1979).
[161] R. E. Weinfeld and T. C. Macasieb, *J. Chromatog.*, **164**, 73 (1979).
[162] P. Jakobsen and A. K. Pedersen, *J. Chromatog.*, **163**, 259 (1979).
[163] M. S. Greenberg and W. J. Mayer, *J. Chromatog.*, **169**, 321 (1979).
[164] Y. G. Yee, P. Rubin and T. F. Blaschke, *J. Chromatog.*, **171**, 357 (1979).
[165] M. A. Evans and R. D. Harbison, *J. Pharm. Sci.*, **66**, 1628 (1977).
[166] K. M. Jensen, *J. Chromatog.*, **111**, 389 (1975).
[167] C. M. Davis, C. J. Meyer and D. C. Fenimore, *Clin. Chim. Acta*, **78**, 71 (1977).
[168] K. H. Dudley in *Trace Organic Analysis*, pp. 381–390. NBS, Washington (1979).
[169] S. A. Wise, S. N. Chesler, F. R. Guenther, H. S. Hertz, L. R. Hilpert, W. E. May and R. M. Parris, *Anal. Chem.*, **52**, 1828 (1980).
[170] J. H. Onley, *J. Assoc. Off. Anal. Chem.*, **60**, 1111 (1977).
[171] R. L. Bong, *J. Assoc. Off. Anal. Chem.*, **60**, 229 (1977).
[172] W. F. Staruszkiewicz, Jr., *J. Assoc. Off. Anal. Chem.*, **60**, 1131 (1977).
[173] G. J. Gleich and W. M. Hull, *J. Allergy Clin. Immunol.*, **66**, 295 (1980).
[174] K. Horn, I. Marschner and P. C. Scriba, *J. Clin. Chem. Clin. Biochem.*, **14**, 353 (1976).
[175] H. P. Thier, *Mitt. Lebensm. Chem. Gerichtl. Chem.*, **30**, 187 (1976).
[176] K. I. Aspila, J. M. Carron and A. S. Y. Chau, *J. Assoc. Off. Anal. Chem.*, **60**, 1097 (1977).
[177] L. R. Hilpert, W. E. May, S. A. Wise, S. N. Chester and H. S. Hertz, *Anal. Chem.*, **50**, 458 (1978).
[178] W. D. MacLeod, P. G. Prohaska, D. D. Gennero and D. W. Brown, *Anal. Chem.*, **54**, 386 (1982).
[179] L. D. Sawyer, *J. Assoc. Off. Anal. Chem.*, **61**, 282 (1978).

[180] B. Borsje, H. A. H. Craenen, R. J. Esser, F. J. Mulder and E. J. de Vries, *J. Assoc. Off. Anal. Chem.*, **61**, 122 (1978).
[181] Analytical Methods Committee, *Analyst*, **103**, 856 (1978).
[182] P. L. Schuler, W. Horwitz and L. Stoloff, *J. Assoc. Off. Anal. Chem.*, **59**, 1315 (1976).
[183] D. W. Fink, *J. Assoc. Off. Anal. Chem.*, **64**, 973 (1981).
[184] A. D. Sauter, P. E. Mills, W. L. Fitch and R. Dyer, *J. High Resol. Chromatog. Chromatog. Commun.*, **5**, 27 (1982).
[185] W. A. Hardwick, *Tech. Q. Master Brew. Assoc. Am.*, **18**, 95 (1981).
[186] M. D. Friesen and L. Garren, *J. Assoc. Off. Anal. Chem.*, **65**, 855 (1982).
[187] M. D. Friesen and L. Garren, *J. Assoc. Off. Anal. Chem.*, **65**, 864 (1982).
[188] H. Egan and T. S. West (eds.) *Collaborative Studies in Chemical Analysis*, Pergamon Press, Oxford (1982).
[189] G. M. Telling, *Proc. Anal. Div. Chem. Soc.*, **16**, 38 (1979).
[190] J. Rado and S. G. Gorbach, *Z. Anal. Chem.*, **302**, 15 (1980).
[191] H. Rohleder and S. G. Gorbach, *Z. Anal. Chem.*, **295**, 342 (1979).
[192] R. J. Henry, *Clinical Chemistry*, Harper & Row, New York (1964).
[193] H. C. Curtius and M. Roth (eds.), *Clinical Biochemistry*, de Gruyter, Berlin (1974).
[194] G. Anido, S. B. Rosalki, E. J. Van Kampen and M. Rubin (eds.), *Quality Control in Clinical Chemistry*, de Gruyter, Berlin (1975).
[195] J. Buettner, R. Borth, J. H. Boutwell, P. M. G. Broughton and R. C. Bowyer, *J. Clin. Chem. Clin. Biochem.*, **18**, 69 (1980).
[196] H. Buettner, *Z. Klin. Chem. Klin. Biochem.*, **5**, 44 (1967).
[197] A. L. Wilson, *Analyst*, **104**, 273 (1979).
[198] Deutsche Forschungsgemeinschaft, Senatskommision zur Prüfung gesundheitsschädlicher Arbeitsstoffe, *MAK, Maximale Arbeitsplatzkonzentrationen*, Verlag Moderne Industrie, Munich (1978).
[199] L. Roth, *Sicherheitsdaten, MAK-Werte (1978)*, Verlag Moderne Industrie, Munich (1978).
[200] Anon, *Anal. Proc.*, **18**, 467 (1981).
[201] R. H. Hill Jr., *Intern. Lab.*, **11**, No. 6, 12 (1981).
[202] P. R. Cooke, *Med. Lab. Sci.*, **37**, 351 (1980).
[203] *Verordnung über den Schutz vor Schäden durch ionisierende Strahlen*, Bundesgesetzblatt, IS 2905, 13 October 1976.

3

Sampling in organic trace analysis

3.1 GENERAL ASPECTS

In sampling, a fraction or portion of a greater quantity of some bulk material is taken in order to facilitate its handling for the analysis. This procedure of reduction of the bulk by selection must not change the ratio of matrix to trace constituent. This relationship must be the same in the sample drawn and in the original bulk.

The source from which the sample is taken may be constant in composition with regard to time, as with many bulk consignments in international trade or local enterprise, e.g. of intermediates or other products of chemical industry, and raw materials. The sampling may then be from ships, trucks, railway vans, packages, bottles etc.

Alternatively, the composition of the source can change with time, and in practice this happens more frequently. The changing composition of the water in a river, and fluctuations in the composition of the exhaust from a factory chimney, are obvious examples. Biological changes in plants and animals also cause variation in composition with time. All samples of fresh foods (vegetables, fish, fruit, meat etc.) are liable to change in matrix and trace concentrations. Chemical changes affecting only one of the constitutents will, of course, influence the compositional ratios and thus the sample and the final result. In control of industrial hazards – such as traces of toxic gases in factories – fluctuations in the concentration of the trace constituent are often observed, whereas the matrix is practically constant in composition. Again, the sample of interest may represent either the composition for a particular situation or an average composition over a period of time. Very often, regulations request data on a time-weighted average basis, which must be achieved by appropriate sampling techniques. There may also be interest in the average (and fluctuations) over greater periods of time with regard to the behaviour of trace constituents in samples for environmental analysis, from wild-life, water ecology etc.

In many instances the trace constituent is separated from the bulk material (or part of it) during the sampling procedure. In such cases the requirement that the concentration ratio of matrix to trace should be unaffected by the sampling product is not met. Such combined sampling and enrichment procedures will also be described here, as they have the advantage that the analyst does not have to handle the whole mass of sample for the whole of the analysis, but deals with only the pre-enriched sample. Such procedures are very helpful for material which is still biochemically active and might transform the constituent to be determined. For example, in the analysis of air, only very rarely will a large sample be taken, stored in a huge container and analysed some time afterwards. Almost always the sampling will include some enrichment step. The same holds for water samples, where sampling/enrichment procedures are increasingly more common than the transportation of large volumes of water from the sampling area to the laboratory.

The sample drawn must often be divided into portions for analysis at different laboratories or for storage for later analyses. The division must be done by a sub-sampling procedure which does not change the composition of the sample and the matrix/trace ratio(s).

Obviously the homogeneity of the bulk is the most critical parameter of the entire sampling procedure. The matrices met in organic trace analysis are often inhomogeneous, for example entire plants (with leaves and roots), tissue specimens, animals, and this contributes to the problems of both the sampling and the trace analysis.

There is a very good survey on accuracy in trace analysis, which deals very extensively with sampling problems [1]. Typically, about 94% of the papers reviewed in it treat problems of inorganic trace analysis, and only 6% are dedicated to problems of organic trace analysis or are of general nature.

3.2 SAMPLING OF SPECIAL MATRICES

3.2.1 Sampling of organic trace constituents in air

It is needless to say that the air we breathe is of special interest to organic trace analysis, with the increasing ecological conscience of society. Consequently, about 8% of all recent publications on organic trace analysis in which a concrete procedure or method is described, are concerned with air as matrix.

There are several institutions which have passed regulations with regard to health standards and MAC-values (maximum allowable concentrations of hazardous material in workplaces) in workshops and industrial areas, with special attention to air. Thus the Environmental Protection Agency (EPA), the (US) National Institute of Occupational Safety and Health (NIOSH) and the American Society for Testing Materials (ASTM) have made recommendations in which the trace analysis of air is involved [2]. Another example is the attention attracted to the release of low molecular-weight chlorinated hydrocarbons (such as carbon tetrachloride, chloroform, trichloroethylene, 1,2-dichloroethane, tetrachloroethylene) into the atmosphere, because some of these substances are potential carcinogens and also might cause damage to the stratosphere ozone layer [3]. The Occupational Safety and Health Administration (OSHA, USA) has published a list of some 170 hazardous gases and vapours, for which air is to be checked [4].

Sampling within the working area is often required. Mobile sampling units or portable equipment are necessary. The use of small tubes filled with 150 mg of activated charcoal is suggested in regulations by OSHA and NIOSH [5, 6]. Plugs of porous polyurethane foam [7] or molecular-sieve columns [8] are also used.

One monitor suggested [9] is about the size of a standard film badge and weighs less than 35 g. It consists of a membrane through which the trace species (vinyl chloride in this case) is free to diffuse. After diffusion it is trapped on activated charcoal. This device meets OSHA requirements for personal monitoring as it permits measurement of a time-weighted average exposure. The membrane considerably reduces the possibility of interference by other air constituents. Such devices have been further discussed [10–13].

A portable battery-operated sampler has been described which uses a tube filled with XAD resin [14]. Breakthrough volumes of organic traces on Tenax GC [15], activated charcoal [16] and impregnated and non-impregnated filters [17] have been reported. Such data are of interest not only in the context of sampling, but also in gas chromatography.

A combination of sampling and qualitative identification is common in industrial health control. For this purpose small tubes filled with an impregnated adsorbent are used, through which the air is pumped, usually manually. The traces react with the impregnation reagents, and the length of the zone, its colour, and other properties can be used for identification and semiquantitative determination. Typical suppliers are Auergesellschaft, Berlin; Dräger-Werke, Lübeck (Draeger Safety, England); Kontron-Technik, Eching bei München.

In laboratory experiments, where mobility is not needed, several principles are used for air-sampling, which are described in some survey articles [18–21]. Sorption–desorption techniques for traces of toxic organic materials in air have been reviewed [22], and a symposium has been held on "Chemical Hazards in the Workplace", at which sampling methods were widely discussed [23].

In these methods a measured gas stream is passed over the adsorbents, which often are cooled. Occasionally, cooling alone is used as a trapping method, as the trapped material can later be volatilized simply by heating. Unfortunately, there is always the danger of formation of aerosols, and this is the more pronounced the more rapidly the gas is cooled. In order to obtain low vapour pressures extremely low temperatures are very often applied, which results in a drastic cooling and even more aerosol formation, instead of quantitative trapping. Sometimes the gas sample is passed through an absorption solution, and the proper selection of such solutions improves the specificity of the trapping procedure (see Table 3.1).

Table 3.1 – Examples of sampling of gases for determination of organic traces.

Trace compound	Matrix	Concentration or amount	Sampling method	Reference
aliphatic alcohols	room air	23–81 ng/l.	cold trap (liquid oxygen)	24
methyl chloride	ambient air	500 ppt	cold trap (−183°C) + glass beads	25
impurities	ambient air	ppt	cold trap	26
ethylene	air over germinating plants	1 ppb	cold trap (+ Porapak Q)	27
acetaldehyde	urban air	1–9 ppb	cold trap	28
aromatic aldehydes	automobile exhaust gases	0.3 ng/l.	cold trap	29
organic compounds	hydrogen (used as coolant gas for electric generators)	1 pg/l.	cold trap (+ Porapak)	30
methyl bromide	air	35 ng/100 ml	cold trap (−78°C) (Tenax GC)	31
lower aliphatic carbonyl compounds	exhaust gases	30 ng–1 μg	cold trap (−186°C)	32
formaldehyde	clean air	0.05 ppb	cold trap	33
volatiles	fish	25 μk/kg	cold trap	34
sym-tetrachloroethane	air		active carbon	35
organic compounds (135)	urban air		Carbochrome K-5	36
vinyl chloride	air	50 ppt	Carborafin 5250	37
acrylonitrile	air	10 ppb	activated carbon	38
vinyl chloride	air	5 ppb	activated charcoal	39
toxic trace gases	air		activated charcoal	40, 41
vinyl chloride	industrial air		membrane + charcoal	42
acrylonitrile	air	0.25 ppm	charcoal	43
acrylonitrile	air	1–50 ppm	charcoal	44
perchlorethylene	air	0.3–3 ppb	charcoal	45
vinyl chloride	air	0.1 ng	charcoal	46
diethylcarbamoyl chloride	air	0.5 ng	charcoal	47
benzene	air	1 ppb	charcoal	48
organic vapours	air	2 ppm	charcoal	49
toxic organic compounds	air		charcoal + membrane	50
volatiles	plants		Tenax GC	51
trimethylamine	air	ppb	Tenax GC	52
trinitrotoluene	air	50–100 ng/l.	Tenax GC	53
N-nitrosodimethylamine	air	μg/l.	Tenax GC	54
chlorinated hydrocarbons	air	1 ng/m³	Tenax GC	55
organic pollutants	air	ppb	Tenax GC	56

organic pollutants	air	100 ppm	Tenax GC	57
N-nitrosodimethylamine	ambient air	ppt	Tenax GC	58
hydrocarbons	ambient air	ppt	Tenax GC	59
organic compounds	human breath		Tenax GC	60
hydrocarbons	jet engine exhaust		Carbosieve B + Tenax	61
phenols	air	ppb	Tenax	62
phthalate esters	air	1 ng/m^3	polyurethane foam	63
triallate	air	0.5 ng/m^3	polyurethane foam	64
pesticides	air	0.1 ng/m^3	polyurethane foam	65
				66, 67
polychlorinated biphenyls	air		polyurethane foam	68
chlorinated compounds	air	ng	polyurethane foam	69
polycyclic aromatic hydrocarbons	air		polyurethane foam	70
polycyclic aromatic hydrocarbons	air	20–1500 ng/m^3	polyurethane foam	71
acrylonitrile	air	100 ppm	Porapak	72
acrylonitrile, acrolein	air	0.1–100 ppm	Porapak	73
di(chloromethyl) ether	air	ppb	Porapak	74
halocarbons	air	30–130 ppt	Porapak	75
organic pollutants	air		silica gel	76, 77
chloroacetyl chloride	air	0.8 ppm	silica gel	78
hydrocarbons (102)	air from workshops		silica gel	79
epichlorhydrin, 2-chloroethanol	air		Amberlite	80
organophosphorus residues	air		XAD-resin	81
chlorinated hydrocarbons	air	0.1 ng/l.	Sepabead GHP-1	82
acetylene	oxygen	0.1–10 ppm	molecular sieve	83
glutaraldehyde	air from hospital	20 $\mu g/m^3$	aluminium oxide	84
various compounds (68)	air		aluminium oxide, Carbopack	85
indoles	air	50 ppt	aluminium oxide	86
chlorpyrifos	air	ppt	Durapak–Carbowax 400	87
acrylonitrile	air	25 $\mu g/m^3$	Chromation–Apiezon	88
1,2-dibromo-3-chloropropane	air	0.07 ppb–20 ppm	Chromosorb	89
benzene	air	2 ppm	rubber	90
		Absorption in solution or on impregnated filters		
formaldehyde	air	0.05 $\mu g/ml$	alkaline H_2O_2	91
formaldehyde	air		1% chromotropic acid (in conc. H_2SO_4)	92
formaldehyde	air	0.1–3.8 ppm	silica, coated with 2,4-dinitrophenylhydrazine	93

Table 3.1 – *continued*

Trace compound	Matrix	Concentration or amount	Sampling method	Reference
aldehydes	air	ng	2,4-dinitrophenylhydrazine	94
aldehydes	air	0.01 ppm	2,4-dinitrophenylhydrazine + HSO_3^-	95
aldehydes	air		2,4-dinitrophenylhydrazine	96
carbonyl compounds	air		2,4-dinitrophenylhydrazine	97
acetone	air	mg/m³	water	98
methyl ethyl ketone	air, factory	mg/m³	water	99
acrolein	air, tobacco smoke	μg	aqueous ethanolic HCl	100
fatty acids	ambient air	3 ppb	1% aqueous NaOH	101
acetic anhydride	workroom air	μg	*m*-aminophenol in toluene–ethyl acetate	102
fatty acids	air	8 ppb	filter impregnated with NaOH	103
isocyanates	air	3 ppt	naphthylmethylamine	104
di-isocyanates	air	7–700 μg/m³	sorbents coated with 9-(2-methylaminoethyl)-anthracene	105
1,6-di-isocyanatohexene	air	15 μg/m³	l-naphthylmethylamine	106
primary aromatic amines	air	10 ng	4-dimethylaminocinnamaldehyde	107
trimethylamine	air	10 ppb	tartaric acid	108
dibutylsulphide	air	0.1–1 ng/m³	slilica gel + Hg (II) acetate	109
S,S,S-tributylphosphorotrithioate				
phenols, cresols	air	3 ppb	10% NaOH	110
pentachlorophenol	air	50 ng/m³	K_2CO_3	111
lindane, DDT	air	4–7 μg/m³	ethanediol	112
aromatic hydrocarbons		8 μg/m³	65% acetic acid	113
		Aerosol filtration		
polycyclic aromatic hydrocarbons	air particulate matter	08–36 ng	glass fibre filter	114, 115, 116
benz(*a*)pyrene	automobile exhaust		filter + condensate	117
benz(*a*)pyrene	air particulate matter	20 ng/m³	high-volume sampler (filter)	118
benzidine, 3,3′-dichlorobenzidine	air	3 μg/m³	glass filter + silica gel	119

For desorption from adsorbents, heating can be applied, during which a stream of pure gas is passed over the adsorbent to flush away the concentrated and desorbed traces. This makes it easy to couple the trapping system with gas chromatography or mass spectrometry, both of which are highly sensitive and rather specific procedures which have greatly contributed to the development of analytical methods for organic traces in air. In several instances it has been proposed to elute the trapped species from the column by using a solvent. The eluate is then analysed.

Special precautions are necessary when labile compounds are sampled. The trace constituent is concentrated by this enrichment procedure and air passed through the sampling unit might oxidize, hydrolyse or otherwise attack the isolated trace species. Thus dimethylamine collected in a sampling unit may react with elevated concentrations of O_3, NO or NO_2 in the presence of humidity, to form *N*-nitrosodimethylamine [120]. The formaldehyde content of air is also rather difficult to determine, as oxidation occurs rather rapidly after sampling.

Aerosols are mostly collected by means of filters, of defined pore size. This type of collection has been highly favoured by trace analysts since the early 1970s, as it is rather simple. Though this technique is sufficient and appropriate for inorganic trace analysis, organic traces frequently have a much higher vapour pressure than inorganic traces, So, in many cases of ecological interest volatile traces can be only partially collected in this way. On the other hand, the method is of interest for determining the aerosol-bound fraction of organic traces.

It is well known that pesticides disappear in time after they have been sprayed on plants, soil etc. The effect is due to chemical transformation or to physical processes (mainly evaporation). Insecticides lost by evaporation obviously have a relatively high vapour pressure, and a certain insecticide content can therefore be expected in air. It is likely that some of the pesticide will be held on dust particles, and another fraction will be present as a free component in the gas phase. Difficulties are encountered in the collection and determination of this fraction (in the ng/m^3 range), since condensation (necessitating cooling to reduce the vapour pressure) results in the formation of aerosols which escape trapping.

High recoveries – even for nanogram quantities – are observed when the air is passed through gauzes coated with poly(ethylene glycol) [121]. Non-polar substances dissolve in the poly(ethylene glycol) with high yield, as has been demonstrated by use of ^{14}C-labelled insecticides. The collected insecticides are eluted and then determined by gas chromatography (see Table 3.2).

The results in Table 3.2 agree with the findings of other authors [122]. The published data obtained by use of only filtration-sampling can be considered too low in all cases where rather volatile pesticides were investigated.

Likewise, the content of benz(*a*)pyrene in dust samples from air has been

Table 3.2 – Differentiation between the gaseous and aerosol-bound fractions of insecticides in air samples [121] (by permission of the copyright holders, Springer Verlag).

Origin of sample	Content (ng/m³) of *p,p'*-DDT		lindane	
	aerosol	gaseous	aerosol	gaseous
Mainz	2–13	191–550	0	7–26
Neustadt	0	0	22	17
Schwarzwald	1	61	0.3	2

found to be lower in summer (2 ng/m^3) than in winter (8 ng/m^3) (the correct value), owing to evaporation of the benzpyrene during sampling at the slightly higher summer temperatures [123]. The critical steps and possible losses during the sampling of polycyclic aromatic hydrocarbons (PAHs) from the atmosphere have been surveyed [124].

The efficiency of Carbopack B and Tenax G [125] and of Porapak Q, Porapak T and Tenax GC for sampling of traces from air has been compared [126].

Finally, it should be mentioned that no sampling is necessary with certain highly specialized techniques of modern instrumental analysis. For example, hydrocarbons in air can be determined directly by laser-induced infrared fluorescence [129] (see also Chapter 7).

"Numerous technical questions will remain concerning the intrinsic accuracy of the act of sampling the atmosphere. Probably, the greatest need is for additional primary standards of substances that can be used to assess the effects of changing in sampling configuration. One of the most serious problems is in the lack of techniques for generation of standard aerosols or the dependable generation of standard mixtures of gases at the levels in the unpolluted atmosphere. While techniques begin to appear for the ppt concentration range not one of them has been successfully standardized against a realistic known concentration in the air" [128].

3.2.2 Sampling of water

Water is a matrix which is analysed in about 12% of all applications of organic trace analysis (calculated from *Analytical Abstracts* for 1978 and 1979). There are several review articles [129–135].

In water analysis individual ('grab') samples are usually taken, from the river, lake, ocean, effluent pipe system etc. In suitable cases continuous sampling methods can be used instead of the discontinuous grab-method [136]. Intermittent automatic sampling is an intermediate form.

In all cases changes in the composition of the water system with time must be taken into consideration. The greatest problem is drawing a sample which is representative for the entire system. Depth and position of sampling, temperature, current of water and other parameters influence the final sample composition. They need to be known for a more precise evaluation of the analytical result.

Sampling is likely to be done with unbreakable plastic materials (bottles, buckets, tubes etc.) It must be kept in mind, though, that non-polar substances tend to become adsorbed very rapidly on the walls of such containers. Sampling for analysis for such substances should preferably be done with glass or metal equipment. Furthermore, the elution of plasticizer material – especially from new plastic containers – can be a great nuisance in trace analysis.

Sampling/enrichment methods play a more and more important role. They have the advantage that less material has to be carried from the sampling station to the laboratory. Also, they often give a better time-weighted average, which it is frequently desirable to know. Consequently, the term 'sampling' in the literature increasingly often means use of such combined methods. Examples can be found from Table 3.3, which demonstrates the extensive use of absorbing materials such as XAD, Porapak and Tenax. The properties of these polymers are described in Table 3.4.

An increasingly common technique in trace enrichment from water is passing the sample through an hplc column packed with a non-polar adsorbent [172]. For example, 2,4-D or 2,4,5-T at the 20-ppb level in water can be adsorbed on a column of Bondapak C_{18} and eluted to give a solution with a concentration of 1 ppm, giving 50-fold enrichment. Analogously, 0.2 ng of PAHs in 1 litre of water can be sampled with enrichment by a factor of 10^4 [173–175]. Diethyl phthalate in ppb concentrations has been separated from water in the sampling procedure [176].

A film of weathered crude oil, floating on the ocean surface, has been collected on Teflon [177]. In another method large bore columns (0.25 in.) were used; their interior wall was coated with SE 30 to give a film about 20 μm thick. The aqueous sample was passed through this system several times, and trace pollutants were fixed [178].

When water in soil is sampled at great depths volatile compounds may be lost owing to the suction which is to be applied. To prevent such losses a Tenax GC trap has been used [179].

Occasionally, lyophilization and cryoconcentration [180] are used for concentrating non-volatile traces from water samples. In cryoconcentration the liquid sample is partially frozen, and the liquid phase is separated. In favourable cases an enrichment of traces ('cryoconcentration') is observed. Such a freeze-concentration has been found to be suitable only for solutions of concentrations below 0.01*M*. At higher concentrations the ice formed will occlude solute. Application to micromolar solutions is possible [181].

For lyophilization, addition of sodium chloride at the start is suggested. The sample is totally frozen and water is sublimed off in vacuum. The resultant dry residue (NaCl + traces) is extracted to separate the trace compound.

Table 3.3 – Use of sampling-enrichment techniques for the isolation of traces of non-volatile organic compounds from water.

Trace	Concentration or amount	Adsorbents	Reference
PAHs	0.1 ppt	Tenax	137
PAHs (60)	20 ppt	Amberlite XAD 2	138
N-methyl-2-methylpyrrolidone	10 ppm	Amberlite XAD 2	139
benzene hexachloride isomers	10 ppt	Amberlite XAD 2	140
several pollutants	ppb	Amberlite XAD 2	141
chlorophenols	100 ppb	Amberlite XAD 2	142
fenitrothion	50 ppb	Amberlite XAD 2	143
PCBs, organochlorine residues	ppt	Amberlite XAD 2	144
chlorinated pesticides	ppt	Amberlite XAD 2	145
organic matter		XAD 2	146
carbamate insecticides		XAD 2	147
chlorinated hydrocarbons (PCBs, insecticides)		XAD 2 or XAD 4 or Tenax	148
phenols, organic acids	0.05 ppm	XAD 4	149
fenitrothion, 5 derivatives	50 ppb	XAD 4 or XAD 7	150
fenitrothion		(XAD 2, XAD 4) XAD 7	151
phenoxyacetic acid herbicides	60 ppt	Amberlite XAD 4	152
non-ionic detergents	10 ppb	Amberlite XAD 4	153
ester of phosphoric acid	10 ppb		154
pollutants (13)		carbon, XAD 4 + XAD 8	155
PAHs (6)	0.1 ppt	polyurethane foam	156
PCBs, DDT		polyurethane foam	157
PAHs	ppb	polyurethane foam, XAD 2, BioRad AG MP 50	158
pesticides, phenols		Spheron MD	159
PCBs	45 ppb	activated carbon, polyurethane foam, XAD 2	160
chlorophenols	ppb	C_{18} chemically bonded reversed phase	161
chlorinated phenols, guaiacols, catechols		C_{18} chemically bonded reversed phase	162
fluridone herbicide	1 ppb	C_{18} chemically bonded reversed phase	163
chlorpyrifos		C_{18} chemically bonded reversed phase	164
nitrosamines	ng/l.	Ambersorb XE 340	165
thiophene, dialkyl sulphides etc.	1 ppm	Porapak R	166
halo-organic compounds	μg/l.	Porapak N	167
N-nitrosamines	0.3 ng	activated carbon	168
chlorinated pesticides		Carbopack B	169
oil	9 ppb	polyester foam	170

Table 3.4 –Materials used for isolation of organic contaminants from water (from [171], by permission of the copyright holders, Elsevier Science Publishing Co).

Material	Composition	Specific surface area (m^2/g, dry polymer)	Thermally stable up to (°C)	Manufacturer
Amberlite				
XAD-2	styrene–divinylbenzene copolymer	290–330	200	Rohm & Haas, Philadelphia, USA
XAD-4	styrene–divinylbenzene copolymer	750		
XAD-7	methacrylate polymers	450	150	
XAD-8	methyl methacrylate polymer	140		
XAD-1	styrene–divinylbenzene copolymer	100		
Ostion SP-1	(similar to XAD-2)	350		Research Institute of Synthetic Resins, Pardubice, Czechoslovakia
Porapak Q	ethylvinylbenzene–divinylbenzene copolymer	630–840	250–300	Waters Associates, Framingham, Mass., USA
Synachrom	styrene, divinylbenzene. ethylvinylbenzene copolymer	520–620	340	Lachema, Brno, Czechoslovakia
Chromosorb 102	styrene–divinylbenzene copolymer	300–400	250	Johns-Manville, New York, USA
Chromosorb 105	polyaromatic type polymers	600–700	200	
Chromosorb 106	polystyrene copolymer	700–800	250	
Tenax	poly(2,6-diphenyl-*p*-phenylene oxide)	19–30	320	Applied Science Labs., State College, Pa., USA*
Spheron MD 30/70	methyl methacrylate–divinylbenzene copolymer	320	230	Laboratory Instruments Prague, Czechoslovakia
Spheron SE	styrene–ethylene dimethacrylate copolymer	70	280	
Amberlyst	anion-exchange resin with trimethylamine groups	25–30		Rohm & Haas, Philadelphia, USA

*World-wide distributor: Chrompack, P.O. Box 3, 4330 AA Middelburg, Netherlands.

3.2.3 Stripping of volatile components from water

Volatile components can be stripped from a water sample by passing a carrier gas through it. The gas flushes the trace compound out of the matrix. After collection (by cold-trap or adsorption) the trace is greatly enriched. The equipment has been described in the literature and compared [182, 183], and a review has been given [184]. Examples of this method, which is a variant of head-space analysis (see Section 3.2.6) are given in Table 3.5.

In almost all cases listed in Table 3.5 the final determination is done by gas chromatography. The trapped substances are recovered by thermal desorption or by elution with a suitable solvent (Nos. 15 and 17).

In another method for the removal of volatile organic compounds, the water sample is sprayed in a jet-system against a glass plate. The resulting fine mist condenses while the more volatile traces remain in the gas phase. Forty trace compounds have thus been determined in a tap water sample [209].

3.2.4 Sampling in clinical chemistry, biochemistry and for pharmaceutical analysis

There are four conditions which must be satisfied for a sample from biological material to be acceptable for organic trace analysis in a clinical laboratory [210].

(1) At the time the sample is collected the patient must be in a physical state appropriate for the contemplated assay. In many instances he is supposed to be in a state of fasting. Hyperventilation or hyperactivity must be avoided.

(2) The sample drawn must be truly representative for the entire test individual (man, animal). This especially applies to blood samples, in which differences are often found according to the point from which the sample is taken, since the concentration of many substances in venous blood is different from that in arterial blood [211, 212]. Sampling conditions, including the point from which the sample is taken, must therefore be precisely indicated.

(3) Circumvention of trauma to body tissues, avoidance of haemolysis and other damage to red blood cells is necessary. Alertness to the possible introduction of contaminants is indicated.

(4) Finally, the specimen drawn must be kept in a manner that preserves its composition with regard to the parameter to be measured. Certain analytes remain stable for long periods, while others require storage at low temperatures or addition to stabilizers. In some instances, no practical method of preservation has so far been found.

Special problems are encountered in the clinical area when the blood of newborns is to be investigated [213]. The blood volume of newborns is only

Table 3.5 – Stripping of volatile compounds by passage of a stream of inert gas through a water sample.

No.	Compound(s) stripped	Concentration or amount	Trap filling	Reference
1	volatile compounds		Tenax GC	185
2	volatile polar compounds		Tenax GC	186
3	volatile micropollutants		Tenax GC	187
4	volatile hydrocarbons	1 µg		188
5	hydrocarbons	10 ng/l.		189
6	volatile impurities		Tenax GC	190, 191
7	halogenated hydrocarbons, esters, ketones	50 ng/l.	Tenax GC	192
8	priority pollutants (36)	0.1 ppb		193
9	chloro- and bromo-derivatives of C_3- and C_4-hydrocarbons	1 µg/l.	Tenax GC	194
10	volatile components from blood (42)		Tenax GC	195
11	chloro-, bromo-derivatives (e.g. $CHCl_3$,CCl_4,trichloroethane)	0.1 ppb	porous polymers	196
12	aromatic hydrocarbons		porous polymers	197
13	derivatives of camphor, terpenes			198
14	trihalomethanes	1 ppb		199
15	volatile contaminants	ppb	charcoal	200
16	pollutants		charcoal	201
17	pesticides, polychlorinated biphenyls	ppb	charcoal	202
18	halocarbons	10 ppt	charcoal (−78°C)	203
19	vinyl chloride	µg/l.	silica gel, Carbosieve	204
20	amines	ppb	Cu(II) salts	205
21	aroma components from wine (19)		absorption in 10% ethanol	206
22	pollutants		none, −196°C	207
23	halogenated hydrocarbons	ppt	none, −196°C	208

about 270 ml and the clinical state of the little patients may have to be monitored several times a day. Thus, the quantity of blood used for clinical analysis must be kept to 10 μl or less per test. Heel pricks are preferred for obtaining the blood samples. They must be performed carefully, and a free flow of blood is essential to avoid contamination with tissue fluids and to prevent haemolysis. To reduce the variability of the sampling procedure, the number of individuals who are allowed to collect blood should be as small as possible.

Sample identification can cause problems, as it is rather difficult to attach a label directly to a capillary blood tube. A convenient way of manipulating, transporting, storing and identifying blood from newborns is to spot the blood onto a filter paper [213, 214]. The dried spots can then be punched out to deliver a fixed volume or amount for clinical analysis. A sequential solvent extraction system is applied to remove fatty acids, cholesterol, amino-acids, and drugs from the filter paper. Direct mass spectrometric analysis is suggested as a fast, sensitive and specific method for determination.

In most investigations the sample material is blood or urine. Though the concentration of the compound of interest in blood can very often be taken as proportional and representative for the rest of the body, this cannot be assumed with urine. Special excretion mechanisms may result in a very drastic change of concentration. Consequently, blood is the preferred sample for scientific studies (and analyses with legal implications). Blood samples are not simple to take. In Germany, only physicians are permitted to puncture veins. Consequently, there is a trend towards use of other body fluids which might be representative for the concentration of drugs and other biochemically active compounds. Saliva is more and more often proposed [215-218], as this material can be obtained by non-invasive methods. Steroids [219], anticonvulsants [220], cannabis alkaloids [221] and carbamazepine [222] have been determined in saliva. Tears have been suggested as a sample source, as they are more stable than saliva [223]. To prove consumption of cannabis, teeth have been swabbed with cotton wool soaked with light petroleum or chloroform(!) [224]. Hair is used as sample material in assessing morphine and cocaine addiction [225].

An excellent survey of sampling problems in clinical chemistry has been given [226].

3.2.5 Techniques using Extrelut®, Sep-Pak®-cartridges and similar materials for sampling/enrichment of traces

In sampling the aqueous systems, e.g. blood, plasma, urine, that are of importance in biochemistry and pharmaceutical analysis, the techniques used are analogous to those for sampling of water, i.e. use of adsorbents to concentrate traces from the matrix. In the Extrelut® technique, plasma or urine is passed through a column filled with the porous adsorbent, which retains the trace compound(s) (plus some other material). By elution with an appropriate solvent separations are achieved (from the co-adsorbed components). Good recoveries and high purification are obtained. The consumption of organic solvents is reduced in comparison to liquid-liquid distribution techniques. Consequently, fewer hazards have to be considered [227]. Some examples can be taken from Table 3.6.

Table 3.6 – Sampling of biological material, with Extrelut®, Sep-Pak C_{18}-cartridges® etc.

No.	Compounds	Concentration or amount	Adsorbent	Reference
1	steroids	10 ng	C_{18}-reversed phase	228
2	steroids		C_{18}-reversed phase	229
3	steroids		C_{18}-reversed phase	230
4	bile acids		C_{18}-reversed phase	231
5	mebendazole + metabolites	20 ng/ml	C_{18}-reversed phase	232
6	ketoconazole	250 ng/ml	C_{18}-reversed phase	233
7	tricyclic antidepressants	ng/ml	C_{18}-reversed phase	234
8	prostacyclin		C_{18}-reversed phase	235
9	drugs		C_{18}-reveresd phase	236
10	steroids		Extrelut	237
11	steroids		Extrelut	238
12	progesterone	ppb	Extrelut	239
13	clonidine (antihypertensive drug)	ppt	Extrelut	240
14	aflatoxin M_1	50 ppt	Extrelut	241
15	drugs		Extrelut	242
16	drugs of abuse		Extrelut	243
17	nicotine	6 ng/ml (smoker) 1 ng/ml (non-smoker)	Extrelut	244
18	cannabinoids		Extrelut	245
19	drugs		XAD 2	246
20	fenitrothion	50 ppb	XAD 7	247
21	drugs		XAD 2	248
22	drugs		XAD 2	249

The substances in Table 3.6 all have been determined in urine or serum samples, with the exception of No. 1 (cell culture incubation media), No. 4 (serum and duodenal extracts), No. 14 (milk), No. 20 (homogenates from chicken liver, clams, pine needles, soil extracts and urine) and No. 21 (extracts from autopsy material). The reversed phases for Nos. 1-9 in Table 3.6 were applied as a Sep-Pak cartridge®. More examples of this enrichment/sampling technique can be found in tables in Chapter 5.

3.2.6 Head-space analysis

Head-space analysis is a sampling technique which can be very useful in organic trace analysis to sample and separate volatile components from solid or liquid matrices.

In principle, the sample is put into a closed vessel. Equilibrium is established in the distribution of the volatile component between the condensed phase and the gas phase, and the gas phase is then analysed.

There are two modifications of this procedure. In the static method the sample is drawn only after equilibrium is obtained. In dynamic sampling the system is purged with carrier gas, which carries away the volatile component, thus disturbing the equilibrium, and this is re-established by transfer of the volatile component from the condensed phase into the gas phase. The components are accumulated from the stripping gas by cryogenic traps or adsorbent-packed traps, as for the analysis of water (Section 3.2.3). The dynamic principle gives better sensitivity, as rather quantitative separation of the volatile component is achieved. The static method is less sensitive, but quicker and simpler [250].

Establishment of the equilibrium can be critical. To improve the system, reagents for salting out the volatile materials for the liquid phase can be used [251]. Increasing the temperature and surface area of the condensed phase are other means of improvement. Calibration, e.g. by addition of a standard concentration of the trace component, is necessary. A few books and review articles offer further information [252–255].

Equipment for head-space analysis – often combined with gas chromatographic equipment – is commercially available. An automated system has been described [26]. An often used technique is condensation of the volatile components on adsorbents which can be heated (rapidly) for desorption (see also Table 3.4), and injection of the desorbed trace components into a gas chromatograph [257, 258]. Injection of nanogram amounts is possible, by collecting traces from the purge gas in the interior of a syringe needle which has been coated with a stationary phase for gas chromatography. The needle is inserted into the hot injection port of the gas chromatograph and the collected traces are thermally desorbed [259]. Examples of the use of the head-space technique in organic trace analysis are given in Table 3.7.

Table 3.7 – Examples of the use of head-space techniques for sampling of organic traces.

Trace compound	Matrix	Concentration or amount	Reference
hydrocarbons	water	100 ng/l.	260
hydrocarbons	water	ppb	261
hydrocarbons	marine biota		262
benzene	water	0.02–20 ppm	263
vinyl chloride	corn oil	1 ppb	264
vinyl chloride	PVC food-packing plastic	>1 ppb	265
vinyl chloride	food	1 ppb	266
acrylonitrile	plastic packaging		267
acrylonitrile	beverages	5 μg/kg	
acrylonitrile	plastic, food		268
halothane metabolites	blood	2 pmole/ml	269
trichloroethane	vinyl chloride polymer		270
halogenated hydrocarbons	water	0.1 μg/1.	271, 272
dichloromethane	polycarbonates	>1 ppm	273
trichloroethanol	urine	>0.5 μg/ml	274
trichloroethylene	oils	>0.4 μg/g	275

Table 3.7 – *continued*

Trace compound	Matrix	Concentration or amount	Reference
chlorinated hydrocarbons	water	0.1 ppb	276
diethyl carbonate	beverages	ppb	277
cyclohexanone	intravenous solutions	mg/l.	278
esters, higher alcohols	beer	mg/l.	279
ethanol	blood		280
ethanol	saliva		281
several monomers	polymers	ppm	282
volatile substances	blood		283
volatile components	waste water		284
volatile compounds	mushrooms		285
volatiles	marihuana, hashish		286
volatiles	waste water	4 μg/l.	287
dimethyl sulphide	white wine	μg/l.	288
volatile compounds	urine		289
biogenic amines	algae	μg/l.	290
volatile compounds (53)	cooked chicken		291
volatiles	milk	ppm	292
organic compounds	conc. HCl or HBr	ppm	293
styrene	food		294
styrene	food	3–10 ppb	295
styrene	fruit	>5 ppb	296
vinyl chloride	ethanediol	>0.5 ppm	297
vinyl chloride	personal monitor	5 ppm	298
vinyl chloride	poly(vinyl chloride)	0.2 ppm	299
vinyl chloride	drugs, cosmetic products	0.1 ppb	300
vinyl chloride	water	1 μg/l.	301
residual volatiles	polymers	ppm	302
volatile organics	water	0.1 μg/l.	303
aromatic hydrocarbons	water, waste water	ppb	304
ethylene oxide	surgical plastics	ng	305
halogenated volatiles	tissue	ng/kg	306
priority pollutants	water	30 μg/l.	307
priority pollutants	water	5 μg/l.	308
residual volatiles	polymers		309
phenol	urine	50 ppm	310
volatile hydrocarbons	blood	4 μg/ml	311
volatile hydrocarbons	food	10 ppb	312
dithiocarbamate pesticides	food		313
acetaldehyde	polyethylene tetraphthalate	50 ppb	314
fruity off-flavour	milk	5 ppb	315
dichlorethylene	tissue	>50 pg	316
flavour compounds	tobacco		317

REFERENCES

[1] P. D, la Fleur (ed.) *Accuracy in Trace Analysis, Sampling, Sample Handling Analysis,* 2 Volumes, NBS, Washington (1976).
[2] B. E. Saltzman and W. R. Burg, *Anal. Chem.,* **49,** 1R (1977).
[3] R. Tatsukawa, T. Okamoto and T. Wakimoto, *Bunseki Kakagu,* **27,** 164 (1978).

[4] J. Diaz-Rueda, H. J. Sloane and R. J. Obremski, *Appl. Spectrosc.*, **31**, 298 (1977).
[5] D. T. Coker, *Proc. Anal. Div. Chem. Soc.*, **14**, 108 (1977).
[6] B. Miller, P. O. Kane, D. B. Robinson and P. J. Whittingham, *Analyst*, **103**, 1165 (1978).
[7] B. C. Turner and D. E. Glotfelty, *Anal. Chem.*, **49**, 7 (1977).
[8] M. Capelloni, L. G. Frederick, D. R. Latshaw and J. B. Wallace, *Anal. Chem.*, **49**, 1218 (1977).
[9] L. H. Nelms, K. D. Reiszner and P. W. West, *Anal. Chem.*, **49**, 994 (1977).
[10] P. W. West, *Intern. Lab.*, **10**, No. 6, 39 (1980).
[11] M. A. Pinches and R. F. Walker, *Ann. Occup. Hyg.*, **23**, 335 (1980).
[12] D. T. Coker, *Analyst*, **106**, 1036 (1981).
[13] R. H. Hill and J. E. Arnold, *Arch. Environ. Contam. Toxicol.*, **8**, 621 (1979).
[14] J. E. Woodrow and J. N. Seiber, *Anal. Chem.*, **50**, 1229 (1978).
[15] K. Kawate, T. Uemura, I. Kifune, Y. Tominaga and K. Oikawa, *Bunseki Kagaku*, **31**, 453 (1982).
[16] J. P. Guenier and J. Muller, *Cah. Notes Doc.*, No. 103, 197 (1981).
[17] W. Funke, J. Koenig and F. Pott, *VDI-Ber.*, **358**, 121 (1980).
[18] B. E. Saltzman and W. R. Burg, *Anal. Chem.*, **49**, 1R (1977).
[19] D. Henschler (ed.), *Analytische Methoden zur Prüfung gesundheitschädlicher Arbeitsstoffe, Vol. 1, Luftanalysen*, Verlag Chemie, Weinheim (1976).
[20] P. Leinster, R. Perry and R. J. Young, *Talanta*, **24**, 205 (1977).
[21] B. A. Rudenko and G. I. Smirnova, *Zh. Analit. Khim.*, **32**, 367 (1977).
[22] Yu. S. Drugov, *Zavodsk. Lab.*, **48**, 3 (1982).
[23] G. Choudhary (ed.), *Chemical Hazards at the Workplace*, American Chem. Soc., Washington (1981).
[24] Y. Hoshika, *Bunseki Kagaku*, **26**, 577 (1977).
[25] D. R. Cronn and D. E. Harsch, *Anal. Lett.*, **9**, 1015 (1976).
[26] B. A. Rudenko and G. I. Smirnova, *Zh. Analit. Khim.*, **32**, 367 (1977).
[27] J. Becka and L. Feltl, *J. Chromatog.*, **131**, 179 (1977).
[28] Y. Hoshika, *J. Chromatog.*, **137**, 455 (1977).
[29] Y. Hoshika and Y. Takata, *Bunseki Kagaku*, **25**, 529 (1976).
[30] D. J. A. Dear, A. F. Dillon and A. N. Freedman, *J. Chromatog.*, **137**, 315 (1977).
[31] T. Dumas, *J. Assoc. Off. Anal. Chem.*, **65**, 913 (1982).
[32] Y. Hoshika, *Analyst*, **106**, 686 (1981).
[33] V. Neitzert and W. Seiler, *Geophys. Res. Lett.*, **8**, 79 (1981).
[34] M. H. Hiatt, *Anal. Chem.*, **53**, 1541 (1981).
[35] S. K. Yasuda and E. D. Loughran, *J. Chromatog.*, **137**, 283 (1977).
[36] B. V. Ioffe, V. A. Isidorov and I. G. Zenkevich, *J. Chromatog.*, **142**, 787 (1977).
[37] R. Ciupe, E. Fasgarasan and L. Pop, *Rev. Chim. (Bucharest)*, **28**, 575 (1977).
[38] R. S. Marano, S. P. Levine and T. M. Harvey, *Anal. Chem.*, **50**, 1948 (1978).
[39] B. Miller, P. O. Kane, D. B. Robinson and P. J. Whittingham, *Analyst*, **103**, 1165 (1978).
[40] D. T. Coker, *Proc. Anal. Div. Chem. Soc.*, **14**, 108 (1977).
[41] J. Diaz-Rueda, H. J. Sloane and R. J. Obremski, *Appl. Spectrosc.*, **31**, 298 (1977).
[42] L. H. Nelms, K. D. Reiszner and P. W. West, *Anal. Chem.*, **49**, 994 (1977).
[43] Y. T. Gagnon and J. C. Posner, *Am. Ind. Hyg. Assoc.*, **40**, 923 (1979).
[44] Anon., *Methods for Determination of Hazardous Substances*, MDHS 1, (Occup. Med. Lab., Edgware Rd., London, N.W.2), (1981).
[45] A. L. Sykes, D. E. Wagoner and C. E. Decker, *Anal. Chem.*, **52**, 1630 (1980).
[46] S. Sadowski and A. Strusinski, *Rocz. Panstw. Zakl. Hig.*, **30**, 413 (1979).
[47] J. P. Anger, J. C. Corbel and G. Paulet, *Toxicol. Eur. Res.*, **3**, 223 (1981).
[48] H. G. Baxter, R. Blakemore, J. P. Moore, D. T. Coker and W. H. McCambley, *Ann. Occup. Hyg.*, **23**, 117 (1980).
[49] R. Ciupe, *Rev. Chim. (Bucharest)*, **32**, 584 (1981).
[50] R. W. Coutant and D. R. Scott, *Environ. Sci. Technol.*, **16**, 410 (1982).
[51] R. A. Cole, *J. Sci. Food Agric.*, **31**, 1242 (1980).
[52] N. Kashihira, K. Kirita, Y. Watanabe and K. Tanaka, *Bunseki Kagaku*, **29**, 853 (1980).

[53] R. W. Bishop, T. A. Ayers and D. S. Rinehart, *Am. Ind. Hyg. Assoc. J.*, **42**, 586 (1981).
[54] T. Matsumura, A. Tanimura, R. Higuchi, N. Yamate and M. Sakata, *Nippon Kagaku Kaishi,* 1410 (1979).
[55] W. N. Billings and T. F. Bidleman, *Environ. Sci. Technol.*, **14**, 679 (1980).
[56] L. Wennrich, T. Welsch and W. Engewald, *J. Chromatog.*, **241**, 49 (1982).
[57] R. H. Brown and C. J. Purnell, *J. Chromatog.*, **178**, 79 (1979).
[58] J. T. Bursey, D. Smith, R. N. Williams, R. E. Berkeley and E. D. Pellizzarri, *Intern. Lab.*, Jan./Feb., 11 (1978).
[59] R. E. Berkeley and E. D. Pellizzarri, *Anal. Lett.*, **11**, 327 (1978).
[60] H. L. Gearhart, S. K. Pierce and D. Payne-Bose, *J. Chromatog. Sci.*, **15**, 480 (1977).
[61] M. S. Black, W. R. Rehg, R. E. Sievers and J. J. Brooks, *J. Chromatog.*, **142**, 809 (1977).
[62] Y. Hoshika and G. Muto, *J. Chromatog.*, **157**, 277 (1978).
[63] H. Yamasaki and K. Kuwata, *Bunseki Kagaku,* **26**, 1 (1977).
[64] R. Grover, L. A. Kerr and S. H. Khan, *J. Agr. Food Chem.*, **29**, 1082 (1981).
[65] R. G. Lewis and K. E. MacLeod, *Anal. Chem.*, **54**, 310 (1982).
[66] R. G. Lewis and M. D. Jackson, *Anal. Chem.*, **54**, 592 (1982).
[67] B. C. Turner and D. E. Glotfelty, *Anal. Chem.*, **49**, 7 (1977).
[68] M. D. Jackson, D. W. Hodgson, K. E. MacLeod and R. G. Lewis, *Bull. Environ. Contam. Toxicol.*, **27**, 226 (1981).
[69] M. Oehme and H. Stray, *Z. Anal. Chem.*, **311**, 665 (1982).
[70] J. L. Lindgren, H. J. Kraus and M. A. Fox, *J. Air Pollut. Control. Assoc.*, **30**, 166 (1980).
[71] K. E. Thrane and A. Mikalsen, *Atmos. Environ.*, **15**, 909 (1981).
[72] Anon, *Methods of Determination of Hazardous Substances,* MDHS 2 (Occup. Med. Hyg. Laborat., Edgware Rd., London, N.W.2), (1981).
[73] D. N. Campbell and R. H. Moore, *Am. Ind. Hyg. Assoc. J.*, **40**, 904 (1979).
[74] G. Mueller, K. Norpoth and S. Z. M. Travenius, *Intern. Arch. Occup. Environ. Health,* **48**, 325 (1981).
[75] J. W. Russell and L. A. Shadoff, *J. Chromatog.*, **134**, 375 (1977).
[76] L. Wennrich, W. Engewald and T. Welsch, *Wiss. Z. Karl-Marx-Univers. Leipzig Math.-Naturwiss. Reihe,* **30** , 82 (1981).
[77] P. A. Rodriguez, C. L. Eddy, G. M. Ridder and C. R. Culbertson, *J. Chromatog.*, **236**, 39 (1982).
[78] P. R. McCullough and J. W. Worley, *Anal. Chem.*, **51**, 1120 (1979).
[79] S. A. Leont'eva, Yu. S. Drugov, N. H. Lulova and N. M. Koroleva, *Zh. Analit. Khim.*, **32**, 1638 (1977).
[80] K. Andersson, J. O. Levin, R. Lindahl and C. A. Nilsson, *Chemosphere,* **10**, 143 (1981).
[81] J. E. Woodrow and J. N. Seiber, *Anal Chem.*, **50**, 1229 (1978).
[82] R. Tatsukawa, T. Okamoto and T. Wakimoto, *Bunseki Kagaku,* **27**, 164 (1978).
[83] M. N. Cappelloni, L. G. Frederick, D. R. Latshaw and J. B. Wallace, *Anal. Chem.*, **49**, 1218 (1977).
[84] W. Wliszczak, F. Meisinger and G. Kainz, *Mikrochim. Acta,* 139 (1977 **II**).
[85] G. Holzer, H. Shanfield, A. Zlatkis, W. Bertsch, P. Juarez, H. Mayfield and H. M. Liebich, *J. Chromatog.*, **142**, 755 (1977).
[86] Y. Hoshika and G. Muto, *Bunseki Kagaku,* **27**, 520 (1978).
[87] R. G. Melcher, W. I. Garner, L. W. Severs and J. R. Vaccaro, *Anal. Chem.*, **50**, 251 (1978).
[88] N. I. Kaznina and N. P. Zinov'eva, *Gig. Sanit.*, 51 (1981).
[89] J. B. Mann, J. J. Freal, H. F. Enos and J. X. Danauskas, *J. Environ. Sci. Health,* **15B**, 519 (1980).
[90] M. Sefton, E. L. Mastracci and J. L. Mann, *Anal. Chem.*, **53**, 458 (1981).
[91] J. M. Lorrain, C. R. Fortune and B. Dellinger, *Anal. Chem.*, **53**, 1302 (1981).
[92] C. W. Lee, Y. S. Fung and K. W. Fung, *Analyst,* **107**, 30 (1982).

[93] R. K. Beasley, C. E. Hoffmann, M. L. Rueppel and J. W. Worley, *Anal. Chem.*, **52,** 1110 (1980).
[94] K. Kuwata, M. Verbori and Y. Yamasaki, *J. Chromatog. Sci.*, **17**, 264 (1979).
[95] J. L. Cheney and C. L. Walters, *Anal. Lett.*, **15,** 621 (1982).
[96] R. Kuntz, W. Lonneman, G. Namie and L. A. Hull, *Anal. Lett.*, **13**, 1409 (1980).
[97] R. A. Smith and I. Drummond, *Analyst,* **104,** 875 (1979).
[98] E. B. Koroleva, L. V. Sakharova, A. L. Kondrasheva and L. B. Sevryugov, *Gig. Sanit.*, 67 (1979).
[99] H. Tyras and A. Blochowicz, *Chem. Anal. (Warsaw)*, **21**, 647 (1976).
[100] B. Nagórski and M. Boguslawska, *Chem. Anal. (Warsaw)*, **22,** 127 (1977).
[101] M. Okabayashi, T. Ishiguro, T. Hasegawa and Y. Shigeta, *Bunseki Kagaku,* **25,** 436 (1976).
[102] C. Prandi and F. Terraneo, *Anal. Chem.*, **50,** 2160 (1978).
[103] T. Saito, T. Takashina, S. Yanagisawa and T. Shirai, *Bunseki Kagaku,* **30,** 790 (1981).
[104] S. P. Levine, J. H. Hoggatt, E. Chaldek, G. Jungclaus and J. L. Gerlock, *Anal. Chem.*, **51,** 1106 (1979).
[105] K. Andersson, A. Gudehn, J. O. Levin and C. A. Nilsson, *Chemosphere,* **11,** 3 (1982).
[106] S. M. Rappaport, *Air Force Off. Sci. Res. Rept.*, AFOSR-TR-81-066 (1981).
[107] D. W. Meddle and A. F. Smith, *Analyst,* **106,** 1088 (1981).
[108] M. Nagase, *Bunseki Kagaku,* **29,** 293 (1980).
[109] B. W. Hermann and J. N. Seiber, *Anal. Chem.*, **53,** 1077 (1981).
[110] I. Kifune, *Bunseki Kagaku,* **28,** 638 (1979).
[111] A. Dahms and W. Metzner, *Holz Roh-Werkst.*, **37,** 341 (1979).
[112] J. Krechniak and W. Foss, *Pr. Cent. Inst. Ochr. Pr.*, **31,** 285 (1981).
[113] A. G. Vitenberg and I. A. Tsibul'skaya, *Gig. Sanit.*, 60 (1982).
[114] J. Mueller and E. Rohbock, *Talanta,* **27,** 673 (1980).
[115] J. Koenig, W. Funcke, E. Balfanz, B. Grosch and F. Pott, *Atmos. Environ.*, **14,** 609 (1980).
[116] W. C. Eisenberg, *J. Chromatog. Sci.*, **16,** 145 (1978).
[117] L. M. Shabad, A. Ja. Chesina, G. A. Smirnow, N. E. Stepina, E. Hüngen, W. Prietsch, N. Jaskullas and M. Naumann, *Z. Ges. Hyg.*, **21,** 869 (1975).
[118] B. Seifert and I. Steinbach, *Z. Anal. Chem.*, **287**, 264 (1977).
[119] R. Morales, S. M. Rappaport and R. E. Hermes, *Am. Ind. Hyg. Assoc. J.*, **40,** 970 (1979).
[120] R. E. Berkley and E. D. Pellizzarri, *Anal. Lett.*, **11,** 327 (1978).
[121] K. Beyermann and W. Eckrich, *Z. Anal. Chem.*, **265,** 4 (1973).
[122] C. W. Stanley, J. E. Barney, M. R. Helton and A. R. Yobs, *Environ. Sci. Technol.*, **5,** 430 (1971).
[123] F. de West and D. Rondia, *Atmos. Environ.*, **10,** 487 (1976).
[124] R. Tomingas, *Z. Anal. Chem.*, **297,** 97 (1979).
[125] P. Ciccioli, G. Bertoni, E. Brancaleoni, R. Fratarcangeli and F. Bruner, *J. Chromatog.*, **126,** 757 (1976).
[126] C. Vidal-Madjar, M. F. Gonnord, F. Benchah and G. Guichon, *J. Chromatog. Sci.*, **16,** 190 (1978).
[127] J. W. Robinson, D. Nettles and P. L. H. Jowett, *Anal. Chim. Acta,* **92,** 13 (1977).
[128] J. P. Lodge, Jr., in *Accuracy in Trace Analysis,* Vol. I, p. 311. NBS, Washington (1976).
[129] M. J. Fishman and D. E. Derman, *Anal. Chem.*, **49,** 139R (1977).
[130] V. Kubelka, J. Mitera, J. Novák and J. Mostecký, *Sci. Papers Prague Inst. Chem. Technol.*, **F22,** 115 (1978).
[131] T. A. Gough, *Analyst,* **103,** 785 (1978).
[132] L. J. Ottendorfer, *Proc. Anal. Div. Chem. Soc.*, **15,** 52 (1978).
[133] J. Rodier, *Analysis of Water,* Wiley, New York (1975).
[134] A. Bjoerseth (ed.), *Analysis of Organic Micropollutants in Water,* Reidel, Dordrecht (1982).
[135] H. Dressler, *Chem. Listy,* **74,** 1245 (1980).
[136] R. Wagner, *Z. Anal. Chem.*, **282,** 315 (1976).
[137] P. W. Jones, *Proc. Anal. Div. Chem. Soc.*, **15,** 158 (1978).

[138] R. Shinohara, M. Koga, J. Shinohara and T. Hori, *Bunseki Kagaku,* **26,** 856 (1977).
[139] M. W. Scoggins and L. Skurcenski, *J. Chromatog. Sci.,* **15,** 573 (1977).
[140] Y. Yamato, M. Suzuki and T. Watanabe, *J. Assoc. Off. Anal. Chem.,* **61,** 1135 (1978).
[141] R. C. Chang and J. S. Fritz, *Talanta,* **25,** 659 (1978).
[142] T. Ramstadt and D. N. Armentrout, *Anal. Chim. Acta,* **102,** 229 (1978).
[143] K. Berkane, G. E. Caissie and V. N. Mallet, *J. Chromatog.,* **139,** 386 (1977).
[144] J. A. Coburn, I. A. Valdmanis and A. S. Y. Chau, *J. Assoc. Off. Anal. Chem.,* **60,** 224 (1977).
[145] E. E. McNeil, R. Otson, W. F. Miles and F. J. M. Rajabalee, *J. Chromatog.,* **132,** 277 (1977).
[146] D. W. Schnare, *J. Water Pollut. Contr. Fed.,* **51,** 2467 (1979).
[147] K. M. S. Sundaram, S. Y. Szeto and R. Hindle, *J. Chromatog.,* **177,** 29 (1979).
[148] N. Picer and M. Picer, *J. Chromatog.,* **193,** 357 (1980).
[149] W. A. Prater, M. S. Simmons and K. H. Mancy, *Anal. Lett.,* **13,** 205 (1980).
[150] G. Volpe and V. N. Mallet, *Intern. J. Environ. Anal. Chem.,* **8,** 291 (1980).
[151] G. Volpe and V. N. Mallet, *J. Chromatog.,* **177,** 385 (1979).
[152] S. Mierzwa and S. Witek, *J. Chromatog.,* **136,** 105 (1977).
[153] P. Jones and G. Nickless, *J. Chromatog.,* **156,** 87 (1978).
[154] C. G. Daughton, D. G. Crosby, R. L. Garnas and D. P. H. Hsieh, *J. Agr. Food Chem.,* **24,** 236 (1976).
[155] P. van Rossum and R. G. Webb, *J. Chromatog.,* **150,** 381 (1978).
[156] D. K. Basu and J. Saxena, *Environ. Sci. Technol.,* **12,** 795 (1978).
[157] G. D. Veith, D. W. Kuehl, F. A. Puglisi, G. E. Glass and J. G. Eaton, *Arch. Environ. Contam. Toxicol.,* **5,** 487 (1977).
[158] J. D. Navratil, R. E. Sievers and H. F. Walton, *Anal. Chem.,* **49,** 2260 (1977).
[159] E. Břízová, M. Popl and J. Čoupek, *J. Chromatog.,* **139,** 15 (1977).
[160] J. Lawrence and H. M. Tosine, *Environ. Sci. Technol.,* **10,** 381 (1976).
[161] C. E. Werkhowen-Goewie, U. A. T. Brinkman and R. W. Frei, *Anal. Chem.,* **53,** 2072 (1981).
[162] L. Renberg and K. Lindström, *J. Chromatog.,* **214,** 327 (1981).
[163] S. D. West and E. W. Day, *J. Assoc. Off. Anal. Chem.,* **64,** 1205 (1981).
[164] W. A. Saner and J. Gilbert, *J. Liq. Chromatog.,* **3,** 1753 (1980).
[165] W. I. Kimoto, C. J. Dooley, J. Carre and W. Fiddler, *Water Res.,* **15,** 1099 (1981).
[166] R. V. Golovnaya, T. A. Misharina and L. A. Semina, *Zh. Analit. Khim.,* **36,** 933 (1981).
[167] R. S. Narang and B. Bush, *Anal. Chem.,* **52,** 2076 (1980).
[168] Y. Zhao, Y. Wang and C. Fu, *Huaxue Tongbao,* 25 (1982).
[169] F. Mangani, G. Crescentini and F. Bruner, *Anal. Chem.,* **53,** 1627 (1981).
[170] L. Zhang, C. Xie, Y. Zhan and Q. Chen, *Fen Hsi Hua Hsueh,* **9,** 376 (1981).
[171] M. Dressler, *J. Chromatog.,* **165,** 167 (1979).
[172] R. W. Edwards, K. A. Nonnemaker and R. L. Cotter, in *Trace Organic Analysis,* pp. 87–94. NBS, Washington (1979).
[173] K. Ogan, E. Katz and W. Slavin, *J. Chromatog. Sci.,* **16,** 517 (1978).
[174] D. Kasiske, K. D. Klinkmueller and M. Sonnenbom, *J. Chromatog.,* **149,** 703 (1978).
[175] J. F. K. Huber and R. R. Becker, *J. Chromatog.,* **142,** 765 (1977).
[176] D. Ishii, K. Hibi, K. Asai and M. Nagaya, *J. Chromatog.,* **152,** 341 (1978).
[177] T. Barth, K. Tjessem and A. Aaberg, *J. Chromatog.,* **214,** 83 (1981).
[178] M. M. Hussein and D. A. M. Mackay, *J. Chromatog.,* **243,** 43 (1982).
[179] A. L. Wood, J. T. Wilson, R. L. Cosby, A. G. Hornsby and L. B. Baskin, *Soil Sci. Soc. Am. J.,* **45,** 442 (1981).
[180] H. Bargnoux, D. Pepin, J. L. Chabrad, F. Vedrine, J. Petit and J. A. Berger, *Analusis,* **5,** 170 (1977).
[181] N. Yonehara and M. Kamada, *Bunseki Kagaku,* **30,** 620 (1981).
[182] T. Ramstadt and T. J. Nestrick, *Water Res.,* **15,** 375 (1981).
[183] D. A. Kalman, R. Dills, C. Perera and F. P. DeWalle, *Anal. Chem.,* **52,** 1993 (1980).
[184] E. Stottmeister and W. Engewald, *Acta Hydrochim. Hydrobiol.,* **9,** 479 (1981).
[185] B. J. Dowty, L. E. Green and J. L. Laseter, *Anal. Chem.,* **48,** 946 (1976).

[186] P. P. K. Kuo, E. S. K. Chian, F. B. DeWalle and J. H. Kim, *Anal. Chem.*, **49,** 1023 (1977).
[187] B. Versino, H. Knöpel, M. de Groot, A. Peil, J. Poelman, H. Schauenberg, H. Vissers and F. Geiss, *J. Chromatog.*, **122,** 373 (1976).
[188] E. W. May, *J. Chromatog. Sci.*, **13,** 535 (1975).
[189] C. D. McAuliffe, *Chem. Techn.*, **1,** 46 (1971).
[190] R. Mindrup, in *Trace Organic Analysis,* pp. 225–229. NBS, Washington (1979).
[191] F. C. McElroy, T. D. Searl and R. A. Brown, in *Trace Organic Analysis,* pp. 7–18. NBS, Washington (1979).
[192] J. Royer and C. Hennequin, *Tech. Sci. Munic.*, **77,** 25 (1982).
[193] J. D. Wilson, *U.S. Environ. Prot. Agency, Off. Res. Dev., Rep.*, EPA 600/4-81-071 (1981).
[194] J. C. Milano and J. L. Vernet, *Analusis,* **9,** 360 (1981).
[195] E. H. Foerster and J. C. Garriot, *J. Anal. Toxicol.*, **5,** 241 (1981).
[196] Ya. L. Khromchenko and B. A. Rudenko, *Zh. Analit. Khim.*, **37,** 924 (1982).
[197] Z. Voznáková, M. Popl and M. Berka, *J. Chromatog. Sci.*, **16,** 123 (1978).
[198] W. Bertsch, E. Anderson and G. Holzer, *J. Chromatog.*, **112,** 701 (1975).
[199] B. D. Quimby, M. F. Delaney, P. C. Uden and R. M. Barnes, *Anal. Chem.*, **51,** 875 (1979).
[200] R. G. Westendorf, *Intern. Lab.*, **12,** No. 7, 32 (1982).
[201] K. Grob and F. Zürcher, *J. Chromatog.*, **117,** 285 (1976).
[202] B. A. Colenutt and S. Thorburn, *Intern. J. Environ. Anal. Chem.*, **7,** 231 (1980).
[203] J. Rivera, M. R. Cuberes and J. Albaiges, *Bull. Environ. Contam. Toxicol.*, **18,** 624 (1977).
[204] T. A. Bellar, J. J. Lichtenberg and J. W. Eichelberger, *Environ. Sci. Technol.*, **10,** 926 (1976).
[205] C. D. Chriswell and J. S. Fritz, *J. Chromatog.*, **136,** 371 (1977).
[206] A. Rapp and W. Knipser, *Chromatographia,* **13,** 698 (1980).
[207] V. Kubelka, J. Mitera, J. Novák, J. Mostecký, *Sbornik Vysoke Skoly Chemicko-Technol. v Praze,* **F20,** 85 (1976).
[208] R. A. Saunders, C. H. Blachly, T. A. Kovacina, R. A. Lamontagne, J. W. Swinnerton and F. E. Saalfeld, *Water Res.*, **9,** 1143 (1975).
[209] C. D. Chriswell, *J. Chromatog.*, **132,** 537 (1977).
[210] W. G. Guder, *Internist,* **21,** 533 (1980).
[211] R. Richterich, *Klinische Chemie,* p. 69. Akademie Verlag, Berlin (1965).
[212] J. D. Pryce and J. Durnford, *Clin. Chem.*, **26,** 1369 (1980).
[213] B. Halpern, in *Trace Organic Analysis,* pp. 373–379. NBS, Washington (1979).
[214] D. N. J. Manning, *Med. Lab. Sci.*, **39,** 257 (1982).
[215] D. D. Breimer and M. Danhof, *Pharm. Intern.*, **1,** 9 (1980).
[216] J. C. Mucklow, *Ther. Drug Monit.*, **4,** 229 (1982).
[217] B. Muehlenbruch, *Pharm. unserer Zeit,* **11,** 41 (1982).
[218] O. R. Idowu and B. Caddy, *J. Forens. Sci. Soc.*, **22,** 123 (1982).
[219] S. J. Gaskell, E. M. H. Finlay and A. W. Pike, *Biomed. Mass Spectrom.*, **7,** 500 (1980).
[220] R. F. Goldsmith and R. A. Ouvrier, *Ther. Drug Monit.*, **3,** 151 (1981).
[221] J. L. Valentine and P. Psaltis, *Anal. Lett.*, **12,** 855 (1979).
[222] J. J. MacKichan, P. K. Duffner and M. E. Cohen, *Br. J. Clin. Pharmacol.*, **12,** 31 (1981).
[223] M. Tondi, R. Mutani, C. Mastropaolo and F. Monaco, *Riv. Ital. Electroencefalog. Neurofisiol. Clin.*, **2,** 85 (1979).
[224] A. Noirfalise and J. Lambert, *Bull. Narc.*, **30,** 65 (1978).
[225] D. Valente, M. Cassini, M. Pigliapochi and G. Vansetti, *Clin. Chem.*, **27,** 1952 (1981).
[226] F. I. Ibbott, in *Accuracy in Trace Analysis,* p. 422. NBS, Washington (1976).
[227] J. Breiter, *Arzneimittelforschung,* **28,** 1941 (1978).
[228] L. C. Ramirez, C. Millot and B. F. Maume, *J. Chromatog.*, **229,** 267 (1982).
[229] G. R. Cannell, J. P. Galligan, R. H. Mortimer and M. J. Thomas, *Clin. Chim. Acta,* **122,** 419 (1982).

[230] C. H. L. Shackleton and J. O. Whitney, *Clin. Chim. Acta,* **107,** 231 (1980).
[231] Y. Ghoos, P. Rutgeerts and G. Vantrappen, *J. Liq. Chromatog.,* **5,** 175 (1982).
[232] R. J. Allan, H. T. Goodman and T. R. Watson, *J. Chromatog.,* **183,** *Biomed. Appl.,* **9,** 311 (1980).
[233] F. A. Andrews, L. R. Peterson, W. H. Beggs, D. Crankshaw and G. A. Sarosi, *Antimicrob. Agents Chemother.,* **19,** 110 (1981).
[234] N. Narasimhachari, *J. Chromatog.,* **225,** *Biomed. Appl.,* **14,** 189 (1981).
[235] V. Skrinska and F. V. Lucas, *Prostaglandins,* **22,** 365, (1981).
[236] E. V. Repique, H. J. Sacks and S. J. Farber, *Clin. Biochem.,* **14,** 196 (1981).
[237] E. Bamberg, E. Moestl, N. E. D. Hassaan, W. Stoekl and H. S. Choi, *Clin. Chim. Acta,* **108,** 479 (1980).
[238] R. Wehner and A. Handke, *Clin. Chim. Acta,* **93,** 429 (1979).
[239] R. Wehner and A. Handke, *J. Chromatog.,* **177,** 237 (1979).
[240] O. P. Edlund, *J. Chromatog.,* **187,** 161 (1980).
[241] R. Gauch, U. Leunenberger and E. Baumgartner, *J. Chromatog.,* **178,** 543 (1979).
[242] P. Hoeltzenbein, G. Bohn and G. Rücker, *Z. Anal. Chem.,* **292,** 216 (1978).
[243] K. Harzer and M. Kaechele, *Beitr. Gerichtl. Mediz.,* **37,** 357 (1979).
[244] H. Mizunuma, Y. Hirayama, S. Sakurai and N. Ikekawa, *Eisei Kagaku,* **28,** 13 (1982).
[245] J. M. Hoellinger, M. Sonnier, M. Hoffelt and N. H. Nguyen, *J. Chromatog.,* **196,** 342 (1980).
[246] M. Bogusz, J. Gierz and J. Bialka, *Vet. Hum. Toxicol.,* **21,** 185 (1979).
[247] P. Hache, R. Marquette, G. Volpe and V. N. Mallet, *J. Assoc. Off. Anal. Chem.,* **64,** 1470 (1981).
[248] M. Bogusz, J. Gierz and J. Bialka, *Arch. Toxicol.,* **41,** 153 (1978).
[249] J. Balkon, B. Donnelly and D. Prendes, *J. Forensic Sci.,* **27,** 23 (1982).
[250] R. Otson, D. T. Williams and P. D. Bothwell, *Environ. Sci. Techn.* **13,** 936 (1979).
[251] S. L. Friant and I. H. Suffet, *Anal. Chem.,* **51,** 2167 (1979).
[252] H. Hachenberg and A. P. Schmidt, *Gas Chromatographic Head Space Analysis,* Heyden, London (1977),
[253] G. Charalambous, *Analysis of Food and Beverages by Headspace Techniques,* Academic Press, New York (1978).
[254] J. Drozd and J. Novák, *J. Chromatog.,* **165,** 141 (1979).
[255] J. Drozd and J. Novák, *Chem. Listy,* **75,** 1148 (1981).
[256] B. Kolb, *J. Chromatog.,* **122,** 553 (1976).
[257] K. E. Murray, *J. Chromatog.,* **135,** 49 (1977).
[258] K. Y. Lee, D. Nurok and A. Zlatkis, *J. Chromatog.,* **158,** 377 (1978).
[259] I. Klimes, W. Stuenzi and D. Lamparsky, *J. Chromatog.,* **136,** 13 (1977).
[260] W. J. Khazal, J. Vejrosta and J. Novák, *J. Chromatog.,* **157,** 125 (1978).
[261] J. Drozd, J. Novák and J. A. Rijks, *J. Chromatog.,* **158,** 471 (1978).
[262] S. N. Chesler, B. H. Gump, H. S. Hertz, W. F. May and S. A. Wise, *Anal. Chem.,* **50,** 805 (1978).
[263] J. Drozd and J. Novák, *J. Chromatog.,* **152,** 55 (1978).
[264] G. W. Diachenko, C. V. Breder, M. E. Brown and J. L. Dennison, *J. Assoc. Off. Anal. Chem.,* **60,** 570 (1977).
[265] J. L. Dennison, C. V. Breder, T. McNeal, R. C. Snyder, J. A. Roach and J. A. Sphon, *J. Assoc. Off. Anal. Chem.,* **61,** 813 (1978).
[266] J. B. H. van Lierop and W. Stek, *J. Chromatog.,* **128,** 183 (1976).
[267] G. B.-M. Gawell, *Analyst,* **104,** 106 (1979).
[268] G. di Pasquale, G. di Iorio, T. Capaccioli and G. R. Verga, *J. Chromatog.,* **160,** 133 (1978).
[269] R. M. Maiorino, I. G. Sipes, A. J. Gandolfi and B. R. Brown, *J. Chromatog.,* **164,** 63 (1979).
[270] J. Gilbert, J. R. Startin and M. A. Wallwork, *J. Chromatog.,* **160,** 127 (1978).
[271] G. J. Piet, P. Slingerland, F. E. de Grunt, M. P. M. van der Heuvel and B. C. J. Zoeteman, *Anal. Lett.,* **11,** 437 (1978).
[272] K. L. E. Kaiser and B. G. Oliver, *Anal. Chem.,* **48,** 2207 (1976).
[273] G. di Pasquale, G. di Iorio and T. Capaccioli, *J. Chromatog.,* **152,** 538 (1978).

[274] J. Lindner and K. Langes, *Mitt. Lebensm. Gerichtl. Chem.*, **28**, 163 (1974).
[275] H. J. Drexler and G. Osterkamp, *J. Clin. Chem. Clin. Biochem.*, **15**, 431 (1977).
[276] E. A. Dietz and K. F. Singley, *Anal. Chem.*, **51**, 1809 (1979).
[277] B. H. van Lierop and H. Nootenboom, *J. Assoc. Off. Anal. Chem.*, **62**, 253 (1979).
[278] G. A. Ulsaker and R. M. Korsnes, *Analyst*, **102**, 882 (1977).
[279] B. Gisler and V. Mäkinen, *Brauwissensch.*, **31**, 177 (1978).
[280] W. Kisser, *Blutalkohol*, **12**, 204 (1975).
[281] A. W. Jones, *Anal. Biochem.*, **86**, 589 (1978).
[282] R. J. Steichen, *Anal. Chem.*, **48**, 1398 (1976).
[283] H. M. Liebich and J. Wöll, *J. Chromatog.*, **142**, 506 (1977).
[284] A. Heyndrickx and C. Van Peteghem, *Europ. J. Toxicol.*, **8**, 275 (1975).
[285] M. Vanhaelen, R. Vanhaelen-Fastre and J. Geeraerts, *J. Chromatog.*, **144**, 108 (1977).
[286] L. V. S. Hood and G. T. Barry, *J. Chromatog.*, **166**, 499 (1978).
[287] E. S. K. Chian, P. P. Kuo, W. J. Cooper, W. F. Cowen and R. C. Fuentes, *Environ. Sci. Technol.*, **11**, 282 (1977).
[288] G. J. Loubser and C. S. du Plaessis, *Vitis*, **15**, 248 (1976).
[289] M. L. McConnell and M. Novotny, *J. Chromatog.*, **112**, 559 (1975).
[290] V. Herrmann and F. Jüttner, *Anal. Biochem.*, **78**, 365 (1977).
[291] R. J. Horvat, *J. Agr. Food Chem.*, **24**, 953 (1976).
[292] V. Palo and J. Hrivnák, *Milchwissensch*, **33**, 285 (1978).
[293] K. W. Grubaugh and G. E. Stobby, *Anal. Chem.*, **50**, 377 (1978).
[294] J. Gilbert and J. R. Startin, *J. Chromatog.*, **205**, 434 (1981).
[295] S. L. Varner and C. V. Breder, *J. Assoc. Off. Anal. Chem.*, **64**, 1122 (1981).
[296] N. Vreden, K. H. van Bracht and L. Matter, *Lebensmittelchem., Gerichtl. Chem.*, **34**, 124 (1980).
[297] G. G. Rusakova and I. A. Kirpa, *Khim. Prom.-st., Ser. Metody Anal. Kontrolya Kach. Prod. Khim., Prom.-sti*, 10 (1979).
[298] N. A. Sebestyen and W. B. Thompson, *Chromatog. Newsl.*, **7**, 29 (1979).
[299] L. Rossi, J. Waibel and C. G. vom Bruck, *Food Cosmet. Toxicol.*, **18**, 527 (1980).
[300] J. R. Watson, R. C. Lawrence and E. G. Lovering, *Can. J. Pharm. Sci.*, **14**, 57 (1979).
[301] V. B. Stein and R. S. Narang, *Bull. Environ. Contam. Toxicol.*, **27**, 583 (1981).
[302] R. P. Lattimer and J. B. Pausch, *Am. Lab.*, **12**, No. 8, 80 (1980).
[303] C. Sadowski and J. Purcell, *Chromatog. Newsl.*, **9**, 52 (1981).
[304] B. V. Stolyarov, *Chromatographia*, **14**, 699 (1981).
[305] C. Bicci and S. Mina, *Farmaco, Ed. Prat.*, **35**, 632 (1980).
[306] G. Alles, U. Bauer and F. Selenka, *Zentralbl. Bakteriol. Mikrobiol. Hyg. Abt. I, Orig. B*, **174**, 238 (1981).
[307] G. R. Umbreit and R. L. Grob, *J. Environ. Sci. Health*, **15A**, 429 (1980).
[308] W. F. Cowen and R. K. Baynes, *J. Environ. Sci. Health*, **15A**, 413 (1980).
[309] R. P. Lattimer and J. B. Pausch, *Intern. Lab.*, **10**, No. 6, 51 (1980).
[310] E. R. Adlard, C. B. Milne and P. E. Tindle, *Chromatographia*, **14**, 507 (1981).
[311] M. Jakubowski, H. Radzikowska-Kintzi and G. Meszka, *Bromatol. Chem. Toksykol.*, **13**, 263 (1980).
[312] R. C. Entz and H. C. Hollifield, *J. Agr. Food Chem.*, **30**, 84 (1982).
[313] Ministry of Agriculture, *Analyst*, **106**, 782 (1981).
[314] M. Dong, A. H. DiEdwardo and F. Zitomer, *J. Chromatog. Sci.*, **18**, 242 (1980).
[315] W. Wellnitz-Ruen, G. A. Reineccius and E. L. Thomas, *J. Agr. Food Chem.*, **30**, 512 (1982).
[316] S. N. Lin, F. W. Y. Fu, J. V. Bruckner and S. Feldman, *J. Chromatog.*, **244**, 311 (1982).
[317] P. Dirinck, J. Veys, M. Decloedt and N. Schlamp, *Tob. Int.*, **182**, 125 (1980).

4

Treatment of samples before analysis

4.1 STORAGE

The problems arising in the storage of samples containing organic trace components are the same as those in inorganic trace analysis, i.e. adsorption on the wall of the storage vessel. Insecticides in water samples, for instance, are very quickly adsorbed on the walls of polyethylene vessels [1]. Further examples are given in Table 4.1.

Table 4.1 – Adsorption from dilute trace solutions.

Adsorbed trace substance	Adsorbing phase	Reference
2,4-D (2,4-dichlorophenoxyacetic acid) 2,4,5-T (2,4,5-trichlorophenoxyacetic acid)	glass	2, 3
basic amino-acids, nucleosides	glass	4
triazolam	glass	5
DDT	particulate matter (from water sample)	6
$CHCl_3$, CCl_4	polyethylene films	7
drugs, proteins	controlled-pore glass surfaces	8
drugs	walls of plastic container	9
$CHCl_3$, CCl_4, ethanol, acetone	polyethylene foil	7
biphenyls	cellulose fibres in paper mill effluents	10
aldrin, DDT	glass, glass wool	11, 11a
pesticides	cellulose triacetate filter	12
tricyclic antidepressants	glass	13, 14
histamine	glass	15
nicotine	glass	16
quinidine	stopper	17
PAHs	aquatic humic acids	18
PAHs	sediments in water sample	19

To avoid losses of this nature it has been recommended that the walls of all equipment should be saturated with a suitably diluted solution of the trace compound to be determined. However, it is better to keep the surface area of the vessels as small as possible in order to minimize the adsorption effect.

Another suggestion is to use an excess of the deuterated analogue of the trace compound which is to be determined [20]. The adsorption of the unlabelled trace is reduced and the losses become smaller. This method is useful only when the excess of the deuterated compound can be determined, e.g. by mass spectrometry.

The adsorption of drugs is said to be decreased when diathylamine is added to the solution containing the drug traces [13]. Addition of aqueous ammonia is reported to prevent the adsorption of nicotine (at the μg/ml level) on the walls of vessels [16]. Histamine, which is adsorbed on glass from aqueous solutions, is said not to be adsorbed from 0.01*M* hydrochloric acid [15]. 2,4-D is claimed to be desorbed by dilute acid [3], but not by hexane or hexane–methylene chloride mixtures [2].

Occasionally, comments are found in the literature which indicate that the addition of a chemically related internal standard or other additive will reduce losses. Unfortunately, these findings cannot be generalized, as the adsorption behaviour depends strongly on the trace compound, the added desorber and the other parameters of the solution/surface system.

The danger of oxidation, hydrolysis, enzymatic or bacterial transformation, and photodecomposition is definitely much greater with organic constituents than with inorganics. The effects are often concentration-dependent which adds to the difficulties of the puzzle. Hence samples are routinely stored at low temperatures to slow down the kinetics of the alteration processes, but such storage can lead to problems of organization if large numbers of samples are to be handled. The storage and filtration of water samples has been reviewed [21].

For the determination of 1–3 ng of PAHs per litre of water, certain precautions must be taken. Tap water directly analysed after addition of traces of 11 PAHs gave recoveries of 86% ± 12%. Complete loss of 7 PAHs was observed, however, when these were added to chlorinated tap water and stored at 5°C for 18 days. To deal with losses from field samples during transportation, a pellet of sodium sulphite spiked with 40 ng of perylene and benzo(*ghi*)perylene as internal standards can be added, to facilitate correction for losses (during transportation the samples have to be stored in the dark) [22].

Methods of dealing with storage problems are summarized in Table 4.2

In clinical chemistry blood is the most often used sample material. It has corresponding importance for the organic trace analyst, and has been rather thoroughly investigated with respect to changes in its constituents during storage, and as a function of temperature [32]. In most cases storage at +4°C is recommended, and at −20°C if low molecular-weight components are to be determined.

Occasionally, stabilizers are added, but may interfere with the following analysis. The trace analyst should therefore know the stabilizers most often

Table 4.2 – Recommendations for storage of solutions containing organic trace components.

Trace	Matrix	Concentration	Recommendation	Reference
volatile compounds	water		storage of water sample at 4° C	23
ATP	plankton (from ocean water)	10 ng/l.	immediate filtration, freeze filter in liquid N_2	24
DDT	water		immediate filtration after sampling to prevent adsorption on particulate matter	25
carbohydrates	ocean water		refrigerate sample (−20° C) in unused(!) polyethylene bottles	26
pesticides	water		lyophilization of sample	27
urea, uric acid, creatinine	serum	ppm	storage at room temperature for up to 4 days	28
acenaphthene, methylindene methylnaphthalene etc.	water	25 μg/l.	filtration (0.4-μm membrane) to eliminate micro-organisms	29
formaldehyde	Hantzsch solution	20 ng/ml	storage in the dark	30
trihalomethanes	water	μg/l.	minimum storage time (results are time-dependent)	31

used in clinical chemistry (Table 4.3), their applications, and any problems arising from their use.

Table 4.3 – Methods used in clinical chemistry to protect trace constituents in a drawn sample [from *Laboratoriums Medizin,* **2,** A+B, 133 (1978) by permission of the copyright holders].

Preventative measure	Effect prevented
addition of EDTA	blood clotting
addition of monoiodoacetate or sodium fluoride	glucose exchange in blood sugar determination
addition of toluene, thymol, boric acid, propan-2-ol or hydrochloric acid	bacterial growth in urine samples
use of non-transparent containers	alteration of light-sensitive components (e.g. porphyrins in urine)
immediate centrifugation of blood after sampling	transfer of enzymes and electrolytes from erythrocytes to serum
covering containers (usually with plastic film)	evaporation
storage at optimal temperature (e.g. 4°C, –20°C)	aging

The problem of instability of samples is particularly serious when constant concentrations or activities of standard reference materials or other calibrated standards must be maintained over long periods. In these cases – and often with serum samples in clinical chemistry – lyophilization is the method of choice for preparing stable standards.

Lyophilization is not free from problems, however. Monosaccharides, amino-sugars and amino-acids are said to be lost, unless glycerol is added [33]. The concentration of volatile components is liable to be changed. Pesticides, for example, are lost when very hard water samples are lyophilized. Acidification prevents such losses [34].

4.2 INFLUENCE OF IMPURITIES DURING STORAGE AND SAMPLE PREPARATION

With the increasing sensitivity of methods for organic trace analysis and the increasing presence of traces of ecological importance, blank values are becoming more and more of a problem in organic trace analysis, just as in inorganic trace analysis.

The difficulty arises because the blank values may be fairly constant or may vary rather widely. Contamination introduced from the reagents, equipment, laboratory atmosphere etc., will often be fairly constant and lead to generally applicable blanks, for a given batch of chemicals, a particular instrument, and so on. Unfortunately it is more often the case that the contamination is more variable, giving inconstant blank values that are difficult to calculate. The influence of contamination is difficult to generalize, as it depends on the efficiency of the separation methods and the specificity of the determination procedure. Relatively non-specific methods such as ultraviolet spectrophotometry tend to be more problematic than methods such as mass spectrometry. Table 4.4 gives some examples of sources of contamination.

Amino-acids are often converted into their fluoroacetylated n-butyl esters for gas chromatographic separation [68]. The method has been used – along with others – for lunar samples [69, 70]. Problems were encountered, though, when the scale of application was changed from the usual mg amounts to μg amounts or carrier-free radioactive solutions of amino-acids [71]. Several additional peaks were recorded which could not be traced to sample contamination from skin, fingerprints, hair etc. The trouble was finally traced to 1–10 ppm of carbonyl compounds in the butanol used, this being sufficient to react with amino groups of the amino-acids during butylation so that only the rest of the amino-acid n-butyl ester is acetylated by the trifluoroacetic anhydride. Side-products are formed and the yield decreases. It is therefore recommended that the purity of the butanol should be checked.

The operator may also be a source of contamination. Grease from the skin can be introduced when equipment is touched with bare fingers [74]. Volatiles from human skin have been studied; they were sampled by placing cotton pads on the axillae, the pads being worn overnight. After removal the pads were placed in glass tubes, warmed to 50°C and flushed with nitrogen. The volatiles were transferred to Tenax, from which they were thermally desorbed and analysed by gas chromatography [75]. About 40 compounds were identified, which partly originated from the air (e.g. benzene, toluene, xylenes, ethylbenzene, trichloroethylene, tetrachloroethylene, trimethylbenzenes, methylene chloride and limonene). Many of these compounds have been identified in studies of urban air [76–79] and have been found in volatiles from human blood [80–82]. Other substances encountered in sweat are known to be constituents of deodorants and cosmetic preparations [83]. It is believed that many of the compounds observed are emissions from man-made sources. These pollutants adhere to hair and skin and can perhaps become concentrated in the unwashed axillary region. Another study has identified C_1–C_3 alcohols, acetone and acetic acid as excretion products [84]. Volatile whole-body effluvia have been found to contain about 135 components [85]. All these findings indicate that sweat can be a significant source of contamination in organic trace analysis.

Information about the composition of modern toilet soaps and deodorants demonstrates [86, 87] that they contain phenols or chlorinated phenols which could possibly interfere with highly sensitive detection methods.

Activated adsorbents for chromatographic purposes attract laboratory vapours. On standing, oxidation and polymerization of these impurities takes place, with formation of several interfering compounds. Filter paper, and plates

Table 4.4 – Examples of contamination from reagents, containers, etc. in organic trace analysis.

Observed problem	Recommended solution	Reference
contamination of solvents and equipment for analysis of phthalate esters	heating of glassware, glass-wool, stationary phases, aluminium foil to 375–600°C in a muffle furnace reduces blank values to about 20–87 ng	35
contamination of silica gel precolumn in analysis of chlorinated hydrocarbons in ocean water	replacement of silica gel by aluminium oxide extracted with dichloromethane and hexane before use	36
glass-wool as source of contamination (up to 0.5 mg/g of glass-wool) for gc-ms analysis of fatty acid traces	extraction of glass-wool for 24 hr (Soxhlet)	37
septa for injection in gc tend to bleed	test septa in advance and use only acceptable types	38
hexachlorobenzene as contaminant in plastic wash-bottles	use bottles made from linear polyethylene or poly(vinyl)-chloride (or Teflon) containing little or no hexachlorobenzene	39
membrane filter contains non-ionic surfactant (polyoxyethylene phenyl ether)	wash filter carefully with distilled water before use	40
impurities in solvent give high blanks and deterioration of steroids in determination of cortisol	use only specially purified solvents	41
0.2–11 μg of cyanogen chloride, as impurity per ml of dichloromethane, will react with primary and secondary amines to form nitriles	avoid use of dichloromethane in this type of analysis	42
water content of solvent disturbs trace determination by nmr	dry solvent with molecular sieve	43
blanks – though very low (4 ng PCBs/1 of water) – must be taken into consideration in use of XAD 2 as extractant of traces from water	determine blank and reproduce procedure exactly	44
foil-lined screw-cap of sample vial was source of phthalate plasticizer contamination	use Teflon foil	45
rubber stoppers used in procedure gave off some volatile material which disturbed gc determination of theophyllin in blood	use glass stoppers	46
impurities in air disturbed determination of about 45 compounds in human breath	allow patient to inhale only purified air	47
acrylonitrile as impurity in acetonitrile	analyse for blank value	48
N-acetylglycine (0.6–55 μg/l.) as impurity in acetic acid	analyse for blank; slow distillation for purification of acetic acid	49
hexamine (7 ppm) as impurity in tetrachloroethylene	analyse for blank	50

Table 4.4 – *continued*

Observed problem	Recommended solution	Reference
0.4 ppm of non-volatile impurities in solvents	analyse blank (by distillation and weighing of residue)	51
squalane as a source of contamination from human skin	touch filters and other equipment only with metal tweezers	52
grease for lubrication of capsule injection system of gc	avoid lubrication	53
di(2-ethylhexyl) phthalate from PVC tubing interfered with gas chromatographic determination of oxprenolol in plasma	replace PVC by Teflon	54
di(2-ethylhexyl) phthalate interfered with formation of prostaglandin derivatives and subsequent determination by gc	replace plastic material by plasticizer-free material	55
di(2-ethylhexyl) phthalate and dibutyl phthalate contaminated intravenous fluids stored in plastic containers made with these plasticizers	use other container material	56
impurities in Tenax GC cartridges for ambient air sampling	extract cartridge in Soxhlet, heat cartridge at 270°C in N_2 atmosphere	57
twelve impurities from Extrelut columns were extracted with ethyl acetate and interfered with gc–ms	extract column with acetone, toluene	58
many impurities have been found in Porapak Q and Tenax GC when operating above their normal working temperature		59
impurities from XAD 2 formed during treatment with methanol	use diethyl ether instead of methanol	60
in flame sealing of glass ampoules PAHs have been formed	chill content of ampoule before sealing	61
tryptophan is converted into norharman by sunlight	protect solution from light	62
water-soluble silanols have been found in blood-collection tubes which have been sterilized by ionic radiation		63
organic impurities have been found in paper and thimbles used for Soxhlet extraction	use glass thimbles instead	64
artefact formation during Soxhlet extraction of soil with acetone and Na_2CO_3 or NH_4HSO_4	be aware of such interferences	65
bottle stoppers of containers for distilled water proved a prime source of contamination	avoid such stoppers	66
impurities in hexane and heptane (0.1–3% in reagent grade material!)	be aware of interference	67

for thin-layer chromatography (tlc) or high-pressure thin-layer chromatography (hptlc) must be stored under strictly controlled conditions.

Contamination from phthalate esters is a rather severe, and hence frequently reported, problem in organic trace analysis. In mass spectrometry an ion with $m/z = 149$ almost always results from these impurities. Migration of the plasticizers from laboratory flooring and wall-paint has been reported [88]; the chemicals, glassware, solvents and filter paper used consequently become contaminated. Special attention has been paid to the contamination of solvents [74, 89, 90]. Filter papers can also become polluted [91].

Glassware to be used in analysis for such organic pollutants should be cleaned by the following steps: washing with detergents, rinsing with 'pure' water, several washings with methanol, acetone or hexane, and drying at 130°C after each washing [92, 93]. Other authors do not recommend the rinsing with organic solvents [94]. Heating for 12 hr at 500°C has been suggested [85]. Equipment thus treated has been found to become contaminated within a few hours merely by standing in the laboratory atmosphere. All glassware should be cleaned briefly before use for phthalate determination at the trace level [95]. A minimum of glassware should be used, and the surface area should be kept as small as possible. Use of plastics should be avoided, as they tend to adsorb these omnipresent plasticizers. This recommendation extends even to Teflon, which otherwise is a non-critical material in organic trace analysis.

Pure water that has come into contact with plastic material tends to pick up phthalates [96–99], as these esters dissolve in both aqueous and non-aqueous media. Contamination has been seen in blood samples stored in poly(vinyl chloride) bags [100]. By use of extreme precautions background levels of about 25 ng of dibutyl phthalate and 50 ng of diethylhexyl phthalate have been obtained in the analysis of biota samples from the open ocean [101]. Polychlorinated dibenzo-*p*-dioxins occur more commonly in the environment than has so far been generally recognized [102]. The implication for the determination of ultratrace concentrations of these highly toxic compounds is obvious.

Nicotine is a widespread compound and has been found in the blank for many analytical systems [102–107]. It has been found in the urine of non-smokers [108]. Indoor environments contaminated with tobacco smoke have been found to contain up to 0.2 μg of *N*-nitrosamines per m^3 of air, depending on the degree of pollution with tobacco smoke [109]. The main-stream smoke – that inhaled by the smoker – was found to contain between 0.1 and 27 ng of dimethylnitrosamine (together with other nitroso compounds), but the side-stream smoke – that passed into the environment – contained between 143 and 415 ng of the same (carcinogenic) compound [110]!

A worldwide questionnaire survey on contamination problems in trace analysis [111] revealed that in 96 laboratories the main sources of contamination were considered to be: laboratory air, reagents, equipment, the investigator. The decomposition step was considered to be most likely to introduce impurities, followed by separation, sampling and storage. The final measurement was found to be the least critical step. Unfortunately, the entire article dealt only with inorganic trace analysis and such a survey is still to be made for organic trace analysis.

4.3 PREPARATION OF SAMPLES FOR ANALYSIS

Heterogeneous material has to be homogenized before a sample is taken. This is done by milling, grinding, mincing, blending etc. The same procedures are used to prepare the sample for better dissolution or digestion, because the larger total surface area obtained by reducing the particle-size makes the attack of reagents more rapid. All these manipulations are done with commercially available homogenizers that are commonly used in chemical laboratories.

A 'cool grinder' (produced by Spex Industries, 3880 Park Avenue, Metuchem, N. J., USA) produces powder from 2-g samples of animal tissue (hair, skin), gum (chewing gum, rubber) or plastics (nylon, polyethylene) within 2 minutes. Materials can be reduced to a powder with a final size of about 100 mesh by impact grinding under liquid nitrogen.

Experience with sampling by shattering cold, brittle biological tissues has been reported [112]. The sample material was chilled with liquid nitrogen. Whole mice have been frozen and homogenized in liquid nitrogen for analysis for *N*-nitrosoatrazine and *N*-nitrosocarbaryl [113]. Whole fruits have been milled to a fine powder after immersion in liquid nitrogen, before the extraction of pesticide traces [114].

Animal tissue can also be homogenized by use of formamide [115, 116]. After its mechanical homogenization, 10 ml of water and 10 ml of formamide are added to about 1 g of sample. On standing at room temperature a clear, colloidal solution is obtained. Sometimes brief heating to 105°C is necessary. As always, the behaviour of organic trace components under such drastic treatment has to be checked.

In analysis for the dithiocarbamate pesticide content of food, it is recommended *not* to use homogenization, as it may cause decomposition of the trace [117]. For determination of oestrogen residues in meat, homogenization in tetrahydrofuran is recommended [118]. In analysis for benzo(*a*)pyrene in marine organisms, alkaline digestion of samples for homogenization is acceptable [119]. (See also Section 5.3.3).

Enzymatic hydrolysis is one method for dissolution of high molecular-weight compounds. After denaturation by brief heating, 1 g of protein is hydrolysed by stirring with 30 mg of trypsin at pH 8.2 in 100 ml of solution for 24 hr at 37°C [120]. Papain and other proteinases have similarly been used [121]. Liver tissue has been digested with alcalase [122]. Collagenase and hyaluronidase have been used for sheep fat [123], and proteinases I, V or VIII for several other tissues [124].

Release of compounds which are bound to proteins can be achieved by enzymes – such as subtilisin – without complete dissolution of the protein. Thus ochratoxin A, at a level of 5 μg/kg, has been set free from pig kidney proteins [125]. Basic or acidic drugs [126, 127] and food dyes [128] have been recovered from proteins by the same treatment.

4.4 EXTRACTION FROM SOLIDS

Extraction – by definition – is the procedure in which a solid material is treated with a solvent in order to dissolve and extract some of the solid phase. Extraction is frequently used as the initial step in the analytical scheme. It is therefore described in this chapter, although it is sometimes used within the analytical scheme. For example, a precipitate, or a residue from an evaporation, may contain a trace compound and other components, and can be extracted with a selective solvent which dissolves the constituent of interest but leaves the rest undissolved (or vice versa).

Extraction is used as the initial step in about 5% of all procedures. Though separation may be achieved in this way there is always the danger of leaving trace material undissolved within the matrix, thus giving rise to losses. Extraction efficiency is difficult to check, as there are no good standards which contain the trace in a reliably implanted form within the solid matrix. Such problems have been mentioned already with respect to samples of oyster tissue (see p. 62). Whenever possible, several methods of extraction or total decomposition should be tried, and should also be checked for their efficiency.

Some information on efficiency can be obtained by means of radioactive tracers incorporated into the authentic intact sample material. In pesticide analysis, for example, plants can be sprayed with radioactively labelled substances. The treated plants are then cultivated for at least one week, by which time incorporation into the cellular tissue should have taken place, especially in the newly grown parts. It can be assumed that the tracer and the inactive material will give the same behaviour under identical growth conditions. Of course, with this system there is always the possibility of changes in metabolism, and this must be checked independently. However, the recovery of radioactivity in the extract gives a good indication of the efficiency of the extraction system.

Extraction has long been used in food analysis for isolation of fats, lipids etc., by Soxhlet extraction. Good yields of the trace components can be expected, along with some dissolution of major components of the matrix. The sample has to be porous and finely ground. Trace components within the interior crystalline phases or fixed in the walls of thick biological membranes are difficult to set free. There is also the possibility of adsorption of the trace species on the powdered matrix material. Extraction of chlorobenzenes from fish or sediments [129] and of benzo(*a*)pyrene, hexachlorobenzene and pentachlorophenol from oyster tissue [130] has been reported. More examples can be found in Tables 5.2 and 5.8.

Table 4.5 – Deproteination methods in organic trace analysis.

Compound	Matrix	Concentration or amount	Deproteination method	Reference
theophylline	plasma		molecular filtration	142
vanilmandelic acid	serum	6 ng/ml	$HClO_4$ precipitation	143
methanesulphonate of *m*-aminobenzoic acid ethyl ester	fish	1–19 ppm	trichloroacetic acid	144
tryptophan	serum	ng	trichloroacetic acid	145
pemoline	serum	0.1 ppm	sulphosalicylic acid	146
diacetoxyscirpenol	corn, mixed feed	> 25 ppb	ammonium sulphate	147
drugs	liver tissue		tungstate–ammonium sulphate	148
histamine, putrescine etc.	tissue		$HClO_4$	149
salicylic acid	serum	> 40 ppb	$HClO_4$	150
taurine	urine		sulphosalicylic acid	151
5-hydroxytryptamine	tissue	> 25 pg	$HClO_4$	152
paraquat	plasma		$CHCl_3$–ethanol (4:1)	153
retinol	serum	> 15 ng	methanol	154
furosemide, 4-chloro-5-sulphamoylanthranilic acid	plasma	10 ng/ml	methanol	155
sedatives (12)	serum	> 10 ng	acetonitrile	156
phenobarbital	serum	> 10 ng	acetonitrile	157
griseofulvin	plasma	50 ng/ml	acetonitrile	158
Azapropazon	plasma	ng	methanol	159

Samples of soft materials are often homogenized by adding an appropriate solvent and mincing, mixing etc. The resulting slurry is filtered or centrifuged and the filtrate, centrifugate or supernatant liquid is used for further analysis. Sometimes, a salt is added to dry the matrix material or to disrupt its cellular structure (for reasons of osmosis) or to increase its surface area after grinding. For example, apples can be macerated by treatment with dichloromethane and anhydrous sodium sulphate, and after filtration the extract analysed for binapacryl, bupirimate and diflubenzuron [131]. For a simple determination of pesticides, fat samples can be ground with anhydrous sodium sulphate to give a powder which is then extracted [132].

Agitation of the finely ground solid material in the extraction medium increases the efficiency. This can be done by vortexing, stirring or ultrasonic vibration. Examples of the last method are the extraction of 0.5-5 ppm of PAHs from rock material or dust [133, 134], and of PAHs from dust [135, 136]. Benzo(*a*)pyrene has been extracted from filters used for collection of ng amounts of the carcinogen from cubic-metre samples of air; ultrasonic agitation greatly speeded up the procedure [137, 138].

4.5 DEPROTEINATION

Methods for deproteination in organic trace analysis generally use precipitation of the major component either by acid or by addition of a water-soluble solvent (see Table 4.5), but there is always the possibility of occlusion or adsorption of traces by the precipitate. In the determination of salicyclic acid (at the 20-μg/g level) in cadaver liver tissue the loss in deproteination has been found (by use of ^{14}C tracer) to be nearly 20% [139]. Usually though, the losses are smaller. The use of acetonitrile for protein precipitation is becoming more common [140] especially when the resulting protein-free solution is to be subjected to high-pressure liquid chromatography (hplc) with ultraviolet detection, as this solvent absorbs only at very short wavelengths. Artefact formation with this solvent during the determination of theophylline has been reported, however [141].

REFERENCES

[1] L. Weil and K. E. Quentin, *Gas-,Wasserfach, Ausg. Wasser-Abwasser*, **111**, 26 (1970).
[2] M. Osadchuk, E. Salahub and P. Robinson, *J. Assoc. Off. Anal. Chem.*, **60**, 1324 (1977).
[3] G. Y. P. Kan, F. T. S. Mah and N. L. Wade, *J. Assoc. Off. Anal. Chem.*, **65**, 763 (1982).
[4] T. Mizutani and A. Mizutani, *Anal. Biochem.*, **83**, 216 (1977).
[5] H. Ko, M. E. Royer, J. B. Hester and K. T. Johnston, *Anal. Lett.*, **10**, 1019 (1977).
[6] A. J. Wilson, *Bull. Envir. Contam. Toxicol.*, **25**, 515 (1976).
[7] S. Badilescu, B. Talpus, R. Schiau and I. Gonciulea, *Rev. Chim. (Bucharest)*, **27**, 979 (1976).

[8] T. Mizutani and A. Mizutani, *J. Pharm. Sci.*, **67**, 1102 (1978).
[9] B. Scales, *Intern. Lab.*, Nov./Dec., 37 (1978).
[10] D. B. Easty and B. A. Wabers, *Anal. Lett.*, **10**, 857 (1978).
[11] V. Leoni, G. Puccetti, R. J. Colombo and A. M. Dovidio, *J. Chromatog.*, **125**, 399 (1976).
[11a] P. R. Musty and G. Nickless, *J. Chromatog.*, **89**, 185 (1974).
[12] D. A. Kurtz, in *Trace Organic Analysis*, pp. 19–32. NBS, Washington (1979).
[13] S. M. Johnson, C. Chan, S. Cheng, J. L. Shimek, G. Nygard and S. K. W. Khalil, *J. Pharm. Sci.*, **71**, 1027 (1982).
[14] M. Zetin, H. R. Rubin and R. Rydzewski, *Am. J. Psychiatry*, **138**, 1247 (1981).
[15] J. Murray and A. S. McGill, *J. Assoc. Off. Anal. Chem.*, **65**, 71 (1982).
[16] O. Grubner, M. W. First and G. L. Huber, *Anal. Chem.*, **52**, 1755 (1980).
[17] D. S. Ooi, W. J. Poznanski and F. M. Smith, *Clin. Biochem.*, **13**, 297 (1980).
[18] E. T. Gjessing and L. Berglind, *Arch. Hydrobiol.*, **91**, 24 (1981).
[19] H. Zawadzke, J. Zerbe and D. Baralkiewicz, *Chem. Anal. (Warsaw)*, **25**, 469 (1980).
[20] R. Self, *Biomed. Mass Spectrom.*, **6**, 315 (1979).
[21] H. Casey and S. M. Walker, *Intern. Environ. Saf.*, 16 (1981).
[22] R. K. Sorrell and K. Reding, *J. Chromatog.*, **185**, 655 (1979).
[23] A. Heyndrickx and C. van Peteghem, *Eur. J. Toxicol.*, **8**, 275 (1975).
[24] L. Hofer-Siegrist, *Schweiz. Z. Hydrol*, **38**, 49 (1976).
[25] A. J. Wilson, *Bull. Environ. Contam. Toxicol.*, **25**, 515 (1976).
[26] H. Hirayama, *Bunseki Kagaku*, **27**, 252 (1978).
[27] H. Bargnoux, D. Pepin, J.-L. Chabard, F. Vedrine, J. Petit and J. A. Berger, *Analusis*, **5**, 170 (1977).
[28] K. Abraham and A. Rössler-Englhardt, *Lab. Med.*, **2**, A+B 133 (1978).
[29] I. Ogawa, G. A. Junk and H. J. Svec, *Talanta*, **28**, 725 (1981).
[30] E. B. Rietz, *Anal. Lett.*, **13**, 1073 (1980).
[31] H. Norin and L. Renberg, *Water Res.*, **14**, 1397 (1980).
[32] N. Kubasik, M. Ricotta, T. Hunter and H. E. Sine, *Clin. Chem.*, **28**, 164 (1982).
[33] R. Dawson and K. Mopper, *Anal. Biochem.*, **84**, 186 (1978).
[34] D. Pepin, J.-L. Chabard, F. Vedrine, J. Petit and J. A. Berger, *Analusis*, **6**, 107 (1978).
[35] R. D. J. Webster and G. Nickless, *Proc. Anal. Div. Chem. Soc.*, **13**, 333 (1976).
[36] J. C. Duinker and M. T. J. Hillebrand, *J. Chromatog.*, **150**, 195 (1978).
[37] D. P. Schwartz, *J. Chromatog.*, **152**, 514 (1978).
[38] V. M. Olsavicky, *J. Chromatog. Sci.*, **16**, 197 (1978).
[39] D. R. Bourke, W. F. Mueller and R. S. H. Yang, *J. Assoc. Off. Anal. Chem.*, **60**, 233 (1977).
[40] A. Otsuki and K. Fuwa, *Talanta*, **24**, 584 (1977).
[41] H. Gleispach, *Chromatographia*, **10**, 40 (1977).
[42] R. A. Franklin, K. Heatherington, B. J. Morrison, P. Sherren and T. J. Ward, *Analyst*, **103**, 660 (1978).
[43] J. B. Alper, *Anal. Chem.*, **50**, 381 (1978).
[44] J. A. Coburn, I. A. Valdmanis and A. S. Y. Chau, *J. Assoc. Off. Anal. Chem.*, **60**, 224 (1977).
[45] D. W. Denney, F. W. Karasek and W. D. Bowers, *J. Chromatog.*, **151**, 75 (1978).
[46] F. Chrzanowski, *J. Pharm. Sci.*, **65**, 735 (1976).
[47] H. L. Gearhart, S. K. Pierce and D. Payne-Bose, *J. Chromatog. Sci.*, **15**, 480 (1977).
[48] H. Plieninger and A. Satyavan, *Liebigs Ann. Chem.*, **535 (1978).**
[49] L. M. S. Onge, S. L. Kittle and P. B. Hamilton, *Anal. Biochem.*, **71**, 156 (1976).
[50] Z. T. Chalaya and O. A. Egoreichenko, *Metody Anal. Kontrolya Proizvod. Khim. Prom.-sti.*, 12 (1977).
[51] B. H. Campbell and L. G. Hallquist, *Anal. Chem.*, **50**, 963 (1978).
[52] E. Wauters, P. Sandra and M. Verzele, *J. Chromatog.*, **170**, 125 (1979).
[53] G. M. Frame, G. A. Flanigan and D. C. Carmody, *J. Chromatog.*, **168**, 365 (1979).
[54] W. D. Hooper and M. T. Smith, *J. Pharm Sci.*, **70**, 346 (1981).
[55] J. Mai, K. E. Kinsella and J. S. Chou, *Lipids*, **15**, 968 (1980).

[56] N. P. H. Ching, G. N. Jham, C. Subbarayan, C. Grossi, R. Hicks and T. F. Nelson, *J. Chromatog.*, **225,** *Biomed. Appl.*, **14,** 196 (1981).
[57] S. A. Hubbard, G. M. Russwurm and S. G. Walburn, *Atmos. Environ.*, **15,** 905 (1981).
[58] M. Ende, P. Pfeifer and G. Spiteller, *J. Chromatog.*, **183,** *Biomed. Appl.*, **9,** 1 (1980).
[59] M. J. Lewis and A. A. Williams, *J. Sci. Food Agric.*, **31,** 1017 (1980).
[60] H. A. James, C. P. Steele and I. Wilson, *J. Chromatog.*, **208,** 89 (1981).
[61] G. R. Sirota, J. F. Uthe, C. J. Musial and V. Zitko, *J. Chromatog.*, **202,** 294 (1980).
[62] M. Kenny, R. F. Lambe, D. A. O'Kelly and A. Darragh, *Clin. Chem.*, **26,** 1511 (1980).
[63] W. E. Jensen, R. T. O'Donnell, I. Rosenberg, D. A. Karlin and R. D. Jones, *Clin. Chem.*, **28,** 1406 (1982).
[64] H. G. Nowicki, C. A. Kieda, R. F. Devine, V. Current and T. H. Schaefers, *Anal. Lett.*, **12,** 769 (1979).
[65] A. L. Lafleur and N. Pangaro, *Anal. Lett.*, **14,** 1613 (1981).
[66] E. J. Sampson and P. H. Culbreth, *Clin. Chem.*, **27,** 1773 (1981).
[67] S. Okada and Y. Iida, *Shitsuryo Bunseki,* **28,** 263 (1980).
[68] C. W. Gehrke, D. Roach, R. W. Zumwalt, D. L. Stalling and L. L. Wall, *Quantitative Gas Liquid Chromatography of Amino Acids in Proteins and Biological Substances, Macro, Semimicro and Micro Methods,* Jackson Anal. Biochem. Lab., Columbia, Mo., (1968).
[69] C. W. Gehrke, R. W. Zumwalt, D. L. Stalling, D. Roach, W. A. Aue, C. Ponnamperuma and K. A. Kvenvolden, *J. Chromatog.*, **59,** 305 (1971).
[70] C. W. Gehrke, *Space Life Sci.*, **3,** 342 (1972).
[71] M. Matucha and E. Smolkova, *J. Chromatog.*, **168,** 255 (1979).
[72] G. von Unruh, G. Remberg and G. Spiteller, *Chem. Ber.*, **104,** 2071 (1971).
[73] G. Remberg, H. Luftmann, G. v. Unruh and G. Spiteller, *Tetrahedron,* **27,** 5853 (1971).
[74] D. H. Hunnemann, *Tetrahedron Lett.*, 1743, (1968).
[75] J. Labows, G. Preti, E. Hoelzle, J. Leyden and A. Kligman, *J. Chromatog.*, **163,** 294 (1979).
[76] J. M. Steyn and H. K. L. Hundt, *J. Chromatog.*, **107,** 196 (1975).
[77] S. Ahuja, *J. Pharm. Sci.*, **65,** 163 (1976).
[78] R. H. Greeley, *Clin. Chem.*, **20,** 192 (1974).
[79] I. A. Zingales, *J. Chromatog.*, **75,** 55 (1973).
[80] C. B. McAllister, *J. Chromatog.*, **151,** 62 (1978).
[81] D. M. Rutherford, *J. Chromatog.*, **137,** 439 (1977).
[82] M. S. Greaves, *Clin. Chem.*, **20,** 141 (1974).
[83] M. Kowblansky, B. M. Scheinthal, G. D. Cravello and L. Chafetz, *J. Chromatog.*, **76,** 467 (1973).
[84] V. P. Savina, N. L. Sokolov and E. A. Ivanov, *Kosm. Biol. Aviakosm. Med.*, **9,** 76 (1976).
[85] R. L. Ellin, R. L. Farrand, F. W. Oberst, C. L. Crouse, N. B. Billups, W. S. Koon, N. P. Muselmann and F. R. Sidell, *J. Chromatog.*, **100,** 137 (1974).
[86] H. Koenig, *Seifen, Oele, Fette, Wachse,* **107,** 532 (1981).
[87] R. Kieffer and H. Scherz, *Z. Anal. Chem.*, **293,** 135 (1978).
[88] D. F. Lee, J. Britton, B. Jeffcoat and R. F. Mitchell, *Nature,* **211,** 521 (1966).
[89] J. Vessman and G. Rietz, *J. Chromatog.*, **100,** 153 (1974); Y. Asakawa and F. Genzida, *J. Sci. Hiroshima Univ., Ser. A-2,* **34,** 103 (1970).
[90] Y. Asakawa, F. Genzida and T. Matsura, *Anal. Lett.*, **2,** 333 (1969).
[91] J. Lam, *Chem. Ind. London,* 1837 (1967).
[92] G. A. Junk, J. J. Richard, M. D. Griesser, D. Witiak, J. L. Witiak, M. D. Arguello, R. Vick, H. J. Svec, J. S. Fritz and G. V. Calder, *J. Chromatog.*, **99,** 745 (1974).
[93] E. E. McNeil, R. Otson, W. F. Miles and F. J. M. Rajabalee, *J. Chromatog.*, **132,** 277 (1977).
[94] K. Grob and F. Zürcher, *J. Chromatog.*, **117,** 285 (1976).
[95] K. S. Anderson and J. Lam, *J. Chromatog.*, **169,** 101 (1979).

[96] C. G. Creed, *Res./Develop.* **27**, 40, (1976).
[97] B. Versino, H. Knöppel, M. deGroot, A. Peil, J. Poelman, H. Schauenburg, H. Vissers and F. Geiss, *J. Chromatog.*, **122**, 373 (1976).
[98] G. A. Junk, H. J. Svec, R. D. Vick and M. J. Avery, *Environ. Sci. Technol.*, **8**, 100 (1974).
[99] S. Mori, *Anal. Chim. Acta*, **108**, 325 (1979).
[100] M. I. Luster, P. W. Albro, K. Chae, G. Clark and J. D. McKinney, *Clin. Chem.*, **24**, 429 (1978).
[101] C. S. Giam, H. S. Cham and G. S. Neff, *Anal. Chem.*, **47**, 2225 (1975).
[102] H. R. Buser, *TrAC*, **1**, 318 (1982).
[103] S. E. Falkman, I. E. Burrows, R. A. Lundgren and B. F. J. Page, *Analyst*, **100**, 99 (1975).
[104] C. Feyerabend, T. Levitt and M. A. H. Russell, *J. Pharm. Pharmacol.*, **27**, 434 (1975).
[105] P. F. Isaac and M. J. Rand, *Nature*, **236**, 308 (1972).
[106] I. E. Burrows, P. J. Corp, C. G. Jackson and B. F. J. Page, *Analyst*, **96**, 81 (1971).
[107] N. Hengen and M. Hengen, *Clin. Chem.*, **24**, 50 (1978).
[108] E. C. Horning, M. G. Horning, D. I. Carroll, R. N. Stillwell and I. Dzidic, *Life Sci.*, **13**, 1331 (1973).
[109] D. Hoffman, K. D. Brunnemann, I. Schmeltz and E. L. Wynder, in *Trace Organic Analysis*, pp. 131–141. NBS, Washington (1979).
[110] K. D. Brunnemann, W. Fink and F. Moser, *Oncology*, **37**, 217 (1980).
[111] A. Mizuike and M. Pinta, *Pure Appl. Chem.*, **50**, 1519 (1978).
[112] J. A. Nichols and L. R. Hageman, *Anal. Chem.*, **51**, 1591 (1979).
[113] I. S. Krull, K. Mills, G. Hoffman and D. H. Fine, *J. Anal. Toxicol.*, **4**, 260 (1980).
[114] W. Pickenhagen, A. Velluz, J. P. Passerat and G. Ohloff, *J. Sci. Food Agric.*, **32**, 1132 (1981).
[115] D. L. Tabern and T. N. Lahr, *Science*, **119**, 739 (1954).
[116] E. M. Pearce, F. Devenuto, W. M. Fitch, H. E. Firschein and U. Westphal, *Anal. Chem.*, **28**, 1762 (1956).
[117] Arbeitsgruppe Pesticide, *Mitt. GDCh-Fachgr. Lebensmittelchem. Gerichtl. Chem.*, **31**, 25 (1977).
[118] H.-J. Stan and F. W. Hogls, *Z. Lebensm. Untersuch. Forsch.*, **166**, 287 (1978).
[119] B. P. Dunn, *Environ. Sci. Technol.*, **10**, 1018 (1976).
[120] R. Clotten and A. Clotten, *Hochspannungselektrophorese*, p. 103. Thieme, Stuttgart (1962).
[121] U. Bergmeyer, *Enzymatische Analyse*, 2 Vols., Verlag Chemie, Weinheim (1974).
[122] S. J. Dickson, W. T. Clearly, A. W. Missen, E. A. Queree and S. M. Shaw, *J. Anal. Toxicol.*, **4**, 74 (1980).
[123] C. Nerenberg, I. Tsina and M. Shaikh, *J. Assoc. Off. Anal. Chem.*, **65**, 635 (1982).
[124] M. Bogusz, J. Bialka and J. Gierz, *Z. Rechtsmed.*, **87**, 287 (1981).
[125] D. C. Hunt, L. A. Philp and N. T. Crosby, *Analyst*, **104**, 1171 (1979).
[126] M. D. Osselton, I. C. Shaw and H. M. Stevens, *Analyst*, **103**, 1160 (1978).
[127] M. D. Osselton, *Forensic Sci. Soc.*, **17**, 189 (1979).
[128] N. Boley, N. T. Crosby and P. Roper, *Analyst*, **104**, 472 (1979).
[129] B. G. Oliver and K. D. Bothen, *Intern. J. Environ. Anal. Chem.*, **12**, 131 (1982).
[130] H. E. Murray, G. S. Neff, Y. Hrung and C. S. Giam, *Bull. Environ. Contam. Toxicol.*, **25**, 663 (1980).
[131] D. J. Austin and K. J. Hall, *Pestic. Sci.*, **12**, 495 (1981).
[132] S. M. Waliszewski and G. A. Szymczynski, *J. Assoc. Off. Anal. Chem.*, **65**, 677 (1982).
[133] Y. L. Tan, *J. Chromatog.*, **176**, 319 (1979).
[134] T. Dutkiewicz, N. Masny, S. Ryborz, J. Masłowski and A. Grabka, *Chem. Anal. (Warsaw)*, **24**, 191 (1979).
[135] E. P. Lankmayer and K. Mueller, *J. Chromatog.*, **170**, 139 (1979).
[136] T. Nielsen, *J. Chromatog.*, **170**, 147 (1979).
[137] D. H. Swanson and J. F. Walling, *Chromatog. Newsl.*, **9**, 25 (1981).
[138] J. Krajewski and K. Lipski, *Chem. Anal. (Warsaw)*, **26**, 431 (1981).

[139] A. Ostrowski, *Arch. Med. Sadowej. Kryminol.*, **31**, 27 (1981).
[140] J. C. Mathies and M. A. Austin, *Clin. Chem.*, **26**, 1760 (1980).
[141] D. A. Jowett., *Clin. Chem.*, **27**, 1785 (1981).
[142] L. C. Franconi, G. L. Hawk, B. J. Sandmann and W. G. Haney, *Anal. Chem.*, **48**, 372 (1976).
[143] S. Takahashi, D. D. Godse, J. J. Warsh and H. C. Stancer, *Anal. Chim. Acta*, **81**, 183 (1977).
[144] J. B. Sills and Ch. W. Luhning, *J. Assoc. Off. Anal. Chem.*, **60**, 961 (1977).
[145] A. M. Krstulovic, P. R. Brown, D. M. Rosie and P. B. Champlin, *Clin. Chem.*, **23**, 1984 (1977).
[146] Y. S. Chu and L. T. Sennello, *J. Chromatog.*, **137**, 343 (1977).
[147] T. R. Romer, T. M. Boling and J. L. MacDonald, *J. Assoc. Off. Anal. Chem.*, **61**, 801 (1978).
[148] L. J. Dusci and L. P. Hackett, *J. Forens. Sci.*, **22**, 545 (1977).
[149] Y. Endo, *Anal. Biochem.*, **89**, 235 (1978).
[150] C. P. Terweij-Groen, T. Vahlkamp and J. C. Kraak, *J. Chromatog.*, **145**, 115 (1978).
[151] M. A. Anzano, J. O. Naewbanij and A. J. Lamb, *Clin. Chem.*, **24**, 321 (1978).
[152] C. Hansson and E. Rosengren, *Anal. Lett.*, **11**, 901 (1978).
[153] J. Knepil, *Clin. Chim. Acta*, **79**, 387 (1977).
[154] R. Frankel, *Mikrochim. Acta*, 359 (1978 **I**).
[155] M. Schäfer, H. E. Geissler and E. Mutschler, *J. Chromatog.*, **143**, 636 (1977).
[156] P. M. Kabra, H. Y. Koo and L. J. Marton, *Clin. Chem.*, **24**, 657 (1978).
[157] P. M. Kabra, B. E. Stafford and L. J. Marton, *Clin. Chem.*, **23**, 1284 (1977).
[158] R. L. Nation, G. W. Peng, V. Smith and W. L. Chiou, *J. Pharm. Sci.*, **67**, 805 (1978).
[159] H. E. Geissler, E. Mutschler and G. Faust-Tinnefeldt, *Arzneim.-Forsch./Drug Res.*, **27**, 1713 (1977).

5

Separation methods and concentration steps

5.1 GENERAL CONSIDERATIONS

Separations play an important – if not *the* most important – role in organic trace analysis. All the sample materials in which great interest is shown (blood, tissue, foods, urine and apparently even systems as 'simple' as air or water) are complicated and made up of a great number of principal and minor components which tend to affect the determination methods in an unpredictable manner. Separations are therefore generally unavoidable (see Table 5.1).

Table 5.1 – Percentage of publications on organic trace analysis (listed in *Analytical Abstracts* 1978–1979) in which separation steps are mentioned.

Number of separation steps necessary	Percentage of publications
0	4
1	32
2	49
>2	15

Except for the few cases which make up the 4% (in Table 5.1) not requiring a separation of the component to be determined, it was always necessary to suggest at least one separation step.

All separation methods try to separate a component A (which is to be determined) from matrix or other constituents (B,C,D . . . . . X) with maximum of yield of A, good separation from B,C, ... X, and efficiency and convenience for the operator. Excessive time consumption, equipment and reagent requirements, and labour cost may render a separation method impracticable.

Table 5.2 – Examples of recoveries (and their standard deviations) in organic trace analysis

Substances	Matrix	Range	Separation	Determination	Recovery (%)	Standard deviation (%)	Reference
		(ppb)					
hexachlorohexane isomers	water	0.2	partition	gc(ecd)	77		1
chlorodibenzo-*p*-dioxins	air-borne dust	0.1	extn./chromat./hplc	gc–ms	73–106	10–13	2
PAHs	fat	0.1–0.9	partn./chromat.	hplc(fluor.)	76–85		3
aflatoxins	milk products	0.05	prot. prep/extn.	tlc(densit.)	93		4
aflatoxins	milk	0.1	extn./chromat.	tlc(densit.)	98		5
aflatoxins	animal tissues	0.1–1	extn./chromat.	tlc(densit.)	90		6
C_1,C_2-halocarbons	fish	>0.1–1	extn.	gc(ecd)	63–82		7
organochlorine pesticides	water	0.1	distillation	gc(ecd)	99		8
dinitrotoluene isomers	sea-water	>0.2	partition	gc(ecd)	91–102		9
isoetharine	plasma	0.9	partition	hplc(electr.)	78		10
nitromethaqualone	blood	1	partition	gc(ecd)	99		11
corticosterone, cortisol	plasma	>2	partition	hplc(uv)	80–85	3–5	12
phencyclidine	serum	>5	partition	gc(N–P)	52		13
propanil	rice, potatoes, water	5	extn./chromat.	hplc(uv)	85–95		14
cyclophosamide	plasma	10	partition	gc(N–P)	97		15
ethmozine	plasma	10	partition	hplc(uv)	90–103	3	16
talinol	plasma	10	partition	hplc(uv)	100	6	17
pentachlorophenol	edible gelatins	>50	extraction	gc(ecd)	83–108		18
carbendazim	onions, cabbage	80	extn./partn.	hplc(uv)	75		19
dodecylbenzenesulphonate	sediment	70	extn./chromat./partn	hplc(uv)	98		20
1,2-dibromo-3-chloropropane	blood, water	0.2–100	partition	gc(ecd)	96	1–6	21
tetrachloroethylene	feeding stuffs	>0.5	extraction	gc(ecd)	92–99	4–12	22
aminocarb	foliage, fish, soil	10	extn./derivat.	gc(ecd)	>65		23
N-nitrosoproline	uncooked meat	>4	extn./chromat.	electrochem.	>90	5–10	24
quaternary ammonium compds.	river water	10–20	partition	hplc	93–106	<20	25
captafol	wheat plants	100	extn./chromat.	gc(ecd)	99–108		26
vitamin D	margarine	100	chromatography	hplc	100	13	27
chlorobenzenes	water	0.1–25	partition	gc(ecd)	80		28

Table 5.2 – *continued*

Substances	Matrix	Range	Separation	Determination	Recovery (%)	Standard deviation (%)	Reference
4-hydroxy-3-methoxyphenyl-ethanediol	cerebrospinal fluid	5–60	partn./derivat.	gc(ecd)	88 ± 3		29
betamethasone, cortisol	plasma	10–200	partition	hplc(uv)	80–85	2	30
fusarium toxin	grains	40–80	extn./partn./chromat.	hplc(uv)	73–81		31
6-mercaptopurine	plasma	3–1800	deproteination	hplc(uv)	94 ± 5		32
chlorinated phenols	urine	10–1000	chromatography	gc(ecd)	>80	10	33
prochloraz	wheat, barley	50–800	extraction	gc(ecd)	78–91		34
clobazam	plasma	10–500	partition	gc(ecd)	90 ± 5	4	35
strychnine	urine, tissue	20–840	extn./chromat.	hplc(uv)	92–102	1	36
tetrahydrocannabinol acid	urine	50–200	partn./derivat.	gc(fid)	66–76		37
sulphamethazine	pork tissue	50–500	extn./partn.	hplc(uv)	78		38
mianserin	plasma	5–500	partn./chromat.	electrochem.	71–76	3–8	39
ethylene oxide	medical devices	100–500	head-space	gc(ecd)	86–99	3	40
bisantrene	blood	125–2000	chromatography	hplc(fluor.)	75		41
dipyridamole	plasma	2–2000	partition	hplc(uv)	87	3–5	42
		(ppm)					
barban	bread, flour	0.1–5	extn./partn./chromat.	hplc(uv)	80		43
N-nitrosodiethanolamine	cosmetic products	0.5–2	extraction	electrochem.	44–96		44
diphenyl ether herbicides	fish	0.1–20	extn./chromat.	gc(ecd)	95–98		45
chlorocholine	wheat, oat grains	1	extn./chromat.	tlc(densit.)	80–90		46
atrazine	soil	1–10	extraction	hplc(uv)	75		47
asulam	wheat	0.1–10	extn./partn.	hplc(uv)	87		48
formaldehyde	shrimp	10	extn./derivat.	hplc(uv)	72		49
paraban	fruit juice, beer	12		tlc(densit.)	100–108	2	50
polychlorinated biphenyls	transformer oil	5–500	partn./chromat.	gc(ecd)	70–80	6	51

In principle a separation can be described by

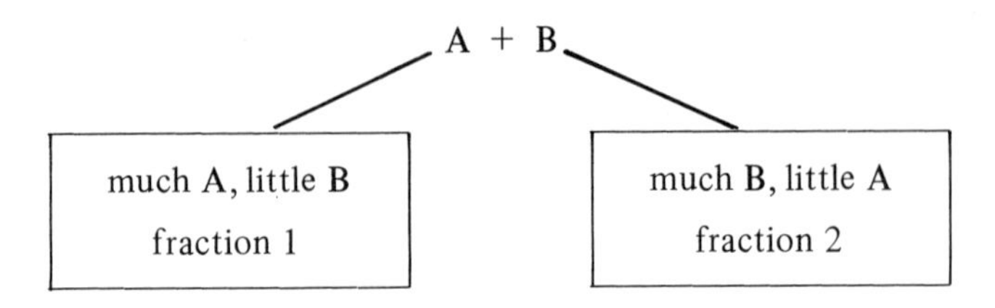

The separation factor β is defined by:

$$\beta = \frac{(c_A/c_B)_{\text{fraction 1}}}{(c_A/c_B)_{\text{fraction 2}}}$$

If there is no separation, then $\beta = 1$. If separation is high, β approaches ∞ or 0. Thermodynamic theory shows that complete separation is impossible, since the concentration of one component cannot equal zero in one fraction, for statistical reasons. In practice, because of the random error inherent in measurement systems, an apparently complete separation may be obtained under carefully controlled conditions, but in organic trace analysis a yield of only 80–90% is not uncommon. Fortunately, such a level is often sufficient for the practical aims of this scientific field. Table 5.2 gives some examples of yields and relative standard deviations. The data are arranged in order of increasing concentration level.

The results of an analytical separation, especially the yields, generally depend on many factors. The nature of the sample and trace analyte can affect the results, through complexity of the matrix, chemical similarity of the analyte to other components, the nature and concentration of the analyte, its tendency to be absorbed, volatilized or lost in other ways, the abundance of the analyte in the environment and reagents (as a source for blanks), etc. The methodology has an effect through the number of separation steps and their nature, the specificity and sensitivity of the method of determination, the need for the use of derivatives, and so on. Last but not least, yields are influenced by the skill of the analyst.

Yields are hence very difficult to judge, and authors often report ranges of yields. However, the examples in Table 5.2 may serve to give an indication of the efficiency of modern organic trace analysis. Very often, there is no need for an extremely high yield, provided the recovery is reproducible and so can be corrected for calculation.

Yields are found to depend on the concentration of the trace to be determined. Thus for propane-1,2-diol dinitrate in blood (after extraction and gc), there was 83% yield at the 10-μg/ml level but only 19% at 10 ng/ml [52]. Aflatoxins in soya sauce, after extraction, clean up on Sep-Pak, thin-layer chromotography and hplc with fluorimetric detection, gave 70% at the ppb-level, but 90% at the 100-ppb level [53]. Acrylamide in carrots, onions and potatoes, determined by extraction, bromination, Florisil column clean-up and gc, gave 80% yield at the 1.5-ng/g level and 98% at the 100-ng/g level in the sample [54]. Prifinium quaternary ammonium ion has been determined in human serum and urine by partition and hplc with ultraviolet detection, the recoveries being 57% for 2.5 ng/ml and 73% for 25–100 ng/ml [55]. Other examples can be found in Table 5.2, e.g. references [16, 25, 31, 37, 44].

The effect of the matrix is shown by the following examples. Determination of chloramphenicol at the 0.1–1 ppm level by extraction and differential pulse polarography gave recoveries of 70% from defatted milk but only 50% from minced meat [56]. Chlormequat residues in cereals, straws, pears, grapes, raisins or milk (measured by extraction, clean up on an alumina column, and tlc with densitometry) gave yields between 70 and 103%, depending on the matrix [57]. Ethmozine at the 0.1–20 ppm level (determined by partition and hplc with ultraviolet detection) was 73% recovered from plasma but >90% from urine [58]. Similarly, penicillinic acid at the 1–50 ppm level is recovered by partition and hplc with ultraviolet detection with 89% yield from plasma and more than 92% yield from urine [59].

Closely related substances can exhibit different yields when determined by the same procedure, at comparable concentrations. An example is the determination of nicotine and cotinine in tissues (extraction, partition and gc–ms) at 2–3 ppb levels; the nicotine yield was 81%, that of cotinine 87% [60]. Noradrenaline and adrenaline, determined in urine by adsorption on alumina and hplc with fluorescence detection at the 20-ppb level, gave yields of 82% and 88% respectively [61]. Chlordiazepoxide and its *N*-desmethyl metabolite have been measured at the 0.5–50 ppb level in mouse brain, by extraction with subsequent hplc and ultraviolet detection, with recoveries of 71% and 81%, respectively [62].

Data from the literature (calculated from *Analytical Abstracts* for 1978 and 1979) indicate that only a few separation procedures are used very extensively in organic trace analysis (Table 5.3).

Table 5.3 – Percentage of publications on organic trace analysis in which a particular separation procedure is mentioned.

Separation procedure	Percentage of publications*
gas chromatography	37
liquid–liquid distribution	32
high-pressure liquid chromatography	16
thin-layer chromatography	9
ion-exchange chromatography	4
other types of chromatography	9

*The sum exceeds 100% since several publications mention more than one separation method.

5.2 EVAPORATION OF SOLVENTS AND DISTILLATION TECHNIQUES

5.2.1 Evaporation of solvents from solutions

After its separation, the trace constituent is almost always present in a relatively large volume of a solvent. Very often, the solvent is evaporated to concentrate the solution. This step is not entirely uncritical, as many organic compounds have an appreciable vapour pressure even at room temperature. Losses have been reported as occurring when solutions containing tricyclic drugs [63] or naphthalenes and biphenyls [64] are evaporated. Often, internal standards can be added and corrections made for any losses. A comparison of common solvent evaporation techniques in organic analysis has been published [65].

Sometimes evaporation is suggested, even though it seems a potential source of loss; for example in the formation of volatile derivatives for gc determination, the reaction solution is sometimes evaporated, the residue dissolved, and the resulting concentrated solution used for the gc. Some examples are given in Table 5.4.

Evaporation is done either in a closed system (distillation) or in an open vessel. Control of external contamination is simpler with the closed system.

For concentrating solutions containing fairly volatile components a 'Kuderna-Danish' concentrator (see Fig. 5.1) is often used [79]. With this equipment

Fig. 5.1 – Kuderna-Danish evaporation equipment.

Table 5.4 – Examples of evaporation of the solvent from a solution containing derivatives prepared for gas chromatography.

Analyte	Amount	Derivative reagent	Internal standard	Solvent evaporated	Temperature for gc (°C)	Reference
diltiazem	20 μg	silylation	+	hexane	270	66
etarphin	10 μg	silylation	+	silylation reagent	236	67
N-hydroxy-2-fluorenylacetamide	>50 pg	trifluoroacetic anhydride	+	cyclohexane	200	68
methylthiouracil	>500 pg	methyl iodide	+	benzene	180	69
PCBs		chlorination with $SbCl_5$	—	pentane	235	70
indomethacin	50 ng	ethyl iodide		hexane	275	71
C_1–C_6 amines (16)		trifluoroacetic anhydride		CH_2Cl_2		72
polyamines (4)		isobutyl chloroformate	—	ether	150	73
hexachlorophen	50 ng	acetic anhydride	+	acetic anhydride	220	74
acenocoumarin	100 ng	pentafluorobenzyl bromide	+	acetone	285	75
tricyclic antidepressants		heptafluorobutyric anhydride	+	ethyl acetate and toluene	205	76
dialkylbarbituric acids (23)		iodopropane	+	methanol	140–200	77
abscisic acid	100 pg	diazomethane	+	ether	197	78

solution volumes are reduced from several hundreds of ml to a few ml. In a further step, micro-equipment of the same pattern can be used for further concentration to just a few μl. Often, the use of a 'keeper', usually a heavy paraffin oil, is desirable to minimize losses [80]. Recoveries of 80-90% have been reported for use of this equipment in evaporation of 100 ml of a hexane solution (containing μg amounts of pesticides with vapour pressures between 10^{-4} and 10^{-7} mmHg) to 0.01 ml (with 'keeper').

In such work the use of solvents with low boiling points is extremely helpful [81]. The evaporation of a methanolic solution of DDT, lindane or DNBP (dinitrobutylphenol) at the 50-ppm and 1-ppm levels, to 1 ml, gave losses of about 2%. After addition of 0.5 ml of poly(ethylene glycol) as keeper all losses were reduced to less than 1%. This keeper retains the trace constituent while the major solvent is distilled off and the concentration of the trace increases. The walls of the still are always wetted by the keeper, which prevents irreversible fixation of the trace on the walls of the equipment. Rinsing and recovery is then simpler [81].

For greater sensitivity it is often necessary to concentrate solutions even more. This can be done by using the equipment [81] shown in Fig. 5.2. The tubes of the vessels act as reflux condensers, which permit better separation between volatile solvent and minor trace constituent, even in the case of compounds with low boiling point. In this way more than 90% recovery of 40 μg of benzene (b.p. 80°C), of 4 μg of lutidine (b.p. about 150°C) and 10 μg of

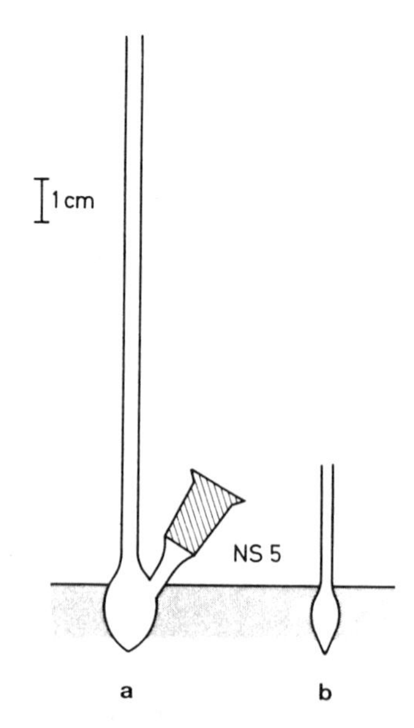

Fig. 5.2 – Micro equipment for evaporation and concentration. (Reproduced from [81], by permission of the copyright holders, Springer Verlag).

diphenyl ether (b.p. 259°C) dissolved in 0.5 ml of dichloromethane could be obtained, in evaporation of 90% of the solvent, with no keeper added. Components with a boiling point higher than 200°C can be recovered quantitatively with such microcolumn equipment (see Fig. 5.3).

Vessels very similar to those in Fig. 5.2 are used in a micro technique for derivative formation [82]. The trace constituent is treated with the derivative-forming reagent, and to complete the reaction the vessel is immersed in hot water, and ethyl acetate is used as solvent. A ring of condensing solvent should form just below the upper end of the little reflux condenser, i.e. the tube. After

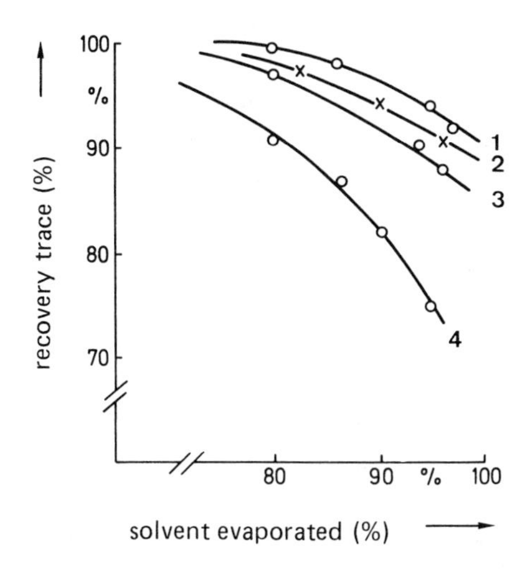

Fig. 5.3 – Recovery of traces in evaporation experiments. Curve 1, 10 μg of benzene (b.p. 80°C); curve 2, 10 μg of diphenyl ether (b.p. 259°C); curve 3, 4 μg of lutidine (b.p. 150°C); curve 4, 40 μg of benzene (b.p. 80°C). (Reproduced from [81] by permission of the copyright holders, Springer Verlag).

completion of the reaction the solvent is gently evaporated, which takes about 10 min per ml. Shortly before dryness is reached the vessel is immersed in ice water. The rest of the solvent vapour condenses, rinsing out the whole reflux column and concentrating the trace constituent into a drop which can be transferred quite simply with a micro syringe. As ethyl acetate forms an azeotrope with water, the residue can be dried by completing the evaporation; this can be done with fairly non-volatile trace constituents (b.p. above 200°C).

A double reservoir evaporation vessel (Fig. 5.4) can be helpful in such operations [83]. The solvents are evaporated by rotary-evaporation under reduced pressure. After reduction of the solution to a volume smaller than that of the lower reservoir, the walls are rinsed with a suitable solvent until the solution reaches the calibration mark. Aliquots are then taken for analysis.

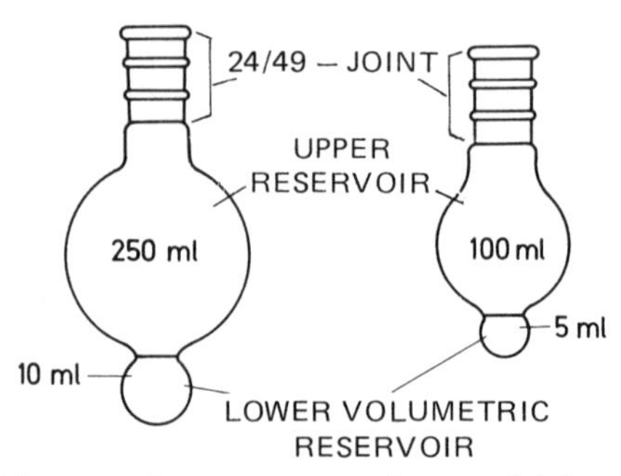

Fig. 5.4 – Double -reservoir rotary evaporation vessel (adapted from T. W. May and D. L. Stalling, *Anal. Chem.*, 1979, **51**, 169 by permission. Copyright 1979 American Chemical Society).

Solutions which contain organic compounds of very low volatility can be brought to dryness much more simply, by means of a gentle stream of filtered air blown onto the surface [84–87] (enough to 'dimple' the surface). The operation is performed at as low a temperature as possible. Any heating necessary is done by means of heating blocks. Naturally, the behaviour of the trace analytes should be checked. This method is often used after liquid–liquid separation steps.

Evaporation to dryness is a step often used in organic trace analysis, for two main purposes:

(1) to prepare a defined volume of sample solution by dissolving the dry residue in an appropriate solvent, so that aliquots can be taken for analysis;

(2) to change the solvent system, e.g. after partition from water with solvent 1, this solvent can be evaporated and the residue taken up with solvent 2, which may be the preferred medium for a subsequent reaction system, etc.

For evaporation in open vessels, any heating should be done from above with an infrared lamp or other heat-source. The solutions are best contained in small conical vessels. The solution does not creep over the top of these little containers, since liquids tend to move away from a place of higher temperature to colder places [88].

The concentration step can be directly coupled to column chromatography [89]. A glass filament (30 cm in length, 1 mm in diameter) is put underneath the column. The effluent is directed onto the outside of the filament, and slowly runs down it. The solvent is evaporated by a stream of gas passed over the surface of the glass thread. Droplets of the concentrated solution are collected at the end of the filament. A 30-fold increase in concentration of *p,p'*-DDT, dieldrin, heptachlor epoxide and others has been obtained, with amounts of about 1 ng. Losses were less than 5%. For aldrin, lindane and some other analytes, losses ranged from 12 to 22%.

5.2.2 Drying of the solution before evaporation

Very often the solution of an organic compound in an organic solvent has to be dried. This is mostly done with anhydrous sodium sulphate, but there is some indication that this reagent contains traces of organic halogen-compounds, detectable by sensitive ecd methods. There may also be losses by adsorption on the drying reagent.

Another, but less often used, technique utilizes distillation of azeotropic mixtures with water (see Table 5.5).

Table 5.5 – Data on azeotropic mixtures with water and some common organic solvents.

Organic compound (%)	Organic compound (%)	b.p. of azeotrope with water (°C)
ethyl acetate (92.1)	–	70.4
propan-1-ol (71.8)	–	87.8
ethanol (18.3)	benzene (74.0)	64.9
ethanol (10.3)	carbon tetrachloride (86.3)	61.8

5.2.3 Steam distillation and co-distillation

The vapour pressure of a mixture of two immiscible solvents is the sum of the vapour pressures of the pure solvents, so the mixture boils at a temperature lower than the b.p. of either component. This principle forms the basis of co-distillation, which can be a useful separation method, though it can also result in the loss of traces during the evaporation of solvents (it is often reported in the literature that the evaporation of aqueous solutions of organic halogen-compounds, e.g. DDT, lindane, leads to losses by co-distillation.)

Co-distillation, mostly with water (in steam distillation), is a useful method for the separation of trace constituents (Table 5.6). The method can be used as a group separation step to separate steam-volatile from involatile components, such as fat or proteins. It must be checked, though, that there is no degradative reaction of the volatile trace constituent(s) at the temperature of distillation. If there is, then steam distillation under reduced pressure can be applied. The substances distilled are recovered from the (usually rather large) volume of steam condensate by liquid–liquid extraction.

Steam distillation can be more effective, rapid and less costly and toxic than solvent extraction with hexane – at least in the separation of traces of DDT from all kinds of materials [127].

Occasionally, a solvent other than water is used. For the separation of acrylonitrile from food-simulating solvents methanol is used for the co-distillation; 10 ppb of the trace can thus be determined [128]. Traces of ethyl methyl ketone or toluene have been separated from waxes and lubricating oil by azeotropic distillation with methanol and cyclohexanone [129].

Table 5.6 – Use of steam distillation to separate traces from matrices.

Trace compounds	Concentration or amount	Matrix	Reference
N-nitrosamines		condensate from 900 cigarettes	90
amines		bacon, fish, beer, frankfurters	91
biphenyl, 2-phenylphenol	ppm	citrus fruits	92, 93
phenols		water	94
volatile phenols		beer, malt, wort	95
biphenyl, 2-hydroxybiphenyl		citrus fruits	96
accelerants		residues from fires (suspected arson)	97
formic acid		blood	98
polychlorobiphenyls		ocean-bottom sediments	99
methanol		food	100
impurities		illicit amphetamine	101
volatile fatty acids		faecal samples	102
pentachloronitrobenzene		vegetables	103
methyl anthranilate		grape juice	104
hydrocarbons	8 ppt	water	105
butylhydroxyanisole	25 ppb	food	106
4-nitrophenol	1 ppm	urine	107
DDT		soil, vegetables	108
volatile nitrosamines	50 ng/l.	water	109
volatile nitrosamines	1 ppb	maize silage, milk	110
aniline	1 ppm	adulterated olive oil	111
primary, secondary amines	ppm	grain, beer	112
phenol	100 μg/l.	urine	113
phenol	100 ng/l.	river water	114
alkylpyridines	>1 ppb	beer	115
sorbic acid	50 ppm	cheese	116
tri-allate, di-allate	50 ppb	milk, plant tissue	117
2,4-dichlorophenol		sediments	118
thiocarbamate herbicides	>9.92 ppm	soil, water, plants	119
dichlobenil	>1 ppb	soil	120
odorous substances		solid swine manure	121
formaldehyde	>0.05 ppm	fish paste	122
acrylonitrile	>10 pg/g	aqueous solutions	123
tetrachloroethylene		food	124
pyrrolidine	picomoles	brain	125
formaldehyde	>0.5 μg/l.	plasma, urine	126

In 'sweep co-distillation' a solvent which is volatile at slightly elevated temperatures is added. When the solvent is distilled off, its vapour carries away the trace as well. Again, the trace must be volatilized without decomposition. Good separations from lipid material are possible with dichloromethane as sweep carrier. The lipids dissolve in the solvent, so no traces are occluded [130, 131]. Examples of the application of the sweep co-distillation method are the determination of pesticides [132, 133] and of diphenyl or *o*-phenylphenol in citrus fruits [134].

5.2.4 Sublimation

Sublimation can be a useful technique for use with already isolated trace compounds. The presence of a solid matrix causes problems, however, as trace constituents that are enclosed (occluded) by the matrix are set free only with some difficulty. PAHs have been sublimed for purification and separation from matrix components [135, 136].

Sublimation has also been suggested for collecting material that has been separated on thin-layer plates [137, 138]. In micro sublimation a cold-finger technique is often applied. In this the trace compound is collected on a cooled condenser positioned in the space above the heated sample, in a closed vessel; the pressure can be reduced if desired. A small steel rod (2 mm in diameter) can serve as the cold finger, for transfer of μg amounts, with yields of more than 80%. The face of the rod is plane and polished and thus can be used as the mirror in infrared microreflectance spectroscopy, making it simple to prepare a sample of the isolated trace compound for infrared examination [139]. Sublimation of thymine has been used to separate this base from other DNA bases, for its determination in biological materials [140].

5.3 LIQUID-LIQUID DISTRIBUTION

5.3.1 General aspects

Liquid-liquid distribution is used in about one third of all separation procedures described in the recent organic trace literature (Table 5.3, p. 000), ranking next to gas chromatography in popularity. Liquid-liquid distribution or 'partition' is very often called 'extraction', although this term should perhaps be reserved for the separation of components from a solid phase. Partition is simple and often rapid and the physicochemical principles of the method hold valid over many orders of magnitude of concentration of the partitioned species. A most important advantage is that, similarly to many inorganic partition systems, liquid-liquid distribution processes in organic mixtures are little affected by the matrix composition. For this reason, the method is ideal as an initial group separation procedure (which can be followed by a 'clean-up' or further separation based on other principles). Thus in almost all of the many hundreds of procedures reported annually for the determination of therapeutic drugs a partition step is used for the initial separation of the trace(s) from biological fluids.

Two main group-separations are very simple to achieve:

(1) separation of polar substances from non-polar;
(2) separation of polar substances into acidic, basic and neutral substances.

The second of these has long been known, and has been used successfully in drug analysis and for forensic purposes. It is the basis of many modern procedures. Very often the partition is repeated, with a change in pH. For example, an acid is partitioned at low pH into an appropriate solvent, from which the acid is stripped by equilibration with an alkaline aqueous solution. From this solution the acidic trace may be transferred into yet another solvent (after acidification) for further purification etc. This method of 'alternate acid–base partitioning' is rather often suggested in the modern literature.

There are several methods for recovering a species from the organic solvent after partitioning:

(1) back-partitioning or 'stripping';
(2) evaporation of the solvent;
(3) adsorption on some adsorbent.

Partitioning – especially as a batch procedure – is used for

(1) group separations;
(2) transfer into smaller volumes to increase the concentration of traces;
(3) transfer of a solute into another medium to permit reactions which are not possible in the original medium or to prevent undesired reactions, e.g. hydrolysis, in the original medium (which is usually an aqueous system).

There is a current trend towards replacing liquid–liquid partitions by reversed phase high-pressure liquid chromatography separations. Such procedures can be automated more easily and on-line combination with measurement methods is possible. There is also less hazard, as less solvent is used than in partition work. Nevertheless the partition methods are still important, for handling a large bulk of sample in preseparations.

In spite of the importance of the technique there are still no general rules for selection of a solvent system for the isolation of a given compound. The most useful approach is still empirical. The partition coefficients are determined for a variety of solvents, and the system is selected which is potentially of greatest value.

For this purpose a series of solvent systems has been recommended, in which the organic phase exhibits increasing polarity [141]:

(1) hexane or cyclohexane/ethanol + water
(2) benzene/methanol + water
(3) chloroform/methanol + water
(4) ethyl acetate/water
(5) butan-1-ol or butan-2-ol/water
(6) phenol/water

Occasionally, the system used is composed of two immiscible organic solvents. In pesticide analysis hexane/acetonitrile, and iso-octane/dimethylformamide are of practical importance [142]. Heptane/90% aqueous ethanol and iso-octane/80% aqueous acetone may also be used. The system hexane/

acetonitrile is of great practical significance in the difficult separation of lipids from many pesticides. These compounds dissolve more readily in the nitrile, while the lipids stay in the hexane phase. For the separation of PAHs, partition between cyclohexane and nitromethane has been suggested [143].

The methods for extraction of pesticides are chosen according to the nature of the pesticide and of the sample [144]. For organophosphorus pesticides in solid matrices with high water content (fruit, vegetables), polar solvents are used, such as acetonitrile, ethyl acetate, acetone or methanol. For samples containing both fat and water (e.g. milk and meat) mixtures of polar and non-polar solvents (methanol-acetonitrile or acetone-methylene chloride) are used. Hexane (and in former times benzene) has been extensively applied for materials with high fat content.

5.3.2 Equipment and techniques

Work with trace amounts requires specialized equipment and techniques, which guarantee:

(1) good separation of the phases after distribution;
(2) a readily visible interphase boundary;
(3) simple operation so that repetitive partitions can be made without great loss of time, reagents and traces;
(4) in special cases a very large phase-volume ratio can be used.

Separatory funnels are often used. They have the disadvantage that the stopcock has to be greased, unless the key is made of Teflon. Since the use of stopcocks is detested by the 'stylish' trace analyst, other equipment is used instead.

Often a test-tube or a centrifuge tube fitted with a ground-glass joint is used. After several inversions or some light shaking of the vessel, equilibrium is attained. Centrifugation can be used for better phase separation, and the layer of interest is removed by a pipette or syringe, or by a suction device (Fig. 5.5).

For partition on the ultramicro scale the two immiscible phases are only stirred, and not shaken [145].A tiny test-tube is used, and a very small glass stirrer [146]. The stirring speed is not critical. The two phases remain separated, with a sharp boundary in between. No droplets of the one phase in the other are formed, and no organic phase tends to adhere to the wall of the vessel (which can cause problems with 'lost' droplets in the usual shaking techniques for equilibration).

For partitioning between very large volumes of one phase, e.g. a water sample, with very small volumes of another solvent, e.g. hexane for separation of pesticides in water analysis, special equipment has been developed [147, 148] (Fig. 5.6). In the pesticide application the flask contained about 980 ml of water and 0.2 ml of hexane and was shaken manually for 2 min. By tilting the flask and carefully adding water through the side-arm, the solvent layer could be

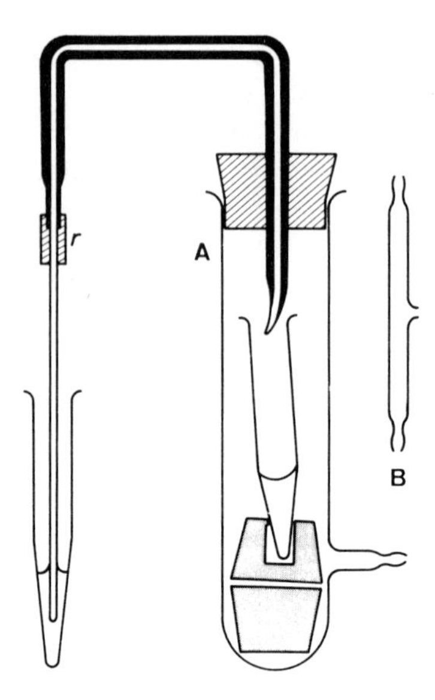

Fig. 5.5 – Suction device for separation of one liquid phase after liquid-liquid distribution (after [88], by permission of the copyright holders, John Wiley Inc.).

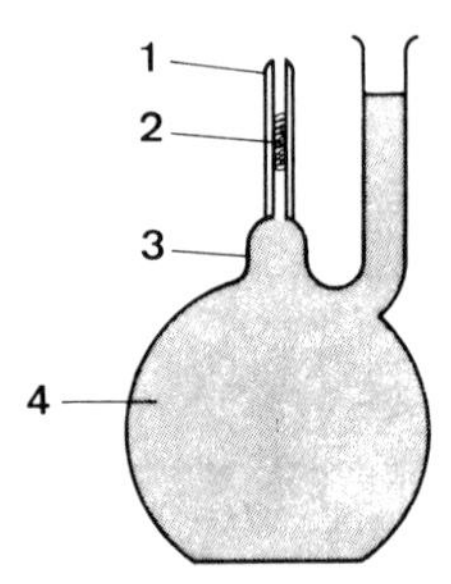

Fig. 5.6 – Micro extraction flask. 1 = Capillary tube; 2 = solvent layer; 3 = modified 1 litre standard flask; 4 = water sample (adapted from [147, 148] by permission of the copyright holders, Elsevier Science Publishing Co., Amsterdam).

held in the central portion and finally displaced into the capillary tube. About 50 μl of hexane were recovered and could be used directly for gas analysis. Recoveries of aldrin, dieldrin or heptachlor epoxide at 10-ppt levels were about 50–60%. Total recovery by means of three successive partitions was 89–93%.

An interesting technique which is occasionally used results in better phase separation. One of the two liquid layers is frozen, and instead of separation of two liquids, a filtration is used. Testosterone has been partitioned from aqueous solution into ether, then the aqueous phase frozen with 'dry ice'-acetone mixture and the ether phase decanted [149]. This technique has been used for some time in inorganic analysis, with high melting-points solvents, such as naphthalene, which can be solidified simply by cooling to room temperature.

5.3.3 Special treatment after partitioning

In some cases a rather drastic treatment of the species partitioned into an organic solvent is recommended. Neutral compounds which have been extracted can be treated with fuming sulphuric acid. This decomposes less stable compounds or sulphonates them into more polar components. Hydrocarbons and some chlorinated pesticides are not changed. Examples are the analysis of Kepone-containing solutions [150], and treatment of extracts containing PCBs and polychlorinated terphenyls [151], or chlorinated compounds [152], e.g. DDT and hexachlorocyclohexane [153].

Another treatment is saponification, which is resisted by hydrocarbons, PAHs and some other very stable compounds in an extracted neutral fraction, e.g. benzo(*a*)pyrene [154, 155].

Sometimes the species to be partitioned is chemically changed by a derivative formation reaction.

5.3.4 Mechanization of liquid–liquid distribution

Some suggestions have been made for automation of this technique. The main difficulties are caused by formation of emulsions [156, 157], and there are also problems connected with automated location of the phase boundary. Centrifugation in combination with a non-wettable Teflon separator has been suggested [15]. Emulsion forming is a general problem in partition systems of course. It can be dealt with by slow filtration (not effective for large volumes, or useful for automation), by addition of neutral salts (e.g. sodium chloride), or of acetone or ethanol, by centrifugation, or by addition of emulsion-breakers.

In the micro-adaptation of an automatic phase separator (Fig. 5.7) [159, 160], the effluent from an hplc-column is mixed with the aqueous solution of a reagent which reacts with basic compounds to give fluorescent products. The excess of reagent remains in the aqueous phase and the fluorescent derivatives are extracted in a coil system, the effluent from which is passed into the separator. About 50% of the organic phase is collected. The rest goes to waste together with all the aqueous phase. This large loss of organic extract was necessary to achieve low noise levels and optimum detector performance. In another example [161] a chloroform phase was separated from an aqueous methanol solution by use of a porous Teflon membrane.

A continuous solvent-extraction apparatus is useful when large volumes of one phase, e.g. water samples, are to be equilibrated with small volumes of an immiscible solvent (Fig. 5.8). The three inverted 250-ml flasks are clamped to a retort stand, which is connected to a vibratory mixer by a bar, so that all three flasks are vigorously shaken. A 10-litre sample is siphoned through the apparatus in about 4 hr and extracted with a total solvent volume of 10 ml of hexane. The extract is finally recovered in a separatory funnel [161].

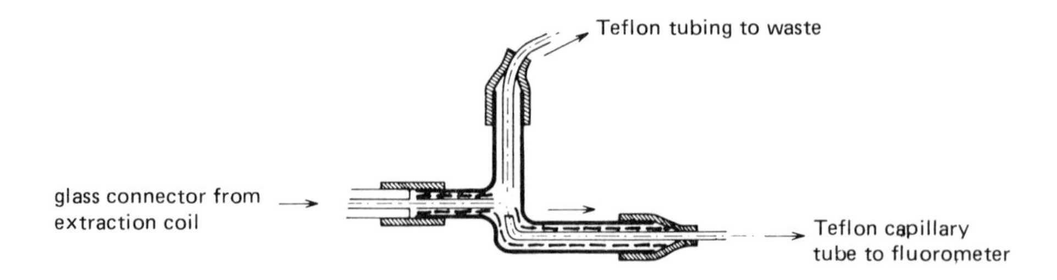

Fig. 5.7 – Modified phase-separator. 2.0-mm o.d. Teflon tubing (dashed lines) was inserted through the lower portion of the glass phase-separator. A hole was cut into it to permit the aqueous phase to escape. A small length of 1-mm i.d. Teflon tubing (o.d. 1.5 mm) was inserted into the first tubing on the entrance side of the phase-separator. Teflon capillary tubing to the fluorometer was inserted into the other end of the 2.0-mm o.d. Teflon tubing, as shown (Reproduced from [159] by permission of the copyright holders, Elsevier Science Publishing Co., Amsterdam).

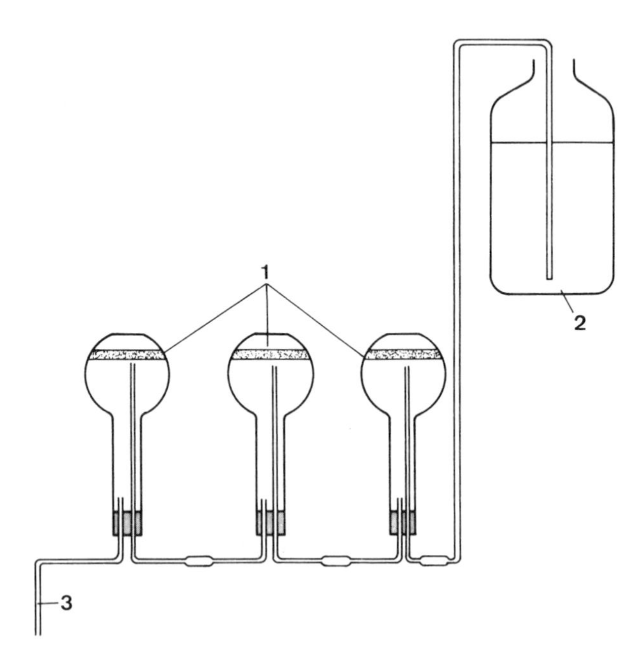

Fig. 5.8 – Continuous solvent extraction apparatus. 1 = Solvent layer (hexane); 2 = sample reservoir; 3 = overflow. (Reproduced from [162], by permission of the copyright holders, Elsevier Science Publishing Co., Amsterdam).

5.3.5 Multistep liquid–liquid distribution processes

In organic preparative chemistry and in biochemistry multistep distribution procedures have been used in which the partition is repeated many times. For practical reasons such multistep distributions are less often applied in trace analysis, as the solutes must be recovered from rather large solvent volumes.

Partition chromatography has found some application in this respect. In this, liquid phase (the stationary phase) is added to a solid support (silica gel, cellulose, cellite, Hyflo Super Cel etc.) in a column. A second phase (the mobile phase) which does not mix with the stationary phase is passed through the column. Added solutes are distributed differently between the two immiscible liquids and traverse the column at different speeds, and are eluted one after another. This multistep distribution system is now less often used, however, as it is replaced by reversed-phase chromatography.

5.4 SEPARATION BY CHROMATOGRAPHY

5.4.1 General aspects

Chromatography is a separation procedure in which a compound is distributed between a mobile phase (liquid or gas) and a stationary phase (solid or liquid). Most separations applied in organic trace analysis are based on chromatographic methods. In three out of every four papers published in this area, chromatography is suggested for the solution of a given problem. There are several textbooks, e.g. [163], and four journals devoted exclusively to chromatography (*Journal of Chromatography, Journal of Chromatographic Science, Chromatographia* and *Journal of High Resolution Chromatography and Chromatographic Communications*).

5.4.2 Types of liquid–solid chromatography

The distribution of a compound between a solid adsorbent phase and a mobile liquid phase is the basis of liquid chromatography (lc). There are various types of interaction and hence of chromatography: adsorption chromatography, ion-exchange chromatography, gel filtration, affinity chromatography.

In adsorption chromatography – by far the most often used technique in organic trace analysis – the sample components are separated on the basis of differences in polarity. The trace substances are adsorbed on the solid phase surface by non-covalent bonds, e.g. hydrogen bonding, non-polar interactions and van der Waals forces.

In ion-exchange chromatography ions interact with a charged solid phase which carries either positive or negative charges. The corresponding oppositely charged type of ion is sorbed from the mobile phase. This interaction is influenced by the presence of other ions, mainly H^+.

In gel filtration a gel is used as the solid phase. It contains numerous pores which larger molecules are unable to enter from the mobile phase, whereas smaller molecules have access to the interior of the pores. When a mobile phase is passed through the gel, elution of these smaller components is retarded relative to that of the larger molecules.

Affinity chromatography is so far practically not used in organic trace analysis of low molecular-weight compounds. In this technique a ligand with specific affinity for a group of components is covalently coupled to the matrix of the immobile phase. When the sample solution is added, only those components with an affinity for the ligand are fixed to the immobile matrix. Desorption is obtained by changing the pH or ionic strength of the eluting solvent.

5.4.3 Theoretical background of chromatographic separations

Chromatographic processes are described in theory by the van Deemter equation (Fig. 5.9), which gives the height of an equivalent theoretical plate (HETP or *H*) in terms of different parameters, such as irregularity of packaging, tortuosity of the channels in the solid phase, the diameter of the stationary-phase particles,

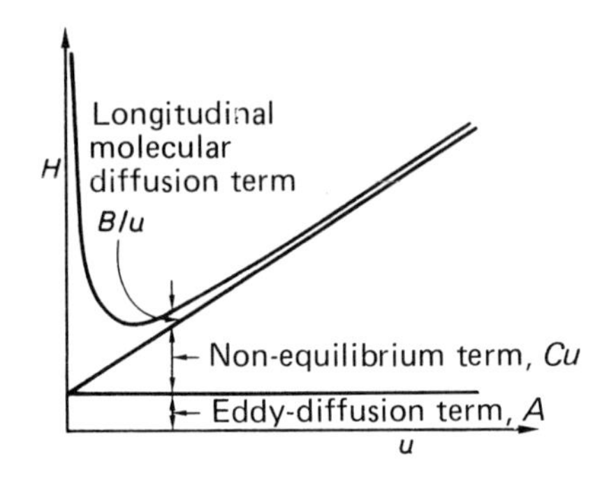

Fig. 5.9 – Schematic graphical representation of the van Deemter equation.

the linear velocity of the mobile phase (u), the diffusion coefficients of the solute molecules in the mobile and solid phases, the distribution coefficient. Good separation efficiency of a chromatographic system is indicated by small values of H. At best, H can become equal to the diameter of the particles of the stationary phase. As a result of the interaction of these different parameters, a minimum plate height H and hence an optimum column efficiency, is observed at certain flow-rates (Fig. 5.9). This optimum value varies with each substance to be chromatographed by the system. The loading of the column also influences H (Fig. 5.10) [164].

Small and uniform adsorbent particles assist in better establishment of equilibrium and separation efficiency, because specific surface area increases with decreasing particle mass. In order to improve column characteristics, very

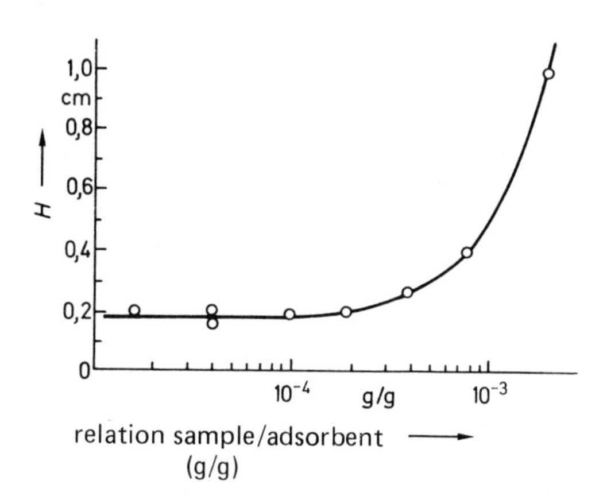

Fig. 5.10 – Relation between HETP and amount of sample applied. (Reproduced from [164], by permission of the copyright holders, Verlag Chemie, Weinheim).

small particles (a few μm in diameter) have been introduced. With such small particles the flow-rate in the column is drastically reduced, so to obtain reasonable analysis times high pressure is applied to transport the mobile phase. The resulting 'high-pressure liquid chromatography' [also called high-performance liquid chromatography (hplc)] is a rapidly expanding field in modern trace analysis, involved in about 16% of all current publications which describe separations.

5.4.4 Principles of column operation

In organic trace analysis the method of elution chromatography is generally used. In this the mixture of solutes to be separated is added to the head of the column. A suitable solvent or mixture of solvents is passed through the column and the separated fractions are eluted separately.

Elution with a solvent of constant composition is called isocratic elution. Good reproducibility is observed with such systems, but fractions which are eluted very slowly show excessive peak broadening, mainly because of longitudinal diffusion. This behaviour can be improved by changing the solvent composition. This can be done discontinuously or continuously, but reproducibility of such gradient elution is poorer than that with isocratic elution. Linear or exponential changes in composition can be achieved, or even more complicated concentration gradients. The selection of solvent is often difficult. A rational procedure for gradient elution has been suggested [165]. The following solvents are successively mixed with the next in the series, to obtain increasingly polar solvent mixtures: n-heptane/carbon tetrachloride/chloroform/ethylene dichloride/2-nitropropane/nitromethane/propyl acetate/methyl acetate/acetone/ethanol/methanol/water.

5.4.5 Adsorbent materials used as stationary phases in liquid chromatography

Rather polar adsorbents, e.g. silica, alumina, cellulose, have long been used. Florisil – a magnesium silicate – is also still often used as a clean-up reagent, mainly in pesticide analysis. For example 10 ppb of polybrominated biphenyls have thus been separated from human fat [166], and pesticides from fat [167, 168]. Pesticides isolated from plants have been cleaned up with silica or alumina [169]. Numerous other examples have been published (see Tables 5.9 and 6.12).

Lately, a chemical modification of these rather polar adsorbent materials has been suggested. Cellulose is used in acetylated form, and silica is changed by fixation of hydrophobic groups through covalent bonds. The original properties of the silica are drastically changed. A 'reversed-phase' (RP) material results, which can be used to separate non-polar compounds, with more or less polar solvents used as mobile phase [170, 171]. The number of publications in this field is increasing rapidly and to-day about two-thirds of all papers on hplc are based on this technique. A review has recently appeared [172]. The method permits the separation of a wide polarity range of organic compounds, from hydrophobic PAHs to amino-acids and sulphonic acids. This versatility is achieved by the control of the mobile phase composition with respect to pH, ionic strength and solvent polarity. Application of the ion-pairing technique [173, 174] results in a further extension of applicability. In this, the analyte ion is paired with a counter-ion to form an uncharged species. By careful choice of the counter-ion, the properties of the various ion-pair compounds produced can be tailored to give good separations. Commonly used counter-ions are alkyl-sulphonates (for bases) and alkyl quarternary ammonium ions (for acids and anions). The hydrophobic interaction of the ion-pairs with the covalently bonded alkyl chains of the reversed-phase material is the basis of the separation. There are some disadvantages, however, as the column material tends to be unstable, and the coating can become abraded, which decreases the separation efficiency.

An alternative to reversed-phase chromatography has been suggested, called 'soap chromatography' [175–177]. It utilizes a column of silica or zirconia, and cationic surfactants dissolved in the eluent phase, which results in a strongly bonded layer of the detergent alkyl chains. The relationship between concentration of surfactant and the distribution coefficient for a given analyte is a function of the alkyl chain length (hydrophobicity) of the surfactant and the polarity (i.e. water content) of the mobile phase [178]. This suggests that many reversed-phase separations can be done with unmodified silica and addition of a surfactant [179–182].

5.4.6 Filling of columns

More and more ready-to-use columns are offered commercially. For group separations or pre-enrichment techniques rather short columns can be purchased, which are used once only and then discarded. Nevertheless, home-made columns are still of interest, as the commercial products are rather expensive.

Two methods are recommended for filling. In one the dry adsorbent is put into the tube and solvent is added. In the other a slurry of the column filling in a solvent is prepared and this suspension is poured or pumped into the tube [183–185]. With wide columns the slurry can be stirred during the packing process. With hplc column materials a liquid of appropriate specific gravity is preferable, to keep the particles suspended and floating during the application. A high-pressure pump is almost always used for the preparation of hplc columns. The methods have been reviewed [186, 187]. The columns must be filled as homogeneously as possible because any inhomogeneity leads to peak broadening and distortion of peak profiles.

No air (or other gas) must be admitted to the adsorbent, once the column has been filled. Gas bubbles would prevent complete wetting of the adsorbent particles. Consequently, precautions must be taken against the columns running dry. Drying leads to cracking of the column filling.

5.4.7 Application of the solute mixture

The volume in which the solute mixture is applied should be as small as possible. With high-pressure systems and other closed-column systems the addition of the solution is often done by injection through a septum, by means of syringes. Somewhat larger volumes are added by means of a sample loop. Injection of volumes up to 0.2 ml is possible [188–190], but enrichment of trace components before chromatography and use of smaller volumes is preferable. Off-line adsorption techniques have been suggested [191–193], but undesired dilution, possible sample contamination from containers, solvents and the laboratory surroundings, and long analysis times can cause problems [194]. Preconcentration at the top of the analytical column before the actual chromatography is therefore used to overcome some of the disadvantages of the off-line strategy [195–197]. Optimum conditions for on-line precolumn enrichment have been determined [194]. Between the precolumn and the analytical column a valve is placed which permits flushing of the sample solvent to waste. Trace compounds are adsorbed in the precolumn and eluted after addition of the eluting solvent. The valve is then switched so that the solution of the trace components is washed from the precolumn onto the analytical main column. Trace analysis of, for example, phthalate esters in ppt concentrations in 1-litre samples of water, is possible with recoveries of about 95–100%. The potential of precolumns for improving the performance of hplc systems has been reviewed [198, 199].

5.4.8 Detection and collection of column effluents

All methods used in instrumental analysis are also employed for registration of column effluent composition. Micro flow-cells with volumes down to 1 μl are used. Tandem methods of detection improve the specificity. In these, two or more measuring principles are combined, either in sequence or in parallel (with stream splitting).

Liquid chromatography can be combined with the highly developed detectors used in gas chromatography by means of 'chain methods'. The column effluent is evaporated on a chain or moving belt which travels through an oven where volatilization or decomposition of the trace constituent is achieved, the vapour being transported in a carrier gas and detected (see p. 257).

If collection is required after the chromatography it is done by use of the standard equipment for this purpose.

5.4.9 Combination of columns, column switching

Advantage can be gained by combining several columns. For example a mixture of A, B, C and D may be applied to a column which separates only A from B, C and D, the entire eluate being fed to another column, which separates B from C and D, which in turn are separated on a third column. Thus, a separation of a complex mixture can be efficiently achieved. Very often, use of a single column for such a mixture would result in excessively long elution times and considerable peak broadening.

Substances which are not eluted or travel only very slowly are dealt with by 'back-washing': they are eluted by reversing the flow of the eluting solvent. This fraction can then be transferred to another column of different characteristics which will resolve it. Microprocessors and programs with electronic control should find considerable use for controlling these complicated operations. A modular system using cassettes has been suggested for gas chromatography [200] or hplc [201]. It has the advantage of using a rather problem-free connection technique based on a Teflon needle-seal. Four combined inlets/outlets are provided with each column cassette. Rapid cassette exchange is possible, and there is a high degree of flexibility, since convenient column back-flushing, improved sample-splitting, simple eluate-splitting, easy column-switching and post-column derivative formation are all provided.

5.4.10 Problems in liquid chromatography

Little is known about the theoretical connection between chemical structure and adsorption behaviour. Predictions of chromatographic performance are therefore difficult to make.

The reproduction of adsorbent characteristics is difficult, though not impossible. Traces of impurities influence the properties of the surface, particularly with the highly efficient and highly sensitive modern column materials for hplc, which have a very small particle diameter and a very high specific surface area. Adsorption of components of the laboratory atmosphere (water, solvents etc.) creates problems, if the storage of adsorbents is not properly controlled.

All column materials tend more or less to 'bleed'. For example, the solubility of amorphous silica is about 10 μg/ml in water at room temperature [202], and increases with temperature to about 250 μg/ml at 200°C, and at pH 10 the dissolution rapidly increases. These factors can lead to bleeding of silica columns. In some cases equilibration of the mobile phase on a precolumn (with 5-μm silica particles) may improve the column stability and result in a more stable base-line in ultraviolet detection at 200 nm [203].

5.4.11 Gel chromatography

As stationary phase a gel is used [204], e.g. of dextran, polyacrylamide, agarose type. Inorganic materials such as silica, certain silicates and porous glass beads are also applied.

Gel chromatography is often used in pesticide analysis as an effective clean-up step. It is useful in the analysis of matrices with a very high lipid content (fat, milk) after partition [205–208], or even without any preseparation, with gel chromatography as the initial step of the analytical scheme [209, 210]. All the examples mentioned describe the determination of PCBs and/or halogen-containing pesticides. Volatile *N*-nitrosamines have been cleaned up on a gel column before gc to determine them at the 0.6 ppb level in dried milk [211]. Because of the significance of the technique for practical pesticide analysis, automatically operating equipment has been devised [212] which permits the treatment of many samples without much attention needed from the analyst.

5.4.12 High-pressure liquid chromatography (hplc)

This technique is considered to be the most important innovation in recent years for trace pesticide analysis [213] and food analysis [214], and in clinical chemistry [215]. It has advantages over gc for the determination of thermally unstable, non-volatile or very polar compounds. It can also be applied to most components normally determined by gc. Theoretically, hplc offers many advantages, and is limited only by the lack of such general detectors as the flame ionization detector (fid) [216].

Theoretically, columns with diameters as small as a few μm are ideally suited for chromatographic separations [217], but in practice many disadvantages arise in the use of such capillary columns. Columns with an internal diameter of about 1–2 mm approach the ideal. With these it is possible to separate amounts of 100 pg injected in 1 μl of sample solution (i.e. a concentration of about 0.1 ppm) by hplc with 10-μm particles as stationary phase. Such microbore columns have very small internal volumes. Consequently, the technique is advantageous when only a small amount of sample material is available, or if extreme sensitivity is to be achieved or the mobile phase is very expensive. The injection of sample volumes of only 1 μl must be done with great precision. Also, the flow-rates produced by the pump must be very constant. The power of the technique is shown by a report [218] of use of a 1-m long column of 3-μm

C_{18}-bonded silica in a 0.3-mm bore fused silica tube, the column having a plate number of more than 10^5.

An interesting new alternative to the normal filled hplc columns is the use of open tubular columns made of fused quartz capillaries having internal diameter of 10–50 μm [219] or 30–100 μm [220], with the stationary phase applied directly to the etched internal surface.

In hplc it is necessary to protect the expensive columns against impurities. Sample solutions and solvents must be filtered before application to the column. Special filter cartridges are usually fitted before the inlet to the column. Very complex sample mixtures need to be preseparated. When a change of solvent is necessary, great differences in polarity should be avoided.

Several types of pumps are available for generating the high pressure needed for solvent transport through the column. Almost all pumping techniques generate fluctuations of pressure, but these should be prevented, mainly to improve detector performance. Damping devices are available for reducing such ripple.

Solvents should be degassed, again mostly for better detector performance.

Detection can be done either on-line or off-line. The most often used on-line systems are ultraviolet detection, fluorescence detection, and measurement of electrochemical properties. Examples are given in Tables 5.7–5.9. Direct coupling to mass spectrometry (ms) is also possible. Generally the solvent is evaporated after the hplc and before the ms. For gc the eluate can be led into the heated injection port of a gas chromatograph, where the solvent is evaporated, and the traces are finally detected by ecd [221].

An internal standard was added in about half the determinations listed in Table 5.7. Examples of a chemically different, but closely related substance by far outweigh the few examples in which an isotopically labelled tracer is used [254, 312]. Examples of pairs of analyte/internal standard are: 4-aminopyridine/2-aminopyridine [223], amitryptilin/loxapine [225], amitryptiline/propranolol [226], amphetamine/aniline [227], cannabinoids/5-phenyl-5-*p*-tolylhydantoin [234], chlordiazepoxides/diazepam [237], chlorpromazine/mesoridazine [242], cinnarizine/chlorbenoxamine [245], clobazam/diazepam [246], cocaine/lignocaine [248], codeine/isopropylcodeine [249], dantrolene/benzanilide [251], diazepam/prazepam [253], dipyridamole/lignocaine [255], ellipticine/11-demethylellipticine [256], ergotamine/ergocristine [257], fluorouracil/5-bromouracil [261], 2-hydroxydesipramine/2-hydroxyimipramine [263], indomethacin/glutethimid [265], indomethacin/indomethacin dimethylamide [267], labetalol/chloroquine [268], labetalol/pericyazine [269], lorazepam/diazepam [270], metapramine/maprotiline [272], methotrexate/aminopterin [275], 8-methoxypsoralen/5-methoxypsoralen [276], metoprolol/4-methylpropranolol [279], morphine/nalorphine [283, 285], morphine/5-quinolinol [286], morphine/3,4-dihydroxybenzylamine [287], norpseudoephedrine/5-hydroxytryptamine [289], papaverine/diphenylhydramine [290], papaverine/mepyramine

Table 5.7 – Use of hplc for the analysis of traces of pharmaceutically used drugs in biological materials.

Substance	Concentration or amount	Determination	Relative standard deviation (%)	Recovery (%)	Reference
adriamycin	1–120 ng/ml	fluorescence			222
4-aminopyridine	2–490 ng/ml	uv			223
amiodarone	23 ppb	uv	2.9		224
amitryptiline, nortriptyline and metabolites	5–500 ng	uv			225
amitryptiline and metabolites	2–250 ng/ml	uv			226
amphetamine	0–2.3 μg/ml	fluorescence	8.8		227
anabolics	>0.5 ppb	tlc			228
anticonvulsants	16–160 ng/l.	uv			229
aspirin and metabolites	1–200 μg/ml	uv	4.5		230
benzodiazepines (15)	>10 ng/ml				231
benzodiazepines and metabolites	5–10 ng/ml	uv	3.8–8.6	91–97	232
11-bromovincamine	20–500 ng/ml	uv			233
cannabinoids	200 ng			80–90	234
captopril	1 pmole	electrochemical			235
celiprolol	>10 ng/ml	uv			236
chlordiazepoxide and metabolites	>20 ng	uv		88	237
chloroquine	25–200 ng/ml	fluorescence			238
chloroquine	>20 ng/ml	uv			239
chloroquine and metabolites	>1 ng/ml	fluorescence		80–88	240
chloroquine and metabolites	5–200 ng/ml	uv	2	85	241
chlorpromazine	1–30 ng/ml	uv			242
cimetidine	25–1000 ng/ml	uv	4.6		243
cimetidine	0.1–4 μg/ml	uv	3		244
cinnarizine	20–100 ng/ml	uv	8.4	67	245
clobazam and metabolites	50–500 ng/ml	uv	1–11	88–100	246
clomiphene	0.4–45 ng/ml	fluorescence			247
cocaine and metabolites	1–100 μg/ml	uv			248
codeine	10–100 μg/l.	fluorescence			249
corticosteroids	50–100 ng	uv or ms		70–80	250
dantrolene and metabolites	0.03–1 μg/ml	uv			251
diazepam and metabolites	2–1000 ng/ml	uv		88	252

Table 5.7 – *continued*

Substance	Concentration or amount	Determination	Relative standard deviation (%)	Recovery	Reference
diazepam and metabolites	ng	uv		95	253
digoxin and metabolites	pg/ml	uv (radiochem.)			254
dipyridamole	0.01–2 μg/ml	uv	4.8	97 ± 2	255
ellipticine	20–500 ng/ml	fluorescence		94	256
ergot alkaloids	>100 pg/ml	fluorescence			257
ethaverine	5–50 ng/ml	uv	2–14		258
5-fluoro-1-(hexylcarbamoyl)uracil and metabolites	>20 ng/ml				259
5-fluoro-l-(hexylcarbamoyl)uracil and metabolites	1–50 ng/ml	uv	3	82–85	260
fluorouracil	0.1–1 μg/ml	uv		74	261
heroin, cocaine	>3 ng	uv			262
2-hydroxydesipramine	10–100 ng	uv	2–5	68	263
indapamide and metabolites	25–125 ng/ml	uv			264
indomethacin	0.5–10 μg/ml	uv		96	265
indomethacin	1–30 ng/ml	fluorescence	4.5		266
indomethacin	>10 ng/ml	uv			267
labetalol	12–250 ng/ml	fluorescence		95± 10	268
labetalol	10–3000 ng	uv			269
lorazepam	2–100 ng/ml	uv	8	66	270
meptazinol	3 ng/ml	fluorescence			271
metapramine	2–300 ng/ml	fluorescence	4		272
methaqualone	$1–10^4$ ng/ml	uv		98	273
methotrexate and metabolites	0.1–10 μg/ml	uv			274
methotrexate	>4 ng	uv	9.7		275
methoxsalen	12–1000	uv	1–5		276
methoxsalen	10–1000 ng/ml	uv			277
metoprolol and metabolites	3–200 ng/ml	fluorescence		83	278
metoprolol	10–300 ng/ml	fluorescence	2–6		279
midazolam and metabolites	15–750 ng/ml	uv		80	280
misonidazol and metabolites	2–25 μg/l.	uv		97–99.8	281

mitomycin C	1–1500 ng/ml	uv		99	282
morphine	>10 ng	fluorescence			283
morphine	20–100 ng/ml	uv			284
morphine	1–40 ng/ml	electrochemical	5.6		285
morphine	25–250 ng/ml	electrochemical	5–9	78	286
morphine	2–50 ng	electrochemical			287
neostigmine	5–400 ng/ml	uv	1.5		288
D-norpseudoephedrine	<300 ng/ml	fluorescence			289
papavarine	2–50 ng/ml	uv	<10	98–107	290
papaverine	10–600 ng/ml	uv		92	291
papaverine	50–1000 ng/ml	uv	7.5	84	292
paracetamol and metabolites	>1 ng	electrochem. or uv	2–5	100	293
paracetamol and metabolites	>0.4 ng	electrochemical	1–3		294
paracetamol		electrochemical			295
penbutolol and metabolites	5–1000 ng/ml	fluorescence			296
phenothiazine + 20 neuroleptics	>0.1 ng/ml	electrochem. or uv	7		297
phenylpropanolamine	5–200 ng	fluorescence	6		298
physostigmine	100 ng/g	uv	90± 7		299
pipothiazine	>0.3 ng/ml	fluorescence			300
podophyllotoxines	50–1000 ng/ml	uv	3		301
prazosin	1–164 ng/ml	uv or fluorescence	1–8		302
prednisolone	25–400 ng/ml	uv			303
prednisone and metabolites	>4 ng/ml	uv	<10	54	304
prednisolone	20–1000 ng/ml	uv	3	93	305
probenecid	1–50 μg/ml	uv	3	98	306
promethazine	1–250 ng/ml	uv		95	307
promethazine	0.2–8 ng/ml	electrochemical	3		308
promethazine and metabolites	1–250 ng/ml	uv			309
propafenone	5–500 ng/ml	uv			310
propafenone	6–270 ng/ml	uv	3–7		311
propafenone	20–250 ng/ml		3	>90	312
propranolol	>2 ng/ml	fluorescence			313
propranolol	2–220 ng/ml	fluorescence			314
propranolol and metabolites	2–150 ng/ml	fluorescence	2		315
propranolol and metabolites	5–300 ng/ml	fluorescence	8	80	316
propranolol and metabolites	5–250 ng/ml	fluorescence			317

Table 5.7 – *continued*

Substance	Concentration or amount	Determination	Relative standard deviation (%)	Recovery	Reference
propranolol and metabolites	>200 pg/ml	fluorescence			318
quaternary ammonium compounds	>1 ng/ml	uv		89–102	319
ranitidine and metabolites	5–5000 ng/ml	uv	3		320
sotalol	0.3–5 μg/ml	fluorescence	3–7		321
sulfinalol	>2 ng/ml	electrochemical	5		322
sulphapyridine	50 ng–40 μg/ml	uv			323
sulphasalazine and metabolites	>50 ng/ml	fluorescence or uv			324
sulphonamides	>10 ppb	electrochemical			325
tetrabenazine and metabolites	>1 ng/ml	fluorescence		74	326
thioguanine	0.2–25 ppm	uv	<3		327
thioridazine and metabolites	50–2000 ng/ml	uv			328
thioridazine and metabolites	20–1000 ng/ml	uv			329
tofisopam	10–800 ng/ml	uv			330
triamterene	40–240 ng/ml	uv	2–9	91–99	331
tricyclic antidepressants and metabolites	>25 ng/ml	uv		>80	332
tricyclic antidepressants	10–800 ng/ml	uv			333
trimethoprim	0.1–1 ppm	uv	5–7	83–90	334
D-tubocurarine	25–500 ng/ml	uv	7	>95	335
verapamil	1–200 ng/ml	fluorescence	4		336
verapamil	1–250 ng/ml	fluorescence	4		337

[291], papaverine/laudanosine [292], pipothiazine/7-methoxypipothiazine [300], prazosin/carbamazepine [302], promethazine/chlorpromazine [307], propranolol/labetalol [131], propranolol/4-methylpropranolol [316], propranolol/ desipramine [317], thioridazine/7-methoxychlorpromazine [328], thioridazine/2-acetylphenothiazine [329].

For the determinations listed in Table 5.7, deproteination was done with perchloric acid [229, 231, 321], trichloroacetic acid [275] and acetonitrile [267, 274, 302, 306, 314, 317, 324, 334]. In all other cases deproteination was not necessary.

Very often, partition was the method of choice for separation of the analyte(s). An organic solvent (immiscible with water) was added to the serum or urine sample for direct liquid–liquid distribution. Other sample preparation methods were adsorption and elution into a reversed-phase column before hplc [232, 267, 284, 305]. Direct determination of the components of interest was made in about a tenth of the applications, by dilution of the sample with mobile phase [229, 230, 235, 262, 264, 273, 275, 281, 293, 294, 302, 314, 324, 327, 334].

The determination is often done by measurement of ultraviolet absorption, fluorescence intensity or an electrochemical property. In less than 5% of all systems analysed in this category was it necessary to make a derivative of the analyte or alter it chemically before hplc or detection. Use of the *o*-phthalaldehyde [227] or dansyl chloride [272] derivatives, or oxidation [283, 326], heating [266] or photochemical reaction [247] are the few examples in which the analytes were not measured directly by means of their own molecular properties.

This indicates the great advantage of the hplc method of determination of traces in pharmacology; the analytes are almost always isolated from serum or urine by a liquid–liquid partition step. The organic phase is used for the hplc separation either directly or after evaporation. Determination by means of the molecular properties of the analyte(s) is usually so easy that use of derivatives is seldom necessary. Metabolites can be measured very often by the same procedure as the parent material. As many therapeutically used drugs have very low volatility and some solubility in water, a method which works at room temperature is obviously useful. Consequently, hplc is beginning to replace gc in this very important area of trace analysis. In fact hplc is a rapidly expanding field of organic trace analysis (which is evidenced by the fact that Tables 5-7–5.9 were compiled entirely from the literature for the period 1980–1982).

Another important area of application of hplc is environmental chemistry and the analysis of food and feeds for organic traces. Table 5.8 gives some examples of this expanding field.

It is obvious that more separation steps are involved than in the application to drug analysis. Typically, the traces are extracted from a solid matrix, and the solution is then evaporated or the trace is distributed into another solvent.

Table 5.8 – Examples of the application of hplc determination of traces in environmental samples or foods.

Substance	Matrix	Concentration or amount	Sample preparation before hplc	Detection	Reference
abscisic, phaseic, indol-3-ylacetic acid	sorghum leaves	1 ng	extn./ion-pair chromatog.	uv	338
aflatoxins	swine tissue	>0.01 ppm	extn.		339
aflatoxins	milk, milk products	0.1–1 ppm	extn./chromatog.		340
aflatoxins	spices	>0.2 ppb	extn.	fluoresc.	341
aflatoxin M_1	milk powder	>0.01 ppb		fluoresc.	342
aflatoxin M_1	dairy products	0.1–1 ppb	extn./chromatog.	fluoresc.	343
aldehydes, ketones	diesel exhaust	>0.01 ppm	trapping in 2,4-dinitrophenylhydrazine soln.	uv	344
alkylbenzenesulphonates	water	5–200 ng	partn.	uv	345
aminotriazole	potatoes, beets	>0.01 ppm	extn./deriv./chromatog.	visible	346
anilazine	potatoes, tomatoes	>0.02 ppm	extn./partn./chromatog.	uv	347
anionic surfactants	water	1–5 ng	chromatog. preconc.	fluoresc.	348
aromatic amines	water, soil	25 pg–5 ng/ml	extn.	electrochem.	349
β-asarone	alcoholic beverages	1 μg/l.	distn./partn.	uv	350
azinphos-methyl	potable water, sea-water	>12 ppb	chromatog. preconcn.	uv	351
benomyl	apple foliage	>0.2 ppm	extn.	uv	352
benzidine, 3,3′-dichlorobenzidine	waste water	2 ppb	none, direct injection	electrochem.	353
benzo(*a*)pyrene	natural and synthetic crudes	20 ng–500 μg/g	chromatog.	fluoresc.	354
benzo(*a*)pyrene	airborne particulates	50 pg–100 ng	extn. of filter	fluoresc.	355
biphenyl, 2-phenylphenol	citrus fruits		steam distn.	uv	356
carbamate pesticides	water	>40 pg		electrochem.	357
carbaryl, l-naphthol	water	0.1–0.5 ppb	chromatog. enrichm.	uv	358
carbofuran + metabolites	water	>1 ppb	direct injection	uv	359
chlorinated dibenzodioxins and dibenzofurans	wood-shavings, liver	25 pg/g	extn./partn./chromatog.	uv and ms	360
chlorinated diphenyl ethers	chicken tissue	>0.25 ppb	extn./chromatog.	uv	361
chlorophenols	urine	ng	chromatog. enrichm.	ms	362
2-chlorophenol	serum	50 ng	partn.	uv	363
chlorsulpheron	soil	>200 pg	extn./partn.	photoconductivity	364

clopidol	chicken tissue, eggs	0.1–0.5 ppm	extn./chromatog.	uv	365
coumarin-based rodenticides	liver, stomach contents	ng–μg	extn./chromatog.	uv	366
curcumin	plasma, urine	0.2–4 μg	partn.	uv	367
cyanuric acid	urine, pool water	0.1 μg/ml	chromatog. enrichm.	uv	368
2,4-dichlorophenoxyacetic acid, dichlobenil	water	0.5–50 ppm	direct injection	uv	369
difenzoquat	water	2–50 ppb	chromatog. enrichm.	uv	370
1,1-dimethylhydrazine	plasma, urine	2–10 ng	direct injection	uv and ms	371
ethylenethiourea	beer	20 ppb	chromatog.	uv	372
fluridone	cotton seed	>0.05 ppm	extn./partn.	uv	373
formaldehyde	clean air	0.1–20 ng/ml	trapping in 2,4-dinitrophenylhydrazine soln.	uv	374
formaldehyde	clean air	>0.03 ppb	trapping in 2,4-dinitrophenylhydrazine soln.	uv	375
formetanate	fresh fruit	0.02–0.5 ppm	extn./partn.	uv	376
furazolidine	swine plasma	>1 ng/ml	partn.	uv or electrochem.	377
furazolidine	turkey tissues	>0.5 ppb	extn./partn.	uv	378
furazolidine	pig liver, kidney	>0.05 mg/kg	extn.	uv	379
glyphosphate + metabolite	straw	>0.2 ng	extn./deriv.	fluoresc.	380
haloalkanes	water	>1 ppm	deriv.	uv	381
indol-3-ylacetic acid	vegetables	0.1–5 ng	partn./chromatog.	fluoresc.	382
isocyanates	air	1–10 $\mu g/m^3$	trapping/deriv.	fluoresc.	383
isocyanates	air	ppt	trapping/deriv.	uv	384
methomyl	fruit crops, water	0.01–1 ppm	extn./partn.	uv	385
4,4′-methylenebis-(2-chloroaniline)	air	>20 pg	trapping (Tenax)	electrochem. (or uv)	386
naphthalene + metabolites	bile from trout, mice	<100 ng		fluoresc.	387
nitrated polycyclic hydrocarbons	aerosols	10–200 pg/m^3		gc or ms	388
nitro-aromatic explosives	gunshot residues	>0.5 pmole	extn. of cotton swab	electrochem.	389
nitrosamines	cosmetic raw materials	ppb	chromatog.	uv	390
N-nitrosamines	food, beverages	μg/kg	chromatog.		391
N-nitrosodiethanolamine	cosmetics	ppb	extn./chromatog.	ms	392
oxfendazole	milk	50 ng/g	extn./partn.	uv	393
pentachlorophenol	muscle, eggs, fat	5 ppb	extn./chromatog.		394
pentachlorophenol	mushrooms	>0.5 μg/kg	steam distn./partn.	uv	395
phenols (germicidal)	serum	40–500 ng/ml	partn.	uv	396

Table 5.8 – *continued*

Substance	Matrix	Concentration or amount	Sample preparation before hlpc	Detection	Reference
phenols	auto exhaust, tobacco smoke	ppb	trapping/deriv./partn.		397
phenols	auto exhaust	1–10 ppm	trapping/partn./chromatog.	uv	398
phenols	water	1–50 ng	partn.	uv	399
polychlorinated biphenyls	silicone fluids	>0.1 ppm	extn.	uv	400
polynuclear aromat. hydrocarb.	oil	100–800 ng	chromatog.	uv	401
polynuclear aromat. hydrocarb.	air particulates		extn./tlc	uv	402
polynuclear aromat. hydrocarb.	oysters, seaweed	25 ppb	saponific./partn./chromatog.	fluoresc.	403
polynuclear aromat. hydrocarb.	drinking water	>7 ng/l.	partn.	uv	404
polynuclear aromat. hydrocarb.	waste waters	>4 ng	direct injection	uv	405
polynuclear aromat. hydrocarb.	beer	>0.5 ppb	partn./chromatog.	uv or fluoresc.	406
pyrazon	water	2 ppb–1 ppm	chromatog./enrichm.	uv	407
satratoxins	cereal grains	>200 ppb	extn./partn./chromatog.	uv	408
stilboestrol	urine, serum	low ppb	partn.	fluoresc.	409
styrene	wine	0.1–2 mg/l.	direct injection	uv	410
sulphamethazine	beef tissues	>0.02 ppm	extn./partn.	uv	411
tetracyclines	honey	>1 ppm	extn.	uv	412
toluene diamines	water from 'boil in bags'	50 ng/l.	partn.	uv	413
1,3,5-triazine herbicides	bacterial cultures	>30 pmole	centrifugation	uv	414
2,4,7-trinitrofluoren-9-one	air (workplace)	>20 ng/ml	trapping/extn.	uv	415
uric acid	water	2 ppb	filtration	uv	416
vitamin E	animal feeds	>9 ppm	extn./partn./chromatog.	uv	417
zeatin	pears, peaches	>5 ng	extn./chromatog./tlc.	uv	418
zearalenone	chicken blood	50–200 ng/ml	protein pptn./partn.	uv	419
zearalenone	urine, liver	>2 ng	extn./partn./chromatog.	uv	420
zearalenone, vomitoxin	foods, feeds	>25 ppb	extn./chromatog.	fluoresc.	421

Separation involves liquid–liquid distribution and clean-up by chromatography. Afterwards the hplc fine separation and determination is run. In only rare cases – with water as matrix – is direct analysis of the aqueous sample solution possible [353, 359, 369, 371, 405, 410].

Again hplc is widely used for the determination of traces of naturally occurring substances in biological material (Table 5.9).

Table 5.9 offers some examples of derivative-formation reactions on pre-columns, e.g. acetylcysteine with *N*-(pyren-l-yl)maleimide [422], amino-acids with 4-fluoro-7-nitrobenzofurazan [424], biogenic amines with phthalaldehyde [427], fatty acids with 9-anthryldiazomethane [440], conversion of fatty acids into *p*-bromophenacyl esters [441], fatty acids (after reaction with $SOCl_2$ to form the acid chlorides) with 1-naphthylamine [442], conversion of sugars into benzyloxime–perbenzoyl derivatives [443], histamine with phthalaldehyde [446], oligosaccharides with benzoyl chloride [462], oxosteroids with dansyl-hydrazine [463], taurine with fluorescamine [477] and conversion of thiamine into thiochrome by treatment with CNBr [478]. Examples of post-column derivative formation are comparatively few: guanidine compounds reacted with ninhydrin [445] and biogenic amines with phthalaldehyde [465].

In view of the great significance of catecholamines in modern research it is not surprising that these substances are often to be determined at concentration levels of a few tenths of a nanogram per ml of blood; hplc has been used very successfully to solve this problem, as catecholamines can be determined very sensitively by electrochemical detection after separation from each other and from other components by hplc [483–498]. A limit of determination of a few pg is reported in almost all these publications, which is sufficient for the determination of these hormones in brain tissue and plasma. Another method for their determination is by fluorescence [499–503]. Both methods have been compared [504] and found to give good agreement. In the fluorescence method limits of determination of a few tenths of a nanogram have been reported. The analytical scheme for determination of catecholamines (and other closely related compounds) almost always consists in extraction of the hormones from the matrix by solvents and subsequent adsorption on alumina and elution before hplc.

Closely related is the group of biogenic amines, which offer many examples of the application of hplc. Again the method is applied primarily to ultraviolet-absorbing or electrochemically active compounds, thus combining separation and detection very efficiently. A third group is formed by the steroids. A survey (with 430 references) on steroid analysis by hplc has been given [505].

5.4.13 Thin-layer chromatography (tlc)

About a tenth of the literature dealing with the separation of traces of organic compounds describes the use of this simple, inexpensive, well-known and rapid technique.

Table 5.9 – Examples of the use of hplc for the determination of traces of naturally occurring substances in biological materials.

Compounds	Matrix	Concentration or amount	Separation	Determination	Reference
N-acetylcysteine	urine	>50 fmole	chromatog./deriv.	fluoresc.	422
N-acetyldopamine	cockroach tissue	>0.1 pmole	extn./chromatog.	electrochem.	423
amino-acids		>5 fmole	deriv.	fluoresc.	424
aromatic amino-acids	serum	>1 ng	deprot.	uv	425
bile acids	serum	>10 ng	chromatog. enrich./ ion-exch.	enzymatically	426
biogenic amines (histamine, noradrenalin, serotonin, dopamine, tyramine)	blood	<100 pg	deprot./deriv./partn.	fluoresc.	427
cortisol	plasma	>5 ng/ml	partn.	uv	428
cortisol	plasma	1–40 ng	partn.	uv	429
7-dehydrocholesterol	rat skin, liver	1 μg	chromatog.	uv	430
24,25-dihydroxyvitamin D_3	plasma	>0.5 ng/ml	partn./chromatog.	uv	431
dopa, 5-hydroxytryptophan	brain tissue	>500 pg	extn.	electrochem.	432
dopa, dopamine, homovanillic acid, 5-hydroxyindol-3-yl-acetic acid	nervous tissue	150 pg	extn./chromatog.	electrochem.	433
dopamine, 5-hydroxytryptamine	serum	0.1–1 ng	deprot.	uv or fluoresc.	434
dopamine, 5-hydroxytryptamine	rat brain	1 ng	extn.	electrochem.	435
dopamine	plasma	>25 pg/ml	adsorp. (alumina)	electrochem.	436
dopamine (lipophilic analogues)	brain	>40 pg	extn./chromatog.	electrochem.	437
dopamine, homovanillic acid	brain tissue homogenates	>50 fmole	extn.	electrochem	438
eugenol	cells from bark		mechanical isolation of oil droplets from cells by micromanipulator	uv	439
fatty acids	complex biological samples	0.2–5 ng	deriv.	fluoresc.	440
fatty acids	mammalian fats	>20–50 pg	deriv.	fluoresc.	441
fatty acids	biological examples	>4 ng	deriv.	fluoresc.	442
L-fucose, D-xylose	algae polysaccharides	> 2 ng	deriv.	uv	443
gibberellins	plant seeds	>10 ng	extn./chromatog.	uv	444
guanidino compounds	serum	>1 pmole		fluoresc.	445

histamine	tissue	> 1 pmole	extn./deriv.	fluoresc.	446
5-hydroxyindol-3-ylacetic acid	urine	2 mg/l.	partn.	uv	447
5-hydroxyindol-3-ylacetic acid	urine	>1 pmole	deprot.	fluoresc.	448, 449
4-hydroxy-3-methoxyphenyl-ethanediol (+ others)	cerebrospinal fluid	>50 pg		electrochem.	450
4-hydroxy-3-methoxyphenyl-acetic acid	cerebrospinal fluid	>40 pg	partn.	electrochem.	451
monoamine metabolites	brain	1–50 ng	extn.	electrochem. or fluoresc.	452
25-hydroxyvitamin D_2	plasma	<100 ng	extn./chromatog.	uv	453, 454
indoles	rumen fluid	1 ng	chromatog. enrichm.	fluoresc.	455
indoles	pineal tissue	>10 ng	extn.	fluoresc.	456
metanephrines	urine	5–30 ng	chromatog. (ion-exch.)	fluoresc.	457
methoxytyramine	brain	15 ng/g	extn./alumina adsorp.	electrochem.	458
nicotinamide	urine	>10 ng	filtration	uv	459
oestriol	urine	>10 ng	partn.	electrochem.	460
oestriol	urine	>5 ng	partn.		461
oligosaccharides	syrup, urine	>0.5 pmole	deriv.	uv	462
17-oxosteroids	syrup, urine	>50 ng/ml	partn./deriv.	fluoresc.	463
phylloquinone	serum	>150 pg	deprot./chromatog.	fluoresc.	464
polyamines	urine, serum	>1 ng	hydrolysis	fluoresc.	465
prostaglandins	synthetic mixture	ng		uv	466, 467
retinoic acid	serum	1–10 ng/ml	deprot./partn.		468
riboflavine	haemodialysate	2–20 ng/ml	chromatog.	fluoresc.	469
serotonin	brain, lung	>100 pg	extn./alumina adsorp.	fluoresc.	470, 471, 472
serotonin	brain, plasma	>10 pg	deprot.	electrochem.	473–476
taurine	urine, cerebrosp. fluid	>0.2 pmole	deprot./ion-exch. chromatog.	fluoresc.	477
thiamine	plasma	3–500 ng/ml	deriv./partn.	fluoresc.	478
thyroxine	serum	2–100 ng/ml	deprot./partn./chromatog.	electrochem.	479
thyroid hormones	serum	ng	deriv.	fluoresc.	480
tocopherols	plasma	3–16 ng	deprot.	fluoresc.	481
vitamin D (+ metabolites)	serum	>2 fmole	partn./chromatog.	protein binding assay	482

In tlc the application of the sample to the plate influences the separation and limit of detection. The spots at the starting line should be as small as possible. To achieve this, micro syringes or capillaries are used. Much better application can be achieved by means of the CAMAG applicator which sprays the solution through a jet onto the adsorbent layer. Sharply defined spots with a diameter of 1-2 mm are obtainable. As the instrument works automatically, volumes of several μl can be applied without much control necessary by the analyst. Quantitative measurements are more accurate and elution is simpler.

Concentration plates [506] consist of a narrow inert layer of porous silica of medium pore size and rather large diameter, with a sharp interface leading to the chromatographic layer. Material applied to the more porous zone is eluted by the solvent and concentrated at the interface. The resulting extremely narrow line of substance provides an optimal start for the chromatography.

Thin-layer chromatography can be combined with practically all other separation or identification techniques. Sample transfer is done by elution of the spots for separation. The adsorbent is separated and then treated with solvent. Microfiltration with slight suction separates the solution from the particles of adsorbent. Equipment for microfiltration of a few μl has been described [507]. After filtration, determination by micro infrared spectrometry is possible [508]. Infrared reflectance spectrometry can be done even without transfer of the separated substances, by direct fuming of the silica layer with hydrofluoric acid [509]; the silica is removed by volatilization as H_2SiF_6, and the organic material from the spots is retained on the aluminium sheet which originally served as carrier for the silica, and is examined *in situ* by infrared reflectance spectrometry. Microgram amounts can be detected.

Combination of tlc with hplc has been suggested [510]. The tlc is done first and the spots are transferred to the hplc equipment. The advantages are that nothing can be lost during the tlc, though it can be in column chromatography if there is insufficient elution. Thin-layer chromatography saves time because several samples can be analysed simultaneously. After preseparation by the more tolerant tlc, the selection of the hplc system is less critical. Finally, it is advantageous to use different mechanisms of separation in the two chromatographic methods, because in that way the overall selectivity is enhanced.

Thin-layer chromatography is used for the detection of ng-amounts or ppb-concentrations of more than 60 drugs in greyhound urine [511], of marijuana metabolites (in the urine of a subject who smoked 1 cigarette containing 16 mg of tetrahydrocannabinol) [512] and of cannabinoids [513]. It is also applied for the determination of aflatoxins [514-518] in milk products and maize, and of zearalenone [519], T_2-toxin [520, 521], Fusarium toxins [522, 523], trychotecene mycotoxins [524, 525] and ochratoxin A [526]. The method is useful for detection of equally small amounts of phenylurea and phenylcarbamate residues [527], triazine herbicides [528, 529], parathion-methyl [530] and photosynthesis herbicides [531]. The last-named were detected by spraying the dry tlc plate with a mixture of spinach chloroplasts and dichlorophenolindophenol. Examples of the application of tlc in food analysis are the identification of trenbolone in veal [532], of furazolidine in chicken, swine or bovine tissues [533] and of aniline and fatty anilides in denatured olive oils [534].

Chromarods® are silica rods covered with an adsorbent layer of silica. They are used – mainly in lipid analysis – in conjunction with the Iatroscan analyser [535–538]. After chromatography the rod is scanned by the analyser, which uses flame-ionization detection. Total lipids (from fish, after extraction) in amounts of 0.5–32 μg [539] and phospholipids [20 μg (extracted from mitochondria)] [540] have been determined.

5.4.14 Ion-exchange chromatography

By ion-exchange chromatography the following separations are possible:

(1) of oppositely charged species;
(2) of ions from uncharged molecules;
(3) within one group of ions, due to differences in charge and ionic size.

Ion-exchange is most often done in aqueous solutions, since many molecules of biochemical interest are highly polar and thus better soluble in polar dissociating solvents. Use of aqueous methanol, acetic acid or other solvents is becoming of increasing importance in separation science [541], but difficulties may arise from solubility of the resin material, since this can cause problems with subsequent highly sensitive methods. Adsorption of trace constituents may be considerable, and this effect is used in ion exclusion chromatography [542].

Pellicular resins with extremely small particle size are needed to give efficient separation of closely related compounds [543]. Several hundreds of components of body fluids have been determined by use of this principle. It has been recommended to use the analysis data in general health-screening programmes.

Examples of the use of ion-exchange resins are the separation of down to 12 ng of histamine in homogenates of rat-brain [544], of putrescine or cadaverine at the 50-ppb level from fish or cheese homogenates [545], of catecholamines from urine (detection limit 1-15 ng) [546], of 0.1–0.5 ppm levels of *N*-nitrosodiethanolamine from cosmetics [547], of surfactants from water (at levels of >20 ppb) [548] and of endogenous plant hormones such as indol-3-ylacetic acid or abscisic acid [549].

5.4.15 Scheme for selection of a liquid chromatographic method

A scheme is given below which helps in the selection of an adequate chromatographic method for the solution of a given trace analysis problem.

Scheme for selection of chromatography [adopted from Varian Information Sheet VEO// INS 2391 (1979), p. 6, by permission].

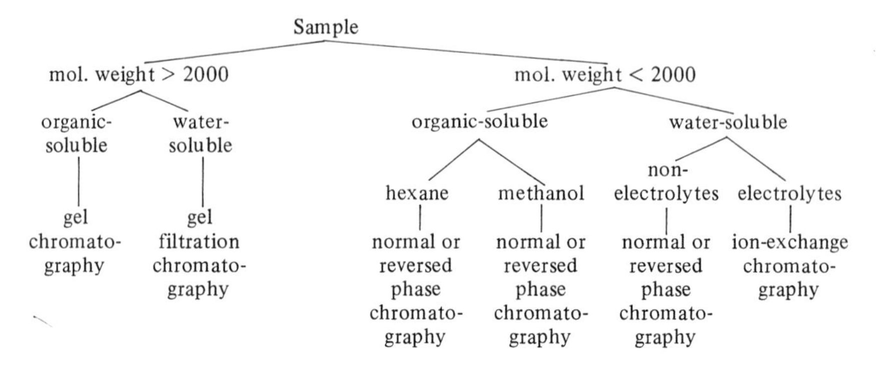

5.5 GAS CHROMATOGRAPHY

5.5.1 General aspects

Gas chromatography is a well established method. About 37% of all literature on separation of organic traces describes its use. Several articles (see Table 5.10) and books on the principles [559-564] and application [565-571] of gc have been published.

Table 5.10 – Review articles on gc.

Topic and reference	Number of references
Gas chromatography [550]	1268
Developments in gas chromatography [551]	211
Use of glass capillary columns [552]	31
Choice of detectors in analysis of pesticides [553]	59
Choice of columns in pesticide analysis [554]	19
Detection of hydrocarbons in the atmosphere [555]	125
Gas chromatography of ^{3}H- and ^{14}C-labelled compounds [556]	431
Analysis of pesticide residues [557]	49
Analysis of drugs in biological samples [558]	3

5.5.2 Types of columns

In gc, filled columns and open columns ('Golay-type') are used. For filled columns an inert support coated with an appropriate liquid phase (Table 5.11) is used. Gaschrom Q and Chromosorb W are the most popular supports, each being mentioned in about 40% of papers on gc. The liquid phases must be stable at temperatures up to 300°C. Separation is based on the boiling point of the trace (usually with SE 30 as liquid phase) or its polarity [with poly(ethylene glycol)]. The columns are made of stainless-steel, glass or silica tubing.

In gc, trace amounts of sample are brought into contact with gram amounts of support and stationary phase, both of which have very large surface area. It is not very surprising that adsorption and losses occur at extremely low trace levels, this being the more pronounced, the more polar the trace compound. Peak broadening and distortion, and unfavourable detection limits are the consequence of such adsorption effects. Special treatment of the stationary phase to deactivate the adsorption sites can improve the detection limit by several orders of magnitude.

Treatment of the support with benzoyl chloride before application of the stationary phase (SE 52) enables determination of ng-amounts of drugs [572]. Analogously, treatment of a chromatographic support consisting of 3% Dexsil 300 on Supelcoport AW DMCS by 2 successive injections of 5 μl of heptafluorobutyric acid once a week, has been reported to deactivate the column to such an extent that about 50 ppb of chloramphenicol in milk could be determined [573].

Open capillaries were introduced about 25 years ago [574]. Stainless-steel, glass or fused quartz capillaries of about 0.25 mm internal diameter are used. They are coated internally with a thin film of stationary phase. The extremely favourable exchange between the mobile and the stationary phase, and the absence of turbulent flow in the column, produce a very high column efficiency, so even very difficult separations can be achieved. One disadvantage is the restriction of the total load to less than a few μg to prevent overloading of the column. The injection of such small amounts necessitates correspondingly sensitive detection, but this is guaranteed by modern detectors. One example may be given (Fig. 5.11). A comparison was made of the performance of a packed column and a glass capillary column for analysis of an extract from a water sample. The same liquid phase (OV-1) was used in both. With the packed column 118 peaks could be identified, whereas with the open column 490 substances could be resolved [575]. This clearly shows that "the future of gc separations of environmental samples lies with capillary columns" [575]. Capillary columns have similarly proved superior to packed columns for the determination of organochlorine compounds [576].

Open columns can be purchased or home-made. The critical step is the conditioning of the surface of the capillary wall. Glass and quartz are currently the preferred materials, and must be roughened so that the liquid film of stationary phase adheres to the wall. If hydrogen chloride gas is passed through soft-glass capillaries, tiny crystals of sodium chloride are formed [577], which serve as the auxiliary supporting phase. In another method a layer of barium carbonate is deposited by applying a barium hydroxide solution to the interior and then passing carbon dioxide over it [578].

Capillary columns are deactivated, like filled columns, in order to prevent excessive adsorption. This is done by treatment with 10% hydrochloric acid at 100°C [579] or by 'bleeding' poly(ethylene glycol) vapours from a precolumn into the capillary [580]. Treatment with hexamethyldisilazane is another possibility [581].

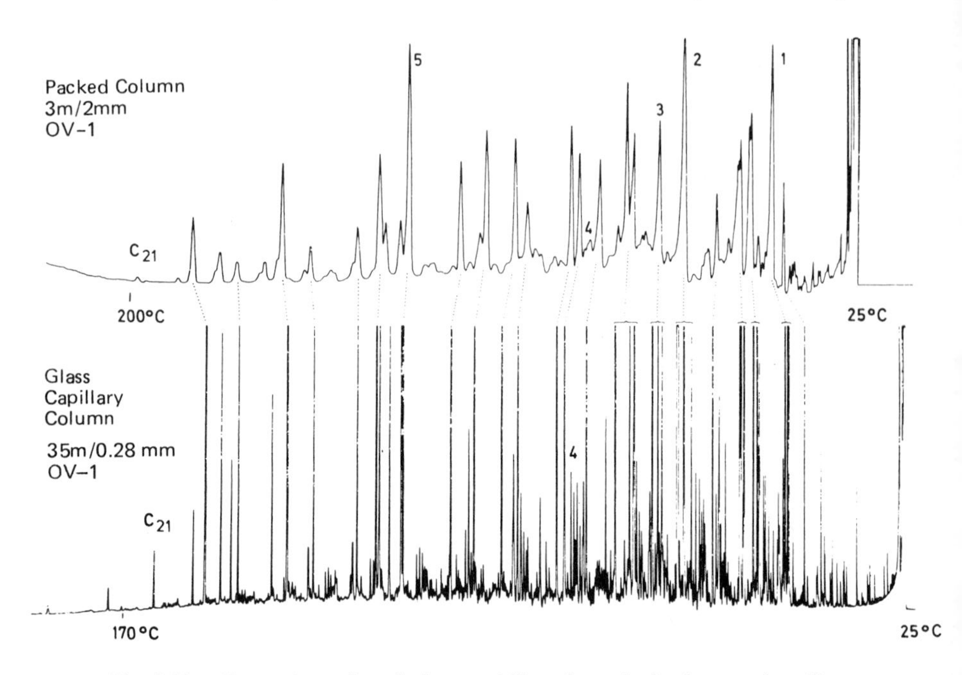

Fig. 5.11 – Comparison of resolution capability of a packed column and capillary column [575]. (Reproduced from the *Journal of Chromatographic Science* by permission of the copyright holders, Preston Publications Inc.).

Fused silica capillary columns are being increasingly used instead of glass capillaries. Fused silica is a highly inert support material, but to avoid tailing, such columns also have to be coated. This creates some problems, as silica is not as easily wetted as glass surfaces by stationary phases [582–587]. Examples of application of fused silica columns in gc are the detection of 190 PAHs (m.w. 300–400) [588], pesticides [589], thermolabile phenylurea herbicides [590], thyroid hormone standard derivatives (with a detection limit of 30 fg!) [591], drugs [592] and pollutants [593, 594].

5.5.3 Stationary phases in gas chromatography

A survey of the literature [595] demonstrates that a few stationary phases are used extensively, while others are applied only rather rarely (see Table 5.11). A survey [596] lists the retention indices of 1318 substances of toxicological interest on SE-30 or OV-1 stationary phases, again indicating that some stationary phases are much preferred to others.

Table 5.11 demonstrates the use of only 10 stationary phases has covered about 82% of all problems so far solved in pesticide analysis. An analyst who keeps columns of QF 1, SE 30, OV 17 and DC 200 available has about a 50% chance of solving a given problem in pesticide analysis, for instance, by using these phases. An analogous picture is seen in the trace gas chromatographic analysis of drugs. Only a few phases, e.g. QF 1, SE 30 and OV 17, are very frequently used. Special cases may require employment of selected rarely used phases, however.

Table 5.11 – Statistical survey (from data given in [595]) of the frequency of use of certain stationary phases in gas chromatography.

Phase material	Percentage of all references
QF 1	15
SE 30	14
OV 17	11
DC 200	10
OV 101	8
OV 1	8
Carbowax 20 M	6
OV 210	6
OV 225	2
XE 60	2
others	18

It is always recommended that at least two columns, working on different principles, should be used. Identification is thus improved and more certain. This necessitates having enough columns readily available.

5.5.4 Special forms of column operation

Many special forms of column operation have been suggested. Precolumns are often used to get rid of some interfering substance or group of substances or a solvent. In the column-switching or 'heart-cutting' techniques, a fraction eluted from one column is led to another column for additional separation.

Packed columns are used as precolumns, because of their high sample capacity, and prevent overloading of the capillary columns, used for the separation because of their high separation efficiency. Column switching is achieved by means of valves set between the two columns and at the beginning and end of the individual columns. It is thus possible to cut out the solvent in the case of identification of phosphorus ester in a wheat flour sample. Another example is the separation of a fraction of interest from a complex mixture, e.g. the methyl esters of oleic acid, linoleic acid and linolenic acid from a mixture of methyl esters of fatty acids [597–600].

One component of a mixture, after separation on the first column, may be transferred to a reactor where some chemical change of the substance takes place (photolysis, electron capture, catalysed reductive dehalogenation etc.) The reaction product (plus residual parent compound) is then flushed into a second column for separation and identification [601–603]. Column bleeding might produce a problem with this technique [604].

The use of column-switching techniques for analysis of complex mixtures, e.g. pesticides and PCBs, has been reviewed [605-607].

A different approach to combining the separation power of two or more stationary phases is the use of mixed-bed columns and homogeneous mixed-phase columns [608].

5.5.5 Bleeding of septa and columns

"Little has been published about ageing of chromatographic columns in general" [608, 609]. Column bleeding will degrade the performance of several types of detectors and severely limit that of gc–ms. It influences the activity of the column and hence the retention data and sensitivity [610–612]. Information on the bleeding of Apiezon L, Carbowax 20 M and OV-101 has been reported [613]. For gc–ms a few phases are preferred because of freedom from bleeding [614]: dimethylsilicones (e.g. OV-101, SP 2100); 50% phenyl-50% methyl silicone (e.g. OV 17, SP 2250); 50% 3-cyanopropyl–50% phenylmethyl silicone (e.g. Silar 5 CP, SP 2300).

The septum of the injection port may bleed and give rise to formation of ghost peaks [615–617]. Ten different types of septa have been tested. One rubbery Teflon septum gave the least bleeding, while all other silicone rubber septa proved less stable [618].

5.5.6 Injection techniques in gas chromatography

Injection is most often done with a syringe. An alternative is use of a sample loop. Solid samples can also be used. Injection can be a rather critical step, however, and several techniques have been proposed.

In high-resolution capillary gc, stream splitting is used to prevent the overloading of the columns that would occur on injection of large volumes. Unfortunately, this results in a considerable loss of the sample, which has often been obtained only after expenditure of much effort on separation and purification. There is also necessarily a loss in sensitivity. To overcome these problems various suggestions have been made, e.g. use of a splitless technique [619, 620] or the use of a 'moving needle system' [621]. A maximum volume of about 5 μl can be handled by these systems. Another suggestion allows injection of volumes up to 0.25 ml [622]. The injector is adjusted to a temperature slightly higher than the boiling point of the solvent. The sample is injected within 40 secs and a valve is then opened which permits flushing of the solvent into a 600:1 stream-splitting device. The sample constituents remain in the glass insert of the injection port. After a few seconds the splitter valve is closed. The injection port is then heated to volatilize the trace components, which are directed to the column. It is a prerequisite, though, that there is no co-distillation of the traces with the solvent.

5.5.7 On-column derivative-formation techniques

Substances which lack the necessary volatility are converted into more volatile derivatives. This can be done in a separate step before injection or 'on-column' by simultaneously injecting the sample solution plus the derivative-forming reagent. As the use of derivatives increases resolution (since less adsorption and retention occurs) these techniques have become very important for the analysis of complex mixtures, e.g. the investigation of narcotic drugs. Furthermore, the analysis time will be considerably shortened, and such economic methods are therefore preferable [623-625]. On-column derivative formation has been suggested for analysis of narcotic drugs [626, 627] to overcome chemical and organizational problems; 1 μl of a silylating reagent is drawn into the syringe, followed by 1 μl of ethyl acetate solution of the drug(s) (250 ng) and the mixture is injected into the gas chromatograph. An example of this technique is demonstrated in Fig. 5.12.

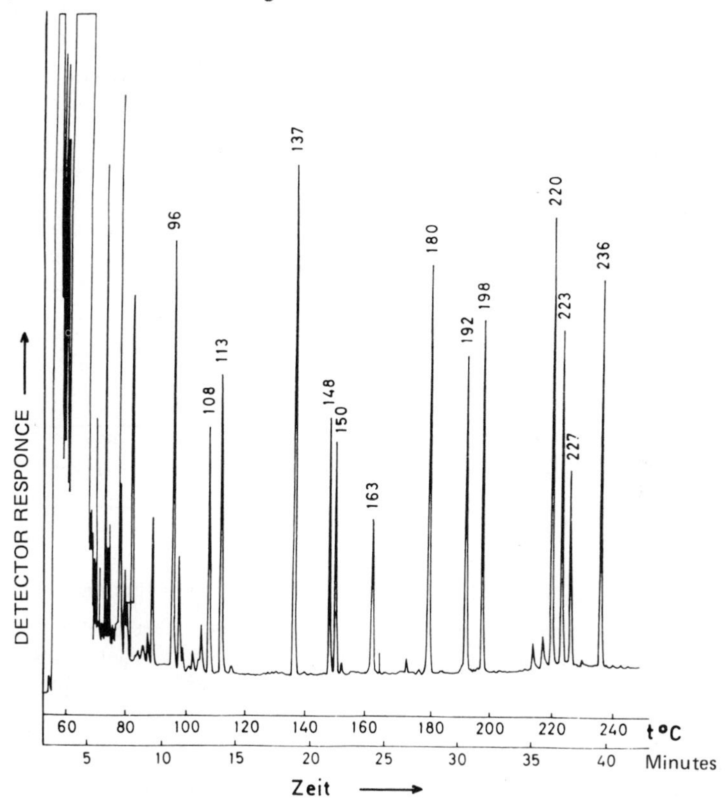

Fig. 5.12 – Chromatogram obtained after flash-heater derivative formation by injecting 1 μl of BSA together with 1 μl of an ethyl acetate solution containing 250 ng of each of the following components: 96 = amphetamine; 108 = ephedrine; 113 = phenmetrazine; 137 = hexadecane; 148 = methyl phenidate; 150 = pethidine; 163 = caffeine; 180 = eicosane; 192 = methadone; 198 = cocaine; 220 = codeine; 223 = ethylmorphine; 227 = morphine; 236 = heroin. Column temperature: 50°C at start, programmed at 5°C/min. (Reproduced from [626], by permission of the copyright holders, Elsevier Science Publishing Co., Amsterdam).

5.5.8 Pyrolysis

Pyrolysis is a sampling technique which is often used with polymer matrices. They are flash-heated to set free any volatile material, which is then analysed by gc. Very often, traces of monomers in polymers are determined in this way [628, 629]. Tryptophan has been determined in proteins [630], saccharine in food [631] and paraquat in river water [632], and mould fungus has been characterized in this way [633]. With ms as the method of detection, capronium chloride in biological material has been determined after pyrolytic decomposition [634]. Bacteria and cells from biological material have been identified [635].

5.5.9 Fraction collection

After separation, the traces are either measured directly (on-line) or collected and analysed separately. Collection of traces after gc separations is much more problematic than with liquid chromatography, but several methods are available:

(1) trapping of the total volume of a fraction;
(2) trapping of the trace constituent in condensed form by
  (*a*) freezing,
  (*b*) passing through an absorption solution,
  (*c*) adsorption on charcoal, molecular sieves, etc.,
  (*d*) co-condensation (or 'raining out').

The simplest way to collect a fraction after gc separation is by eluting all of it into an evacuated vessel. Such a discontinuous operation is not optimal, however, as many receivers are needed, and these often need to be heated to prevent partial condensation of the collected effluent. The chromatographic process has to be stopped between collection of the fractions, to permit change-over of receivers.

Trapping of the trace constituent by condensation is usually tried, if possible. The efficiency of trapping by freezing decreases with decreasing sample mass. It can be calculated that a sample of 100 μg of benzene can be trapped with nearly 100% efficiency at −78°C whereas 1 μg would not be collected at all [636]. Furthermore, many trace constituents give non-ideal behaviour if they form aerosols on cooling. The effect is even more pronounced with drastic freezing, as more condensation nuclei are formed in less time, with the consequence that the aerosol particles become smaller and even more difficult to isolate from the gas flow.

Thus additional aids have been suggested, such as filter pads or a cooled absorption liquid. Figure 5.13 shows an example in which the gas from the chromatograph outlet is passed into a capillary which contains 2 μl of carbon tetrachloride as collection aid. The trapping solution containing the trace compound is then transferred to a thin-layer chromatographic plate.

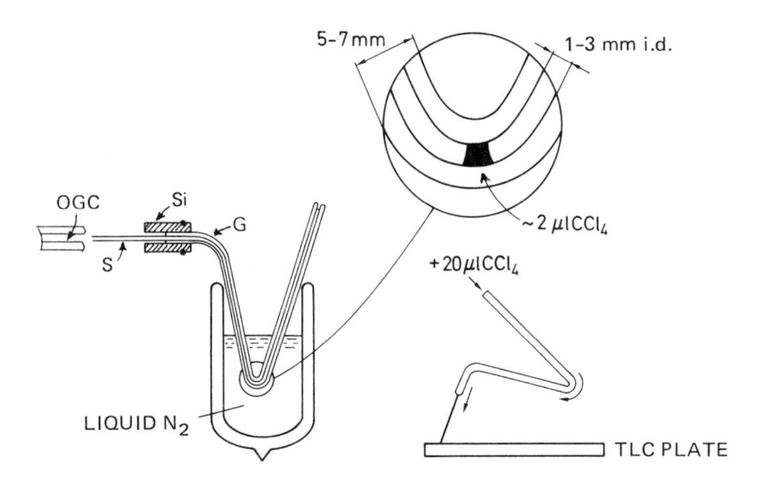

Fig. 5.13 – Micropreparative collection system and its application. *OGC*, outlet of the gas chromatograph; *S*, metal capillary sealed into the glass capillary *G; Si*, silicone rubber tube. (Reproduced from L. S. Ettre and W. H. McFadden (eds.), *Ancillary Techniques of Gas Chromatography*, p. 319, by permission of the copyright holders, Wiley-Interscience).

Powdered potassium bromide or a cooled potassium bromide plate may serve as a target for condensation of traces (10 μg) which are then investigated by infrared spectroscopy [637].

Good recoveries – even of ng amounts – have been reported [638] for use of a cooled microcolumn filled with the same column material as the analytical column, and connected to its exit. The trace material thus trapped was then removed either by elution or heating and injected into another gc system for further investigation.

'Raining out' (co-condensation) gives good recoveries. The most elegant method is to use argon as carrier gas, and a trap cooled with liquid nitrogen, where the carrier gas is condensed. Any trace present is 'co-precipitated' with the droplets of liquid argon formed. On warming, the argon is volatilized, leaving most of the trace constituent behind, ready for further analysis [639].

An analogous method feeds the vapour or a solvent into the carrier gas stream after this has left the column [81]. The mixture is then cooled and the condensing solvent collects condensable trace matter with about 80-90% recovery (see Figs. 5.14 and 5.15). Water, acetone and dichloromethane are good solvents for this purpose.

5.5.10 Carbon skeleton technique

In this technique the sample components are catalytically hydrogenated in a precolumn before gc analysis. Hydrogen is used as both reducing agent and

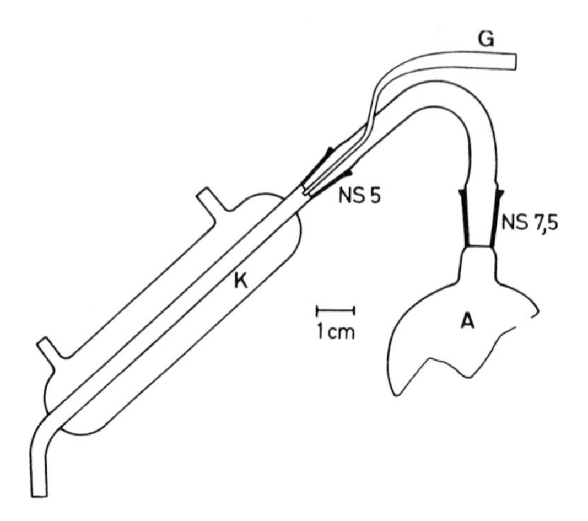

Fig. 5.14 – Equipment for collection of a gas chromatographic fraction by 'co-condensation'. Reproduced from K. Beyermann, *Z. Anal. Chem.*, **230,** 414 (1967) by permission of the copyright holders, Springer Verlag.

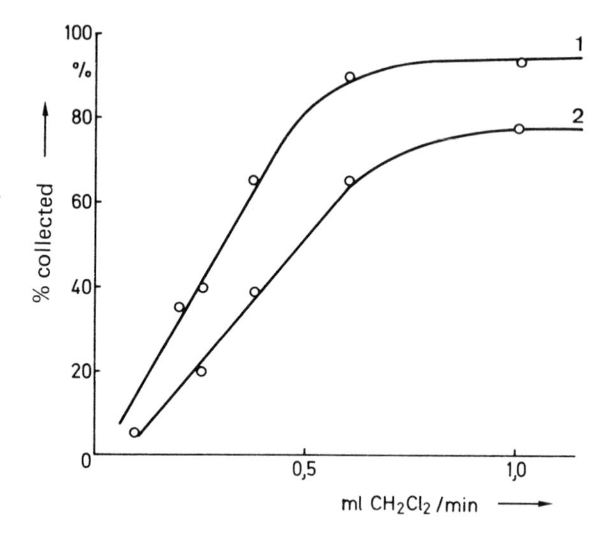

Fig. 5.15 – Recovery of traces in collection experiments with the equipment shown in Fig. 5.14: (1) 0.4 μg of anthracene; (2) 0.1 μg of benzpyrene. (Reproduced from [81] by permission of the copyright holders, Springer Verlag).

carrier gas. Platinum or palladium on a carrier serves as hydrogenation catalyst. The chromatograms are very often greatly simplified, as many compounds with different functional groups and the same carbon skeleton yield the same parent hydrocarbon [640–648]. Equipment for this technique is commercially available. The carbon skeleton method has proved very useful in the analysis of complex mixtures, e.g. Arochlor (see Fig. 5.16) [646].

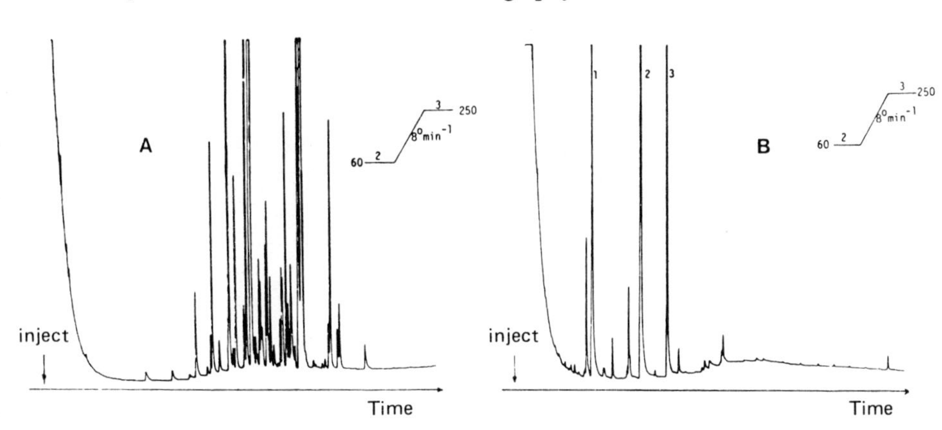

Fig. 5.16 – Capillary GC traces of a mixture of Arochlor 1248 (10 ppm), D88 (a PCN) (10 ppm), aldrin (4.75 ppm), dieldrin (5 ppm), heptachlor (5.25 ppm), p,p′-DDT (4.25 ppm), p,p′-DDE (4.75 ppm) and p,p′-TDE (4.50 ppm) before catalysis (A) and after catalysis (B). Peaks: 1 = naphthalene, 2 = biphenyl, 3 = diphenylethane. Injection, 1 μl. (Reproduced from [646], by permission of the copyright holders, Elsevier Science Publishing Co., Amsterdam).

Several approaches were used in the Arochlor work [646]. In one the catalyst was placed at the top of the gc column, within the injection port (see Fig. 5.17). To differentiate chlorinated and dechlorinated mixtures on the same column, a twin-oven arrangement was designed (see Fig. 5.18).

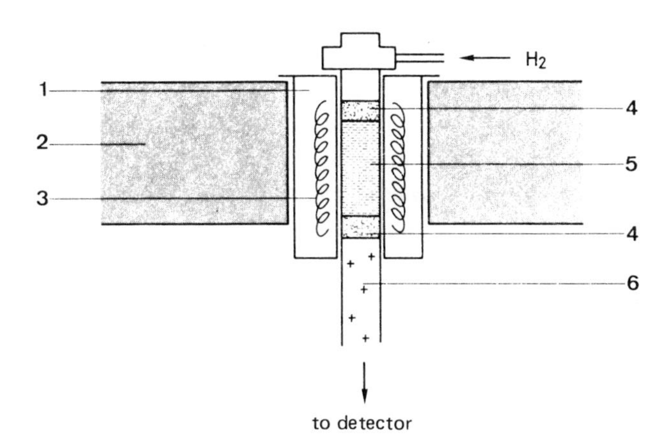

Fig. 5.17 – Arrangement for on-column catalytic hydrodechlorination using injection-point heater as reactor oven. 1 = injection block; 2 = gas chromatograph oven wall; 3 = injection block water; 4 = glass wool; 5 = catalyst; 6 = column packing. (Reproduced from [646], by permission of the copyright holders, Elsevier Science Publishing Co., Amsterdam).

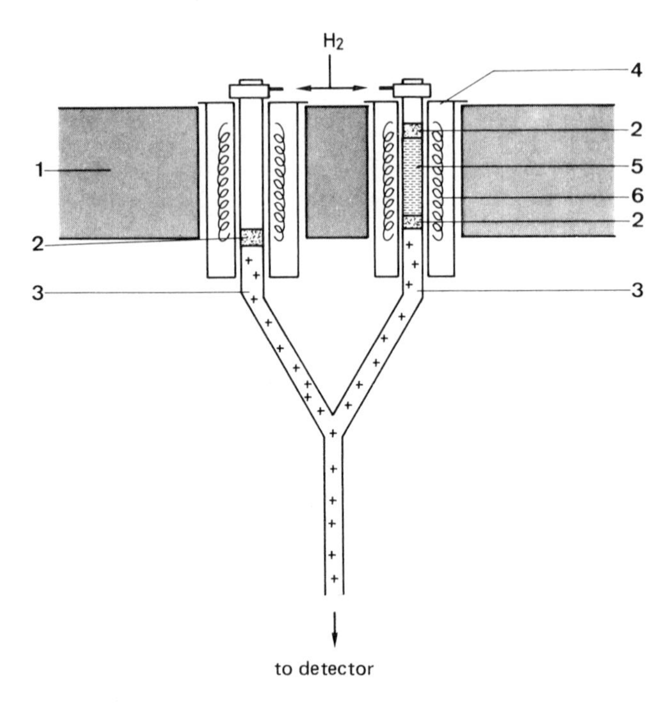

Fig. 5.18 – Twin oven arrangement for with and without catalysis determination of hydrocarbons such as naphthalene and biphenyl. 1 = gas chromatograph oven wall; 2 = glass wool; 3 = column packing; 4 = injection block; 5 = catalyst; 6 = injection block heater. (Reproduced from [646], by permission of the copyright holders, Elsevier Science Publishing Co., Amsterdam).

The direct use of the catalyst 'on-column' necessitates use of hydrogen as carrier gas. If another carrier gas is more useful or the potential explosive danger of hydrogen is to be minimized, a precolumn (off-line) can be used for hydrodechlorination (see Fig. 5.19). The products are trapped and injected afterwards.

5.6 ZONE MELTING

Zone melting is based on the properties of a molten liquid in contact with its crystals at the melting point. Traces are distributed between the molten phase and the solid phase. A distribution coefficient is defined, which is unity in unfavourable cases, with no difference in concentration between the two phases. With a favourable distribution coefficient, however, repeated melting and freezing in a zone-melting apparatus gives enrichment of the traces and purification of the major component (which is often the primary purpose). The system has also the advantages that it is closed, and no reagent is added, so risk of contamination is reduced.

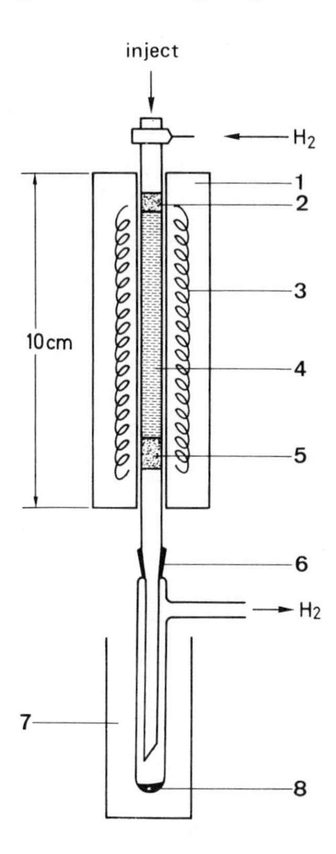

Fig. 5.19 – Catalysis oven and trap arrangement for hydrodechlorination of samples (≈ 50 μl) prior to analysis by capillary GC. 1 = catalysis oven block; 2 = glass wool; 3 = catalysis oven heater; 4 = catalyst; 5 = glass wool; 6 = ground glass joint; 7 = cold trap; 8 = catalysed sample. (Reproduced from [646], by permission of the copyright holders, Elsevier Science Publishing Co., Amsterdam).

Micro zone-melting has been developed [649-651]. The method has been used to concentrate traces of quinones or aldehydes from dilute solutions: 0.05 ppm of anthracene in phenanthrene has been recovered with 95% yield [652].

5.7 PRECIPITATION AND CO-PRECIPITATION

In organic analysis precipitation and co-precipitation are rarely used, owing to the relatively high solubility of organic compounds. Occasionally, though, co-precipitation offers an alternative to other separation methods. Thus ng-amounts of amines can be co-precipitated with creatinine tetraphenylborate from 100 ml of solution, with high yield [653], Barbiturates have been co-precipitated with mercuric barbiturate [654, 655].

5.8 SEPARATION WITH MEMBRANES

In dialysis, molecules of higher molecular weight are separated from those of lower m.w. by differences in their permeation through membranes. Single-step dialysis and multistep equipment based on the counter-current principle have been described [656]. Dialysis is very often used as the major separation method in automated equipment of the AutoAnalyzer® type.

Examples of the use of dialysis are rare, although some potential seems to exist for organic trace analysis. Mycotoxins in concentrations of 3–4000 ppb in animal feedstuffs have been separated in a membrane clean-up [657]. Ochratoxin A has been determined at the 1-μg/kg level in pig kidney tissue by using an enzymatic digestion concurrently with dialysis from water–methane [658]. For isolation of metabolites and conjugates of newly developed drugs dialysis is suggested – together with partition methods and chromatography – to separate the metabolites for structure determination (by ms) [659].

Dialysis is one of the methods of choice for separating protein-bound fractions from non-bound fractions, e.g. in blood after injection of a drug that exhibits protein binding. Other methods for the determination of the free fraction are ultrafiltration or gel filtration. These methods have been compared for the determination of 300 ng/ml of disopyramide [660].

Such methods of separation of certain species ('speciation') will become more and more important, since non-bound components have very different bioavailability and pharmocological and toxicological activity. Such speciation analysis will have to be performed at trace levels.

Ion-exchange membranes combine the properties of ion-exchange resins and diffusion membranes. They provide interesting possibilities for separation of charged species from oppositely charged species or from non-charged molecules. Ion-exchange membranes are much thicker than dialysis membranes. Diffusion is consequently much slower and of sufficient speed only for small ions. The method has its merits for desalination of solutions. The speed of transfer of ions through the barrier can be increased considerably by application of a D.C. voltage. The advantage of ion-exchange membranes over ion-exchange resins is their considerably smaller surface area, which permits better control of adsorption of non-electrolytes. The method has proved useful for eliminating inorganic ionic constituents from swimming-pool water. The urea concentration (a few ppm) was determinable afterwards by micro infrared reflectance spectrometry [661]. The desalination of ocean water for the determination of monosaccharides at the pg-level is achieved by the same principle [662].

'Pervaporation' is a technique in which an aqueous solution is separated from a vacuum by a membrane. Some volatile organic compounds diffuse through this membrane into the vacuum compartment, which is directly connected to the ionization source of a mass spectrometer [663].

5.9 DIASOLYSIS

Diasolysis is a combination of liquid-liquid distribution and dialysis, obtained by separating two immiscible liquid phases by a membrane which is semipermeable to one solute [664]. The use of silicone rubber membranes has been recommended [665]. They swell when treated with suitable organic solvents and such a swollen membrane forms a barrier to compounds of higher molecular weight. If the swelling agent is non-polar, polar substances are not dissolved within the membrane, which thus becomes impenetrable to these compounds. It is possible to apply these systems to solutions which in liquid-liquid distribution would form emulsions on shaking. With diasolysis no shaking is necessary, as the extractant is slowly pumped through the interior of a silicone rubber tube, the walls of which act as the membrane although 0.5 mm thick. In this way 50 ng of DDT can be recovered from 1 litre of water with a yield of 85%, by diasolysis into a few ml of toluene. Other pesticides have been isolated from a slurry of spinach to which the radioactively labelled compounds had been added.

5.10 'ADSUBBLE' METHODS AND 'SOLVENT SUBLATION' TECHNIQUES

Adsorptive bubble methods (abbreviated to 'adsubble') comprise the various ways of separating dissolved or suspended materials by means of adsorption or attachment to the surface of bubbles rising through the liquid which contains the species to be separated [666].

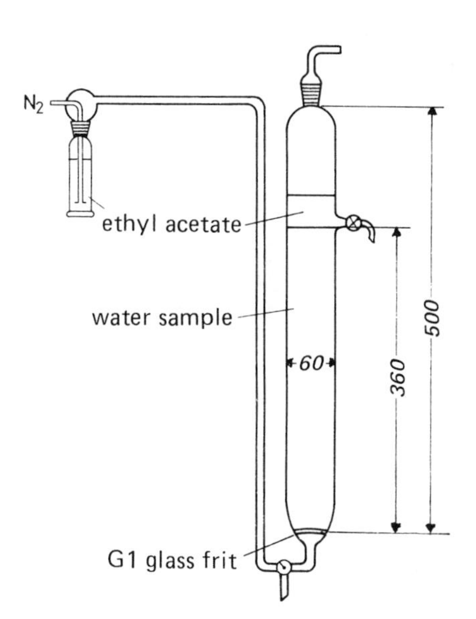

Fig. 5.20 – Equipment for solvent sublation technique [667]. (From *Tenside-Detergents,* Vol. 8 (1971), p. 61, Hanser Publishers, Munich).

In the trace analysis of surface-active compounds 'solvent sublation' is used. Air is passed through a two-phase system of ethyl acetate and water [667], and surface-active components concentrate at the phase boundary. They are separated by collecting the ethyl acetate (see Fig. 5.20). By tlc and micro infrared spectrometry, as little as 0.2 ppb of detergents can then be detected [668]. Similar applications have been described [669–671].

5.11 ELECTROPHORESIS

In electrophoresis, charged particles or molecules are separated in an electric field, according to their mobility. The method is very useful in differentiation of macromolecular ions and even of whole cells.

Isotachophoresis is an electrophoretic technique in which sample components are separated according to differences in the net mobility. Usually a polyacrylamide gel in a column is used as carrier. Spacer ions are added to the analyte mixture. Sample components and spacers are dissolved in the terminal electrolyte, and the leading electrolyte is initially applied through the polyacrylamide gel. On application of the electric field separation of the sample constituents takes place by migration between the leading and terminal electrolytes. After completion of the separation the separated zones and spacer ions migrate with the same velocity in direct contact with each other into an elution chamber where they are detected. Isotachophoresis gives a high resolving power because of its concentrating effect. Zone boundaries are actively sharpened during the separation, and this prevents diffusional broadening [672–677] (see Fig. 5.21).

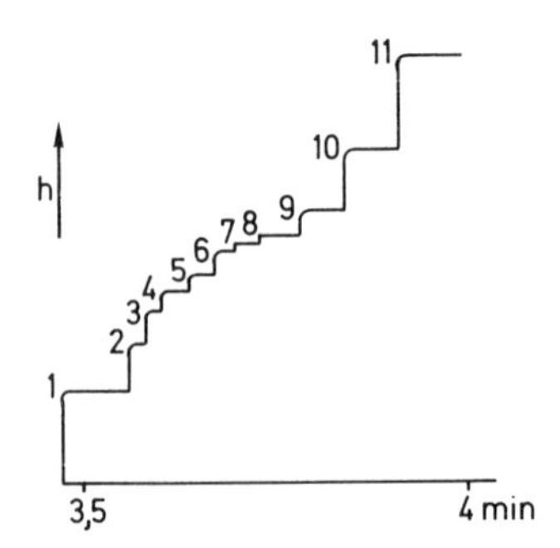

Fig. 5.21 – Example of an isotachophoretic separation of the acids from citrate cyclus. (Each acid was present in 2–4 nmole quantity. 1, chloride, 2, oxalacetate; 3, acetoxalate; 4, fumarate; 5, ketoglutarate; 6, citrate; 7, malate; 8, isocitrate; 9, lactate; 10, succinate; 11, acetate (as terminal electrolyte). Signal height h is plotted as a function of time. (Adapted from [678], by permission of the copyright holders, Vogel-Verlag, Würzburg).

Few applications in organic trace analysis have been reported so far. Aspartic acid, asparagin, glutamic acid and glutamine have been isolated in 5 μl of serum filtrate [679]; 50 ng of valproic acid [680], EDTA (with a limit of detection of 10 ppm) [681], and histamine (50 ppm) [682] have been determined. Isotachophoresis has been used to separate paraquat and diquat from other components of plant extracts before mass spectrometry [683].

5.12 PLASMA ELECTROPHORESIS (GASEOUS ELECTROPHORESIS)

Plasma chromatography of gaseous electrophoresis [684] was first described in 1970 [685]. The method is based on the reaction between a trace molecule and a reactant ion, e.g. $(H_2O)_2H^+$ or $(H_2O)_2O^-$, generated at atmospheric pressure in the gas phase. The reactant ions are formed by direct ionization with a ^{63}Ni source, which emits β-radiation. The ion–molecule complexes formed drift into a separation compartment. In it a potential is applied between two grid electrodes. The ions formed are separated in the field, depending on several parameters. The drift time increases with molecular or ionic weight [686, 687], but also depends on the size and shape of the ions [688, 689]. Ion–molecule complexes arrive at the spectrometer detector and a 'plasmagram' is recorded. The equipment can be purchased [690]. Coupling with gas chromatography has been suggested. Trace levels of organic species outgassed from epoxy polymers have been determined [691].

5.13 CHECKING THE EFFICIENCY OF SEPARATION STEPS BY RADIOCHEMICAL METHODS

The radiochemical tracer method is invaluable for checking separation procedures. The specific activities of ^{14}C-labelled compounds permit the detection of ng amounts. Their behaviour in separation operations and other analytical steps can be studied quite easily.

Any labelled compounds purchased should be checked for chemical identity. During storage the compounds can decompose by radiolysis or some other change in the chemical nature of the substance may have happened. Use of results without due precautions and critical appraisal may lead to misinterpretations. The check for purity is best done by thin-layer chromatography of the labelled compound mixed with some inactive authentic material; the radioactivity should be found at the same place as the authentic substance.

For estimating losses during analytical separations the isotope dilution method is extremely useful. To the mixture of analytes is added a small amount (a few ng) of a labelled specimen of the compound to be determined in the mixture. This compound is separated as usual. As the labelled compound behaves exactly like the non-labelled material the final yield can be calculated from radioactivity measurements. A correction for losses is made quite simply. It is a prerequisite, though, that the activity must become homogeneously distributed within the sample directly after addition of the tracer. With multiphase systems, e.g. soil or biological tissue, it can become difficult to guarantee homogeneity of the distribution, which is the most critical step of the entire procedure.

The isotope dilution method requires two final measurements:

(1) determination of the activity to calculate the yield;
(2) determination of the mass of the isolated mixture of labelled and non-labelled compound.

In some instances the second determination is difficult to do with sufficient sensitivity. The double isotope dilution technique can be tried in such cases. It has been successfully used (amongst other applications) for steroid analysis. For this purpose a serum sample is acetylated with ^{14}C-labelled acetic anhydride. Of course, many compounds beside steroids form acetyl derivatives in that way, and must be separated. Losses will occur during this operation, and to estimate these a tritium-labelled steroid acetyl ester is added. After separation, measurement of the ^{3}H-content permits correction for the yield, and the ^{14}C-activity is a measure of the amount of steroid present. Extremely sensitive determinations are possible.

5.14 TREATMENT OF SAMPLE MATERIALS TO ALTER THEIR SEPARATION CHARACTERISTICS

5.14.1 General considerations

Many chemical modifications are recommended for influencing the behaviour of the sample components during separation. For example, volatility can be changed, mainly to increase it to improve gas chromatographic separations. Solubility may be altered, which will affect liquid–liquid distribution processes and adsorption mechanisms.

A change in the physico-chemical properties of a given analyte is based on alterations of

(1) its polarity,
(2) its ionic nature (or absence of ionic features),
(3) the molecular weight and dimensions,
(4) the molecular shape.

Molecular weight, dimensions and shape are difficult to change. Either fragmentation (and consequent loss of individual characteristics) or attachment of another group (to form a bigger molecule) is possible. However, if the same group is added to all the analytes, the resulting molecular weights will be relatively closer together, and the chromatographic behaviour more similar. Separation can even become more difficult with this strategy. On the other hand, solubility or mobility are influenced by such an increase in molecular dimensions, and new group separations may become possible.

Most alterations aim at a change of polarity or ionic nature. The latter is changed simply by variation in pH if there are ionic groups within the analyte. This principle has long been used in very effective separation schemes in partition, ion-exchange etc.

Polarity is more difficult to alter. Mostly it is done by masking the polar group(s) by reaction to give a less polar derivative. In about 16% of all publications on organic trace analysis during the years 1975–1980 such a derivative-formation step is mentioned, which indicates the importance of this procedure. In about 10% of all publications the object was to improve the volatility of trace compounds to enable use of gas chromatographic separation. Other purposes are the introduction of a chromophore – often absorbing in the ultraviolet region – or the attachment of an electrophore for subsequent electrochemical determination.

Derivative-formation techniques and reagents are described in several handbooks [692–697].

5.14.2 Continuous and discontinuous techniques for forming derivatives

In principle, chemical modification of a compound can be done at various stages in the analytical procedure:

(1) pre-column (i.e. before separation of the trace compounds); this is done on-line, continuously, e.g. as part of a chromatographic system, or off-line, in a batch procedure, separately from the chromatographic step;
(2) on-column (see Section 5.5.7, p. 174);
(3) post-column (i.e. after the trace compounds have been separated); again, the reaction can be performed on-line (in a reactor included in a chromatographic system) or off-line (after fraction collection).

Each approach has its own advantages and disadvantages [698, 699]. One advantage of pre-column techniques lies in the fact that the reaction is not dependent on the mobile phase [700]. Another is that the kinetics of the reaction are not critical as long as they are reproducible. A large excess of reagent can often be used, and may be eliminated during the following chromatography or before it. On the negative side, there is greater risk of artefact formation. Post-column reactions are not very widely used, so far. The critical problem in these is the interdependence between the mobile phase and the reaction medium. The design of post-column reactors is important. There are tubular or capillary reactors, bed reactors and air-segmented streams such as are used in AutoAnalyzers®. Band broadening is a critical complication, but can be reduced through improved technical design of the reactors. Fluorescent markers have been introduced, which do not exhibit fluorescence by themselves [701, 702]. Electroanalytical detection has been used [703, 704].

A typical post-column system is shown in Fig. 5.22 [705]. It was designed for hplc separation of carbamates. All the *N*-methylcarbamates are hydrolysed on-line with sodium hydroxide in a 3-m glass coil heated to 100°C, to form methylamine, which is then reacted with *o*-phthalaldehyde and 2-mercapto-ethanol to form a fluorescent product. Nanogram and subnanogram quantities can be determined. Of 235 pesticides tested only *N*-methylcarbamates reacted.

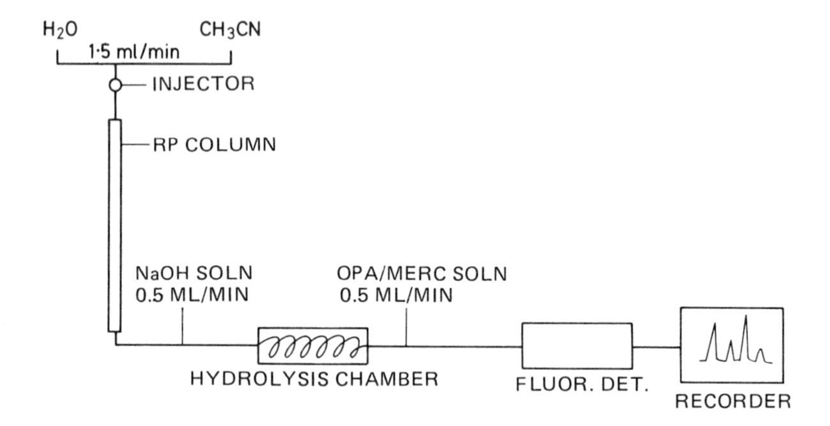

Fig. 5.22 – Post-column fluorometric system for hplc. OPA = *o*-phthalaldehyde; MERC = 2-mercaptoethanol. Hydrolysis chamber: 3-m coil, 100°C. Fluorometric detector: excitation wavelength 340 nm, emission wavelength 455 nm. (Reproduced from [705], by permission of the copyright holders, Elsevier Science Publishing Co., Amsterdam).

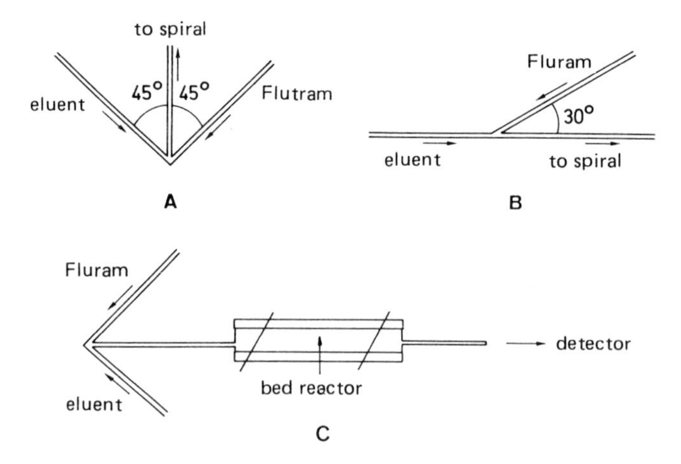

Fig. 5.23 – A, B, tubular reactors; C, bed reactor [712]. (Reproduced from the *Journal of Chromatographic Science*, by permission of the copyright holders, Preston Publications Inc.).

The air-segmented reactor is widely known through its use in AutoAnalyzers® [706]. It is based on the segmentation of flowing liquid streams with air bubbles introduced into them at certain intervals. This reduces axial diffusion of the sample zones and minimizes band broadening, but there is some carry-over from one segment to another. The method is useful for relatively slow post-column reactions (reaction times greater than 5 min.)

Tubular reactors are recommended for faster reactions (completed within 30 sec or less). The principle has been applied successfully to reactions with fluorescamine ('Fluram') and *o*-phthalaldehyde, which are both selective reagents for primary amines [707, 708]. The reagent is added to the eluate containing the amine [702].

For reactions requiring reaction times between about 30 seconds and several minutes, bed reactors are preferred [709–711] (see Fig. 5.23).

5.14.3 Examples of the application of derivative formation

Table 5.12 gives a survey which indicates the great variety of reagents used. Their frequency of appearance in this table is an indicator of their importance (and popularity among analysts). The working range of the method is also given in Table 5.12, since it depends on the determination step.

Diazomethane – although highly toxic and a potential carcinogen – is often used to make derivatives of traces. Methods for preparation of radioactive diazomethane have been described [836].

Photolysis is mainly suggested for chemical change of halogen-containing compounds, e.g. pesticides (Table 5.13).

A few examples of on-column reactions are given in Table 5.14.

A review of these alkylation techniques [861] indicates that the main field of application is to acidic compounds. Problems arise from isomerization of heat-sensitive compounds on injection, e.g. phenobarbital to form two compounds. Polyenoic fatty acids can be isomerized to conjugated polyenoic fatty acids. On the other hand, the chemical degradation of cocaine has proved rather advantageous for its qualitative analysis [862].

PCBs form a group of environmental contaminants with hundreds of members. The gas chromatographic determination is accordingly very difficult. Techniques have therefore been suggested which lead towards one compound, having the carbon skeleton of all PCBs, by dehalogenation [863] or by formation of the totally halogenated compound [864], e.g. in perchlorination by use of PCl_5 as chlorinating reagent.

Bromination is achieved (e.g. of aromatic amines) by adding aqueous bromine solution to the trace compounds at room temperature. After 60 min the excess of bromine is reduced with sulphurous acid. The bromination products are extracted and investigated by gas chromatography. A few nanograms of aromatic amines have been determined in 1 litre of water [865].

Table 5.12 – Examples of derivative-formation reactions used in organic trace analysis.

Reagent	Substrate (trace)	Concentration or amount	Purpose	Reference
acetic anhydride–pyridine	chloramphenicol	ng	gc-ecd	713
	5-chloro-7-iodo-8-quinolinol		gc	714
	hexachlorophene	ppm	gc-ecd	715
acetic anhydride	chloro-, nitrophenols	μg/l.	gc	716
7-acetoxy-4-bromomethylcoumarin	carboxylic acids		fluoresc.	717
bromo-2,3,4,5,6-pentafluorotoluene	bentazon	ng	gc	718
ethyl iodide	Ag-salts of carboxylic acids	μg	gc-ecd	719
butylamine	benzaldehyde, tolualdehyde	ng	gc	720
benzyl bromide	tetra-n-butylammonium salts of carboxylic acids	ppb	gc	721
butyl iodide	Ag-salts of sulphonic acids	ppb	gc	721
butaneboric acid	pyridoxine	10 ng	gc	722
bromomethylpentafluorobenzene	Cl-methylphenoxy acids	ng	gc	723
bromomethylmethoxycoumarin	carboxylic acids	ng	fluoresc.	724
	2,4-D	ng	fluoresc.	725
butanol	nitriloacetic acid	ppb	gc-ms-fid	726
quinoline-8-sulphonic acid chloride	putrescine, spermine, spermidin	ng	hplc	727
	aliphatic biogenic amines	ng	hplc	728
dansyl chloride	oestrogens	ppb	tlc-fluoresc.	729
	amines, phenolic amines, amino-acids	ng	tlc-fluoresc.	730
	labetol diastereomers		ms	731
	herbicides with amino groups		tlc-fluoresc.	732
	phenols (after cleavage from organophosphates)	ng	tlc-fluoresc.	733
	amines, phenols, alcohols		hplc-fluoresc.	734
2,4-dinitrophenylhydrazine	carbonyl compounds		uv	735
3,5-diaminobenzoic acid	turimycin (macrolid antibiotics)		fluoresc.	736
diazomethane	esters of phosphoric acid and thiophosphoric acid	ppb	gc	737
	hexachlorophene	ppb	gc-ecd	738, 739
	trichlorophenoxyacetic acid		gc-ecd	740
	5-phenyl-2,4-oxazolidinedione	ng	gc	741, 742
	trichloroacetic acid, oxalic acid	ppb	gc-ms	743
	2,4-D	ppb	gc-ms	744
	prostaglandins	ng	gc-ms	745

diazoethane	dopamine, norepinephrine	10 ng	gc	746
dicyclohexyl-(7-methoxycoumarin-4-yl) methylpseudourea	oxocarboxylic acids		fluoresc.	747
dimethylsulphate	2,4-D	ppb	gc-ecd	748
4-fluoro-3-nitrobenzotrifluoride	tranexamic acid	ng	gc-ecd	749
fluorescamine	histidine, histamine	ng	tlc-fluoresc.	750
	dopamine, norepinephrine		hplc-fluoresc.	751
flophemesyl chloride	alcohols	fg	gc-ecd	752
flophemesyl chloride	alcohols, *N*-nitrosodiethanolamine	low pg	gc	753
hexafluoracetylacetone	guanidines		gc-ecd	754
heptafluorobutyric anhydride	salsolinol	ng	gc-ecd	755
	lofepramine	ng	gc-ms	756
	morphine	ppb	gc-ecd	757
	nomifensine (psych. drug)	ppb	gc-ecd	758
	carbofuran	ppb	gc-ms	759
	ephedrines	ng	gc-ecd	760
	thyroid hormones	pg	gc-ms	761
	androsterone	ppb	gc-ecd	762
	carbofuram, diethylstilboestrol	ppb	gc-ecd	763
	50 amino-acids		gc-fid	764
	thyroxine, tri-iodothyronine	pg	gc-ecd	765
	carbamate insecticides	ng	gc-coul. detn.	766
	pilocarpine	ng	gc-ecd	767
methyl iodide	oryzalin	ppb	gc-ecd	768
	methyl thiouracil	pg	gc-ecd	769
	urea and carbamate herbicides	ppb	gc-ecd	770
	3-hydroxy-l-nitrosopyrrolidine		gc-ms	771
	cytokinins	ng	gc-ms	772
9-(methylaminomethyl)anthracene	isocyanates	0.1 $\mu g/m^3$	hplc	773
4-(6-methylbenzthiazol-2-yl) phenylisocyanate	alcohols, amines	100 pg	fluoresc.	774
7-chloro-4-nitrobenzo-2-oxo-1,3-diazole (NBD-chloride)	hydroxyphenylamines		tlc-fluoresc.	775
	amphetamine, metamphetamine	pg	tlc-fluoresc.	776
	tolbutamide, glibornuride	ppm	fluoresc.	777
	nitrosamines	ng	hplc-fluoresc.	778
	N-nitrosoproline	ng	hplc-fluoresc.	779

Table 5.12 – *continued*

Reagent	Substrate (trace)	Concentration or amount	Purpose	Reference
4-nitrobenzoyl chloride	digitalis glycosides		chromat.-uv	780, 781
	amphetamines	ng	hplc-uv	782
	saccharides	ng	hplc-uv	783
o-phthalaldehyde	dopa, dopamine, norepinephrine	ng	hplc-fluor.	784
	histamine	ppm	fluoresc.	785
pentafluorobenzaldehyde	aliphatic primary amines	pg	gc-ecd	786
	phenylpropylamine		gc-ecd	787
pentafluorobenzoyl chloride	*o*-phenylphenol	ppb	gc-ecd	788
pentafluorobenzyl bromide	captan	ppb	gc-ecd	789
	phenoxyalkanoic acid herbicides		gc	790
	carboxylic acids, phenols	ppb	gc-ecd	791
	cocaine, benzoylecgonine	ppb	gc-ecd	792
pentafluorobenzylhydroxylamine	keto steroids	ng	gc-ecd	793
	prostaglandins	ng	gc-ecd-hplc	794
phenyldiazomethane	aliphatic acids		gc-fid	795
phenyl isothiocyanate	amines		hplc	796
pentafluoropropionic anhydride	diethylstilbeostrol	ppm	gc-ecd	797
	hydroxyindoleacetic acid		gc-ms	798
	meperidine, propoxyphene (narcotics)		gc	799
	fenfluramine, norfenfluramine	ng	gc-ecd	800
	hydroxyphenylacetic acids		gc-ecd	801
	3-methoxy-4-hydroxyphenylglycol	ppb	gc-ecd	802
butyl isocyanate	amines (from herbicides)	ng	gc-ms	803
phosphorus tribromide	ancymidol	ppb	gc-ecd	804
	fluridone (aquatic herbicide)	10 ppb	gc	805
silylating reagents	fatty acids	pg	gc-ms	806
	bufuralol	ng	gc-ecd	807
	acids		gc	808, 809
	indolalkylamines	pg	gc-ms	810
	gestagens, oestrogens	ng	gc-fid	811

	phenols		gc-fid	812, 813
	histamine, imidazolacetic acid	ng	gc-ms	814
	melatonine	ppt	gc-ms	815
	iodofenphos	ppb	gc-ecd	816
	noludar, persedon (narcotics)	ng	gc	817
	46 amino-acids		gc-ms	818
	2,4-D	ppm	gc-fid	819
	coprostanol, cholesterol	ng	gc-ms	820
	tetrahydrocannabinols	ng	gc-ms	821
	hydroxylated steroids		gc-ms	822
	derivatives of cholesterol	ng	ms	823
	aromatic carboxylic acids		gc-fid	824
tosyl chloride	secondary amines		gc-coul. detn.	825
2,2,2-trichloroethyl chloroformate	amines, primary, secondary	pg	gc	826
trifluoroethanol–BF_3	phthalic acids	pg	gc-ecd	827
trifluoracetic anhydride	16 aliphatic amines		gc	828
	aminochlorobenzophenones	ppb	gc-ms	829
	biogenic amines	ppb	gc-ms	830
	monosaccharides	pg	gc-ecd	831
	2,4-D	ppt	gc-ecd	832
	4-aminobutyric acid	pg	gc-ms-ecd	833
	N-hydroxy-2-fluorenylacetamide	pg	gc-ecd	834
	acids, phenols		gc	835

Table 5.13 – Photolysis or photodecomposition methods.

Compound	Method	Reference
PCBs	photoisomerization	837
	photodechlorination	838
	photolysis	839, 840
	TiO_2-photodechlorination	841–843
mirex	photolysis	839, 844
PBB*	photolysis	845, 846
endosulfan	photolysis	847
heptachlor	photolysis	848
polychloronaphthalenes	photolysis	849
TCDD	photolysis	850

*PPB = polybrominated biphenyls.

Table 5.14 – Examples of on-column methods.

Compound	Reagent	Reference
aromatic acids	trimethylphenylammonium acetate	851
barbiturates	trimethylphenylammonium acetate	852
S-containing carbamates	trimethylphenylammonium acetate	853
diphenylhydantoin, carbamazepine, primidone	trimethylphenylammonium acetate	854
anticonvulsives	trimethylphenylammonium acetate	855
anticonvulsives	tetramethylammonium hydroxide	856
carbamates	trimethylanilinium hydroxide	857
phenytoin	trimethylanilinium hydroxide	858
barbiturates	trimethylanilinium hydroxide	859
phenobarbital	tetraethylammonium hydroxide	860

By proper choice of derivative-formation technique the reaction can be improved. Dansylation is achieved by reaction in an aprotic solvent (dimethylformamide, acetonitrile, acetone, ethyl acetate) [866]. The reaction medium is prepared by heating potassium fluoride (which has been solubilized with 18-crown-6 in the solvent) together with the substrate compound. The non-solvated fluoride ions activate the hydrogen atoms of amino or hydroxy groups for dansylation, on heating for 3 hr at 65°C. No extractive separation of the reagent from the dansylation product is necessary. The solution can be chromatographed directly.

In ion-pair formation no covalent bonds are formed. The reaction cannot be called a derivative-formation method *in sensu strictu*. This important method is mentioned here, nevertheless (see Table 5.15).

Extractive alkylation is a mixed two-stage technique in which an organic compound – usually an acid – is partitioned into a solvent by the ion-pair technique. In a second step it is alkylated in the organic layer. Tetrabutylammonium, tetrapentylammonium or tetrahexylammonium ions are used as

Table 5.15 – Ion-pair formation as a tool in separation of organic trace compounds.

Trace compounds	Counter-ion	Purpose	Reference
carboxylic acids, phenols, sulphonic acids	protonated long-chain aliphatic amines	hplc	867
chlorimipramine		hplc	868
diminazene (reduced to 4-aminobenzamidine)		hplc	869
substituted pyridines	di(ethylhexyl) phosphate	chromatogr.	870
catecholamines	C_{10}–C_4-alkyl sulphates	chromatogr.	871
alkaloids	tetraphenylborate	precipitation	872
anionic surfactants	Bindschedlers Green	liq. liq. distr.	873
anionic surfactants	Methylene Blue	liq. liq. distr.	874
polyoxyethylene alkyl ether non-ionic surfactants	picrate	photometry	875
codeine	picrate	liq. liq. distr.	876
phenolic acids	quaternary ammonium	liq. liq. distr.	877
methylguanidine	hexanitrodiphenylamine	liq. liq. distr.	878
drugs (basic)	methanesulphonate	chromatogr.	879
drugs (acidic)	tetrabutylammonium	chromatogr.	
catecholamines	aliphatic sulphonates	chromatogr.	880
alkaloids	picrate	liq. liq. distr.	881
anionic surfactants	pentanesulphonate	hplc	882
8 guanidino compounds	hexanesulphonate	hplc	883
chlorpromazine	heptanesulphonate	hplc	884
norepinephrine, dopamine, serotonin	heptanesulphonate	hplc	885
apovincaminic acid	tetrabutylammonium	hplc	886
anions	tetrabutylammonium	hplc	887
tertiary amines, e.g. chlorpheniramine	coumarin-6-sulphonate	partition, fluoresc.	888
probenecid, metabolites	tetrapentylammonium	partition	889

counter-ions in the partition process. Methyl iodide in dichloromethane is often used as the alkylating reagent. Gas chromatography is used for the subsequent separation and determination. Examples are the determination of carboxylic acids [890–892], phenols [893–894], barbituric acids [895, 896], sulphonamides [897–899], 5-chloro-7-iodo-8-hydroxyquinoline [900], chlorthalidone [901], furosemide [902], hydrochlorothiazide [903] and pemoline [904].

Formation of charge-transfer complexes is also not a derivative-formation reaction in the strict sense. It can be useful for changing the chromatographic behaviour of aromatic hydrocarbons by the addition of caffeine, picric acid or 9-oxo-2-trinitrofluorene [905]; 1 ng of benzo(*a*)pyrene in 1 kg of food can be determined with recoveries of about 95% and good reproducibility in this way. In this case the behaviour in liquid–liquid distribution was influenced by complex formation with caffeine in formic acid solution [906]. Analogously, polycyclic aromatic hydrocarbons in concentrations of more than 0.2 ppm in vegetable oil have been separated after caffeine complexation [907].

Finally, a technique will be mentioned in which a group is reacted or more than one reaction is used in order to improve separation behaviour. For instance, aromatic nitro compounds have been isolated from air and reduced to anilines with zinc powder and hydrochloric acid, the amines then being reacted with heptafluorobutyric anhydride and the products analysed by gc [908]. *N*-Nitrosamines have been denitrosated, e.g. by irradiation at 365 nm, and the amines formed determined after dansylation [909, 910]. Phenylurea herbicides are hydrolysed by heating at 165°C to form anilines which are then converted into derivatives for gc determination [911, 912]. Traces of isocyanates are converted into amines [913]. Pentachlorophenol (extracted from sawdust) is oxidized to form chloranil by warming with concentrated nitric acid and the oxidation product is detected by tlc [914]. Groups are sometimes split from parent compounds, e.g. 4-nitroaniline from nicarbazin [915], or 4-bromo-2,5-dichlorophenol from bromophos [916]. 2,2-Dimethyl-2,3-dihydrobenzofuran-7-ol is produced from the thiocarbamidate compound carbofuran [917]. Piperazine is split from triforine [918]. From pemoline, fenozolone or thozalinone, a common hydrolysis product (i.e. 5-phenyloxazolidine-2,4-dione) can be split [919], thus enabling determination of all three compounds. Different esters from the same acid (in residue analysis) are hydrolysed and the acid is determined. Carbon disulphide can be split from dithiocarbamate fungicides and determined [920].

5.15 SAMPLE TRANSFER

Very often, material isolated during the separation scheme is to be transferred from one vessel to another, or at the end of a separation procedure the fraction isolated is to be introduced into a measuring system – a critical step which can become difficult with a few micrograms or even nanograms of solid samples.

Modern developments in instrumental analysis enable the inspection of small particles and areas about 1 μm in diameter by means of molecule-specific methods (MOLE, a laser-raman microprobe, or infrared microscopes). Application to investigation of nanogram amounts is possible, but these have to be transferred quantitatively into the equipment. There is still the missing link of such a transfer method.

Transfer is done (on-line) by pumping the solution of a gas sample into the vessel or to the place where the final determination takes place. An off-line transfer uses pipettes (see Fig. 5.24) or microsyringes. A transfer device with as small a surface area as possible is preferable. Therefore, the syringes usually used are less ideal than pipettes with almost spherical bulbs and hence optimal volume to surface area ratio. Fixed-volume pipettes (0.1 or 0.2 μl) are commercially available (CAMAG).

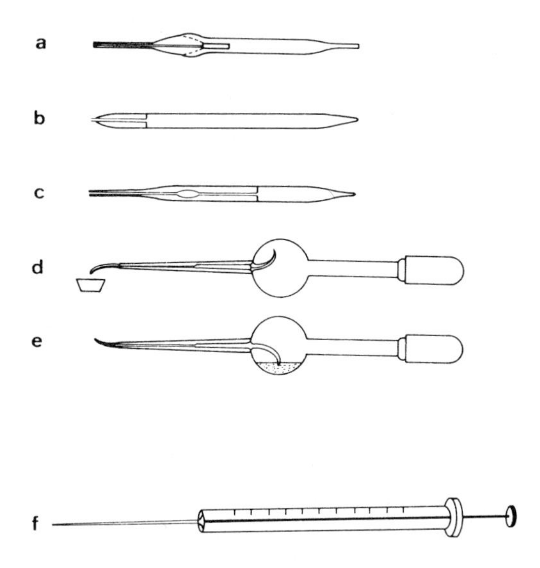

Fig. 5.24 – Transfer pipettes: (a) and (b) self-filling; (c) self-adjusting; (d) Grunbaum, used for transfer; (e) Grunbaum, used for dispensing; (f) Hamilton syringe-type.

Aliquots can be taken by use of micropipettes. Micro vessels for this purpose have been described which give the possibility of working with volumes of 0.1 ml with relative standard deviations of only a few per cent (see Fig. 5.25).

For transfer of solid particles fine tungsten or polyethylene needles are used [921, 922].

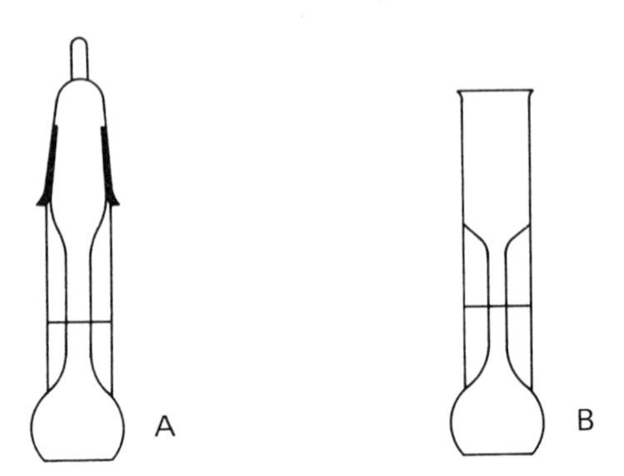

Fig. 5.25 – Micro standard flasks. A = standard flask, 0.2 ml, glass stoppered; B = standard flask, 0.2 ml.

As solid systems are difficult to handle, and solutions exhibit a surface tension that limits the size of droplets that can be formed, transfer of small amounts in the gas phase can be helpful, but quantitative recovery is difficult to guarantee. On the other hand, the method can be rather simple. For example 1-5 μg of DNBP has been transferred successfully after sublimation and condensation on a plane 2-mm diameter mirror for infrared reflectance spectrometry [923].

REFERENCES

[1] B. Stachel, K. Baetjer, M. Cetinkaya, J. Dueszeln, U. Lahl, K. Lierse, W. Thiemann, B. Gabel, R. Kozicki and A. Podbielski, *Anal. Chem.,* **53,** 1469 (1981).
[2] L. L. Lamparski and T. J. Nestrick, *Anal. Chem.,* **52,** 2045 (1980).
[3] A. Van Heddeghem, A. Huyghebaert and H. deMoor, *Z. Lebensm. Unters. Forsch.,* **171,** 9 (1980).
[4] F. Y. Tripet, C. Riva and J. Vogel, *Mitt. Geb. Lebensmittelunters. Hyg.,* **72,** 367 (1981).
[5] P. Lafont and M. G. Siriwidarna, *J. Chromatog.,* **219,** 162 (1981).
[6] R. D. Stubblefield and O. L. Shotwell, *J. Assoc. Off. Anal. Chem.,* **64,** 964 (1981).
[7] E. B. Ofstad, H. Drangsholt and G. E. Carlsberg, *Sci. Total Environ.,* **20,** 205 (1981).
[8] M. Godefroot, M. Stechele, P. Sandra and M. Verzele, *J. High Resolut. Chromatog. Chromatog. Commun.,* **5,** 75 (1982).
[9] A. Hashimoto, H. Sakino, E. Yamagami and S. Tateishi, *Analyst,* **105,** 787 (1980).
[10] G. B. Park, R. F. Koss, J. Utter, B. A. Mayes and J. Edelson, *J. Pharm. Sci.,* **71,** 932 (1982).
[11] M. Van Boven, P. Daenens and G. Vandereycken, *J. Chromatog.,* **182,** *Biomed. Appl.,* **8,** 435 (1980).
[12] M. Alvinerie and P. L. Toutain, *J. Pharm. Sci.,* **71,** 816 (1982).
[13] S. Stavchansky and A. Loper, *J. Pharm. Sci.,* **71,** 194 (1982).
[14] M. A. Alawi, *Z. Anal. Chem.,* **312,** 536 (1982).
[15] N. van den Bosch and D. de Vos, *J. Chromatog.,* **183,** *Biomed. Appl.,* **9,** 49 (1980).
[16] C. C. Whitney, S. H. Weinstein and J. C. Gaylord, *J. Pharm. Sci.,* **70,** 462 (1981).
[17] H. Poetter, M. Huelm and K. Richter. *J. Chromatog.,* **241,** 189 (1982).
[18] T. Stijve, *Dtsch. Lebensm. Rundsch.,* **77,** 249 (1981).

[19] C. F. Aten, J. B. Bourke and R. A. Marafioti, *J. Agr. Food Chem.*, **30**, 610 (1982).
[20] Y. Tanaka and H. Nakanishi, *Bunseki Kagaku*, **30**, 569 (1981).
[21] P. E. Kastl, E. A. Hermann and W. H. Braun, *J. Chromatog.*, **213**, 156 (1981).
[22] T. Rihs and W. Herzog, *Mitt. Geb. Lebensmittlunters. Hyg.*, **73**, 88 (1982).
[23] C. W. Stanley and L. M. Delphia, *J. Agr. Food Chem.*, **29**, 1038 (1981).
[24] K. Hasebe and J. Osteryoung, *Talanta*, **29**, 655 (1982).
[25] V. T. Wee and J. M. Kennedy, *Anal. Chem.*, **54**, 1631 (1982).
[26] M. G. Beyer, *Z. Lebensm. Unters Forsch.*, **173**, 368 (1981).
[27] P. J. Van Niekirk and S. C. C. Smit, *J. Am. Oil Chem. Soc.*, **57**, 417 (1980).
[28] B. G. Oliver and K. D. Bothen, *Anal. Chem.*, **52**, 2066 (1980).
[29] M. Stanley and F. Iohan, *Res. Commun. Psychol. Psychiatry Behav.*, **5**, 419 (1980).
[30] M. C. Petersen, R. L. Nation and J. J. Ashley, *J. Chromatog.*, **183**, *Biomed. Appl.*, **9**, 131 (1980).
[31] T. Suzuki, M. Kurisu, N. Nose and A. Watanabe, *Shokuhin Eiseigaku Zasshi*, **22**, 197 (1981).
[32] R. A. DeAbreu, J. M. van Baal, T. J. Schouten, E. D. A. M. Schretlen and C. H. M. M. De Bruyn, *J. Chromatog.*, **227**, *Biomed. Appl.*, **16**, 526 (1982).
[33] T. R. Edgerton, R. F. Moseman, E. M. Lores and L. H. Wright, *Anal. Chem.*, **52**, 1774 (1980).
[34] L. Somerville, *Meded Fac. Landbouwet., Rijksuniv. Gent*, **45**, 841 (1980).
[35] R. Riva, G. Tedeschi, F. Albani and A. Baruzzi, *J. Chromatog.*, **225**, *Biomed. Appl.*, **14**, 219 (1981).
[36] T. Egloff, A. Niederwieser, K. Pfister, A. Otten, B. Steinmann, W. Steiner and R. Gitzelmann, *J. Clin. Chem. Clin. Biochem.*, **20**, 203 (1982).
[37] J. D. Whiting and W. W. Manders, *J. Anal. Toxicol.*, **6**, 49 (1982).
[38] A. B. Vilim, L. Larocque and A. I. Macintosh, *J. Liq. Chromatog.*, **3**, 1725 (1980).
[39] R. F. Suckow, T. B. Cooper, F. M. Quitkin and J. W. Stewart, *J. Pharm. Sci.*, **71**, 889 (1982).
[40] K. Mori, S. Nishida and H. Harada, *Eisei Kagaku*, **26**, 107 (1980).
[41] K. Lu, N. Savaraj, M. T. Huang, D. Moore and T. L. Loo, *J. Liq. Chromatog.*, **5**, 1323 (1982).
[42] J. Rosenfeld, D. Devereaux, M. R. Buchanan and A. G. Turpie, *J. Chromatog.*, **231**, *Biomed. Appl.*, **20**, 216 (1982).
[43] J. F. Lawrence, L. G. Panopio and H. A. McLeod, *J. Chromatog.*, **195**, 113 (1980).
[44] S. G. Vohra and G. W. Harrington, *Food Cosmet. Toxicol.*, **19**, 485 (1981).
[45] K. Ishikawa, S. Suzuki, N. Sato, K. Takatsuki and K. Sakai, *Shokuhin Eiseigaku Zasshi*, **22**, 56 (1981).
[46] M. Goldhagen and W. Meier, *Mitt. Geb. Lebensmittelunters. Hyg.*, **71**, 248 (1980).
[47] N. M. J. Vermeulen, Z. Apostolides, D. J. J. Potgieter, P. C. Nel and N. S. H. Smit, *J. Chromatog.*, **240**, 247 (1982).
[48] J. F. Lawrence, L. G. Panopio and H. A. McLeod, *J. Agr. Food Chem.*, **28**, 1323 (1980).
[49] T. Radford and D. E. Dalsis, *J. Agr. Food Chem.*, **30**, 600 (1982).
[50] J. Sherma and S. Zorn, *Intern. Lab.*, **12**, No. 6, 68 (1982).
[51] R. J. Gordon, J. Szita and E. J. Faeder, *Anal. Chem.*, **54**, 478 (1982).
[52] S. D. Erk, C. H. Jarboe, P. E. Newton and C. Pflederer, *J. Chromatog.*, **240**, 117 (1982).
[53] R. D. Wei, S. C. Chang and S. S. Lee, *J. Assoc. Off. Anal. Chem.*, **63**, 1269 (1980).
[54] A. Arikawa and M. Shiga, *Bunseki Kagaku*, **29**, T33 (1980).
[55] Y. Tokuma, Y. Tamura and H. Noguchi, *J. Chromatog.*, **231**, *Biomed. Appl.*, **20**, 129 (1982).
[56] J. J. van der Lee, W. P. van Bennenkom and H. J. de Jong, *Anal. Chim. Acta*, **117**, 171 (1980).
[57] T. Stijve, *Dtsch. Lebensm. Rundsch.*, **76**, 234 (1980).
[58] V. K. Piotrovski and V. I. Metelitsa, *J. Chromatog.*, **231**, *Biomed. Appl.*, **20**, 205 (1982).
[59] P. K. Chan, M. Y. Siraj and A. W. Hayes, *J. Chromatog.*, **194**, 387 (1980).

[60] J. A. Thompson, M. S. Ho and D. R. Petersen, *J. Chromatog.*, **231,** *Biomed. Appl.*, **20,** 53 (1982).
[61] N. Nimura, K. Ishida and T. Kinoshita, *J. Chromatog.*, **221,** *Biomed. Appl.*, **10,** 249 (1980).
[62] H. B. Greizerstein and I. G. McLaughlin, *J. Liq. Chromatog.*, **3,** 1023 (1980).
[63] P. K. Sonsalla, T. A. Jennison and B. S. Finkle, *Clin. Chem.*, **28,** 1401 (1982).
[64] C. E. Higgins and M. R. Guerin, *Anal. Chem.*, **52,** 1984 (1980).
[65] M. D. Erickson, M. T. Giguere and D. A. Whitacker, *Anal. Lett.*, **14,** 841 (1981).
[66] V. Rovei, M. Mitchard and P. L. Morselli, *J. Chromatog.*, **138,** 391 (1977).
[67] S. P. Jindal and P. Vestergard, *J. Pharm. Sci.*, **67,** 260 (1978).
[68] C. Razzouk, G. Lhoest, M. Roberfroid and M. Mercier, *Anal. Biochem.*, **83,** 194 (1977).
[69] L. Laitem and P. Gaspar, *J. Chromatog.*, **140,** 266 (1977).
[70] R. Vanderpoorten and R. Van Renterghem, *Milchwissensch.*, **32,** 644 (1977).
[71] K. Møller-Jensen, *J. Chromatog.*, **153,** 195 (1978).
[72] G. A. Jungclaus, *J. Chromatog.*, **139,** 174 (1977).
[73] M. Makita, S. Yamamoto, M. Miyake and K. Masamoto, *J. Chromatog.*, **156,** 340 (1978).
[74] W. E. Dodson, E. E. Tyrala and R. E. Hillman, *Clin. Chem.*, **23,** 944 (1977).
[75] G. Bianchetti, R. Latini and P. L. Morselli, *J. Chromatog.*, **124,** 331 (1976).
[76] R. G. Jenkins and R. O. Friedel, *J. Pharm. Sci.*, **67,** 17 (1978).
[77] J. F. Menez, F. Berthou, D. Picart, L. Bardou and H. H. Floch, *J. Chromatog.*, **129,** 155 (1976).
[78] S. A. Quarrie, *Anal. Biochem.*, **87,** 148 (1978).
[79] F. Gunther, R. C. Blinn, M. J. Kolbezen, J. H. Barclay, W. D. Harris and H. S. Simon, *Anal. Chem.*, **23,** 1835 (1951).
[80] R. G. Lewis, in *Accuracy in Trace Analysis,* NBS, Washington (1976).
[81] K. Beyermann, A. Kessler and P. W. Ungerer, *Z. Anal. Chem.*, **251,** 289 (1970).
[82] W. Dünges, *Anal. Chem.*, **45,** 963 (1973); *Chromatographia,* **9,** 571 (1976).
[83] T. W. May and D. L. Stalling, *Anal. Chem.*, **51,** 169 (1979).
[84] G. M. Frame and G. A. Flanigan, *Anal. Lett.*, **11,** 603 (1978).
[85] R. D. Wanchope, *Anal. Chem.*, **47,** 1879 (1975).
[86] M. Beroza and M. C. Bowman, *Anal. Chem.*, **39,** 1200 (1967).
[87] M. Beroza, M. C. Bowman and B. A. Bierl, *Anal. Chem.*, **44,** 2411 (1972).
[88] A. A. Benedetti-Pichler, *Microtechniques of Inorganic Analysis,* Wiley, New York (1950).
[89] J. Solomon, *Anal. Chem.*, **51,** 1861 (1979).
[90] H.-J. Klimisch, L. Stadler and S. Brahm, *Z. Lebensm. Unters. Forsch.*, **162,** 131 (1976).
[91] G. M. Singer and W. Lijinsky, *J. Agr. Food Chem.*, **24,** 550 (1976).
[92] A. Tanaka, N. Nose, T. Suzuki, A. Hirose and A. Watanabe, *Analyst,* **103,** 851 (1978).
[93] H. Pyysalo, A. Kiviranta and S. Lahtinen, *J. Chromatog.*, **168,** 512 (1979).
[94] M. Uchiyama and M. Yamaguchi, *Bunseki Kagaku,* **26,** 872 (1977).
[95] H. Kieninger and D. Boeck, *Brauwissensch.* **30,** 357 (1977).
[96] E. Lord, N. G. Bunton and N. T. Crosby, *J. Assoc. Publ. Analysts,* **16,** 25 (1978).
[97] I. C. Sone, J. N. Lomonte, L. A. Fletcher and W. T. Lowry, *J. Forens. Sci.*, **23,** 78 (1978).
[98] J. Angerer, *Mikrochim. Acta,* 405 (1977 **II**).
[99] K. Shimokawa, H. Takada and N. Watanabe, *Bunseki Kagaku,* **27,** 452 (1978).
[100] A. Baron, J. Calvez and J. F. Drilleau, *Ann. Fals. Exp. Chim.*, **71,** 29 (1978).
[101] A. M. van der Ark, A. B. E. Theeuwen and A. M. A. Verweij, *Pharm. Weekbl.*, **112,** 977 (1977).
[102] J. B. Zijlstra, J. Beukema, B. G. Wolthers, B. M. Byrne, A. Groen and J. Dankert, *Clin. Chim. Acta,* **78,** 243 (1977).
[103] E. Kröller, *Deutsch. Lebensm. Rundsch.*, **70,** 69 (1974).

[104] D. J. Casimir, J. C. Moyer and L. R. Mattick, *J. Assoc. Off. Anal. Chem.*, **59**, 269 (1976).
[105] K. Grob, *J. Chromatog.*, **84**, 255 (1973).
[106] S. Dilli and K. Robards, *Analyst*, **102**, 201 (1977).
[107] D. E. Ott, *J. Assoc. Off. Anal. Chem.*, **62**, 93 (1979).
[108] G. D. Veith and L. M. Kiwus, *Bull. Environ. Contam. Toxicol.*, **17**, 631 (1977).
[109] G. Hartmetz and J. Slemrova, *Bull. Environ. Contam. Toxicol.*, **25**, 106 (1980).
[110] K. Moehler and E. S. M. El-Refai, *Z. Lebensm. Unters. Forsch.*, **172**, 449 (1981).
[111] H. J. Jeuring, W. Verwaal, A. Brands, K. van Ede and J. W. Dornseiffen, *Z. Lebensm. Unters. Forsch.* **175**, 85 (1982).
[112] B. D. Ripley, B. J. French and L. V. Edgington, *J. Assoc. Off. Anal. Chem.*, **65**, 1066 (1982).
[113] P. B. Van Roosmalen, J. Pundham and I. Drummonds, *Intern. Arch. Occup. Environ. Health*, **48**, 159 (1981).
[114] P. J. Rennie, *Analyst*, **107**, 327 (1982).
[115] T. L. Peppard and S. A. Halsey, *J. Chromatog.*, **202**, 271 (1980).
[116] H. Werner and F. Jensen, *Dtsch. Lebensm. Rundsch.*, **76**, 431 (1980).
[117] L. W. Cook, F. W. Zach and J. R. Fleeker, *J. Assoc. Off. Anal. Chem.*, **65**, 215 (1982).
[118] G. Y. P. Kan, F. T. S. Mah, L. N. Wade and M. L. Bothwell, *J. Assoc. Off. Anal. Chem.*, **64**, (1981).
[119] I. S. Kofman and V. I. Kofanov, *Gig. Sanit.*, 41 (1979).
[120] F. Herzel, *J. Chromatog.*, **193**, 320 (1980).
[121] A. Yasuhara and K. Fuwa, *Agric. Biol. Chem.*, **44**, 2491 (1980).
[122] K. Kido, T. Sakuma and T. Watanabe, *Eisei Kagaku*, **26**, 224 (1980).
[123] T. Ramstad and L. W. Nicholson, *Anal. Chem.*, **54**, 1191 (1982).
[124] B. Zimmerli, H. Zimmermann and F. Mueller, *Mitt. Geb. Lebensmittelunters. Hyg.*, **73**, 71 (1982).
[125] Y. Okana, T. Miyata, K. Iwasaki, K. Takahama, T. Hitoshi, Y. Kasé, I. Matsumoto and T. Shinka, *Anal. Biochem.*, **115**, 254 (1981).
[126] G. F. Dregval, F. N. Kunznetsova and N. M. Brodskaya, *Gig. Sanit.*, 46 (1979).
[127] M. Cooke, G. Nickless, A. C. Povey and D. J. Roberts, *Sci. Total Environ.*, **13**, 17 (1979).
[128] T. McNeal, W. C. Brumley, C. Breder and J. A. Spohn, *J. Assoc. Off. Anal. Chem.*, **62**, 41 (1979).
[129] B. S. Rawat and G. Prasad, *Anal. Chem.*, **53**, 1144 (1981).
[130] M. Eichner, *Chemie f. Labor u. Betrieb*, **28**, 299 (1977).
[131] J. Pflugmacher and W. Ebing, *Z. Anal. Chem.*, **263**, 120 (1973).
[132] M. Eichner, *Z. Lebensm. Unters. Forsch.*, **167**, 245 (1978).
[133] J. Pflugmacher and W. Ebing, *Landwirtsch. Forsch.*, **32**, 82 (1979).
[134] L. Bertlig, J. Delventhal and I. Tietz, *Dtsch. Lebensm. Rundsch.*, **74**, 41 (1978).
[135] J. Jäger, *Chem. Listy*, **70**, 875 (1976).
[136] H. Matsushita, K. Arashidani and H. Hayashi, *Bunseki Kagaku*, **25**, 412 (1976).
[137] E. Stahl, *Dünnschichtchromatographie*, p. 82. Springer Verlag, Heidelberg (1967).
[138] A. Colmsjö and U. Stenberg, *J. Chromatog.*, **169**, 205 (1978).
[139] K. Beyermann and D. Hasenmaier, unpublished results.
[140] F. D. Gauchel, D. Gauchel, K. Beyermann and R. K. Zahn, *Z. Anal. Chem.*, **259**, 177 (1972).
[141] T. P. King and L. C. Craig, *Methods of Biochem. Anal.*, **10**, 201 (1962).
[142] M. Beroza, M. N. Inscoe and M. C. Bowman, *Residue Reviews*, **30**, 1 (1969).
[143] Y. L. Tan, *J. Chromatog.*, **176**, 319 (1979).
[144] P. F. Blanchet, *J. Chromatog.*, **179**, 123 (1979).
[145] W. Dünges, *Anal. Chem.*, **49**, 442 (1977).
[146] A. A. Benedetti-Pichler, *Introduction to the Microtechnique of Inorganic Analysis*, p. 135. Wiley, New York (1950).
[147] K. Grob, K. Grob, Jr. and G. Grob, *J. Chromatog.*, **106**, 299 (1975).
[148] D. A. J. Murray, *J. Chromatog.*, **177**, 135 (1979).

[149] R. C. Cochran, K. J. Darney and L. L. Ewing, *J. Chromatog.*, **173**, 349 (1979).
[150] A. P. Borsetti and J. A. G. Roach, *Bull. Environ. Contam. Toxicol.*, **20**, 241 (1978).
[151] J. Jan and S. Malnersic, *Bull. Environ. Contam. Toxicol.*, **19**, 772 (1978).
[152] E. B. Ofstad, E. G. Lunde and H. Drangsholt, *Intern. J. Environ. Anal. Chem.*, **6**, 119 (1979).
[153] P. P. Singh and R. P. Chawla, *Talanta*, **29**, 231 (1982).
[154] Y. Saito, H. Sekita, M. Takeda and M. Uchiyama, *J. Assoc. Off. Anal. Chem.*, **61**, 129 (1978).
[155] J. P. Hanus, H. Guerrero, E. R. Biehl and Ch. T. Kenner, *J. Assoc. Off. Anal. Chem.*, **62**, 29 (1979).
[156] *Technicon AutoAnalyzer Bibliography 1957-1967*, Technicon Corp., Ardsley, N.Y. (1968).
[157] C. J. O. R. Morris and P. Morris, *Separation Methods in Biochemistry*, Pitmans, London (1964).
[158] D. W. Sutton and D. G. Vallis, *J. Radioanal. Chem.*, **2**, 377 (1969).
[159] J. F. Lawrence, U. A. T. Brinkman and R. W. Frei, *J. Chromatog.*, **185**, 473 (1979).
[160] R. J. Reddingius, G. J. de Jong, U. A. T. Brinkman and R. W. Frei, *J. Chromatog.*, **205**, 77 (1981).
[161] J. Kawase, A. Nakae and M. Yamanaka, *Anal. Chem.*, **51**, 1641 (1979).
[162] D. A. J. Murray, *J. Chromatog.*, **177**, 135 (1979).
[163] G. Schwedt, *Chromatographische Trennmethoden*, Thieme, Stuttgart (1979).
[164] Bock, R., *Methoden der analytische Chemie*, Vol. 1, Verlag Chemie, Weinheim (1974).
[165] R. P. W. Scott and P. Kucera, *Anal. Chem.*, **45**, 749 (1973).
[166] L. A. Hu, G. A. S. Ansari, M. T. Moslen and E. S. Reynolds, *J. Chromatog.*, **241**, 419 (1982).
[167] F. Quaranta, L. Sardi and V. Vigneri, *Relata Techn.*, **12**, 61 (1980).
[168] J. D. Tessari, L. F. Griffin and M. J. Aaronson, *Bull. Environ. Contam. Toxicol.*, **25**, 59 (1980).
[169] A. Ambrus, J. Lantos, E. Visi, I. Csatlos and L. Sarvari, *J. Assoc. Off. Anal. Chem.*, **64**, 733 (1981).
[170] J. J. Kirkland, *J. Chromatog. Sci.*, **9**, 206 (1971).
[171] R. E. Majors, *Anal. Chem.*, **44**, 1722 (1972).
[172] N. H. C. Cooke and K. Olson, *J. Chromatog. Sci.*, **18**, 512 (1980).
[173] C. Horváth and S. R. Lipsky, *Nature*, **211**, 748 (1966).
[174] S. Eksborg and G. Schill, *Anal. Chem.*, **45**, 2092 (1973).
[175] M. T. Gilbert and R. A. Wall, *J. Chromatog.*, **149**, 341 (1978).
[176] J. H. Knox and G. R. Laird, *J. Chromatog.*, **122**, 17 (1976).
[177] Y. Ghaemi and R. A. Wall, *J. Chromatog.*, **174**, 51 (1979).
[178] C. Horváth, W. Melander, I. Molnár and P. Molnár, *Anal. Chem.*, **49**, 2295 (1977).
[179] D. P. Wittmer, N. O. Nuessle and W. G. Haney, *Anal. Chem.*, **47**, 1422 (1975).
[180] J. Korpi, D. P. Wittmer, B. J. Sandmann and W. G. Haney, *J. Pharm. Sci.*, **65**, 1087 (1976).
[181] R. Gloor and E. L. Johnson, *J. Chromatog. Sci.*, **15**, 413 (1977).
[182] J. C. Kraak, K. M. Jonker and J. F. K. Huber, *J. Chromatog.*, **142**, 671 (1977).
[183] Y. Yamauchi and J. Kumanotani, *J. Chromatog.*, **210**, 512 (1981).
[184] J. C. Liao and J. L. Ponzo, *J. Chromatog. Sci.*, **20**, 14 (1982).
[185] J. N. Grover, R. D. Hartley and M. M. Smith, *Lab. Pract.*, **31**, 110 (1982).
[186] T. J. N. Webber and E. H. McKerrel, *J. Chromatog.*, **122**, 243 (1976).
[187] M. Broquire, *J. Chromatog.*, **170**, 43 (1979).
[188] B. L. Karger, M. Martin and G. Guiochon, *Anal. Chem.*, **46**, 1640 (1974).
[189] W. Strubert, *Chromatographia*, **6**, 205 (1973).
[190] J. F. K. Huber, J. A. R. J. Hulsman and C. A. M. Meyers, *J. Chromatog.*, **62**, 79 (1971).
[191] K. Grob, *J. Chromatog.*, **84**, 255 (1973).
[192] A. Tateda and J. S. Fritz, *J. Chromatog.*, **152**, 329 (1978).

[193] S. K. Maitra, T. T. Yoshikawa, J. L. Hansen, I. Nilson-Ehle, W. J. Palin, M. C. Schotz and L. B. Guze, *Clin. Chem.*, **23**, 2275 (1977).
[194] H. P. M. Van Vliet, Th. C. Bootsman, R. W. Frei and U. A. Th. Brinkman, *J. Chromatog.*, **185**, 483 (1979).
[195] W. E. May, S. N. Chesler, S. P. Cram, B. H. Gump, H. S. Hertz, D. P. Eragonio and S. M. Dyszel, *J. Chromatog. Sci.*, **13**, 535 (1975).
[196] R. W. Frei, *Intern. J. Environ. Anal. Chem.*, **5**, 143 (1978).
[197] J. Lankelma and H. Poppe, *J. Chromatog.*, **149**, 587 (1978).
[198] G. J. DeJong and J. Zeeman, *Chromatographia*, **15**, 453 (1982).
[199] R. W. Frei and U. A. T. Brinkman, *TrAC*, **1**, 45 (1981).
[200] R. E. Kaiser, *Chromatographia*, **7**, 688 (1974).
[201] E. Oelrich, H. Preusch, E. Wilhelm and D. Theurkauf, *J. Chromatog. Sci.*, **17**, 289 (1979).
[202] R. B. Sosman, *The Phases of Silica*, Rutgers University Press, New Brunswick, N. J. (1965).
[203] J. G. Atwood, G. J. Schmidt and W. Slavin, *J. Chromatog.*, **171**, 109 (1979).
[204] H. Determann, *Gelchromatographie*, Springer Verlag, Heidelberg (1967).
[205] T. R. Schwartz and R. G. Lehmann, *Bull. Environ. Contam. Toxicol.*, **28**, 723 (1982).
[206] W. Specht and M. Tillkes, *Z. Anal. Chem.*, **301**, 300 (1980).
[207] K. E. MacLeod, R. C. Hanisch and R. G. Lewis, *J. Anal. Toxicol.*, **6**, 38 (1982).
[208] M. G. Beyer, *Z. Lebensm. Unters. Forsch.*, **173**, 275 (1981).
[209] H. Steinwandter, *Z. Anal. Chem.*, **312**, 342 (1982).
[210] A. L. Smrek and L. L. Needham, *Bull. Environ. Contam. Toxicol.*, **28**, 718 (1982).
[211] D. C. Havery, J. H. Hotchkiss and T. Fazio, *J. Dairy Sci.*, **65**, 182 (1982).
[212] J. Pflugmacher and W. Ebing, *J. Chromatog.*, **160**, 213 (1978).
[213] W. Thornburg, *Anal. Chem.*, **51**, 196R (1979).
[214] J. A. Yeransian, K. G. Sloman and A. K. Foltz, *Anal. Chem.*, **51**, 105R (1979).
[215] M. A. Evenson and G. D. Camack, *Anal. Chem.*, **51**, 35R (1979).
[216] J. A. Hackwell and C. Self, *Proc. Anal. Div. Chem. Soc.*, **15**, 307 (1978).
[217] G. Görlitz and I. Halász, *Talanta*, **26**, 773 (1979).
[218] F. J. Yang, *J. Chromatog.*, **236**, 265 (1982).
[219] R. Tijsen, J. P. A. Bleumer, A. L. C. Smit and M. E. van Kreveld, *J. Chromatog.*, **218**, 137 (1981).
[220] F. J. Yang, *J. High Resolut. Chromatog., Chromatog. Commun.*, **3**, 589 (1980).
[221] I. S. Krull and D. Bushee, *Anal. Lett.*, **13**, 1277 (1980).
[222] S. Shinozawa and T. Oda, *J. Chromatog.*, **212**, 323 (1981).
[223] Y. Shinohara, R. D. Miller and N. Castagnoli, *J. Chromatog.*, **230**, *Biomed. App.*, **19**, 363 (1982).
[224] G. C. A. Storey and D. W. Holt, *J. Chromatog.*, **245**, 377 (1982).
[225] R. F. Suckow and T. B. Cooper, *J. Chromatog.*, **230**, *Biomed. Appl.*, **19**, 391 (1982).
[226] G. A. Smith, P. Schulz, K. M. Giacomini and T. F. Blaschke, *J. Pharm. Sci.*, **71**, 581 (1982).
[227] B. Kinberger, *J. Chromatog.*, **213**, 166 (1981).
[228] F. Smets and M. Vandewalle, *Z. Lebensm. Unters. Forsch.*, **175**, 29 (1982).
[229] J. C. Kraak, F. Smedes and J. W. A. Meijer, *Chromatographia*, **13**, 673 (1980).
[230] R. H. Rumble, M. S. Roberts and S. Wanwimolruk, *J. Chromatog.*, **225**, *Biomed. Appl.*, **14**, 252 (1981).
[231] C. Violon, L. Pessemier and A. Vercruysse, *J. Chromatog.*, **236**, 157 (1982).
[232] T. J. Good and J. S. Andrews, *J. Chromatog. Sci.*, **19**, 562 (1981).
[233] R. R. Brodie and L. F. Chasseaud, *J. Chromatog.*, **228**, *Biomed. Appl.*, **17**, 413 (1982).
[234] T. Daldrup, *Beitr. Gerichtl. Medizin*, **38**, 67 (1980).
[235] D. Perret and P. L. Drucy, *J. Liq. Chromatog.*, **5**, 97 (1982).
[236] J. N. Buskin, R. A. Upton, F. Soergel, R. L. Williams, E. Lang and L. Z. Benet, *J. Chromatog.*, **230**, *Biomed. Appl.*, **19**, 454 (1982).

[237] T. B. Vree, A. M. Baars, Y. A. Hekster and E. van der Kleijn, *J. Chromatog.*, **224**, *Biomed. Appl.*, **13**, 519 (1981).
[238] B. Oosterhuis, M. van den Berg and C. J. van Boxtel, *Pharm. Weekbl. Sci. Ed.*, **3**, 263 (1981).
[239] M. A. Staiger, D. P. Nguyen and F. C. Churchill, *J. Chromatog.*, **225**, *Biomed. Appl.*, **14**, 139 (1981).
[240] G. Alvan, L. Ekman and B. Lindstrom, *J. Chromatog.*, **229**, *Biomed. Appl.*, **18**, 241 (1982).
[241] N. D. Brown, B. T. Poon and J. D. Chulay, *J. Chromatog.*, **229**, *Biomed. Appl.*, **18**, 248 (1982).
[242] K. K. Midha, J. K. Cooper, I. J. McGilveray, A. G. Butterfield and J. W. Hubbard, *J. Pharm. Sci.*, **70**, 1043 (1981).
[243] G. W. Mihaly, S. Cockbain, D. B. Jones, R. G. Hanson and R. A. Smallwood, *J. Pharm. Sci.*, **71**, 590 (1982).
[244] B. Lorenzo and D. E. Drayer, *J. Lab. Clin. Med.*, **97**, 545 (1981).
[245] H. K. L. Hundt, L. W. Brown and E. C. Clark, *J. Chromatog.*, **183**, *Biomed. Appl.*, **9**, 378 (1980).
[246] A. Brachet-Liermain, C. Jarry, O. Faure, M. Guyot and P. Loiseau, *Ther. Drug. Monit.*, **4**, 301 (1982).
[247] P. J. Harman, G. L. Blackman and G. Philipou, *J. Chromatog.*, **225**, *Biomed. Appl.*, **14**, 131 (1981).
[248] M. A. Evans and T. Morarity, *J. Anal. Toxicol.*, **4**, 19 (1980).
[249] I. W. Tsina, M. Fass, J. A. Debban and S. B. Matin, *Clin. Chem.*, **28**, 1137 (1982).
[250] E. Houghton, M. C. Dumasia and J. K. Wellby, *Biomed. Mass Spectrom.*, **8**, 558 (1981).
[251] Y. Katogi, N. Tamaki and M. Adachi, *J. Chromatog.*, **228**, *Biomed. Appl.*, **17**, 404 (1982).
[252] S. N. Rao, A. K. Dhar, H. Kutt and M. Okamoto, *J. Chromatog.*, **231**, *Biomed. Appl.*, **20**, 341 (1982).
[253] N. Ratnaraj, V. D. Goldberg, A. Elyas and P. T. Lascelles, *Analyst*, **106**, 1001 (1981).
[254] M. H. Gault, M. Ahmed, N. Tibbo, L. Longerich and D. Sugden, *J. Chromatog.*, **182**, *Biomed. Appl.*, **8**, 465 (1980).
[255] C. Williams, C. S. Huang, R. Erb and M. A. Gonzalez, *J. Chromatog.*, **225**, *Biomed. Appl.*, **14**, 225 (1981).
[256] G. Bykadi, K. P. Flora, J. C. Cradock and G. K. Poochkian, *J. Chromatog.*, **231**, *Biomed. Appl.*, **20**, 137 (1982).
[257] P. O. Edlund, *J. Chromatog.*, **226**, *Biomed. Appl.*, **15**, 107 (1981).
[258] R. R. Brodie, L. F. Chasseaud, L. M. Walmsley, H. H. Soegtrop, A. Darragh and D. A. O'Kelley, *J. Chromatog.*, **182**, *Biomed. Appl.*, **8**, 379 (1980).
[259] K. Mori, H. Kobe, H. Namekawa, H. Misono, Y. Yokoyama and T. Kobari, *Chemotherapy (Tokyo)*, **29**, 314 (1981).
[260] O. Nakajima, Y. Yoshida, I. Takako, Y. Takemasa, Y. Imamura and Y. Koyama, *J. Chromatog.*, **225**, *Biomed. Appl.*, **14**, 91 (1981).
[261] C. Aubert, C. Luccioni, P. Coassolo, J. P. Sommadossi and J. P. Cano, *Arzneim. Forsch.*, **31 (II)**, 2048 (1981).
[262] G. K. Poochikian and J. C. Cradock, *J. Pharm. Sci.*, **69**, 637 (1980).
[263] S. H. Y. Wong, T. McCauley and P. A. Kramer, *J. Chromatog.*, **226**, *Biomed. Appl.*, **15**, 147 (1981).
[264] P. Pietta, A. Calatroni and A. Rava, *J. Chromatog.*, **228**, *Biomed. Appl.*, **17**, 377 (1982).
[265] H. B. Greizerstein and I. G. McLaughlin, *J. Liq. Chromatog.*, **5**, 337 (1982).
[266] W. F. Bayne, T. East and D. Dye, *J. Pharm. Sci.*, **70**, 458 (1981).
[267] H. Kubo, Y. Kobayashi and T. Kinoshita, *Bunseki Kagaku*, **31**, 462 (1982).
[268] B. Oosterhuis, M. van den Berg and C. J. Van Boxtel, *J. Chromatog.*, **226**, *Biomed. Appl.*, **15**, 259 (1981).
[269] T. F. Woodman and B. Johnson, *Ther. Drug Monit.*, **3**, 371 (1981).

[270] L. M. Walmsley and L. F. Chasseaud, *J. Chromatog.*, **226**, *Biomed. Appl.*, **15**, 155 (1981).
[271] T. Frost, *Analyst*, **106**, 999 (1981).
[272] J. P. Sommadossi, M. Lemar, J. Necciari, Y. Sumirtapura, J. P. Cano and J. Gaillot, *J. Chromatog.*, **228**, *Biomed. Appl.*, **17**, 205 (1982).
[273] R. A. Hux, H. Y. Mohammed and F. F. Cantwell, *Anal. Chem.*, **54**, 113 (1982).
[274] M. L. Chen and W. L. Chiou, *J. Chromatog.*, **226**, *Biomed. Appl.*, **15**, 125 (1981).
[275] D. A. Cairnes and W. E. Evans, *J. Chromatog.*, **231**, *Biomed. Appl.*, **20**, 103 (1982).
[276] J. G. Monbaliu, M. T. Rosseel and M. G. Bogaert, *J. Pharm. Sci.*, **70**, 965 (1981).
[277] M. J. Herfst, P. M. Edelbroek and F. A. deWolf, *Clin. Chem.*, **26**, 1825 (1980).
[278] D. B. Pautler and W. J. Jusko, *J. Chromatog.*, **228**, *Biomed. Appl.*, **17**, 215 (1982).
[279] M. T. Rosseel, F. M. Belpaire, I. Bekaert and M. G. Bogaert, *J. Pharm. Sci.*, **71**, 114 (1982).
[280] J. Vasiliades and T. H. Sahawneh, *J. Chromatog.*, **225**, *Biomed. Appl.*, **14**, 226 (1981).
[281] P. G. Meering and R. A. A. Maes, *J. Chromatog.*, **225**, *Biomed. Appl.*, **14**, 407 (1981).
[282] J. Den Hartigh, W. J. van Oort, M. C. Y. M. Bocken and H. M. Pinedo, *Anal. Chim. Acta*, **127**, 47 (1981).
[283] P. E. Nelson, S. L. Nolan and K. R. Bedford, *J. Chromatog.*, **234**, 407 (1982).
[284] J. O. Svensson, A. Randers, J. Sawe and F. Sjoqvist, *J. Chromatog.*, **230**, *Biomed. Appl.*, **19**, 427 (1982).
[285] J. E. Wallace, S. C. Harris and M. W. Peek, *Anal. Chem.*, **52**, 1328 (1980).
[286] H. Vandenberghe, S. M. MacLeod, H. Chinyanga and S. J. Soldin, *Ther. Drug Monit.*, **4**, 307 (1982).
[287] K. Ishikawa, J. L. Martinez and J. L. McCaugh, *J. Chromatog.*, **231**, *Biomed. Appl.*, **20**, 255 (1982).
[288] M. G. M. DeRuyter, R. Cronnelly and N. Castagnoli, Jr., *J. Chromatog.*, **183**, *Biomed. Appl.*, **9**, 193 (1980).
[289] J. Meyer and P. Portmann, *Pharm. Helv. Acta*, **57**, 12 (1982).
[290] T. R. Patel and J. L. Lach, *Drug Dev. Ind. Pharm.*, **7**, 317 (1981).
[291] G. Hoogewijs, Y. Michotte, J. Lambrecht and D. L. Massart, *J. Chromatog.*, **226**, *Biomed. Appl.*, **15**, 423 (1981).
[292] S. R. Gautam, A. Nahum, J. Baechler and D. W. A. Bourne, *J. Chromatog.*, **182**, *Biomed. Appl.*, **8**, 482 (1980).
[293] J. M. Wilson, J. T. Slattery, A. J. Forte and S. D. Nelson, *J. Chromatog.*, **227**, *Biomed. Appl.*, **16**, 453 (1982).
[294] M. Hamilton and P. T. Kissinger, *Anal. Biochem.*, **125**, 143 (1982).
[295] F. X. Walsh, P. J. Langlais and E. D. Bird, *Clin. Chem.*, **28**, 382 (1982).
[296] N. Bernard, G. Cuisinaud and J. Sassard, *J. Chromatog.*, **228**, *Biomed. Appl.*, **17**, 355 (1982).
[297] S. H. Curry, E. A. Brown, O. Y. P. Hu and J. H. Perrin, *J. Chromatog.*, **231**, *Biomed. Appl.*, **20**, 361 (1982).
[298] W. D. Mason and E. N. Amick, *J. Pharm. Sci.*, **70**, 707 (1981).
[299] D. J. De Wildt, A. J. Porsius and H. H. van Rooy, *J. Chromatog.*, **225**, *Biomed. Appl.*, **14**, 381 (1981).
[300] Y. Le Roux, J. Gaillot and A. Bieder, *J. Chromatog.*, **230**, *Biomed. Appl.*, **19**, 401 (1982).
[301] J. J. M. Holthuis, W. J. van Oort and H. M. Oinedo, *Anal. Chim. Acta*, **130**, 23 (1981).
[302] E. T. Lin, R. A. Baughman and L. Z. Benet, *J. Chromatog.*, **183**, *Biomed. Appl.*, **9**, 367 (1980).
[303] N. R. Scott, J. Chakraborty and V. Marks, *Anal. Biochem.*, **108**, 266 (1980).
[304] M. L. Rocci and W. J. Jusko, *J. Chromatog.*, **224**, *Biomed. Appl.*, **13**, 221 (1981).
[305] H. Kubo, K. Tsutsumi and T. Kinoshita, *Bunseki Kagaku*, **30**, 656 (1981).
[306] H. Haraguchi, M. Oshigiri and M. Hata, *Igaku no Ayumi*, **116**, 702 (1981).
[307] R. B. Patel and P. G. Welling, *J. Pharm. Sci.*, **71**, 529 (1982).

[308] J. E. Wallace, E. L. Shimek, S. Stavchansky and S. C. Harris, *Anal. Chem.*, **53**, 960 (1981).
[309] R. B. Patel and P. G. Welling, *Clin. Chem.*, **27**, 1780 (1981).
[310] S. R. Harapat and R. E. Kates, *J. Chromatog.*, **230**, *Biomed. Appl.*, **19**, 448 (1982).
[311] E. Brode, R. Sachse and H. D. Hoffmann, *Arzneim. Forsch.*, **32**(I), 1 (1982).
[312] E. Brode, *J. Clin. Chem. Clin. Biochem.*, **20**, 39 (1982).
[313] O. H. Drummer, J. McNeil and W. J. Louis, *J. Pharm. Sci.*, **70**, 1030 (1981).
[314] H. Winkler, W. Ried and B. Lemmer, *J. Chromatog.*, **228**, *Biomed. Appl.*, **17**, 223 (1982).
[315] M. Lo and S. Riegelman, *J. Chromatog.*, **183**, *Biomed. Appl.*, **9**, 213 (1980).
[316] M. T. Rosseel and M. G. Bogaert, *J. Pharm. Sci.*, **70**, 688 (1981).
[317] F. Albani, R. Riva and A. Baruzzi, *J. Chromatog.*, **228**, *Biomed. Appl.*, **17**, 362 (1982).
[318] M. W. Lo, B. Silber and S. Riegelman, *J. Chromatog. Sci.*, **20**, 126 (1982).
[319] J. E. Greving, J. H. G. Jonkman, H. G. M. Westenberg and R. A. De Zeeuw, *Pharm. Weekbl., Sci. Ed.*, **2**, 833 (1980).
[320] G. W. Mihaly, O. H. Drummer, A. Marshall, R. A. Smallwood and W. J. Louis, *J. Pharm. Sci.*, **69**, 1155 (1980).
[321] M. A. Lefebre, J. Girault, M. C. Saux and J. B. Fourtillan, *J. Pharm. Sci.*, **69**, 1216 (1980).
[322] G. B. Park, R. F. Koss, S. K. O'Neil, G. P. Palace and J. Edelson, *Anal. Chem.*, **53**, 604 (1981).
[323] E. M. Sharp, S. M. Wallace, K. W. Hindmarsh and M. A. Brown, *Can. J. Pharm. Sci.*, **15**, 35 (1980).
[324] P. N. Shaw, L. Aarons and J. B. Houston, *J. Pharm. Pharmacol.*, **32**, 67 (1980).
[325] M. A. Alawi and H. A. Ruessel, *Chromatographia*, **14**, 704 (1981).
[326] M. S. Roberts, H. M. Watson, S. McLean and K. S. Millingen, *J. Chromatog.*, **226**, *Biomed. Appl.*, **15**, 175 (1981).
[327] H. Breithaupt, and G. Goebel, *J. Chromatog. Sci.*, **19**, 496 (1981).
[328] T. Skinner, R. Gochnauer and M. Linnoila, *Acta Pharmacol. Toxicol.*, **48**, 223 (1981).
[329] C. D. Kilts, K. S. Patrick, G. R. Breese and R. B. Mailman, *J. Chromatog.*, **231**, *Biomed. Appl.*, **20**, 377 (1982).
[330] A. Gesztesi, T. Sido-Lenarth, M. Sajgo, A. Fazekas, *Zentralbl. Pharm. Pharmakother. Labor. Diagnostik*, **120**, 642 (1981).
[331] G. Yakatan and J. E. Cruz, *J. Pharm. Sci.*, **70**, 949 (1981).
[332] R. Dixon and D. Martin, *Res. Commun. Chem. Pathol. Pharmacol.*, **33**, 537 (1981).
[333] J. T. Streator, L. S. Eichmeier and M. E. Caplis, *J. Anal. Toxicol.*, **4**, 58 (1980).
[334] S. L. Ali and H. Moeller, *Z. Anal. Chem.*, **311**, 514 (1982).
[335] A. Meulemans, J. Mohler, D. Henzel and P. Duvaldestin, *J. Chromatog.*, **226**, *Biomed. Appl.*, **15**, 255 (1981).
[336] T. M. Jaouni, M. B. Leon, D. R. Rosing and H. M. Fales, *J. Chromatog.*, **182**, *Biomed. Appl.*, **8**, 473 (1980).
[337] E. Watson and P. A. Kapur, *J. Pharm. Sci.*, **70**, 800 (1981).
[338] R. C. Durley, T. Kannangara and G. M. Simpson, *J. Chromatog.*, **236**, 181 (1982).
[339] D. M. Miller, D. M. Wilson, R. D. Wyatt, J. K. McKinney, W. A. Crowell and B. P. Stuart, *J. Assoc. Off. Anal. Chem.*, **65**, 1 (1982).
[340] T. Goto, M. Manabe and S. Matsuura, *Agric. Biol. Chem.*, **46**, 801 (1982).
[341] M. Awe and J. L. Schranz, *J. Assoc. Off. Anal. Chem.*, **64**, 1377 (1981).
[342] J. M. Fremy, T. Cariou and C. Terrier, *Ann. Fals. Expert Chim. Toxicol.*, **74**, 465 (1981).
[343] J. M. Fremy and B. Boursier, *J. Chromatog.*, **219**, 156 (1981).
[344] G. Creech, R. T. Johnson and J. O. Stoffer, *J. Chromatog. Sci.*, **20**, 67 (1982).
[345] T. Matsueda, Y. Osaki and S. Shigee, *Bunseki Kagaku*, **31**, 59 (1982).
[346] H. Loekke, *J. Chromatog.*, **200**, 234 (1980).
[347] J. F. Lawrence and L. G. Panopio, *J. Assoc. Off. Anal. Chem.*, **63**, 1300 (1980).
[348] F. Smedes and J. C. Kraak, *J. Chromatog.*, **247**, 123 (1982).

[349] J. R. Rice and P. T. Kissinger, *Environ. Sci. Techn.*, **16**, 263 (1982).
[350] G. Micali, P. Curro and G. Calabro, *J. Chromatog.*, **194**, 245 (1980).
[351] R. J. Bushway, *J. Liq. Chromatog.*, **5**, 49 (1982).
[352] M. Chiba and D. F. Veres, *J. Assoc. Off. Anal. Chem.*, **63**, 1291 (1980).
[353] D. N. Armentrout and S. S. Cutie, *J. Chromatog. Sci.*, **18**, 370 (1980).
[354] B. A. Tomkins, R. R. Reagan, J. E. Caton and W. H. Griest, *Anal. Chem.*, **53**, 1213 (1981).
[355] H. Matsushita, T. Shiozaki, Y. Kato and S. Goto, *Bunseki Kagaku*, **30**, 362 (1981).
[356] K. Nakashima, T. Nakagawa and S. Era, *Shokuhin Eiseigaku Zasshi*, **22**, 233 (1981).
[357] J. L. Anderson and D. J. Chesney, *Anal. Chem.*, **52**, 2156 (1980).
[358] R. J. Bushway, *J. Chromatog.*, **211**, 135 (1981).
[359] P. H. Cramer, A. D. Drinkwine, J. E. Going and A. E. Carey, *J. Chromatog.*, **235**, 489 (1982).
[360] J. J. Ryan and J. C. Pilon, *J. Chromatograph.*, **197**, 171 (1980).
[361] W. H. Newsome and J. B. Shields, *J. Chromatog.*, **247**, 171 (1982).
[362] L. H. Wright, T. R. Edgerton, S. J; Arbes and E. M. Lores, *Biomed. Mass Spectrom.*, **8**, 475 (1981).
[363] L. L. Needham, R. H. Hill and S. L. Sirmans, *Analyst*, **105**, 811 (1980).
[364] E. W. Zahnow, *J. Agr. Food Chem.*, **30**, 854 (1982).
[365] K. Otsuka, K. Horibe, A. Sugitani and F. Yamada, *Shokuhin Eiseigaku Zasshi*, **22**, 462 (1981).
[366] D. E. Mundy and A. F. Machin, *J. Chromatog.*, **234**, 427 (1982).
[367] S. Singh, M. Khanna and J. P. S. Sarin, *Indian Drugs*, **18**, 207 (1981).
[368] T. V. Briggle, L. M. Allen, R. C. Duncan and C. D. Pfaffenberger, *J. Assoc. Off. Anal. Chem.*, **64**, 1222 (1981).
[369] W. J. Connick and J. M. Simoneaux, *J. Agr. Food Chem.*, **30**, 258 (1982).
[370] I. Ahmad, *J. Assoc. Off. Anal. Chem.*, **65**, 1097 (1982).
[371] E. S. Fiala and C. Kulakis, *J. Chromatog.*, **214**, 229 (1981).
[372] R. C. Massey, P. E. Key and D. J. McWeeny, *J. Chromatog.*, **240**, 254 (1982).
[373] S. D. West, *J. Agr. Food Chem.*, **29**, 624 (1981).
[374] D. C. Lowe, U. Schmidt, D. H. Ehhalt, C. G. B. Frischkorn and H. W. Nuernberg, *Environ. Sci. Technol.*, **15**, 819 (1981).
[375] H. Tuss, V. Neitzert, W. Seiler and R. Neeb, *Z. Anal. Chem.*, **312**, 613 (1982).
[376] J. F. Lawrence, L. G. Panopio, D. A. Lewis and H. A. McLeod, *J. Agr. Food Chem.*, **29**, 722 (1981).
[377] H. S. Veale and G. W. Harrington, *J. Chromatog.*, **240**, 230 (1982).
[378] W. Winterlin, G. Hall and C. Mourer, *J. Assoc. Off. Anal. Chem.*, **64**, 1055 (1981).
[379] G. F. Ernst and A. vander Kaaden, *J. Chromatog.*, **198**, 526 (1980).
[380] H. Roseboom and C. J. Berkhoff, *Anal. Chim. Acta*, **135**, 373 (1982).
[381] A. M. Cheh and R. E. Carlson, *Anal. Chem.*, **53**, 1001 (1981).
[382] R. Budini, S. Girotti, A. M. Pierpaoli and D. Tonelli, *Microchem. J.*, **27**, 365 (1982).
[383] L. H. Kormos, R. L. Sandridge and J. Keller, *Anal. Chem.*, **53**, 1122 (1981).
[384] P. A. Goldberg, R. F. Walker, P. A. Ellwood and H. L. Hardy, *J. Chromatog.*, **212**, 93 (1981).
[385] M. A. Alawi and H. A. Ruessel, *Z. Anal. Chem.*, **309**, 8 (1981).
[386] C. J. Purnell and C. J. Warwick, *Analyst*, **105**, 861 (1980).
[387] M. M. Krahn, D. W. Brown, T. K. Collier, A. J. Friedman, R. G. Jenkins and D. C. Malins, *J. Biochem. Biophys. Methods*, **2**, 233 (1980).
[388] M. Oehme, S. Mano and H. Stray, *J. High Resol. Chromatog., Chromatog. Commun.*, **5**, 417 (1982).
[389] K. Bratin, P. T. Kissinger, R. C. Briner and C. S. Bruntlett, *Anal. Chim. Acta*, **130**, 295 (1981).
[390] I. E. Rosenberg, J. Gross, T. Spears and P. Rahn, *J. Soc. Cosmet. Chem.*, **31**, 237 (1980).
[391] R. C. Massey, C. Crews and D. J. McWeeny, *J. Chromatog.*, **241**, 423 (1982).
[392] M. Takeuchi, K. Mizuishi and H. Harada, *Tokyo-toritsu Eisei Kenkyusho Kenkyu Nempo*, 86 (1980).

[393] I. W. Tsina and S. B. Martin, *J. Pharm. Sci.*, **70**, 858 (1981).
[394] D. E. Mundy and A. F. Machin, *J. Chromatog.*, **216**, 229 (1981).
[395] R. Schoenhaber, A. Schwarz and D. Kranl, *Dtsch. Lebensm. Rundsch.*, **78**, 254 (1982).
[396] L. L. Needham, R. Hill, D. C. Paschal and S. L. Sirmans, *Clin. Chim. Acta*, **107**, 261 (1980).
[397] K. Kuwata, M. Uebori and Y. Yamazaki, *Anal. Chem.*, **53**, 1531 (1981).
[398] P. Roumeliotis, W. Liebald and K. K. Unger, *Intern. J. Environ. Anal. Chem.*, **9**, 27 (1981).
[399] P. A. Realini, *J. Chromatog. Sci.*, **19**, 124 (1981).
[400] H. M. Klimisch, D. N. Ingebrigtson and N. Donald, *Anal. Chem.*, **52**, 1675 (1980).
[401] W. J. Sonnefeld, W. H. Zoller, W. E. May and S. A. Wise, *Anal. Chem.*, **54**, 723 (1982).
[402] D. R. Choudhury and B. Bush, *Anal. Chem.*, **53**, 1351 (1981).
[403] H. Obana, S. Hori and T. Kashimoto, *Bull. Environ. Contam. Toxicol.*, **26**, 613 (1981).
[404] R. Fischer, *Z. Anal. Chem.*, **311**, 109 (1982).
[405] L. Szepesy and M. Czencz, *Period. Polytech., Chem. Eng.*, **24**, 123 (1980).
[406] F. L. Joe, E. L. Roseboro and T. Fazio, *J. Assoc. Off. Anal. Chem.*, **64**, 641 (1981).
[407] I. Ahmad, *J. Environ. Sci. Health*, **17B**, 253 (1982).
[408] M. E. Stack and R. M. Eppley, *J. Assoc. Off. Anal. Chem.*, **63**, 1278 (1980).
[409] A. T. R. Williams, S. A. Winfield and R. C. Belloli, *J. Chromatog.*, **235**, 461 (1982).
[410] W. R. Sponholz and P. Lamberty, *Z. Lebensm. Unters. Forsch.*, **171**, 451 (1980).
[411] D. Seymour and B. D. Rupe, *J. Pharm. Sci.*, **69**, 701 (1980).
[412] U. Juergens, *Z. Lebensm. Unters. Forsch.*, **173**, 356 (1981).
[413] R. C. Snyder and C. V. Breder, *J. Chromatog.*, **236**, 429 (1982).
[414] P. Beilstein, A. M. Cook and R. Huetter, *J. Agr. Food Chem.*, **29**, 1132 (1981).
[415] M. J. Seymour, *J. Chromatog.*, **236**, 530 (1982).
[416] L. Brown, J. Braven, M. M. Rhead, R. Evens and E. I. Butler, *Water Res.*, **16**, 1409 (1982).
[417] H. Cohen and M. R. Lapointe, *J. Assoc. Off. Anal. Chem.*, **63**, 1254 (1980).
[418] E. A. Stahly and D. A. Buchanan, *J. Chromatog.*, **235**, 453 (1982).
[419] G. V. Turner, T. D. Phillips, N. D. Heidelbaugh and L. H. Russell, *J. Environ. Sci. Health*, **17**, 197 (1982).
[420] L. J. James, L. G. McGirr and T. K. Smith, *J. Assoc. Off. Anal. Chem.*, **65**, 8 (1982).
[421] H. Schweighardt, J. Boehm, J. Leibetseder, M. Schuh and E. Glawischnig, *Chromatographia*, **13**, 447 (1980).
[422] B. Kagedal and M. Kallberg, *J. Chromatog.*, **229**, *Biomed. Appl.*, **18**, 409 (1982).
[423] D. Omar and L. L. Murdock, *J. Chromatog.*, **224**, *Biomed. Appl.*, **13**, 310 (1981).
[424] Y. Watanabe and K. Imai, *Anal. Biochem.*, **116**, 471 (1981).
[425] A. F. Fell, G. P. Anderson, B. J. Clark and J. M. Rattenbury, *J. Pharm. Pharmacol.*, **33**, (Suppl.) 88 (1981).
[426] S. Onishi, S. Itoh and Y. Ishida, *Biochem. J.*, **204**, 135 (1982).
[427] T. P. Davis, C. W. Gehrke, C. H. Williams, C. W. Gehrke, Jr. and K. O. Gehrhardt, *J. Chromatog.*, **228**, *Biomed. Appl.*, **17**, 113 (1982).
[428] S. Bouqiet, A. M. Brisson and J. Gombert, *Ann. Biol. Clin.*, **39**, 189 (1981).
[429] T. Matsuzawa, N. Sugimoto and I. Ishiguro, *Anal. Biochem.*, **115**, 250 (1981).
[430] K. Takada, R. Ito, K. Tokunaga, T. Kobayashi and T. Takao, *J. Chromatog.*, **216**, 385 (1981).
[431] B. E. Dryer and D. B. P. Goodman, *Anal. Biochem.*, **114**, 37 (1981).
[432] S. P. Arneric, D. B. Goodale, J. R. Flynn and J. P. Long, *Brain Res. Bull.*, **6**, 407 (1981).
[433] B. H. Westerink and T. B. A. Mulder, *J. Neurochem.*, **36**, 1449 (1981).
[434] H. Svendsen and T. Greibrokk, *J. Chromatog.*, **213**, 429 (1981).
[435] C. D. Kilts, G. R. Breese and R. B. Mailman, *J. Chromatog.*, **225**, *Biomed. Appl.*, **14**, 347 (1981).

[436] D. S. Goldstein, G. Z. Feuerstein, I. J. Kopin and H. R. Kaiser, *Clin. Chim. Acta,* **117,** 113 (1981).
[437] M. G. P. Feenstra, J. W. Homan, D. Dijkstra, T. B. A. Mulder, H. Rollema, B. H. C. Westerink and A. S. Horn, *J. Chromatog.,* **230,** *Biomed. Appl.,* **19,** 271 (1982).
[438] O. Magnusson, L. B. Nilsson and D. Westerlund, *J. Chromatog.,* **221,** *Biomed. Appl.,* **10,** 237 (1980).
[439] Y. Hashimoto, K. Kawanishi, H. Tomita, Y. Uhara and M. Moriyasu, *Anal. Lett.,* **14,** 1525 (1981).
[440] N. Nimura and T. Kinoshita, *Anal. Lett.,* **13,** 191 (1980).
[441] J. Halgunset, E. W. Lund and A. Sunde, *J. Chromatog.,* **237,** 496 (1982).
[442] M. Ikeda, K. Shimada and T. Sakaguchi, *Chem. Pharm. Bull.,* **30,** 2258 (1982).
[443] B. Kloareg, *J. Chromatog.,* **236,** 217 (1982).
[444] G. W. M. Barendse, P. H. van de Werken and N. Takahashi, *J. Chromatog.,* **198,** 449 (1980).
[445] Y. Hiraga and T. Kinoshita, *J. Chromatog.,* **226,** *Biomed. Appl.,* **15,** 43 (1981).
[446] G. Skofitsch, A. Saria, P. Holzer and F. Lembeck, *J. Chromatog.,* **226,** *Biomed. Appl.,* **15,** 53 (1981).
[447] P. S. Draganac, S. J. Steindel and W. G. Trawick, *Clin. Chem.,* **26,** 910 (1980).
[448] J. P. Garnier, C. Dreux and B. Bousquet, *Feuill. Biol.,* **22,** 105 (1981).
[449] J. P. Garnier, B. Bousquet and C. Dreux, *J. Autom. Chem.,* **4,** 65 (1982).
[450] P. J. Langlais, W. J. McEntee and E. D. Bird, *Clin. Chem.,* **26,** 786 (1980).
[451] A. M. Krstulovic, L. Bertani-Dziedzic, S. Bautista-Cerqueira and S. E. Gitlow, *J. Chromatog.,* **227,** *Biomed. Appl.,* **16,** 379 (1982).
[452] A. J. Cross and M. H. Joseph, *Life Sci.,* **28,** 499 (1981).
[453] H. Turnbull, D. J. H. Trafford and H. L. J. Makin, *Clin. Chim. Acta,* **120,** 65 (1982).
[454] D. J. H. Trafford, D. A. Seamark, H. Turnbull and H. L. J. Makin, *J. Chromatog.,* **226,** *Biomed. Appl.,* **15,** 351 (1981).
[455] A. M. Craig and M. J. Savage, *Altex Chromatogram,* **3,** 1,4 (1980).
[456] G. M. Anderson, J. Young, D. J. Cohen and S. N. Young, *J. Chromatog.,* **228,** *Biomed. Appl.,* **17,** 155 (1982).
[457] G. P. Jackman, *Clin. Chim. Acta,* **120,** 137 (1982).
[458] F. Ponzio, G. Achilli and S. Algeri, *J. Neurochem.,* **36,** 1361 (1981).
[459] R. W. McKee, Y. A. Kang-Lee. M. Panaqua and M. E. Swendseid, *J. Chromatog.,* **230,** *Biomed. Appl.,* **19,** 309 (1982).
[460] O. Horoshima, S. Ikenoya, O. Satoru, M. Ohmae and K. Kawabe, *Chem. Pharm. Bull.,* **28,** 2512 (1980).
[461] B. Nadjem, K. Abdeddiam and M. H. Guermouche, *Analusis,* **10,** 83 (1982).
[462] P. F. Daniel, D. F. DeFeudis, I. T. Lott and R. H. McCluer, *Carbohydr. Res.,* **97,** 161 (1981).
[463] T. Kawasaki, M. Maeda and A. Tsuji, *J. Chromatog.,* **226,** *Biomed. Appl.,* **15,** 1 (1981).
[464] M. F. Lefevre, R. W. Frei and A. H. M. T. Scholten, *Chromatographia,* **15,** 459 (1982).
[465] R. C. Simpson, H. Y. Mohammed and H. Veening, *J. Liq. Chromatog.,* **5,** 245 (1982).
[466] D. M. Desiderio, M. D. Cunningham, J. A. Trimble and C. B. Stout, *J. Liq. Chromatog.,* **4,** 1261 (1981).
[467] S. Inayama, H. Hori, T. Shibata, Y. Ozawa, K. Yamagami, M. Imazu and H. Hayashida, *J. Chromatog.,* **194,** 85 (1980).
[468] A. P. De Leenheer and W. E. Lambert, *Intern. J. Vitam. Nutr. Res.,* **51,** 180 (1981).
[469] Y. H. Mohammed, H. Veening and D. A. Dayton, *J. Chromatog.,* **226,** *Biomed. Appl.,* **15,** 471 (1981).
[470] D. E. Mais, P. D. Lahr and T. R. Bosin, *J. Chromatog.,* **225,** *Biomed. Appl.,* **14,** 27 (1981).
[471] G. P. Jackman, V. J. Carson, A. Bobick and H. Skews, *J. Chromatog.,* **182,** *Biomed. Appl.,* **8,** 277 (1980).
[472] S. Hori, K. Ohtani, S. Ohtani, K. Kayanuma and T. Ito, *J. Chromatog.,* **231,** *Biomed. Appl.,* **20,** 161 (1982).

[473] A. J. Falkowski and R. Wei, *Anal. Biochem.*, **115**, 311 (1981).
[474] Z. Lackovic, M. Parenti and N. H. Neff, *Eur. J. Pharmacol.*, **69**, 347 (1981).
[475] A. Shum, M. J. Sole and G. R. van Loon, *J. Chromatog.*, **228**, *Biomed. Appl.*, **17**, 123 (1982).
[476] L. Semerdjian-Rouquier, L. Bossi and B. Scatton, *J. Chromatog.*, **218**, 663 (1981).
[477] T. Stabler and A. L. Siegel, *Clin. Chem.*, **27**, 1771 (1981).
[478] M. Hisaoka, T. Terao, T. Kojima and T. Morioka, *Sankyo Kenkyusho Nempo*, **32**, 98 (1980).
[479] I. D. Hay, T. M. Annesley, N. S. Jiang and C. A. Gorman, *J. Chromatog.*, **226**, *Biomed. Appl.*, **15**, 383 (1981).
[480] M. Nishikimi, N. Yamamoto and Y. Shizuta, *Biochem. Intern.*, **2**, 7 (1981).
[481] J. Lehmann and H. L. Martin, *Clin. Chem.*, **28**, 1784 (1982).
[482] M. T. Parviainen, K. E. Savolainen, P. H. Korhonen, E. M. Alhava and J. S. Visakorpi, *Clin. Chem. Acta*, **114**, 233 (1981).
[483] Y. Maruyama, T. Oshima and E. Nakajima, *Life Sci.*, **26**, 1115 (1980).
[484] T. Nabeshima, M. Hiramatsu, S. Noma, M. Ukai, M. Amano and T. Kameyama, *Res. Commun. Chem., Pathol. Pharmacol.*, **35**, 421 (1982).
[485] F. Smedes, J. C. Kraak and H. Poppe, *J. Chromatog.*, **231**, *Biomed. Appl.*, **20**, 25 (1982).
[486] C. L. Davies and S. G. Molyneux, *J. Chromatog.*, **231**, *Biomed. Appl.*, **20**, 41 (1982).
[487] D. A. Jenner, M. J. Brown and F. J. M. Lhoste, *J. Chromatog.*, **224**, *Biomed. Appl.*, **13**, 507 (1981).
[488] E. Watson, *Life Sci.*, **28**, 493 (1981).
[489] D. S. Goldstein, F. Feuerstein, J. L. Izzo, I. J. Kopin and H. R. Keiser, *Life Sci.*, **28**, 467 (1981).
[490] I. N. Mefford, *J. Neurosci. Methods*, **3**, 207 (1981).
[491] T. H. Mueller and K. Unsicker, *J. Neurosci. Methods*, **4**, 39 (1981).
[492] S. Oka, M. Sekiya, H. Osada, K. Fujita, T. Kato and T. Nagatsu, *Clin. Chem.*, **28**, 646 (1982).
[493] I. N. Mefford, M. Gilberg and J. D. Barchas, *Anal. Biochem.*, **104**, 469 (1980).
[494] A. M. Krstulovic, A. W. Dziedzic, L. Bertani-Dziedzic and D. DiRico, *J. Chromatog.*, **217**, 523 (1981).
[495] R. C. Causon, M. E. Carruthers and R. Rodnight, *Anal. Biochem.*, **116**, 223 (1981).
[496] R. L. Michaud, M. J. Bannon and R. H. Roth, *J. Chromatog.*, **225**, *Biomed. Appl.*, **14**, 335 (1981).
[497] L. R. Hegstrand and B. Eichelmann, *J. Chromatog.*, **22**, *Biomed. Appl.*, **11**, 107 (1981).
[498] I. M. Mefford, M. M. Ward, L. Miles, B. Taylor, M. A. Chesney, D. L. Keegan and J. D. Barchas, *Life Sci.*, **28**, 477 (1981).
[499] Y. Yui, Y. Itokawa and C. Kawai, *Anal. Biochem.*, **108**, 11 (1980).
[500] K. Arashidani and T. Nakamura, *J. UOEH*, **2**, 439 (1980); *Chem. Abstr.*, **94**, 170382j (1981).
[501] G. M. Anderson, J. Young, P. I. Jatlow and D. J. Cohen, *Clin. Chem.*, **27**, 2060 (1981).
[502] A. Koike, M. Kimura and E. Nakamura, *Jikeikai Med. J.*, **28**, 193 (1981).
[503] A. Yamatodani and H. Wada, *Clin. Chem.*, **27**, 1983 (1981).
[504] R. C. Causon and M. E. Carruthers, *J. Chromatog.*, **229**, *Biomed. Appl.*, **18**, 301 (1982).
[505] E. Heftmann and J. T. Lin, *J. Liq. Chromatog.*, **5**, (Suppl. 1), 121 (1982).
[506] H. Halpaap and H. F. Krebs, *J. Chromatog.*, **142**, 823 (1977).
[507] A. A. Benedetti-Pichler, *Introduction to the Microtechnique of Inorganic Analysis*, Wiley, New York (1950).
[508] K. Beyermann and E. Roeder, *Z. Anal. Chem.*, **230**, 347 (1967).
[509] K. Beyermann and J. Dietz, *Z. Anal. Chem.*, **256**, 349 (1971).
[510] P. Kissinger, in E. Reid (ed.), *Blood Drugs and other Analytical Challenges*, Horwood, Chichester (1978).
[511] D. W. Hill, T. R. Kelley, S. W. Matiuck, K. J. Langner and D. E. Phillips, *Anal. Lett.*, **15**, 193 (1982).

[512] S. L. Kanter, L. E. Hollister and M. Musumeci, *J. Chromatog.,* **234,** 201 (1982).
[513] H. Neuninger, *Arch. Kriminol.,* **167,** 99 (1981).
[514] Arbeitsgruppe "Toxine", *Mitt Geb. Lebensmittelunters. Hyg.,* **73,** 362 (1982).
[515] P. Lafont, M. Siriwardana, J. Jacquet, M. Gaillardin and J. Sarfati, *Lait,* **61,** 275 (1981).
[516] H. V. Della Rosa and E. C. F. Moraes, *Rev. Farm. Bioquim. Univ. S. Paulo,* **17,** 270 (1981).
[517] O. L. Shotwell, W. R. Burg and T. Diller, *J. Assoc. Off. Anal. Chem.,* **64,** 1060 (1981).
[518] V. I. Mochalov, L. N. Semenova and L. A. Eremina, *Molochn. Prom-st.,* 13 (1981).
[519] V. V. Ermakov and N. A. Kostyunina, *Vopr. Pitan.,* 19 (1981).
[520] A. Bata and R. Lasztiti, *Elalmiszervizsgalati Kozl.,* **26,** 223 (1980).
[521] R. Schmidt, A. Bieger, E. Ziegenhagen and K. Dose, *Z. Anal. Chem.,* **308,** 133 (1981).
[522] H. Kamimura, M. Nishijima, K. Yasuda, K. Saito, A. Ibe, T. Nagayama, H. Ushiyama and Y. Naoi, *J. Assoc. Off. Anal. Chem.,* **64,** 1067 (1981).
[523] T. Ilus, M. L. Niku-Paavola and T. M. Enari, *Eur. J. Appl. Microbiol. Biotechnol.,* **11,** 244 (1981).
[524] A. Sano amd S. Takitani, *Maikotokishin,* **12,** 26 (1980).
[525] A. Sano, Y. Asabe, S. Takitani and Y. Ueno, *J. Chromatog.,* **235,** 257 (1982).
[526] E. Asensio, I. Sarmiento and K. Dose, *Z. Anal. Chem.,* **311,** 511 (1982).
[527] Y. D. Ha and K. G. Bergner, *Dtsch. Lebensm. Rundsch.,* **76,** 390 (1980).
[528] J. Sherma and N. T. Miller, *J. Liq. Chromatog.,* **3,** 901 (1980).
[529] M. A. Klisenko, S. Kudela, D. B. Girenko and M. S. Petrosyan, *Zh. Analit. Khim.,* **36,** 1383 (1981).
[530] S. U. Bhaskar and N. V. Nadakumar, *J. Assoc. Off. Anal. Chem.,* **64,** 1312 (1981).
[531] J. F. Lawrence, *J. Assoc. Off. Anal. Chem.,* **63,** 758 (1980).
[532] D. Schopper and B. Hoffmann, *Arch. Lebensmittelhyg.,* **32,** 141 (1981).
[533] J. P. Heotis, J. L. Mertz, R. J. Herrett, J. R. Diaz, D. C. van Hart and J. Olivard, *J. Assoc. Off. Anal. Chem.,* **63,** 720 (1980).
[534] E. Vioque and A. Vioque, *Gras. Aceites (Seville),* **33,** 162 (1982).
[535] J. C. Sipos and R. G. Ackman, *J. Chromatog. Sci.,* **16,** 443 (1978).
[536] P. Van Tornout, R. Vercaemst, H. Caster, M. J. Lievens, W. de Keersgieter, F. Soetewey and M. Rosseneu, *J. Chromatog.,* **164,** 222 (1979).
[537] L. Vandamme, V. Blaton and H. Peeters, *J. Chromatog.,* **145,** 151 (1978).
[538] R. G. Ackman and A. D. Woyewoda, *J. Chromatog. Sci.,* **17,** 514 (1979).
[539] H. R. Harvey and J. S. Patton, *Anal. Biochem.,* **116,** 312 (1981).
[540] S. M. Innis and M. T. Clandinin, *J. Chromatog.,* **205,** 490 (1981).
[541] N. E. Skelly and R. H. Stehl, in *Recent Developments in Separation Science,* N. N. Li (ed.), Vol. 1, p. 141. CRC Press, Cleveland (1972).
[542] R. M. Wheaton and W. C. Bauman, *Ann. N.Y. Acad. Sci.,* **57,** 159 (1953).
[543] C. D. Scott, *Anal. Chem.,* **43,** 2121 (1971).
[544] S. J. Lewis, M. R. Fennessy, F. J. Laska and D. A. Taylor, *Agents Actions,* **10,** 197 (1980).
[545] U. Pechanek, G. Blaicher, W. Pfannhauser and H. Woidich, *Z. Lebensm. Unters. Forsch.,* **171,** 420 (1980).
[546] K. Olek, S. Uhlhaas and P. Wardenbach, *J. Clin. Chem. Clin. Biochem.,* **18,** 567 (1980).
[547] Y. Fukuda, Y. Morikawa and I. Matsumoto, *Anal. Chem.,* **53,** 2000 (1981).
[548] T. Saito and K. Hagiwara, *Z. Anal. Chem.,* **312,** 533 (1982).
[549] B. Andersson and K. Andersson, *J. Chromatog.,* **242,** 353 (1982).
[550] S. P. Cram and T. H. Risby, *Anal. Chem.,* **50,** 213 R (1978).
[551] S. Ettre, *J. Chromatog.,* **112,** 1 (1975).
[552] G. Schomburg, R. Dielmann, H. Husmann and F. Weeke, *J. Chromatog.,* **122,** 55 (1976).
[553] W. A. Aue, *J. Chromatog. Sci.,* **13,** 329 (1975).
[554] W. A. Aue, *Intern. J. Environ. Anal. Chem.,* **5,** 1 (1977).

[555] P. Leinster, R. Perry and R. J. Young, *Talanta,* **24,** 205 (1977).
[556] M. Matucha and E. Smolkova, *J. Chromatog.,* **127,** 163 (1976).
[557] J. Sherma, *Chemist-Analyst,* **65,** (1976).
[558] B. Scales, *Am. Lab.,* **10,** No. 10, 93 (1978).
[559] R. E. Kaiser, *Chromatographie in der Gasphase,* Bibliogr. Institut, Mannheim (1975).
[560] K. Metzner, *Gas-Chromatographische Spurenanalyse,* Akad. Verlagsgesellsch., Leipzig (1976).
[561] J. C. Giddings, E. Grushka, J. Cazes and P. R. Brown, *Advances in Chromatography,* Vol. 14, Dekker, New York (1976).
[562] L. S. Ettre, *Gas-Chromatographie mit Kapillarsäulen,* Vieweg, Braunschweig (1976).
[563] T. S. Ma and A. S. Ladas, *Organic Functional Group Analysis by Gas Chromatography,* Academic Press, London (1976).
[564] W. Jennings, *Gas Chromatography with Capillary Columns,* Academic Press, London (1978).
[565] B. J. Gudzinowicz and M. J. Gudzinowicz, *Analysis of Drugs and Metabolites by Gas Chromatography and Mass Spectrometry,* 7 Volumes. Dekker, New York (1977–1980).
[566] E. Reid (ed.), *Blood Drugs and Other Analytical Challenges,* Horwood, Chichester (1978).
[567] V. G. Berezkin, V. R. Alishoyev and I. B. Nemirovskaya, *Gas Chromatography of Polymers,* Elsevier, Amsterdam (1977).
[568] G. J. Dickes and P. V. Nicholas, *Gas Chromatography in Food Analysis,* Butterworths, London (1976).
[569] C. E. Jones and C. A. Cramers, *Analytical Pyrolysis,* Elsevier, Amsterdam (1977).
[570] R. W. May, E. F. Pearson and D. Scothern, *Pyrolysis-Gas Chromatography,* Chemical Society. London (1977).
[571] H. Hachenberg and A. P. Schmidt, *Gas Chromatographic Head Space Analysis,* Heyden, London (1977).
[572] H. V. Street, W. Vycudilik and G. Machata, *J. Chromatog.,* **168,** 117 (1979).
[573] J. M. Wal, J. C. Peleran and G. Bories, *J. Chromatog.,* **168,** 179 (1979).
[574] M. J. E Golay, *Anal. Chem.,* **29,** 928 (1957).
[575] L. H. Keith, *J. Chromatog. Sci.,* **17,** 48 (1979).
[576] J. C. Caprais and M. Marchand, *Analusis,* **9,** 140 (1981).
[577] H. T. Badings, J. J. G. van der Pol and D. G. Schmidt, *Chromatographia,* **13,** 17 (1980).
[578] K. Grob, G. Grob and K. Grob, Jr., *Chromatographia,* **10,** 181 (1977).
[579] M. L. Lee, D. L. Vassilaros, L. V. Phillips, D. M. Hercules, H. Azumaya, J. W. Jorgenson, M. P. Maskarinec and M. Novotny, *Anal. Lett.,* **12,** 191 (1979).
[580] R. C. M. De Nijs, J. F. Franken, R. P. M. Dooper, J. Rijks, H. J. J. M. De Ruwe and F. L. Schulting, *J. Chromatog.,* **167,** 231 (1978).
[581] K. Grob, G. Grob and K. Grob, Jr., *J. High Resolut. Chromatog.,* **2,** 31 (1979).
[582] G. Schomburg, H. Husmann and H. Behlau, *Chromatographia,* **13,** 321 (1980).
[583] G. Schomburg, H. Husmann and H. Behlau, *J. Chromatog.,* **203,** 179 (1981).
[584] W. G. Jennings, *J. High Resolut. Chromatog. Chromatog. Commun.,* **3,** 601 (1980).
[585] R. C. M. Nijs and R. P. M. Dooper, *J. High Resolut. Chromatog. Chromatog. Commun.,* **3,** 583 (1980).
[586] H. Saito, *J. Chromatog.,* **243,** 189 (1982).
[587] R. E. Kaiser (ed.), *Capillary Chromatography,* Huethig, Heidelberg (1981).
[588] T. Romanowski, W. Funcke, J. Koenig and E. Balfanz, *J. High Resolut. Chromatog. Chromatog. Commun.,* **4,** 209 (1981).
[589] M. G. Proske, M. Bender, G. Schomburg and E. Huebinger, *J. Chromatog.,* **240,** 95 (1982).
[590] K. Grob, Jr., *J. Chromatog.,* **208,** 217 (1981).
[591] J. A. Corkill and R. W. Glese, *Anal. Chem.,* **53,** 1667 (1981).
[592] L. L. Plotczyk, *J. Chromatog.,* **240,** 349 (1982).
[593] A. D. Sauter, L. D. Betowski, T. R. Smith, V. A. Strickler, R. G. Beimer, B. N. Colby and J. E. Wilkinson, *J. High Resolut. Chromatog. Chromatog. Commun.,* **4,** 366 (1981).

[594] G. T. Hunt and M. P. Hoyt, *J. High Resolut. Chromatog. Chromatog. Commun.*, **5**, 291 (1982).
[595] W. Ebind, *Gaschromatographie der Pflanzenschutzmittel, Mitteilung aus der Biologischen Bundesanstalt für Land- und Forstwirtschaft*, Berlin, No. 190 (1979).
[596] R. E. Ardrey and A. C. Moffat, *J. Chromatog.*, **220**, *Chromatog. Review*, **25**, 195 (1981).
[597] W. Blass, K. Riegner and H. Hulpke, *J. Chromatog.*, **172**, 67 (1979).
[598] G. Schomburg, H. Husmann and F. Weeke, *J. Chromatog.*, **112**, 205 (1975).
[599] G. Schomburg and H. Husmann, *Chromatographia*, **8**, 517 (1975).
[600] G; Schomburg, R. Dielmann, H. Husmann and F. Weeke, *J. Chromatog.*, **122**, 55 (1976).
[601] W. A. Aue and S. Kapila, *Anal. Chem.*, **50**, 536 (1978).
[602] S. Kapila and W. A. Aue, *J. Chromatog. Sci.*, **15**, 569 (1977).
[603] W. A. Aue, V. Paramasigamani and S. Kapila, *Microchim. Acta*, 193 (1978 I).
[604] V. Paramasigamani and W. A. Aue, *J. Chromatog.*, **168**, 202 (1979).
[605] F. Seefeld and W. Tunkel, *Chem. Tech. (Leipzig)*, **33**, 470 (1981).
[606] V. Paramasigamani, S. Kapila and W. A. Aue, *J. Chromatog. Sci.*, **18**, 191 (1980).
[607] D. R. Deans, *J. Chromatog.*, **203**, 19 (1981).
[608] J. A. G. Dominguez, E. F. Sanchez, J. G. Munoz and M. J. Molera, *J. Chromatog. Sci.*, **17**, 281 (1979).
[609] P. L. Fitzgerald, T. N. Gallaher, F. A. Paloscay and J. J. Leary, *J. Chromatog. Sci.*, **15**, 119 (1977).
[610] N. F. Ives and L. Giuffrida, *J. Assoc. Off. Anal. Chem.*, **53**, 973 (1970).
[611] J. J. Franken, R. C. M. de Nijs and F. L. Schulting, *J. Chromatog.*, **144**, 253 (1977).
[612] W. A. Aue, *Intern. J. Environ. Anal. Chem.*, **5**, 6 (1977).
[613] V. Paramasigamani and W. A. Aue, *J. Chromatog.*, **168**, 202 (1979).
[614] R. J. Leibrand, *J. Chromatog. Sci.*, **13**, 556 (1975).
[615] V. M. Oslavicky, *J. Chromatog. Sci.*, **16**, 197 (1978).
[616] P. Welsh, *Ind. Res.*, **19**, 72 (1977).
[617] J. E. Purcell, H. D. Downs and L. S. Ettre, *Chromatographia*, **8**, 605 (1975).
[618] D. M. Ottenstein and P. H. Silvis, *J. Chromatog. Sci.*, **17**, 389 (1979).
[619] K. Grob and G. Grob, *J. Chromatog. Sci.*, **7**, 584 1969; **7**, 587 (1969).
[620] M. Novotny and A. Zlatkis, *J. Chromatog. Sci.*, **8**, 346 (1970).
[621] P. M. J. van den Berg and Th. P. H. Cox, *Chromatographia*, **5**, 301 (1972).
[622] W. Voigt, K. Jacob and H. W. Obwexer, *J. Chromatog.*, **174**, 437 (1979).
[623] E. Bailey, M. Fenoughty and L. Richardson, *J. Chromatog.*, **131**, 347 (1977).
[624] A. G. de Boer, J. Rösr-Kaiser, H. Bracht and D. D. Breimer, *J. Chromatog.*, **145**, 105 (1978).
[625] M. T. Rosseel, M. G. Bogaert and M. Claeys, *J. Pharm. Sci.*, **67**, 802 (1978).
[626] A. S. Christophersen and K. E. Rasmussen, *J. Chromatog.*, **168**, 476 (1979).
[627] A. S. Christophersen and K. A. Rasmussen, *J. Chromatog.*, **174**, 454 (1979).
[628] C. E. Roland Jones and C. A. Cramers, *Analytical Pyrolysis*, Elsevier, Amsterdam (1977).
[629] R. W. May, E. F. Pearson and D. Scothern, *Pyrolysis-Gas Chromatography*, Chemical Society, London (1977).
[630] N. D. Danielson and L. B. Rogers, *Anal. Chem.*, **50**, 1680 (1978).
[631] S. S. Szinai and T. A. Roy, *J. Chromatog. Sci.*, **14**, 327 (1976).
[632] A. J. Cannard and W. J. Criddle, *Analyst*, **100**, 848 (1975).
[633] G. Blomquist, E. Johansson, B. Söderström and S. Wold, *J. Chromatog.*, **173**, 7, 19 (1979).
[634] K. Ohya and M. Sano, *Biomed. Mass Spectrom.*, **4**, 241 (1977).
[635] H. L. C. Meuzelaar, P. G. Kistemaker and M. A. Posthumus, *Biomed. Mass Spectrom.*, **1**, 312 (1974).
[636] D. A. Leathard and B. C. Shurlock, *Identification Techniques in GC*, p. 195. Wiley, New York (1970).
[637] T. H. Parliment, *Microchem. J.*, **20**, 492 (1975).
[638] J. W. Amy, E. M. Chait, W. E. Baitinger and F. W. McLafferty, *Anal. Chem.*, **37**, 1265 (1965).

[639] S. C. Brooks and V. C. Godefroi, *Anal. Biochem.*, **7,** 135 (1964).
[640] M. Beroza, *Anal. Chem.*, **34,** 1801 (1962).
[641] R. I. Asai, F. A. Gunther, W. E. Westlake and Y. Iwata, *J. Agr. Food Chem.*, **19,** 396 (1971).
[642] B. Zimmerli, *J. Chromatog.*, **88,** 65 (1974).
[643] M. Cooke, G. Nickless, A. M. Prescott and D. J. Roberts, *J. Chromatog.*, **156,** 293 (1978).
[644] M. Cooke and D. J. Roberts, *Bull. Environ. Contam. Toxicol.*, **23,** 285 (1979).
[645] M. Cooke, K. D. Khallef, G. Nickless and D. J. Roberts, *J. Chromatog.*, **178,** 183 (1979).
[646] M. Cooke, G. Nickless and D. J. Roberts, *J. Chromatog.*, **187,** 47 (1980).
[647] M. Cooke and R. D. James, *J. Chromatog.*, **193,** 437 (1980).
[648] D. J. Roberts, M. Cooke and G. Nickless, *J. Chromatog.*, **213,** 73 (1981).
[649] H. Schildknecht and A. Mannl, *Angew. Chem.*, **69,** 634 (1957).
[650] H. Röck, *Naturwissensch.*, **43,** 81 (1956).
[651] P. Süe, J. Pauly and A. Nouaille, *Bull. Soc. Chim. France,* **593** (1958).
[652] M. Furusawa and M. Tachibana, *Bull. Chem. Soc. Japan,* **54,** 2968 (1981).
[653] K. Beyermann and F. Schredelseker, *Z. Anal. Chem.*, **256,** 279 (1971).
[654] P. Hoeltzenbein, G. Bohn and G. Ruecker, *Z. Anal. Chem.*, **298,** 45 (1979).
[655] M. Hellmann, *Z. Anal. Chem.*, **300,** 44 (1980).
[656] L. C. Craig and T. P. King, *J. Am. Chem. Soc.*, **79,** 3729 (1957).
[657] B. A. Roberts and D. S. P. Patterson, *J. Assoc. Off. Anal. Chem.*, **58,** 1178 (1975).
[658] D. C. Hunt, L. A. Philp and N. T. Crosby, *Analyst,* **104,** 1171 (1979).
[659] D. Arndts and K. L. Rominger, *Arzneimitt.-Forsch.*, **28,** 1951 (1978).
[660] R. L. G. Norris, J. T. Ahokas and P. J. Ravenscroft, *J. Pharm. Methods,* **7,** 7 (1982).
[661] K. Beyermann and J. Dietz, *Z. Anal. Chem.*, **271,** 362 (1974).
[662] G. Eklund, B. Josefsson and C. Roos, *J. Chromatog.*, **142,** 575 (1977).
[663] H. Eustache and G. Histi, *J. Membr. Sci.*, **8,** 105 (1981).
[664] H. Brintzinger and H. Beier, *Kolloid-Z.*, **79,** 324 (1937).
[665] W. Albrecht and K. Beyermann, *Z. Anal. Chem.*, **286,** 102 (1977).
[666] R. Lemlich, *Recent Developments in Separation Science,* Vol. I, 113 (1972).
[667] R. Wickbold, *Tenside,* **8,** 61 (1971).
[668] E. Kunkel, *Mikrochim. Acta,* 227 (1977 **II**).
[669] C. Divo, S. Gafa, T. LaNoce, A. Paris, C. Ruffo and M. Sana, *Riv. Ital. Sostanze Grasse,* **57,** 329 (1980).
[670] G. Loglio, U. Tesei, P. C. Legittimo, E. Racnalelli and R. Cini, *Ann. Chim. (Rome),* **71,** 251 (1981).
[671] IUPAC Commission on Microchemical Techniques and Trace Analysis, *Pure Appl. Chem.*, **54,** 1555 (1982).
[672] F. M. Everaerts, J. L. Beckers and Th. E. P. E. M. Verheggen, *Isotachophoresis – Theory, Instrumentation,* Elsevier, Amsterdam (1976).
[673] B. J. Radola and D. Graesslin (eds.), *Electrofocussing and Isotachophoresis,* de Gruyter, Berlin (1977).
[674] F. E. P. Mikkers, F. M. Everaerts and J. A. F. Peek, *J. Chromatog.*, **168,** 293 (1979).
[675] F. E. P. Mikkers, F. M. Everaerts and T. P. E. M. Verheggen, *J. Chromatog.*, **169,** 1, 11 (1979).
[676] P. Delmotte, *J. Chromatog.*, **165,** 87 (1979).
[677] F. M. Everaerts, T. P. E. M. Verheggen and F. E. P. Mikkers, *J. Chromatog.*, **169,** 21 (1979).
[678] P. Bocek, M. Deml and J. Janák, *Laborpraxis,* May 1979, 18.
[679] D. V. Robinson and M. Rimpler, *J. Clin. Chem. Clin. Biochem.*, **16,** 1 (1978).
[680] F. E. P. Mikkers, T. P. E. M. Verheggen, F. M. Everaerts, J. Hulsman and C. Meijers, *J. Chromatog.*, **182,** *Biomed. Appl.*, **8,** 496 (1980).
[681] Y. Ito, M. Toyoda, H. Suzuki and M. Iwaida, *J. Assoc. Off. Anal. Chem.*, **63,** 1219 (1980).
[682] K. Rubach, P. Offizorz and C. Breyer, *Z. Lebensm. Unters. Forsch.*, **172,** 351 (1981).
[683] E. Kenndler and D. Kaniansky, *J. Chromatog.*, **209,** 306 (1981).

[684] H. E. Revercomb and E. A. Mason, *Anal. Chem.*, **47**, 970 (1975).
[685] F. W. Karasek, *Res. Develop.*, **21**, 34 (1970).
[686] F. W. Karasek, H. Kim and S. Rokushika, *Anal. Chem.*, **50**, 2013 (1978).
[687] J. M. Preston, F. W. Karasek and S. H. Kim, *Anal. Chem.*, **49**, 1746 (1977).
[688] J. C. Tou and G. U. Boggs, *Anal. Chem.*, **48**, 1351 (1976).
[689] T. W. Carr, *J. Chromatog. Sci.*, **15**, 85 (1977).
[690] M. J. Cohen and F. W. Karasek, *J. Chromatog. Sci.*, **8**, 330 (1970).
[691] W. T. Carr, in *Trace Organic Analysis*, pp. 697–703. NBS, Washington (1979).
[692] A. E. Pierce, *Silylation of Organic Compounds*, Pierce Co., Rockford, Ill. (1968).
[693] J. F. Lawrence and R. W. Frei, *Chemical Derivatization in Liquid Chromatography*, Elsevier, Amsterdam (1976).
[694] K. Blau and G. King, *Handbook of Derivatives for Chromatography*, Heyden, London (1977).
[695] D. R. Knapp, *Handbook of Analytical Derivatization Reactions*, Wiley, New York (1979).
[696] J. Drozd, *Chemical Derivatization in Gas Chromatography*, Elsevier, Amsterdam (1981).
[697] R. W. Frei and J. F. Lawrence (eds.), *Chemical Derivatization in Analytical Chemistry*, Plenum Press, New York (1981).
[698] R. W. Frei and W. Santi, *Z. Anal. Chem.*, **277**, 303 (1975).
[699] T. Jupille, *J. Chromatog. Sci.*, **17**, 160 (1979).
[700] R. W. Frei, *J. Chromatog.*, **165**, 75 (1979).
[701] R. W. Frei, L. Michel and W. Santi, *J. Chromatog.*, **126**, 665 (1976).
[702] R. W. Frei, L. Michel and W. Santi, *J. Chromatog.*, **142**, 261 (1977).
[703] R. S. Deelder, M. G. F. Kroll, A. J. B. Beeren and J. H. M. van den Berg, *J. Chromatog.*, **149**, 669 (1978).
[704] P. T. Kissinger, *Anal. Chem.*, **49**, 447A (1977).
[705] R. T. Krause, *J. Chromatog.*, **185**, 615 (1979).
[706] L. F. Skeggs, *Am. J. Clin. Pathol.*, **28**, 311 (1957).
[707] S. Udenfriend, S. Stein, S. Bohlen, P. Dairman, W. Leimgruber and M. Weigele, *Science*, **178**, 871 (1972).
[708] M. Roth, *Anal. Chem.*, **43**, 880 (1971).
[709] R. S. Deelder, M. G. F. Kroll and J. H. M. van den Berg, *J. Chromatog.*, **125**, 307 (1976).
[710] T. D. Schlabach, S. H. Chang, K. M. Goodring and F. E. Regnier, *J. Chromatog.*, **134**, 91 (1977).
[711] K. M. Jonker, H. Poppe and J. F. K. Huber, *Chromatographia*, **11**, 123 (1978).
[712] R. W. Frei and A. H. M. T. Scholten, *J. Chromatog. Sci.*, **17**, 152 (1979).
[713] K. Sasaki, M. Takeda and M. Uchiyama, *J. Assoc. Off. Anal. Chem.*, **59**, 1118 (1976).
[714] C. T. Chen, K. Samejima and Z. Tamura, *Chem. Pharm. Bull.*, **24**, 97 (1976).
[715] W. E. Dodson, E. E. Tyrala and R. E. Hillman, *Clin. Chem.*, **23**, 944 (1977).
[716] J. Mathew and A. W. Elzerman, *Anal. Lett.*, **14**, 1351 (1981).
[717] H. Tsuchiya, T. Hayashi, H. Naruse and N. Takagi, *J. Chromatog.*, **234**, 121 (1982).
[718] J. D. Gaynor and D. MacTavish, *J. Agr. Food Chem.*, **29**, 626 (1981).
[719] R. Gloor and H. Leidner, *Chromatographia*, **9**, 618 (1976).
[720] Y. Hoshika and Y. Takata, *Bunseki Kagaku*, **25**, 529 (1976).
[721] D. Klockow, W. Bayer and W. Faigle, *Z. Anal. Chem.*, **292**, 385 (1978).
[722] M. S. Chauhan and K. Dakshinamurti, *J. Chromatog.*, **237**, 159 (1982).
[723] H. Agenian and A. S. Y. Chau, *Analyst*, **101**, 732 (1976).
[724] W. Dünges, A. Meyer, K. E. Müller, R. Pietschmann, C. Plachetta, R. Sehr and H. Tuss, *Z. Anal. Chem.*, **288**, 361 (1977).
[725] W. Dünges, *Chromatographia*, **9**, 624 (1976).
[726] D. T. Williams, F. Benoit, K. Muzika and R. O'Grady, *J. Chromatog.*, **136**, 423 (1977).
[727] I. Pigulla and E. Röder, *Z. Anal. Chem.*, **293**, 404 (1978).
[728] E. Röder, I. Pigulla and J. Troschütz, *Z. Anal. Chem.*, **288**, 56 (1977).
[729] H. J. Stan and F. W. Hohls, *Z. Lebensm. Untersuch. Forsch.*, **166**, 287 (1978).
[730] B. A. Davis, *J. Chromatog.*, **151**, 252 (1978).

[731] D. W. Selby and R. P. Munden, *Proc. Anal. Div. Chem. Soc.*, **14**, 296 (1977).
[732] J. Pribyl and F. Herzel, *Z. Anal. Chem.*, **286**, 95 (1977).
[733] J. F. Lawrence, C. Renault and R. W. Frei, *J. Chromatog.*, **121**, 343 (1976).
[734] E. Johnson, A. Abu-Shumays and S. R. Abbott, *J. Chromatog.*, **134**, 107 (1977).
[735] S. Selim, *J. Chromatog.*, **136**, 271 (1977).
[736] G. Hesse, W. Forberg, H. Fricke and G. Bradler, *Pharmazie*, **32**, 472 (1977).
[737] C. G. Daughton, D. G. Crosby, R. L. Garnas and D. P. H. Hsieh, *J. Agr. Food Chem.*, **24**, 236 (1976).
[738] O. W. van Auken, M. Hulse and C. L. Durocher, *J. Assoc. Off. Anal. Chem.*, **60**, 1087 (1977).
[739] O. W. van Auken and M. Hulse, *J. Assoc. Off. Anal. Chem.*, **60**, 1081 (1977).
[740] C. R. Nony, M. C. Bowman, C. L. Holder and J. F. Young, *J. Pharm. Sci.*, **65**, 1810 (1976).
[741] N. P. E. Vermeulen, D. de Roode and D. D. Breimer, *Pharm. Weekbl.*, **112**, 637 (1977).
[742] N. P. E. Vermeulen, D. de Roode and J. D. Breimer, *J. Chromatog.*, **137**, 333 (1977).
[743] W. H. Braun, *J. Chromatog.*, **150**, 212 (1978).
[744] C. H. van Peteghem and A. M. Heyndrickx, *J. Agr. Food Chem.*, **24**, 635 (1976).
[745] A. Ferretti and V. P. Flanagan, *Anal. Lett.*, **11**, 195 (1978).
[746] M. S. Chauhan and K. Dakshinamurti, *J. Chromatog.*, **227**, *Biomed. Appl.*, **16**, 323 (1982).
[747] S. Goya, A. Takadate and H. Fujino, *Yakugaku Zasshi*, **102**, 63 (1982).
[748] V. D. Chmil and M. A. Klisenko, *Zh. Analit. Khim.*, **32**, 592 (1977).
[749] J. Vessman and S. Strömberg, *Anal. Chem.*, **49**, 369 (1977).
[750] H. Nakamura, *J. Chromatog.*, **131**, 215 (1977).
[751] K. Imai, M. Tsukamoto and Z. Tamura, *J. Chromatog.*, **137**, 357 (1977).
[752] P. M. Burkinshaw, E. D. Morgan and C. F. Poole, *J. Chromatog.*, **132**, 548 (1977).
[753] C. F. Poole, W. F. Sye, S. Singhawangcha, F. Hsu, A. Zlatkis, A. Arfwidson and J. Vessman, *J. Chromatog.*, **199**, 123 (1980).
[754] T. Kawabata, H. Ohshima, T. Ishibashi, M. Matsui and T. Kitsuwa, *J. Chromatog.*, **140**, 47 (1977).
[755] P. J. O'Neill and R. G. Rahwan, *J. Pharm. Sci.*, **66**, 893 (1977).
[756] K. Matsubayashi, H. Hakusui and M. Sano, *J. Chromatog.*, **143**, 571 (1977).
[757] G. Nicolau, G. Lear, B. Kaul and B. Davidow, *Clin. Chem.*, **23**, 1640 (1977).
[758] E. Bailey, M. Fenoughty and P. L. Richardson, *J. Chromatog.*, **131**, 347 (1977).
[759] R. A. Chapman and J. R. Robinson, *J. Chromatog.*, **140**, 209 (1977).
[760] E. T. Lin, D. C. Brater and L. Z. Benet, *J. Chromatog.*, **140**, 275 (1977).
[761] B. A. Petersen and P. Vouros, *Anal. Chem.*, **49**, 1304 (1977).
[762] T. Fehér, L. Bodrogi, K. G. Fehér and E. Poteczin, *Chromatographia*, **10**, 86 (1977).
[763] J. F. Lawrence and J. J. Ryan, *J. Chromatog.*, **130**, 97 (1977).
[764] R. J. Siezen and R. H. Mague, *J. Chromatog.*, **130**, 151 (1977).
[765] B. A. Petersen, R. N. Hanson, R. W. Giese and B. L. Karger, *J. Chromatog.*, **126**, 503 (1976).
[766] J. F. Lawrence, *J. Chromatog.*, **123**, 287 (1976).
[767] S. W. Dziedzic, S. E. Gitlow and D. L. Krohn, *J. Pharm. Sci.*, **65**, 1262 (1976).
[768] F. F. Sieck, W. S. Johnson, A. F. Cockerill, D. N. B. Mallen, D. J. Osborne and S. J. Barton, *J. Agr. Food Chem.*, **24**, 617 (1976).
[769] L. Laitem and P. Gaspar, *J. Chromatog.*, **140**, 266 (1977).
[770] J. F. Lawrence, *J. Assoc. Off. Anal. Chem.*, **59**, 1061 (1976).
[771] N. P. Sen, W. F. Miles, S. Seaman and J. F. Lawrence, *J. Chromatog.*, **128**, 169 (1976).
[772] H. Young, *Anal. Biochem.*, **79**, 226 (1977).
[773] C. Sango and E. Zimerson, *J. Liq. Chromatog.*, **3**, 971 (1980).
[774] R. Wintersteiger, G. Gamse and W. Pacha, *Z. Anal. Chem.*, **312**, 455 (1982).
[775] G. L. Dadisch and P. Wolschann, *Z. Anal. Chem.*, **292**, 219 (1978).
[776] T. J. Hopen, R. C. Briner, H. G. Sadler and R. L. Smith, *J. Forensic Sci.*, **21**, 842 (1976).

[777] R. Becker, *Arzneimittelforsch.*, **27**, 102 (1977).
[778] H. J. Klimisch and D. Ambrosius, *J. Chromatog.*, **121**, 93 (1976).
[779] J. H. Wolfram, J. I. Feinberg, R. C. Doerr and W. Fiddler, *J. Chromatog.*, **132**, 37 (1977).
[780] F. Nachtmann, H. Spitzy and R. W. Frei, *Anal. Chem.*, **48**, 1576 (1976).
[781] F. Nachtmann, H. Spitzy and R. W. Frei, *J. Chromatog.*, **122**, 293 (1976).
[782] C. R. Clark, J. D. Teague, M. M. Wells and J. H. Ellis, *Anal. Chem.*, **49**, 912 (1977).
[783] F. Nachtmann and K. W. Budna, *J. Chromatog.*, **136**, 279 (1977).
[784] P. M. Froehlich and T. D. Cunningham, *Anal. Chim. Acta*, **97**, 357 (1978).
[785] W. F. Staruszkiewicz, E. M. Waldron and J. F. Bond, *J. Assoc. Off. Anal. Chem.*, **60**, 1125 (1977).
[786] Y. Hoshika, *Anal. Chem.*, **49**, 541 (1977).
[787] L. Neelakantan and H. B. Kostenbauder, *J. Pharm. Sci.*, **65**, 740 (1976).
[788] N. Nose, S. Kobayashi, A. Tanaka, A. Hirose and A. Watanabe, *J. Chromatog.*, **125**, 439 (1976).
[789] J. H. Onley, *J. Assoc. Off. Anal. Chem.*, **60**, 679 (1977).
[790] H. Agemian and A. S. Y. Chau, *J. Assoc. Off. Anal. Chem.*, **60**, 1070 (1977).
[791] B. Davis, *Anal. Chem.*, **49**, 832 (1977).
[792] M. J. Kogan, K. G. Verebey, A. C. DePace, R. B. Resnick and S. J. Mule, *Anal. Chem.*, **49**, 1965 (1977).
[793] K. T. Koshy, D. G. Kaiser and A. L. VanDerSlik, *J. Chromatog. Sci.*, **13**, 97 (1975).
[794] F. A. Fitzpatrick, M. A. Wynalda and D. G. Kaiser, *Anal. Chem.*, **49**, 1032 (1977).
[795] E. K. Doms, *J. Chromatog.*, **140**, 29 (1977).
[796] F. T. Noggle, *J. Assoc. Off. Anal. Chem.*, **63**, 702 (1980).
[797] J. R. King, C. R. Nony and M. C. Bowman, *J. Chromatog. Sci.*, **15**, 14 (1977).
[798] D. D. Godse, J. J. Warsh and H. C. Stancer, *Anal. Chem.*, **49**, 915 (1977).
[799] D. R. Wilkinson, F. Pavlokowski and P. Jenson, *J. Forensic Sci.*, **21**, 564 (1976).
[800] S. Caccia and A. Jori, *J. Chromatog.*, **144**, 127 (1977).
[801] B. A. Davis, D. A. Durden, P. Pun-Li and A. A. Boulton, *J. Chromatog.*, **142**, 517 (1977).
[802] A. E. Halaris, E. M. Demet and M. E. Halari, *Clin. Chim. Acta*, **78**, 285 (1977).
[803] J. Slemrova and I. Nitsche, *J. Chromatog.*, **135**, 401 (1977).
[804] S. D. West and E. W. Day, *J. Assoc. Off. Anal. Chem.*, **60**, 904 (1977).
[805] S. D. West and R. O. Burger, *J. Assoc. Off. Anal. Chem.*, **63**, 1304 (1980).
[806] G. Phillipou, D. A. Bigham and R. F. Seamark, *Lipids*, **10**, 714 (1975).
[807] R. J. Francis, P. B. East, S. J. McLaren and J. Laman, *Biomed. Mass Spectrom.*, **3**, 281 (1976).
[808] R. H. Horrocks, E. J. Hindle, P. A. Lawson, D. H. Orrell and A. J. Poole, *Clin. Chim. Acta*, **69**, 93 (1976).
[809] S. I. Goodman, P. Helland, O. Stokke, A. Flatmark and E. Jellum, *J. Chromatog.*, **142**, 497 (1977).
[810] M. Donike, R. Gola and L. Jaenicke, *J. Chromatog.*, **134**, 385 (1977).
[811] H. O. Günther, *Z. Anal. Chem.*, **290**, 389 (1978).
[812] H. Kieninger and D. Boeck, *Brauwiss.*, **30**, 357 (1977).
[813] F. Drawert, V. Lessing and G. Leupold, *Chem. Mikrobiol. Technol. Lebensm.*, **5**, 65 (1977).
[814] N. Mahy and E. Gelpi, *J. Chromatog.*, **130**, 237 (1977).
[815] B. R. Wilson, W. Snedden, R. E. Silman, I. Smith and P. Mullen, *Anal. Biochem.*, **81**, 283 (1977).
[816] M. C. Ivey and D. D. Oehler, *J. Agr. Food. Chem.*, **24**, 1049 (1976).
[817] M. Lauwereys and A. Vercruysse, *Chromatographia*, **9**, 520 (1976).
[818] K. R. Leimer, R. H. Rice and C. W. Gehrke, *J. Chromatog.*, **141**, 355 ((1977).
[819] M. Arjmand and R. O. Mumma, *J. Agr. Food Chem.*, **24**, 574 (1976).
[820] R. Tatsukawa, J. Itoh and T. Wakimoto, *J. Agr. Chem. Soc. Japan*, **51**, 315 (1977).
[821] J. Rosenfeld, *Anal. Lett.*, **10**, 917 (1977).
[822] H. Miyazaki, M. Ishibashi, M. Itoh, K. Yamashita and T. Nambara, *J. Chromatog.*, **133**, 311 (1977).

[823] A. Sanghvi, M. Galli-Kienle and G. Galli, *Anal. Biochem.*, **85**, 430 (1978).
[824] K. Olek and P. Wardenbach, *J. Clin. Chem. Clin. Biochem.*, **15**, 657 (1977).
[825] G. M. Singer and W. Lijinsky, *J. Agr. Food Chem.*, **24**, 550 (1976).
[826] N. O. Ahnfelt and P. Hartvig, *Acta Pharm. Suec.*, **17**, 307 (1980).
[827] R. Takeshita, E. Takabatake, K. Minagawe and Y. Takizawa, *J. Chromatog.*, **133**, 303 (1977).
[828] G. Jungcalus, *J. Chromatog.*, **139**, 174 (1977).
[829] M. C. Sanchez, K. Colome and E. Gelpi, *J. Chromatog.*, **126**, 601 (1976).
[830] V. Herrmann and F. Jüttner, *Anal. Biochem.*, **78**, 365 (1977).
[831] G. Eklund, B. Josefson and C. Roos, *J. Chromatog.*, **142**, 575 (1977).
[832] S. Mierzwa and S. Wotek, *J. Chromatog.*, **136**, 105 (1977).
[833] R. Schmid and M. Karobath, *J. Chromatog.*, **139**, 101 (1977).
[834] C. Razzouk, G. Lhoest, M. Roberfroid and M. Mercier, *Anal. Biochem.*, **83**, 194 (1977).
[835] P. Stantscheff, *Z. Anal. Chem.*, **292**, 39 (1978).
[836] S. Fittkau, *Pharmazie*, **35**, 646 (1980).
[837] W. J. Trotter, *J. Assoc. Off. Anal. Chem.*, **58**, 461 (1975).
[838] M. Mansour and H. Parlar, *J. Agr. Food Chem.*, **26**, 483 (1978).
[839] D. Friedman and P. Lombardo, *J. Assoc. Off. Anal. Chem.*, **58**, 703 (1975).
[840] W. H. Dennis and W. J. Cooper, *Bull. Environ. Contam. Toxicol.*, **16**, 425 (1976).
[841] J. H. Carey, J. Lawrence and H. M. Tosine, *Bull. Environ. Contam. Toxicol.*, **16**, 697 (1976).
[842] R. G. Lewis, R. G. Hanish, K. E. Macleod and G. W. Sovocool, *J. Agr. Food Chem.*, **24**, 1030 (1976).
[843] J. Mes, D. J. Davies and W. Miles, *Bull. Environ. Contam. Toxicol.*, **19**, 546 (1978).
[844] R. H. Lane, R. M. Grodner and J. L. Graves, *J. Agr. Food Chem.*, **24**, 192 (1976).
[845] D. R. Erney, *J. Assoc. Off. Anal. Chem.*, **58**, 1202 (1975).
[846] W. J. Trotter, *Bull. Environ. Contam. Toxicol.*, **18**, 726 (1977).
[847] T. B. Putnam, D.D. Bills and L. M. Libbey, *Bull. Environ. Contam. Toxicol.*, **13**, 662 (1975).
[848] P. M. Ward, *J. Assoc. Off. Anal. Chem.*, **60**, 673 (1977).
[849] L. O. Ruzo, N. J. Bunce, S. Safe and O. Hutzinger, *Bull. Environ. Contam. Toxicol.*, **14**, 341 (1975).
[850] J. R. Plimmer, *Bull. Environ. Contam. Toxicol.*, **20**, 87 (1978).
[851] J. Královsky and P. Matoušek, *J. Chromatog.*, **147**, 404 (1978).
[852] M. Mráz and V. Šedivec, *Collection Czech. Chem. Commun.*, **42**, 1347 (1977).
[853] R. H. Bromilow and K. A. Lord, *J. Chromatog.*, **125**, 495 (1976).
[854] C. V. Abraham, *Microchem. J.*, **21**, 272 (1976).
[855] W. Beyer, W. Truppe and W. Mlekusch, *Z. Med. Labortechnik*, **17**, 267 (1976).
[856] A. Sengupta and M. A. Peat, *J. Chromatog.*, **137**, 206 (1977).
[857] R. G. Wien and F. S. Tanaka, *J. Chromatog.*, **130**, 55 (1977).
[858] K. K. Midha, I. J. McGilveray and D. L. Wilson, *J. Pharm. Sci.*, **65**, 1240 (1976).
[859] J. de Graeve and J. Vanroy, *J. Chromatog.*, **129**, 171 (1976).
[860] A. Kumps and Y. Mardens, *Clin. Chim. Acta*, **62**, 371 (1975).
[861] W. C. Kossa, J. MacGee, S. Ramachandran and A. J. Webber, *J. Chromatog. Sci.*, **17**, 177 (1979).
[862] R. H. Hammer, J. L. Templeton and H. L. Panzik, *J. Pharm. Sci.*, **63**, 1963 (1974).
[863] A. De Kok, R. B. Geerdink, R. W. Frei and U. A. T. Brinkman, *Intern. J. Environ. Anal. Chem.*, **9**, 301 (1981).
[864] M. A. T. Kerkhoff, A. De Vries, R. C. C. Wegman and A. W. M. Hofstee, *Chemosphere*, **11**, 165 (1982).
[865] R. C. C. Wegman and G. A. L. de Korte, *Intern. J. Environ. Anal. Chem.*, **9**, 1 (1981).
[866] B. A. Davis, *J. Chromatog.*, **151**, 252 (1978).
[867] C. P. Terwij-Groen and J. C. Kraak, *J. Chromatog.*, **138**, 245 (1977).
[868] B. Mellström and G. Tybring, *J. Chromatog.*, **143**, 597 (1977).
[869] H. G. Fouda, *J. Chromatog. Sci.*, **15**, 537 (1977).
[870] E. Soczewinski, D. Ratajewicz and M. Rojowska, *Acta Polonia Pharm.*, **34**, 181 (1977).

[871] C. Horvath, W. Melander, I. Molnar and P. Molnar, *Anal. Chem.*, **49**, 2295 (1977).
[872] A. F. Casy, *Proc. Anal. Div. Chem. Soc.*, **14**, 292 (1977).
[873] K. Toei, H. Miyata, S. Motomizu and J. Tetsumoto, *Bunseki Kagaku*, **27**, 138 (1978).
[874] K. Oba, K. Miura, H. Sekiguchi, R. Yagi and A. Mori, *Water Res.*, **10**, 149 (1976).
[875] L. Favretto, B. Stancher and F. Tunis, *Analyst*, **103**, 955 (1978).
[876] B. Karlberg, P. A. Johansson and S. Thelander, *Anal. Chim. Acta*, **104**, 21 (1979)..
[877] J. Kuczynski and E. Soczewinski, *Chem. Anal. (Warsaw)*, **23**, 421 (1978).
[878] S. Eksborg, B. A. Persson, L. G. Allgen, J. Bergström, L. Zimmermann and P. Fürst, *Clin. Chem. Acta*, **82**, 141 (1978).
[879] B. A. Persson and P. O. Lagerström, *J. Chromatog.*, **122**, 305 (1976).
[880] T. P. Moyer and N. S. Jiang, *J. Chromatog.*, **153**, 365 (1978).
[881] J. M. Huen, R. W. Frei, W. Santi and J. P. Thevenin, *J. Chromatog.*, **149**, 359 (1978).
[882] C. P. Terweij-Groen, J. C. Kraak, W. M. A. Niessen, J. F. Lawrence, C. E. Werkhoven-Goewie, U. A. T. Brinkman and R. W. Frei, *Intern. J. Environ. Anal. Chem.*, **9**, 45 (1981).
[883] M. D. Baker, H. Y. Mohammed and H. Veening, *Anal. Chem.*, **53**, 1658 (1981).
[884] D. Stevenson and E. Reid, *Anal. Lett.*, **14**, 741 (1981).
[885] J. J. Warsh, A. Chiu and D. D. Godse, *J. Chromatog.*, **228**, *Biomed. Appl.*, **17**, 131 (1982).
[886] M. Kozma, P. Pudleiner and L. Vereczkey, *J. Chromatog.*, **241**, 177 (1982).
[887] D. W. Fritz and R. D. Strahm, *J. High Resolut. Chromatog., Chromatog. Commun.*, **4**, 584 (1981).
[888] L. L. Dent, J. T. Stewart and I. L. Honigberg, *Anal. Lett.*, **14**, 1031 (1981).
[889] B. E. Roos, G. Wickstrom, P. Hartvig and J. L. G. Nilsson, *Eur. J. Clin. Pharmacol.*, **17**, 223 (1980).
[890] O. Gyllenhaal, H. Brötell and P. Hartvig, *J. Chromatog.*, **129**, 295 (1975).
[891] J. E. Greving, J. H. G. Jonkman, F. Fiks, R. de Zeeuw, L. E. van Bork and N. G. M. Orie, *J. Chromatog.*, **142**, 611 (1977).
[892] A. Arbin, *J. Chromatog.*, **144**, 85 (1977).
[893] H. Brötell, H. Ersson and O. Gyllenhaal, *J. Chromatog.*, **78**, 293 (1973).
[894] P. Hartvig and C. Fagerlund, *J. Chromatog.*, **140**, 170 (1977).
[895] O. Gyllenhaal, H. Brötell and B. Sandgren, *J. Chromatog.*, **122**, 471 (1976).
[896] M. Garle and I. Petters, *J. Chromatog.*, **140**, 165 (1977).
[897] O. Gyllenhaal, U. Tjärnlund, H. Ehrsson and P. Hartvig, *J. Chromatog.*, **156**, 275 (1978).
[898] P. Hartvig, P. Gyllenhaal and M. Hammarlund, *J. Chromatog.*, **151**, 232 (1978).
[899] C. Fagerlund, P. Hartvig and B. Lindström, *J. Chromatog.*, **168**, 107 (1979).
[900] P. H. Degen, W. Schneider, P. Vuillard, U. P. Geiger and W. Riess, *J. Chromatog.*, **117**, 407 (1976).
[901] M. Ervic and K. Gustavii, *Anal. Chem.*, **46**, 39 (1974).
[902] B. Lindström and M. Molander, *J. Chromatog.*, **101**, 219 (1974).
[903] B. Lindström and M. Molander, *J. Chromatog.*, **114**, 459 (1975).
[904] D. J. Hoffman, *J. Pharm. Sci.*, **69**, 445 (1979).
[905] A. Berg and J. Lam, *J. Chromatog.*, **16**, 157 (1964).
[906] C. Gertz, *Z. Lebensm. Unters. Forsch.*, **173**, 208 (1981).
[907] L. Kolarovic and H. Traitler, *J. Chromatog.*, **237**, 263 (1982).
[908] K. Morita, K. Fukamachi and H. Tokiwa, *Bunseki Kagaku*, **31**, 255 (1982).
[909] F. A. Medvedev, D. B. Melamed and Y. L. Kostyukovskii, *Biomed. Mass Spectrom.*, **7**, 354 (1980).
[910] W. I. Kimoto and W. Fidler, *J. Assoc. Off. Anal. Chem.*, **65**, 1162 (1982).
[911] A. De Kok, I. M. Roorda, R. W. Frei and U. A. T. Brinkman, *Chromatographia*, **14**, 579 (1981).
[912] A. H. M. T. Scholten, B. J. De Vos, J. F. Lawrence, U. A. T. Brinkman and R. W. Frei, *Anal. Lett.*, **13**, 1235 (1980).
[913] G. Skarping, C. Sango and B. E. F. Smith, *J. Chromatog.*, **208**, 313 (1981).
[914] H. H. Ting and M. P. Quick, *J. Chromatog.*, **195**, 441 (1980).
[915] N. Nose, Y. Hoshino, Y. Kikuchi, F. Yamada and S. Kawauchi, *Shokuhin Eiseigaku Zasshi*, **22**, 496 (1981).

[916] S. Traore and J. J. Aaron, *Talanta,* **28,** 765 (1981).
[917] T. R. Nelson and M. H. Gruenauer, *J. Chromatog.,* **212,** 366 (1981).
[918] W. H. Newsome, *J. Agr. Food Chem.,* **30,** 778 (1982).
[919] W. Gielsdorf, *J. Clin. Chem. Clin. Biochem.,* **20,** 65 (1982).
[920] K. Makino, N. Kashihira, K. Kirita and Y. Watanabe, *Bunseki Kagaku,* **31,** 417 (1982).
[921] W. C. McCrone and J. G. Delly, *The Particle Atlas,* 2nd Ed., 4 Vols., Ann Arbor (1973).
[922] W. C. McCrone, *Z. Anal. Chem.,* **282,** 275 (1976).
[923] D. Hasenmaier, *Dissertation,* Mainz (1980).

6

Methods for the determination or detection of compounds

6.1 GENERAL ASPECTS

In the analysis of organic substances meaningful results are nearly always obtainable only by molecule-specific methods. Elemental analysis, at best, gives only hints of composition in the trace concentration range. For this reason, an extremely effective tool for the trace analysis of inorganic components, viz. radiochemical activation analysis, is almost useless for the trace analysis of organic compounds.

A look at the literature shows a picture similar to that for separation methods. Only a few procedures are extensively used, though many methods have been suggested (see Table 6.1).

Table 6.1 – Percentage of publications which describe certain determination methods of organic trace analysis (calculated from *Analytical Abstracts* for 1978 and 1979).

Method	Percentage of publications
gas chromatography	37
mass spectrometry	14
ultraviolet spectrophotometry	11
thin-layer densitometry	8
fluorescence, phosphorescence	8
radioimmunoassay, enzymatic methods	6
electrochemical methods	4
infrared spectrometry	3

The current trend is towards more use of radioimmunoassay and fluorescence (mainly in combination with hplc) and towards tlc densitometry.

The methods used can be assessed according to their merits in the following respects:

(1) capacity to measure molecular properties directly and specifically;
(2) tolerance towards other substances and interferences;
(3) sensitivity, working range of concentration, limit of detection;
(4) mode of operation, viz. destructive (e.g. fid) or non-destructuve (e.g. photometry);
(5) capacity for automation.

The concentrations and working areas of the methods most often used are given below. This classification is schematic, as the limits of determination can be improved for several of the methods. The scheme demonstrates that the range from 1 to 100 ng/ml is covered by most of the methods, which thus offer means of determination which would have been impossible 20 years ago.

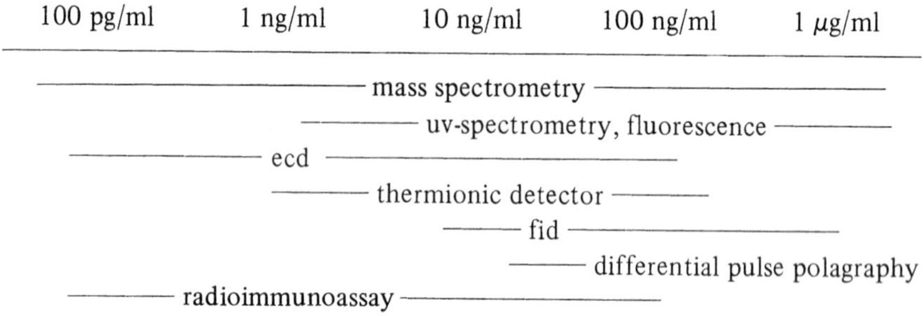

100 pg/ml   1 ng/ml   10 ng/ml   100 ng/ml   1 μg/ml

——————— mass spectrometry ———————
——— uv-spectrometry, fluorescence ———
——————— ecd ———————
——— thermionic detector ———
—— fid ———————
—— differential pulse polagraphy
——— radioimmunoassay ———————

6.2 GENERAL REMARKS ON SPECTROSCOPIC METHODS

There are two principal modes of operation for spectroscopic methods. In both, the primary energy I_1 interacts with the sample components. In absorption methods the energy I_2 which has not been absorbed by the sample is measured. In secondary emission methods the energy I_s which results from fluorescence or phosphorescence processes is monitored. In absorption techniques in trace analysis comparatively weak interaction with the primary energy takes place. Thus, the incident radiation and the radiation transmitted through the sample have about the same energy. This calls for a very accurate determination of small differences in two energy beams in trace analysis. In secondary emission, no effect is seen with zero concentrations of the analyte (under ideal conditions), but even a very small concentration of an emitting substance gives a measurable increase in secondary emission, which can be compared with the zero effect more simply and easily than absorption signals can be compared. Consequently, secondary emission methods have a much lower limit of detection than absorption methods.

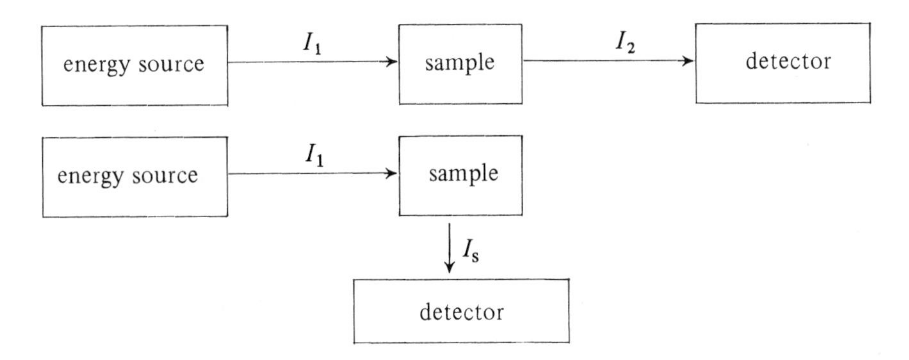

There are some practical and experimental limitations, though, which partly offset these advantages. First, not all molecules are capable of giving secondary emission, which limits the general applicability of this method. On the other hand, it adds to the specificity. Secondly, general (Rayleigh) scattering by the sample often interferes.

6.3 SPECTROMETRY WITH ULTRAVIOLET AND VISIBLE LIGHT

This technique measures molecular properties indirectly, as features of the chromophoric groups are detected. The wavelength of maximal absorption depends on the chemical nature of the chromophore. The method is non-destructive and the species can be additionally determined by another method, except in cases when a reagent has been added to give a reaction product which is measured spectrophotometrically. Fairly dilute solutions ($\sim 10^{-5}M$) can be investigated, but appropriate selection of solvent is necessary. Other sample components can be tolerated provided they do not absorb radiation in the same wavelength region, but there are techniques for dealing with certain cases of spectral overlap. Measurement can be done on-line by using flow-through cuvettes.

"The 254 nm uv-absorbance detector is the 'Chevrolet' of hplc detectors. It is simple, cheap, rugged, reliable and sensitive. A low-pressure mercury arc is used, which is an inexpensive light source that concentrates most of its output energy in a narrow spectral band [1]. The use of this detector calls for chromophores with absorption maxima at or near 254 nm. Chromophores which absorb light at this wavelength can be attached to non-absorbing molecules with great advantage. A molar extinction coefficient of about 10^4 is desirable for trace analysis with ng-amounts. The most common approach involves the attachment of a substituted (favorably nitro-group) aromatic chromophore by a derivatization reaction to the substance to be determined" [2]. Highly specialized procedures permit the determination of ng-amounts and less [3].

6.4 INFRARED ABSORPTION

6.4.1 General aspects

Infrared spectrometry is – next to mass spectrometry – the procedure of the highest molecular specificity in organic trace analysis. Except for two enantiomers, no two chemical compounds have the same ir spectrum. Measurement in the fingerprint region permits a reliable identification and determination of organic substances. Unfortunately, the limit of detection is not very low. Even μg amounts are determinable only by use of refined techniques. There is also great interference from other constituents. A trace which is to be investigated by ir methods must be rather pure and its solution must be fairly concentrated. For example, 0.1% solutions can be investigated in microcuvettes (1 μl volume). The method is non-destructive, but recovery is difficult, as very often the trace compound is incorporated into KBr pellets for measurement and is difficult to recover from them. On-line measurement is possible with a gas stream as carrier, provided it does not absorb ir radiation in the same region. Such methods require special gas cells or 'light pipes'.

Fourier analysis is a new instrumental method by which the signal to noise ratio can be improved. The spectrum is scanned repetitively and the data are averaged. As many repetitions are required, very rapid scanning is necessary, which demands very good response by the detectors and the optical equipment. Determination of ng-amounts by Fourier-ir analysis has been reported [4–6]. The latest developments in instrumentation and application of ir have been reviewed (621 references) [7].

6.4.2 Developments and improvements in ir instrumentation

A computerized ir microspectrophotometer allows identification of samples as small as 20 μm (Nanospec™/20 IR, produced by Nanometrics, Sunnyvale, California, USA). New ir detectors (triglycine sulphate, mercury cadmium telluride, indium antimonide) have sensitivities several hundred times better than the conventional detectors. The response time of modern detectors is very good.

An interesting approach is the use of tunable lasers as the light-source. The output is varied in the 50–200 cm^{-1} range by change of the operating temperature (of the laser) between 15 and 100 K, at a tuning rate of 4 cm^{-1}/deg (LS-3 System of Spectra Physics, Bedford, USA). In principle, such a system can operate without the usual prisms or gratings. Application has been suggested in low-level trace-gas detection, e.g. 170 ppb of SO_2, by use of a path-length of 300 m.

By combination of ir spectrometers with computerized data-handling it is possible to add or subtract spectra, store and retrieve spectra, and do difference spectrometry, with the facility of automatic scaling for compensation work. Spectral averaging over time, spectral accumulation and digital smoothing are also possible. Figure 6.1 gives an example of such an application.

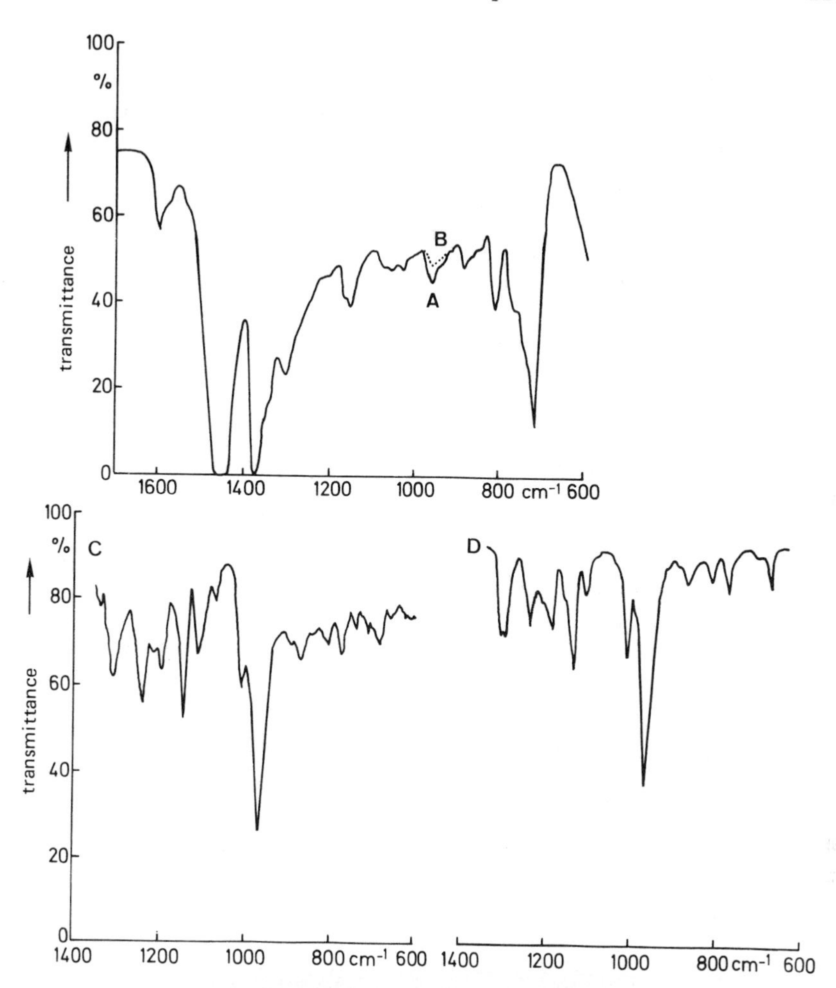

Fig. 6.1 – Example of electronic subtraction of two spectra (by courtesy of Perkin-Elmer): (A) standard lubricant (200-μm path-length cell); (B) sample A with 400 ppm tricresyl phosphate added; (C) computer difference (A – B) with scale expansion and digital smoothing; (D) reference spectrum of tricresyl phosphate.

6.4.3 Sample preparation

Microsampling is a promising development for ir, in which reduction of the beam area is used [8]. Small KBr pellets, 1 mm in diameter, are used, and give good spectra with about 1 μg of sample material.

Aqueous solutions are difficult to analyse, as they dissolve KBr. The solvent has to be evaporated first, and it is then difficult to transfer the isolated trace constituent into the KBr powder. Grinding is necessary in order to obtain crystals sufficiently small for excessive scattering to be avoided.

Other materials have been used for transmission spectrometry of aqueous solutions after evaporation, e.g. sapphire [9] or diamond [10]. In the author's experience sample carriers consisting of a tiny plate (3 mm diameter, 1 mm thickness) made of ZnSe or TlI–TlBr (KRS 5) are very useful for evaporation of even very polar solvents.

Aqueous solutions can be analysed by attentuated total reflection (ATR) or multiple ATR, but the sensitivity is not very high. Microgram amounts of herbicides have been detected [11]. By electronic time-averaging the limit of detection can be improved.

Diffuse reflectance spectroscopy in the infrared became possible after application of the Fourier transform technique, which overcomes the difficulty of detection of scattered radiation at very low intensity [12, 13]. The detection limits are in the μg range [14, 15]. The method is used to measure ir spectra of chromatographically separated fractions, after evaporation of the solvent on KBr.

6.4.4 Micro specular reflectance

This is an ir method which is free from many of the disadvantages of the KBr technique. The sample material is deposited on a tiny mirror (diameter 2 mm) by evaporation of a drop of solution. Naturally, the analyte volatility must be low. Any solvent can be used, including water [16], since it is evaporated. This makes possible the investigation of many compounds of biochemical significance, which are often soluble only in water or very polar solvents. Proteins can be determined in μg quantities [17]. Sample application is very simple (see Fig. 6.2). After addition of the solution and evaporation of the solvent the mirror is fixed in its holder in the reflectance unit and the ir spectrum is recorded. The sensitivity is good, since the light travels through the sample layer twice, being reflected back by the mirror. Spectra are easily obtainable with μg amounts, and

Fig. 6.2 – Micro mirror for infrared reflectance spectroscopy. (Reproduced from [16], by permission of the copyright holders, Springer Verlag).

have about the same quality as the spectra obtained for mg quantities by the standard procedures (see Fig. 6.3). The method is readily combined with separation methods, such as tlc [18] or gc [19]. The determination of non-volatile traces in the nanogram range is achieved by time-averaging [20].

The disadvantage of the method – as with all ir procedures – lies in the necessity to have the trace compound in practically pure form. Also, band broadening is occasionally seen.

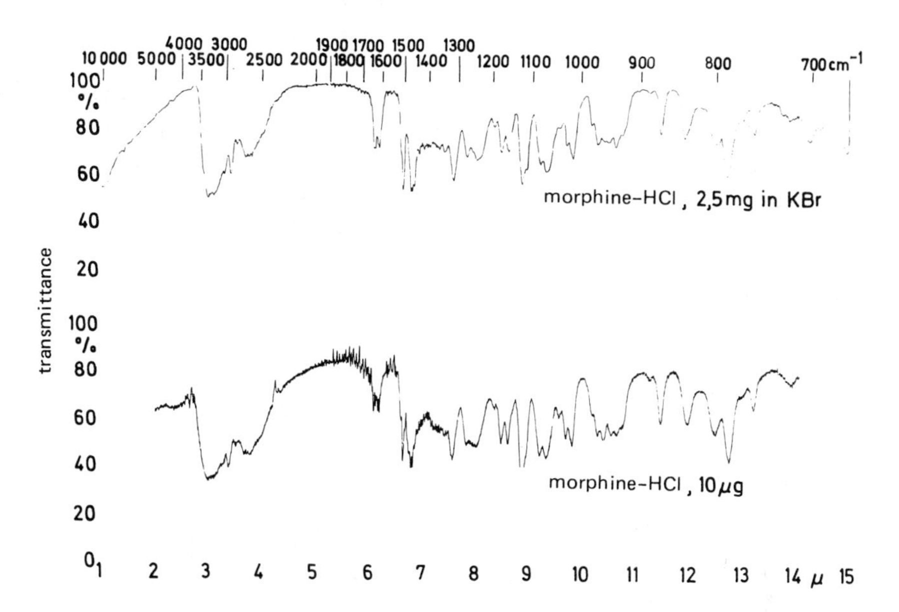

Fig. 6.3 – Infrared spectra of morphine-HCl. Top: 2.5 mg in 750 mg of KBr, conventional pellet technique. Bottom: 10 μg on top of a 3-mm micro mirror, reflectance method. (Reproduced from [16], by permission of the copyright holders, Springer Verlag).

6.4.5 Influence of temperature in ir spectrometry of traces

Cooling the sample to the temperature of liquid nitrogen or below leads to a reduction of molecular randomness and hence to an enhancement in resolution of ir spectra [21–23]. So-called cryogenic ir has been successfully used for the identification of petroleum fractions [24] which could not be differentiated at room temperature.

6.4.6 Matrix isolation spectroscopy

In matrix isolation techniques sample components are mixed with a 10^4–10^8-fold amount of a diluent. Often, an inert gas – e.g. N_2, Ar or He – is used and sample and gas are deposited from the gaseous state at very low temperatures. All sample molecules have diluent molecules or atoms as near neighbours. As the interaction of the trace molecules with each other is zero and with matrix components weak, perturbation of the spectroscopic behaviour is minimized. Very sharp spectra result [25–29]. The method is used in ir spectroscopy (see Fig. 6.4 [29]). Even more impressive results can be seen in fluorescence spectroscopy when the matrix isolation technique is used (see Figs. 6.5 and 6.6 [29]).

The method has been used in combination with gc. The separated compounds are collected individually on cooled mirrors. Under carefully selected collection conditions each matrix contains sample molecules of one species only, isolated by deposition along with the carrier gas molecules [30]. Sharp bands in the ir are observed. It is possible to detect as little as 0.16 ng by use of strong ir bands; weak bands allow detection of about 80 ng of analyte [31].

The gases used (N_2, Ar, He) have the advantages that they can be used as the carrier gas in the gc as well as the trapping medium in the deposition technique and that they do not absorb ir radiation. They have the disadvantage that they require very low temperatures for condensation. Other solvents, which have been used in low-temperature fluorescence, have therefore been suggested [32, 33].

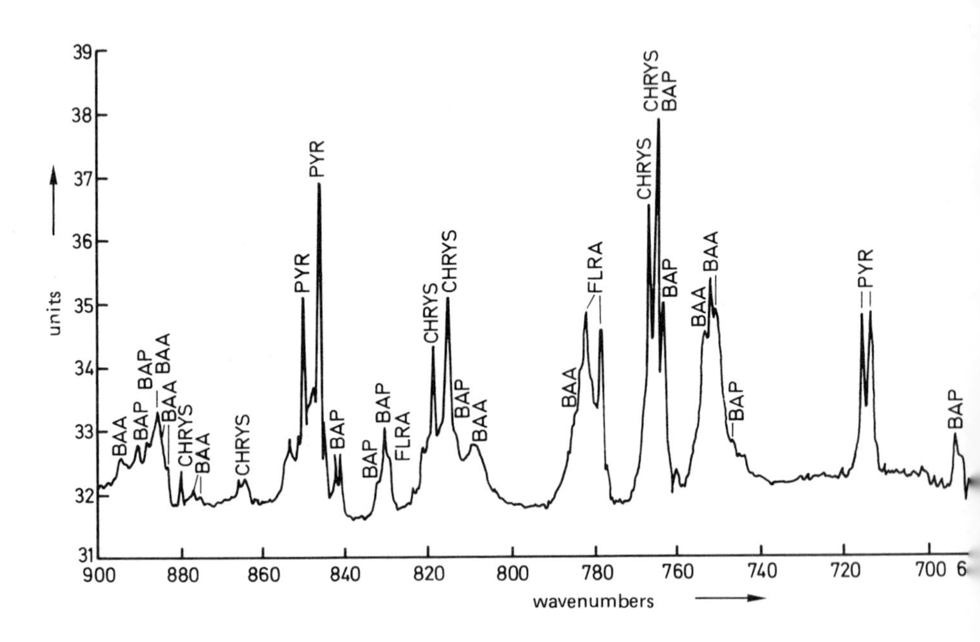

Fig. 6.4 – Matrix isolation Fourier-transform infrared spectrum of a synthetic mixture of five polycyclic aromatic hydrocarbons (25 μg of each) at the temperature of liquid N_2. BAA, benz(*a*)anthracene; CHRYS, chrysene; PYR, pyrene; FLRA, fluoranthene; BAP, benz(*a*)pyrene. [Reproduced by permission from E. L.Wehry and G. Mamantov, *Anal. Chem.*, **51**, 643A (1979). Copyright 1979 American Chemical Society].

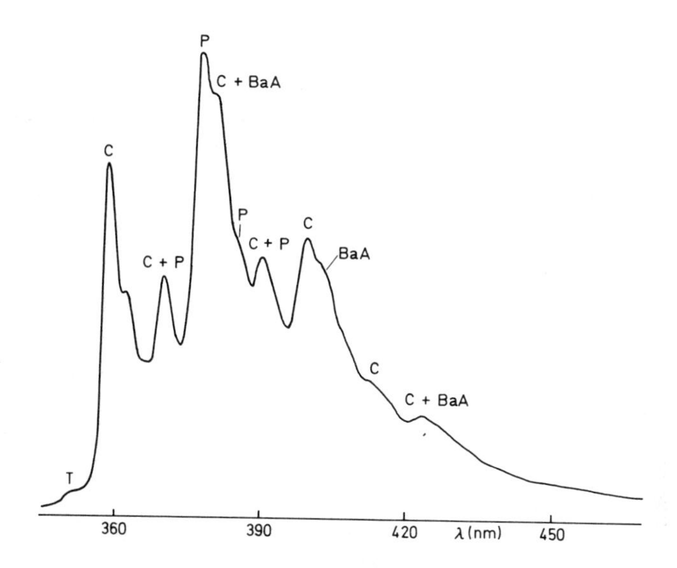

Fig. 6.5 – Room temperature fluorescence spectrum of an acetone solution of PAHs (10^{-6}M). P, pyrene; BaA, benz(*a*)anthracene; C, chrysene; T, triphenylene. [Reproduced by permission from E. L. Wehry and G. Mamantov, *Anal. Chem.*, **51,** 643A (1979). Copyright 1979 American Chemical Society].

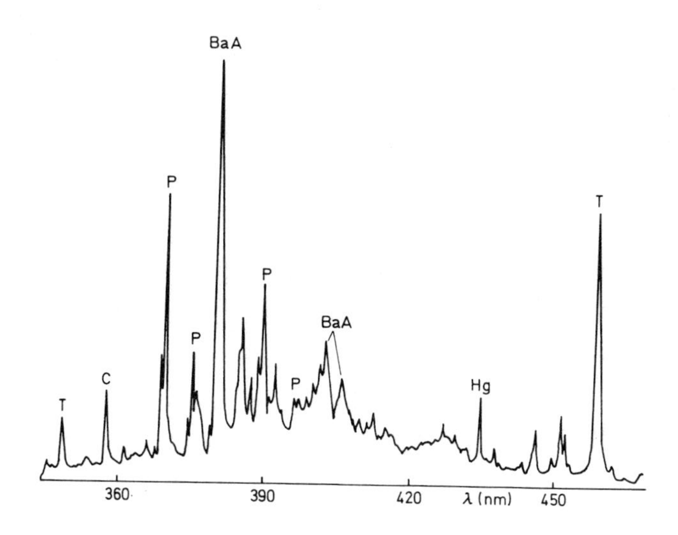

Fig. 6.6 – Fluorescence spectrum, matrix isolation mode, liquid-N_2 temperature, 100 ng of BaA, P and C, 4 μg of T (for symbols see Fig. 6.5), BaA, T and C are isomers! [Reproduced by permission from E. L. Wehry and G. Mamantov, *Anal. Chem.*, **51,** 643A (1979). Copyright 1979 American Chemical Society].

6.4.7 Coupling of ir with gc

As with all chromatographic separation and detection systems, measurement in the gc–ir technique can be done both on-line and off-line. On-line, the effluent can be constantly monitored. An alternative is on-line measurement with stopped flow; in this the continuous operation is interrupted and the signal from the static system is recorded. Off-line detection uses a detection system that is separate from the chromatographic column, and requires trapping of the effluent or collection in a separable gas cuvette.

Obviously, an on-line system with no interruption offers the greatest ease of operation. Consequently, much effort has been made to develop and improve such combinations. The gc effluent is passed through a cuvette or light-pipe for inspection by ir. Such light-pipes are often designed with reflecting walls, so that the light-path is multiplied several times. The volume of the cuvette and its associated transfer lines should be small enough to prevent significant peak broadening. Some factors to consider in optimizing light-pipe volumes with respect to gc peak volumes have been discussed in the literature [34, 35]. The volume of the cell should be equal to or slightly smaller than the volume of carrier gas corresponding to the peak-width at half-height. The investigator has to decide when the measurement of a given component is to start and finish.

Evidently, a successful determination requires a very rapid scanning of the entire ir spectrum within seconds. In the example of Fig. 6.7 even a scanning time as short as 4 sec could become problematic, since an ir band scanned at the beginning of the first second will be smaller than if it were scanned at the end of the next second, because of its somewhat lower concentration within the light-pipe. An ir band which happens to be recorded at the peak maximum will

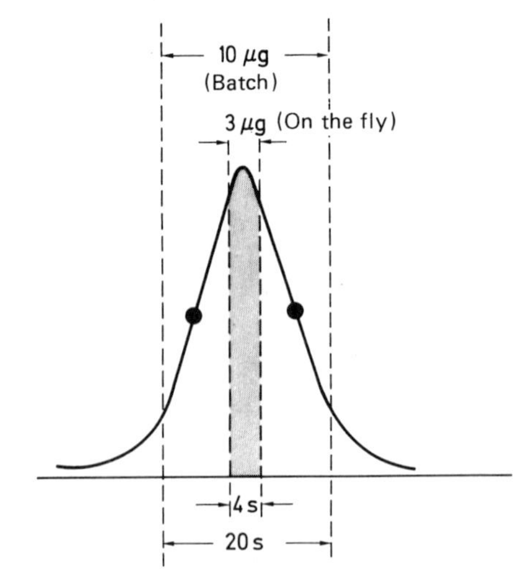

Fig. 6.7 – Quantities of material available to a spectrometer by gc 'batch' and 'on the fly' methods. [Adapted from L. S. Ettre and W. H. McFadden (eds.), *Ancillary Techniques of Gas Chromatography*, p. 229, by permission of the copyright holders, Wiley–Interscience Inc.].

have its full height whereas another band registered at the end of the scanning time will be attenuated. One solution to this problem is multiple scanning and time averaging of the data collected during the elution time. Fast optical and electronic equipment is mandatory for such work.

The data collection is started automatically by triggering the infrared-inspection and data-storage in the computer system whenever the gc peak-height exceeds a preselected threshold ('on the fly method'). Alternatively, the ir system can be set to collect data continuously in groups of preselected numbers of scans ('every mode method'). This latter technique results in a large number of spectra, which are recorded as a function of time. A combination of both methods is the 'every threshold' approach, in which data are collected not over the entire chromatogram, but only on those parts which exceed a certain threshold [36].

The routine application of Fourier-transform gc–ir analysis requires the ability to determine quickly which of the thousands of interferograms collected contain spectral information [37]. There are three possibilities.

(1) Use of a separate detector along with the ir-spectrometer. Unless a non-destructive detector (so far, only low-sensitivity thermal conductivity detectors are available) is used, the gc procedure has to be run twice, once for collection of the interferograms, the other for measurement. The problem is to duplicate the instrument performance sufficiently well.
(2) Use of a stream splitter to divert part of the effluent to a gc detector and part to the ir cell. However, the use of the splitter and extra tubing can degrade the chromatographic performance of the system.
(3) Use of the ir spectrometer as a detector, with fast computing equipment and low spectral resolution.

The combination of gc and ir has been described in several publications [38–47]. A real breakthrough was achieved, however, only after the introduction of Fourier-transform ir (see Table 6.2).

Table 6.2 – Application of Fourier-transform ir in combination with gc

Problem or result	Reference
achievement of ng sensitivity	49
100 ng of isobutyl methacrylate (as 100 ppm mixture with CH_2Cl_2)	50
double beam FT–ir for improved sensitivity	51, 52
50 ng of isobutyl methacrylate determined 'on the fly'	53
analysis of waste water effluents	54, 55
analysis of shale oil acids and bases	56
isobutyl methacrylate, limit of detection 460 ng	57
PAHs, limit of detection 720 ng	58
21 priority pollutants detected	59
isobutyl methacrylate, valeraldehyde etc., limit of detection 20 ng	60
55 organic compounds	61

The off-line strategy is much simpler in terms of technology. It requires more manipulation, though, with the attendant danger of loss and contamination. Trapping is most often used. Fractions are collected on a heat-stable adsorbent such as Tenax. The traces are desorbed by passing a current of gas through the Tenax collector and into the ir gas cuvette, or – more simply – by placing the Tenax + trace in the cuvette, which is then heated to desorb the analyte [62].

Equipment is also available for 'on the fly' and stopped flow action. The gc equipment and the ir spectrometer can be used separately, if desired.

6.4.8 Combination of ir with liquid chromatography

In thin-layer chromatography the separated analytes are usually eluted from the spots and investigated after evaporation of the solvent. This method requires some manipulations, and losses and contamination are always possible. Therefore, many efforts have been made to measure ir spectra directly *in situ*. Scattering has been minimized by treatment of the adsorbent layer with Fluorolube, an ir mulling oil [63]; its refractive index closely matches that of the adsorbent. FT-ir spectra of microgram quantities of pesticides have been obtained.

For combination with column chromatography two approaches have been used. In one, the column effluent is investigated directly in a micro ir-cell, but good compensation of the solvent bands is necessary and any change in solvent composition, flow etc. results in problems. Also, most window materials – such as KBr, NaCl – do not tolerate polar solvents. Examples are the detection of dioctyl phthalate and diethyl phthalate (in iso-octane as solvent) [64], ng amounts of glycerides, alkanes and alkenes by gradient elution [65], polyoxyethylene surfactants in chloroform as solvent [66] and of 525 ng of paraffin oil [67]. Use of FT-ir improves the signal-to-noise ratio, and this is essential for direct ir analysis of column eluates, but solvent band compensation is always necessary [68].

In the other method, to circumvent these problems the solvent is evaporated on KBr powder and the diffuse reflectance spectrum measured [67]. Detection is in the submicrogram range, again with the use of FT-ir methodology [69].

6.4.9 Application of ir methods in organic trace analysis

Anionic surfactants must be identified in waste waters in concentrations of a few ng/ml or even less. Methods for separation include partition of their Methylene Blue salts [70], solvent sublation and tlc [71, 72], and partition and tlc [73]. Infrared spectroscopy is capable of demonstrating alterations in traces of anionic surfactants during analytical operations [74].

In gas analysis, ir techniques are used for the identification of commonly encountered tear gases [75], and for detection of traces of methane in medicinal gases [76]. Ethylene and isoprene have been determined in the gas phase of cigarette smoke [77]. Toxic gases in tissue specimens have been released by maceration, collected by the head-space method and analysed by ir [78]. OSHA lists some 170 hazardous chemicals which are gases or vapours and must be determined in the atmosphere. They are adsorbed on charcoal (carried by the exposed worker) and eluted with carbon disulphide, which is then inspected by ir [79, 80]. Trace impurities in gases are determined by condensation of the matrix gas and the impurities. Such condensates are analysed by ir at low temperature. In this way ppm concentrations of Freon 12 and vinyl chloride have been detected [81].

The purity of butane-1,4-diol has been checked after gc separation [82], and commercial chloroform inspected for the highly toxic chloromethyl methyl ether by FT–ir spectroscopy [83]. In the latter case the signal-averaging capability and high sensitivity of the FT–ir system were essential in overcoming the strong absorbance of chloroform and detecting 1 ppm of the ether.

Infrared methods are very useful for identification of drugs and their derivatives, in legal cases [84, 85]. In a systematic approach for detection of explosive residues ir techniques play an important role [86].

Petroleum oils and vegetable or animal oils must be determined in the presence of each other in water samples [87], besides other non-polar compounds, e.g. PAHs. For this purpose samples are extracted with carbon tetrachloride and analysed by ir methods [88, 89]. Accordingly, the persistence of petroleum in marine sediments [90] and in rapid weathering processes of fuel oils in natural water [91] has been determined. Residues of 2-chloroacetanilides in the range of 0.1 ng per ml of water have been determined by partition and gc–ir [92].

Pesticide traces in human viscera (after partition and tlc) [93] have been determined. Differentiation of the α-isomer (main ir peak at 1200 cm^{-1}) and β-isomer (2950 and 4112 cm^{-1}) of endosulfan after extraction from tissue, partition and tlc is possible [94]. Application of different techniques – including ir – for detection of ppm concentrations of chlorophenoxy acid herbicides has been demonstrated [95]. Benzo(*a*)pyrene has been determined in plant material after extraction, column chromatography and saponification of the eluate, followed by several partition steps before the final chromatography [96].

6.4.10 Advantages of ir methods

For identification of alkyl-9-fluorenones in diesel-exhaust particulate matter, high-resolution gas chromatography has been used, with detection by flame ionization, FT–ir and mass spectrometry. Any one type of detection would have been insufficient for confident identification. The mass spectra, for example, indicated the presence of alkyl benzo(*c*)cinnolines, which in fact were absent [97]. The results of this study illustrate the need to use complementary analytical techniques for analysis of complex unknowns.

In forensic chemistry ir is considered to be a very reliable method. When compared with two colorimetric methods, four tlc and four gc systems, ir proved to be the only reliable method for identification of heroin [98]. Infrared spectroscopy can also be very useful for objective presentation of the results of hplc separations, which without such a specific monitoring method are less readily accepted for forensic purposes, although there are definitive advantages with this separation system in legal chemistry [99].

A comparison of the capability of gc–FT-ir, hplc–FT-ir and gc–ms has been made, in which an industrial waste water effluent was analysed for chloronitrobenzenes [100]. It became evident that gc–ms could not differentiate between the isomeric forms of chloronitrobenzene, whereas gc–FT-ir could. The information content of gc–FT-ir generally matches or exceeds the information content of gc–ms analysis. However, the combined information from both techniques enables positive identification not afforded by either alone. The sensitivity of gc–FT-ir is sufficient for the analysis of many real samples and is no longer a significant limitation. The limiting factor is the lack of an adequate reference library of vapour-phase ir spectra. On the other hand, gc–ms provides better reconstructed gas chromatograms, a larger reference library and faster spectrum production. For gc–ms computer-search systems exist, but at present FT-ir libraries still have to be searched by hand. Theoretically, the sensitivity of gc–ms is about 2–3 orders of magnitude greater than that of gc–FT-ir. However, the background levels of environmental samples can often limit detection by gc–ms to about 10 ng of a component. Thus, for these very important samples the differences in limit of detection are not very significant.

The necessity for combination of several methods has been emphasized [101]. The versatility of a system consisting of gc–ir, collection of peaks of interest, and analysis by nmr, was illustrated by several examples, including some where gc–ms was not specific.

6.5 RAMAN SPECTROSCOPY

Up to the present, Raman methods have not been mentioned very often in organic trace analysis. About 0.3% of all papers in this area, published in 1978–1979, suggested this method. There seems to be some future in it, though, and recent instrumental developments have improved the possibilities for its application.

When light passes through a transparent medium a small fraction of the radiation is scattered by the molecules. The scattering is mainly elastic (Rayleigh scattering) but part is inelastic (Raman scattering). The inelastic scattering carries information about the vibrational and rotational energy levels of the molecules. Information from Raman spectra complements that from ir investigations.

The experimental advantage over ir consists in the possibility of using excitation light of shorter wavelength, which ought to be strictly monochromatic. The laser technique has two additional advantages. Laser sources have a very high energy output, so the Raman spectra become correspondingly stronger. Also the laser can be focused on a very small area, thus enabling the investigation of small samples. With modern photon-counting techniques high sensitivities and low detection limits can be obtained [102]. A 'Molecular Optics Laser Examiner' Raman microprobe (MOLE) is commercially available. Samples of a few picograms can be inspected [103]. The sample area is about 1 μm in diameter. Aqueous samples can be investigated, as with all Raman experiments. A micro cell has been developed [104] which gives a detection limit of 1 ng of β-carotene per μl. The cell is made of small glass melting-point capillaries with an internal diameter such that two tunable lasers can be used for excitation. When it was used in combination with chromatography, no signal distortion was observed for flow-rates between 0.1 and 10 ml/min.

The examples of application focus on environmental pollution-control problems. Phenolic compounds have been examined in water at 50–300 ppb levels [105]. A detection limit of 10–100 pg of PAHs is indicated [106]. After chemical derivative formation phenols [107] and amines [108] have been determined in the ppb–ppm region. The possibilities of Raman resonance spectroscopy for detection of water pollutants have been reviewed [109].

In life sciences the determination of carotenoid concentrations in marine phytoplankton is another example of the potentials of Raman spectroscopy. Analyses at the $10^{-8}M$ level can be done with an error of about 10%. Separation of the carotenoids is not necessary and the spectra can be obtained *in situ* [110]. Catecholamines have been determined (at the $10^{-5}M$ level) as their aminochromes [111].

6.6 FLUORESCENCE AND PHOSPHORESCENCE

Chemical systems, on excitation by electromagnetic radiation, re-emit the radiation either at the same wavelength or at a modified wavelength (photoluminescence). With fluorescence the luminescence process ceases almost immediately after irradiation is stopped. Phosphorescence endures for an easily detectable length of time [112–115]. Though phosphorescence is less widely used in trace analysis fluorescence is of constantly increasing importance. There are many substances which exhibit native fluorescence. Non-fluorescent compounds can often be coupled with reagents which fluoresce. Fluorescence measures molecular properties more specifically than uv spectrophotometry but with less specificity than ir spectroscopy. Fluorescence can be excited at discrete wavelengths which are specific for the given molecule. The wavelength of the emitted light depends on the chemical structure of the substance. As only a minority of organic substances gives native fluorescence a certain selectivity is guaranteed, and can often be improved by use of pH-control, solvent effects, etc. Very often ng amounts and sometimes less can be determined by specialized techniques [116].

Quenching of fluorescence intensity by interfering substances is one of the greatest disadvantages of fluorescence spectroscopy. Therefore, the analytes must either be separated or the interference must otherwise be dealt with. The optical properties of many fluorescent substances depend on pH, solvent nature, temperature etc.

Fluorescence, and especially phosphorence, effects depend on temperature [117]. In many cases the band-width decreases at low temperatures and peak sharpening is observed (see also Figs. 6.5 and 6.6, p. 229). An increase in the working range results, as the limit of detection can be improved by up to two orders of magnitude. In combination with time averaging it is possible to determine 4 ng of benz(*a*)pyrene by recording the decay of the phosphorescence intensity with time [118, 119].

In phosphorescence the sensitivity can often be enhanced by addition of certain substances to the sample. Thus, ethanol increases the phosphorescence of barbiturates, chlorpromazine, sulphamethazine and other drugs, sometimes by a factor of more than 100 [120]. Phosphorescence measurements at room temperature are also possible. Several reviews have been dedicated to this topic [121-125].

Fluorescence can be combined with tlc. Quantitative measurements *in situ* on the plate are possible, with good accuracy. This possibility – together with an increasing use of fluorescence detection with hplc – is responsible for the popularity fluorescence methods find in trace anlysis. Phosphorescence can also be coupled with tlc [126], especially at low temperatures. Spraying with a number of solvents improves the limits of detection of the method. Several applications of both methods are given in Table 6.3.

6.7 OPTOACOUSTIC SPECTROSCOPY

Optoacoustic spectroscopy (OAS) or photoacoustic spectroscopy (PAS) is a promising field of development in organic trace analysis [147, 148].

In this method, light-energy interacts with matter, provoking changes of pressure in it, which are recorded by use of a microphone. The method is to be classified, therefore, as a secondary emission method, with all its inherent advantages (of potentially high sensitivity) and disadvantages (e.g. non-linear calibration curves and some difficulties in quantitative work.) By use of lasers as the light-source rather low limits of detection can be obtained: 1 ppm of ethylene is detectable in flowing air, and the limit of detection in a static sample approaches 10 ppb [149].

The method seems to have some possibilities in surface analysis, as the penetration of primary energy is not very deep and emission energy is released only from the surface layers. This can be used for direct inspection of organic compounds adsorbed on powdered materials, e.g. on thin-layer adsorbents [150]. The determination of 0.2 μg of fluorescein on silica has been demonstrated [151]. Foodstuff dyes have been determined quantitatively after tlc

Table 6.3 – Examples of the application of fluorescence and phosphorescence methods.

Trace	Matrix	Level	Separation	Comments	Reference
butaperazine	plasma	50–300 ng/ml	partition	native fluorescence	127
cortisol	serum	<50 ng	partition	derivative formation	128
flecainide	plasma	>50 ng	partition	native fluorescence	129
fluoranthene		>20 fg(!)		laser-induced fluorescence	130
histamine	canned fish	8–300 ppm	partition	derivative formation (phthalaldehyde)	131
histamine	tissue	1 ng/ml	ion-exchange	derivative formation (phthalaldehyde)	132
histamine	urine	13 ng/ml	partition/ion-exchange	derivative formation (phthalaldehyde)	133
indol-3-ylacetic acid	plants	0.1–1 ng	partition		134
naphthacene	anthracene	>0.1 ppm			135
PAHs	coal	1 pg/g		laser, 4 K, Shpol'skii glass	136
PAHs		10 pg	chromatography	laser-induced fluorescence	137
perylene	cigarette smoke	2 μg	partition	Shpol'skii solvents, liquid N_2	138
PAHs		50 ng/ml		Shpol'skii matrices, 77 K	139
quinine				laser-induced fluorescence	140
quinolines		>30 ppm		fluorescence and phosphorescence compared	141
serotonin	biological material	1–100 ng	partition		142
aromatic amines	silica gel	1–10 ng		room-temperature phosphorescence	143
epinephrine		0.2 ng/ml		derivative formation, phosphorescence	144
PAHs	coal product	1 ng	partition	room-temperature phosphorescence	145
22 PAHs		0.5 ng/ml		77 K phosphorescence	146

[152]. Substances separated by electrophoresis have been determined directly on the electropherograms. The optoacoustic method is more sensitive than transmission densitometry and can also be used on colourless spots [153]. The limits of detection for 40 PAHs were found to be 2 ng/ml in laser-excited fluorescence and 8 ng/ml in photoacoustic spectroscopy [154].

A single-beam spectrometer has been described for use in the near infrared [155]. Use of an ir laser and the optoacoustic method is claimed to give an improvement by a factor of 10^3 in gc detection (relative to 'classical' ir-gc detection). Limits of detection of 20 pg have been reported [156]. If a Fourier-transform infrared spectrometer is coupled with a photoacoustic cell [157], spectra of powdered samples can be obtained, which are similar to those obtained by diffuse reflectance techniques [158].

6.8 MASS SPECTROMETRY

6.8.1 General aspects

Mass spectrometry (ms) is a very important field of organic trace analysis. About 14% of all publications in this area (during 1975–1980) mentioned ms as a tool or were dedicated to ms. Several review articles have been published (see Table 6.4), and there are also books (e.g. [181]), especially on application to the life sciences [182–184] chapters in monographs [185, 186], and conference proceedings [187, 188]. Mass spectrometry of priority pollutants has been described [189], as well as practical aspects of mass spectrometry [190]. More recent developments in ms have been summarized [191].

Table 6.4 – Review articles on mass spectrometry.

Field of application, topic of survey	References
Comparison of quadrupole and magnet ms	159
Design considerations of ms sources	160
Combination gc–ms	161, 162
Recent literature on ms (1263 references)	163
Trends in ms (methods, equipment)	164
Field desorption (fd–ms)	165
Measurement of hydrocarbons in atmosphere (125 references)	166
gc–ms analysis of water and air	167
Analysis of human breath and urine by gc–ms (98 references)	168
gc–ms in pesticide analysis	169
Environmental applications of ms (349 references)	170
ms in analysis of food and consumer goods (425 + 266 references)	171
Analysis of effluents (178 references)	172
N-Nitroso compounds (200 references)	173
Derivative formation (79 references)	174
Multiple selected-ion monitoring	175
Recent advances	176, 177
Application for determination of traces in food (1978 references)	178
Negative chemical ionization, applied in environmental chemistry	179
Mass spectrometry of chlorinated dibenzo-*p*-dioxins (152 references)	180

6.8.2 Principles of mass spectrometry

In mass spectrometry the sample molecules are converted into ions in a vacuum or a gas atmosphere. Several excitation processes are used in order to form the ions, which are then separated and characterized. The resolution is based on differences in the ion-paths in a magnetic field or an electrostatic field or both. Partition can also be based upon variations in ion velocity in a field-free space.

There are several methods of measurement – using one or several photomultipliers – and of recording the results. The registration modes have an influence on the sensitivity and detection limits of the entire ms process. The most often used excitation methods [160] will be presented first.

Electron impact (EI)

The EI source is currently the most commonly used. Electrons are emitted by heating a rhenium or tungsten filament cathode, and accelerated by a potential difference. Ions are formed by collision of the electrons with sample molecules. Most organic compounds have ionization potentials of 7–20 eV. An EI source operating at the conventional 70 eV provides enough energy to ionize all molecules. The ions fragment in a way characteristic of the parent molecule. The positive ions formed are extracted from the ionization chamber and accelerated by a high voltage. Further separation is achieved in the mass spectrometer.

Chemical ionization (CI)

Many compounds do not form a molecule-ion in an EI system. To cope with this problem the CI technique was developed. In it a reagent gas (Ar, H_2O, NH_3, D_2O, H_2, CH_4, O_2, at 0.2–2 mmHg) is ionized, and the ionic species (such as CH_5^+ from CH_4) reacts with the sample molecules. From a molecule MH, species such as M_2^+ or M^+ may be produced. The ion–molecule reaction is much gentler than that obtained by EI ionization. The quasi-molecular ion formed is more stable than the ion produced in EI. Generally, spectra obtained by the CI mode are simpler than those from EI excitation (see Fig. 6.8).

Negative chemical ionization

Negative chemical ionization is becoming more often used. It is based on the capture of electrons, produced in the ion-source according to the reaction $CH_4 = CH_4^+ + e^-$, at atmospheric pressure. The electrons are accepted by electron-capturing groups or atoms in the molecule, which thus becomes rather specifically charged and excited. Halogen-containing pesticides are excited about 10–100 times more sensitively than in normal ms. Linearity of response over 3 orders of magnitude is obtained. If Freon (CF_2Cl_2) is used as reagent gas, chloride ions can become attached, giving an $(M+Cl)^-$ ion. This has the advantage that no electron-capturing atom is required in the molecule to be excited. Recent developments have been reviewed [192, 193]. In assay for dopamine the use of negative-ion ms led to a 10–1000-fold improvement in the detection

a
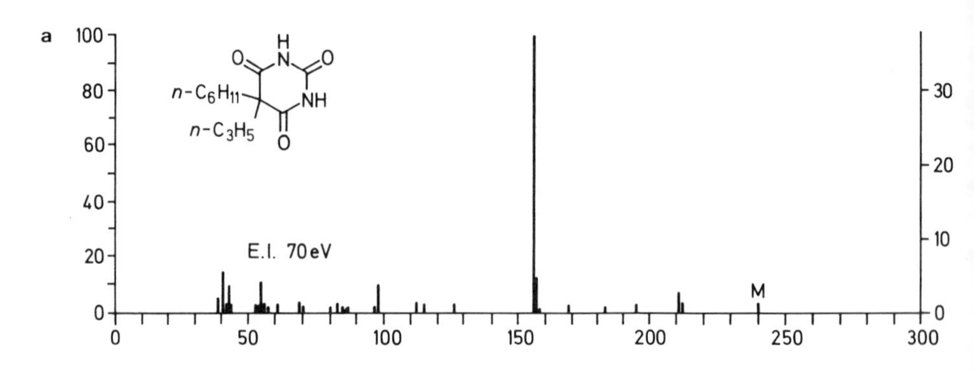

b
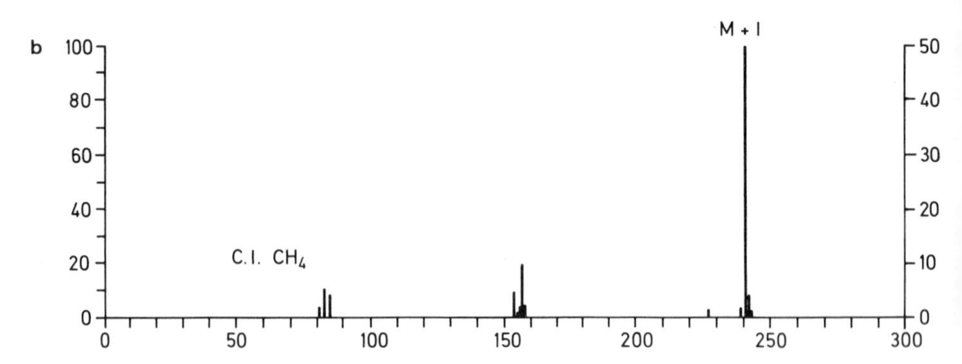

Fig. 6.8 – Comparison of mass spectra obtained by the EI mode (top) (70 eV) and the CI mode (CH_4 as reagent gas) (from [160]). The formula of the barbiturate investigated is shown (m.w. = 240). (Reproduced from *Journal of Chromatographic Science* by permission of the copyright holders, Preston Publications Inc.).

limit in comparison with electron impact or positive chemical ionization [194]. The detection limit of a few pg/ml for polybrominated biphenyls is lower than that for electron-capture detection by a factor of about 20 [195]. Organophosphate pesticides have been determined by chloride attachment, with CH_2Cl_2 as reagent gas [196].

Field Ionization (FI)

A very sharp tungsten edge or a fine tungsten wire (10 μm) at the entrance to the ion gun is held at a high potential (8–14 kV). The resultant high field gives positive ions for gaseous samples, which enter the ion source.

Field desorption (FD)

Field desorption is a specialized form of field ionization in use for molecules which cannot be vaporized. In such cases, an activated emitter wire is dipped into the sample solution under study or droplets of the solution are added to the wire and evaporated. The emitter wire is then gradually heated while it is at high voltage [197–199]. Such a wire should have high ionization efficiency, large surface area and strength, and should be easy to make. Carbon, nickel, cobalt and silicon are used as materials for making emitters [200]. The dipping method for sample loading is not quantitative, and it is better to freeze microlitre droplets onto the emitter [201]. Field desorption is used for non-volatile trace compounds which are of significance in biochemistry, e.g. for phospholipids [202], guanosine, creatine, choline chloride and peptides [203] in amounts of 0.1–1 μg, pesticides of carbamate, thiocarbamate or phenylurea type [204], mycotoxins [205] and oligosaccharides [206].

Atmospheric pressure ionization (API)

Ion-molecule reactions occur in an inert gas weak plasma at 1 atm pressure. Ionization energy is provided by a β-emitting ^{63}Ni source or by a high-voltage corona discharge. The gas drifts through a tiny hole (25–50 μm diameter) into the ms vacuum [207]. Trichlorophenoxyacetic acid has been routinely determined at the 20-ppm level in blood by use of atmospheric ionization [208].

Laser desorption mass spectrometry

In this version, a micro volume of the sample is excited to the ionic state by a focused laser beam [209]. The ions formed are analysed by mass spectrometry [210–212].

Other excitation modes

Besides these standard methods, other excitation modes are occasionally used or are in development. Combination of excitation modes is becoming more popular. It is often necessary to run one sample by EI and another by CI, and it is inconvenient to change the sources in the spectrometer. Sometimes, two excitation modes are used for a given compound. The CI or FD spectrum might show a molecular ion while the fragmentation pattern of an EI spectrum might contribute to the identification of the compound [213].

The sensitivity of the different excitation modes depends on the experimental conditions. As a general rule 1–10 ng can be registered by the EI and CI modes. FI or FD spectra have been obtained with 15–100 ng. For API a sensitivity of 300 fg has been reported in some cases.

The limit of detection of an ms method also depends on the registration procedure. In single-ion monitoring (see p. 242) sensitivities in the low pg range have been observed for EI or CI excitation.

Combination of ms with other techniques – mainly gc – is of greatest importance. Such interfacing directly influences the action of the source. Consequently, the source should be carefully selected, and a proper ion-source design is very important for gc–ms [214]. All the excitation methods mentioned above are used in gc–ms combinations, with the exception of FD, which was developed for solids.

Separation of ions

Several principles are applied. In a time of flight spectrometer the ions formed are accelerated by a very brief potential pulse. The ions then travel through a field-free space about 1 m in length, where they become separated according to their different speeds (which depend on the mass and charge of the ions).

In focusing mass spectrometers the difference in the radial path through the magnetic field is a function of the m/z ratio of the charged particles (mass m, charge z), and ions with a given m/z ratio can be focused on the detector entrance-slit by varying the accelerating electric field or the magnetic field. This is the most widely used technique.

In quadrupole mass spectrometry separation is achieved by allowing the charged particles to travel in a longitudinal channel formed by 4 electrodes; the diagonally opposite electrodes carry the same charge. In addition a high-frequency a.c. field is superimposed. Under these conditions the ions follow an oscillating path, the amplitude of which increases with the path-length. When the amplitude is large enough the ions contact the charged electrodes which form the wall of the channel and become eliminated from the beam. Only those ions which have not been sufficiently excited will leave the system at the exit. The whole system thus operates as a mass filter.

The different types of separation have several advantages and disadvantages. Quadrupole filters are preferable in the field of low-resolution ms analysis [215]. Magnetic instruments are necessary for precise determinations of masses in high-resolution ms.

Monitoring

There are different ways of monitoring the separated ions. Today registration by means of photomultipliers is almost always applied. In principle, one photomultiplier can be set at a given point of the mass spectrometer and the electromagnetic conditions changed so that the different ions separated reach the detector consecutively. As information on many types of ions is gained this kind of registration is called multiple-ion detection (MID). On the other hand, several multipliers can be used, set at different points on the registration plane of the spectrometer. A corresponding number of types of ions can thus be registered simultaneously. Finally, one detector can be used, set at the point of the registration plane corresponding to a single desired m/z ratio. This method is called single-ion monitoring or the selected-ion mode (SIM). The speed and reproducibility of registration and thus of the entire mass spectrometric measurement depend on the registration mode. Naturally, SIM gives less information than MID.

6.8.3 Further developments in ms instrumental design

One pronounced trend in further development is the use of higher resolution spectrometers. The signal/noise ratio is much improved, and smaller concentrations of traces are detectable (see Fig. 6.9).

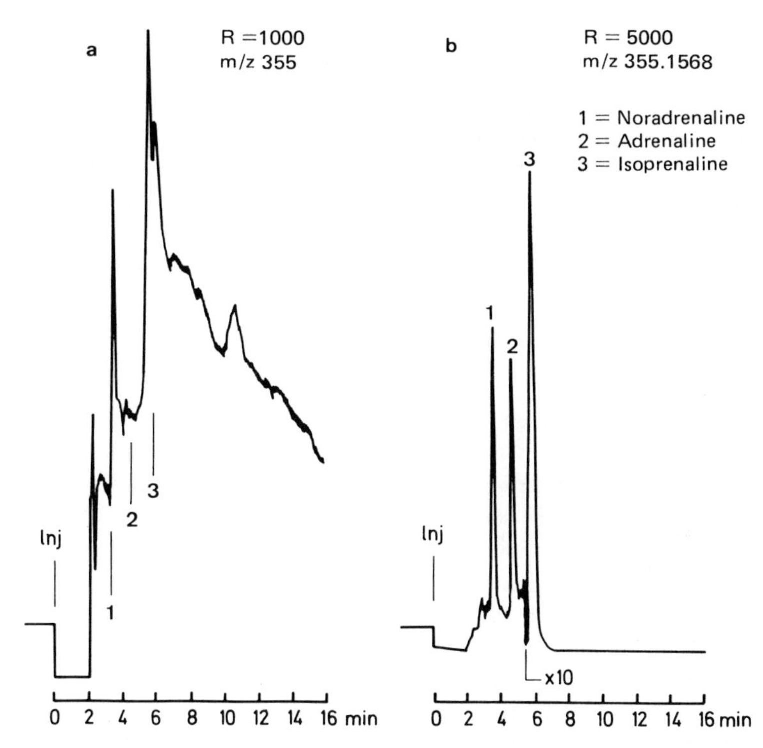

Fig. 6.9 – Detection of catecholamines with low resolution (a), registration of $m/z = 355$; high resolution (b), registration of m/z 355.1568. Detection of 0.2 ng of adrenaline per ml and 0.7 ng of noradrenaline per ml. ([Reproduced from K. Jacob, W. Vogt, M. Knedel and G. Schwertferger, *J. Chromatog.*, **146**, 221 (1978), by permission of the copyright holders, Elsevier Science Publishing Co., Amsterdam].

In organic trace analysis the gc–ms combination is very successful. The separation is achieved by gas chromatography and the mass spectrometer detects the separated components very effectively. The limitation lies in the necessity for the sample material or its derivatives to be volatile. This can be a severe limitation if derivatives have to be made, because there is then the possibility of introduction of additional errors, the analysis takes longer, and additional skilled personnel are needed. These drawbacks have initiated consideration of other approaches, such as using one mass spectrometer purely for separation and a second one for detection. The resulting method has been called ms–ms (or sometimes tandem ms). The ms–ms method has been reviewed [216–219]. It has its merits for the determination of traces at low concentrations in complex relatively non-volatile mixtures [220]. Rapid analysis without extraction of the sample or derivative formation has been demonstrated for the characterization of nutmeg constituents [221], an example of the potential of ms–ms in food analysis. The most interesting application for trace analysis, however, is the determination of highly toxic components at extremely low concentrations, e.g. dioxins [222, 223] and PAHs [224].

The selectivity of mass spectrometry is improved if the ions can be made to undergo fragmentation outside the ion-source. For this purpose a double-focusing spectrometer is used in which the magnetic sector is between the ion-source and the electric sector. In such an instrument the parent molecular ions can be separated from all fragments formed in the source. At the interfocus between the sectors a small reaction cell filled with reagent gas is installed, and collisions take place between the reagent gas and the separated ions, to form new fragments which are then separated within the electric sector. The method has been named collision-activation mass-analysed kinetic-energy spectrometry (CA-MIKES) or direct analysis of daughter ions (DADI) [225–227]. The spectra have little background noise.

In the 'linked scan' method the electric field and the magnetic field in a double-focusing mass spectrometer are varied simultaneously. Ions which undergo fragmentation between the source and the first analyser are specifically detected [228, 229].

In summarizing, it can be said that mass spectrometry is highly specific. Individual molecules or fragments of molecules or ions are recognizable, ng-quantities are detectable and sometimes even pg-amounts or less can be traced. Quantitative measurement is more difficult, since mass spectrometry is primarily a qualitative tool. If combination with other methods of detection is desired, the ms must be the final step. Interference from the matrix and other foreign material can become a problem. As long as the other substances exhibit different volatility from the analyte a separation is often possible by stepwise heating of the sample. The limit of detection deteriorates with the presence of other materials.

The combination with gas chromatography as a separation method is extremely useful. With open columns – having about 8×10^4 theoretical plates per metre – investigations of ppb concentrations are possible with ms. The greatest problem is the enormous number of compounds which can be simultaneously detected in a sample. As a good gc–ms combination is capable of producing one mass spectrum per second, a large number of data are produced which must be fed into a collection and data-processing system of high capacity and high speed. Often enough, the reliable identification of components in a complex mixture is very difficult and requires either a good deal of clean-up and separation, or a mass spectrometer with high resolution. Since the necessary instrumentation is very expensive and requires a lot of maintenance, in order to make the system economically viable the spectrometer must be used constantly, and by many research workers, which creates organizational problems in organic trace analysis if cross-contamination is to be avoided.

6.8.4 Combination of ms with other methods

Table 6.5 shows the extent to which ms is combined with various separation methods in organic trace analysis. The most often mentioned combination is gc–ms, which contributes over a tenth of the entire literature on organic trace analysis.

Table 6.5 – Frequency of publications on mass spectrometry in the literature.

Topic	Percentage of entire literature on organic trace analysis
general aspects of ms in trace analysis	2
gc–ms-coupling	12
coupling of ms with tlc or liquid chromatography	1
direct injection	0.5
total	about 14

6.8.5 Combination of mass spectrometry with gas chromatography

There are several ways to combine gc with ms, and these have been reviewed [23]. All of them have to take into consideration that gc operates at atmospheric pressure, whereas ms – with the exception of API – requires a vacuum. The introduction of the gas sample and carrier gas is very often combined with a step to reduce the amount of carrier gas. Some methods are shown in Fig. 6.10.

The interface is a critical point. The gas-flow is influenced by the type of gas chromatographic column used and whether a single-step or two-step separator is used. Yields depend on the molecular weight of the trace component. Platinum has been recommended as the interface material [232], while other authors favour glass [233-235].

The advantages of a gc–ms combination are fully gained only when handling of data by computer is available [231]. Two signals are usually obtained – total ion current and the ms signal. These data are processed by the computer. Data stored on magnetic tape or discs can be reproduced at any desired time. This can be of great value, especially when the analysis cannot be repeated. The data-processing system can give the total ion current, the mass spectrum at any retention time of interest and the 'mass chromatogram', i.e. the recording of the gas chromatogram with respect to a selected single mass. With these computer techniques it becomes possible to check a gc peak for homogeneity with respect to composition. Several spectra are recorded during elution and studied for changes in the fragmentation pattern. This means that many spectra have to be recorded, a task which cannot be fulfilled without the aid of a computer.

Spectrum refining can be performed by subtracting mass spectra from each other in order to eliminate interfering masses, e.g. from overlapping peaks and from column bleeding.

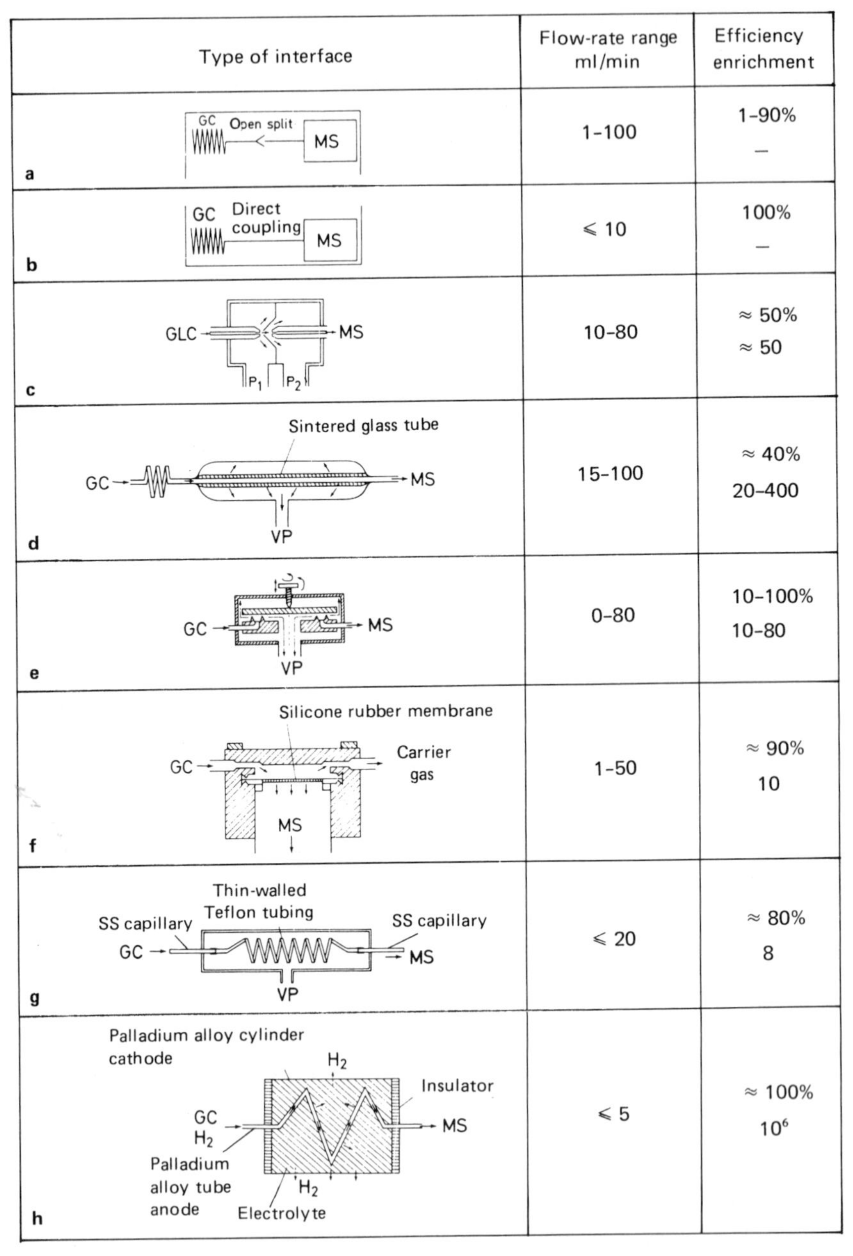

Type of interface	Flow-rate range ml/min	Efficiency enrichment
a	1–100	1–90% —
b	⩽ 10	100% —
c	10–80	≈ 50% ≈ 50
d	15–100	≈ 40% 20–400
e	0–80	10–100% 10–80
f	1–50	≈ 90% 10
g	⩽ 20	≈ 80% 8
h	⩽ 5	≈ 100% 10^6

Fig. 6.10 – Schematic survey of the different interfacing techniques used in gc-ms. (a) Open split coupling; (b) vacuum coupling; (c) jet separator; (d) frit separator; (e) slit separator; (f) membrane separator; (g) Teflon separator; (h) electrolytic silver-palladium separator. (Reproduced from [231] by permission of the copyright holders, Elsevier Science Publishing Co., Amsterdam).

6.8.6 Coupling of liquid chromatography with mass spectrometry

There are several methods of coupling liquid chromatography with ms [236, 237]. Most often used is direct introduction of the sample [238-241], with jet enrichment if necessary [242, 243]. Increasing use is being made of mechanical transfer by a moving belt or wire [244-246]. Less often applied, so far, are high-capacity ion sources, operated at atmospheric pressure [247], a semi-permeable silicone membrane [248], or chemical modification of the sample at the interface, e.g. by reduction to a hydrocarbon [249]. Evaporation of the effluent by laser energy, transportation of the molecular beam into the source and excitation has also been suggested [250, 251].

6.8.7 Application of mass spectrometry to the analysis of waters, effluents etc.

Mass spectrometry is well suited for the analysis of traces in aqueous systems after preseparation and concentrations, as indicated in Table 6.6.

Table 6.6 – Examples of the application of mass spectrometry (after preseparation) for the determination of traces in water and effluents.

Trace	Level	Separations before ms	Reference
hydrocarbons		partition/gc	252
hydrocarbons		head-space gc	253
terphenyls	0.1 ppb	partition/gel chromatog./gc	254
PAHs	50 ng	partition/lc	255
hydrocarbons		gc	256
alkanes		partition/gel chromatog./gc	257
60 compounds	20 ppt	XAD2 resin/gc	258
PAHs	0.5 ng	partition/lc/gc	259
PAHs	ng/l.	XAD 2 resin/gc	260
vinyl chloride	μg/l.	purging/gc	261
β-chloro-ethers		gc	262
organohalogens	μg/l.	gc	263
pentachlorophenol	200 ng	partition/gc	264
organohalogens	1 ppb	gc	265
chlorinated compounds	0.1-10 ppb	partition/gc	266
halogenated unknowns		gc	267
vinyl chloride	0.01-10 ppb	purging	268
vinyl chloride	ppt	gc	269
polychlorinated naphthalenes	130 μg/kg	partition/lc/gc	270
chlorinated pesticides	0.05 ppt DDT	XAD 2/gc	271
PCBs, DDT	20 pg	polyurethane/gc	272
kepone		gc	273
acrylamide	0.1 ng/ml	gc	274
alkylbenzenesulphonates	μg/l.	partition/gc	275
glutamic acid	ng	ion-exchange/gc	276

Coprostanol is a characteristic sterol found in the faeces of man and higher animals. This compound can serve as an indicator for faecal pollution. It is extracted with n-hexane, silylated, separated by gc and determined by ms. This determination is about 6 orders of magnitude more sensitive than that by gc-fid. The limit of detection in water is 1 ng/l. [277].

Geosmin (1,10-dimethyl-*trans*-9-decalol) is a metabolite of some actinomycetes. It gives a characteristic smell to water, with an odour threshold (for man) of 10 ng/l. For determination 1 litre of water is partitioned with 50 ml of CH_2Cl_2 after addition of 300 g of NaCl. The separated organic layer is dried (Na_2SO_4), concentrated (in a Kuderna-Danish evaporator) and subjected to gc–ms. Recovery for 30 ng/l. is about 69% [278].

6.8.8 Application of mass spectrometry to determination of organic traces in air, tobacco smoke and particulate matter from air

Table 6.7 gives examples of the determination of traces in air by mass spectrometry.

Table 6.7 – Mass spectrometric determination of traces in air or tobacco smoke.

Compound	Level	Separation before ms	Reference
55 aromatic hydrocarbons	0.02–1 μg/m³	gc	279
PAHS	30 ppm		280
PAHs	10 ng	tlc/hplc/gc	281
PAHs	ng	gc	282
PAHs	ng/m³	partition/gc	283
PAHs (about 26)		partition/chromatog./gc	284
PAHs (about 50)		partition/chromatog./gc	285
PAHs		partition/chromatog./gc	286
PAHs		gc	287
PAHs (about 100)		gc	288
PAHs (about 150)		gc	289
PAHs *		gel/tlc/gc	290
PAHs * (about 150)		partition/chromatog./gc	291
PAHs *		gel/gc	292
naphthalenes *	ng–2 μg	partition/chromatog./gc	293
chrysene, benzopyrene	0.1–50 ng	extraction	294
halocarbons		Tenax/gc	295
methyl bromide	2 ng/m³	gc	296
halocarbons	0.1 ng/l.	gc	297
vinyl chloride	0.05 ppm	gc	298
chlorofluorocarbons	220 ppt	gc	299
methyl chloride	500 ppt	gc	300
chlorofluorocarbons	5 ppt	gc	301

*In tobacco smoke. All other cases are for air samples.

6.8.9 Application of mass spectrometry in life sciences

Gas chromatography–mass spectrometry is used for profiling various components of human body fluids in healthy and diseased states [302], with special reference to organic acids. A number of metabolic disorders can be diagnosed by means of this modern technique. The obvious aim is improved diagnosis of such diseases.

Micro-organisms are identified by means of their cellular fatty acids. By use of derivative formation and ir as an additional confirmatory method it is possible to differentiate *Pseudomonas aeruginaosa, P. cepacia* and *P. maltophilia.*

Trace amounts of volatile compounds are emitted from mushrooms and can affect insect chemosensory behaviour. For the isolation and analysis of such interesting compounds, mushrooms are macerated and kept in a head-space equipment. The volatilized compounds are isolated from the gas phase by adsorption on a Tenax column. Desorption is done by flash-heating of the column, and is followed by gc–ms [303].

Table 6.8 gives examples of the application of gc–ms in trace analysis of biological material. In all cases mentioned, at least one derivative-formation step was necessary to obtain compounds of the necessary volatility, from the originally highly polar traces. Furthermore, the table demonstrates that an internal standard was used in the majority of cases. Very often, the deuterium-labelled analogue ('D' in the 'internal standard' column in Table 6.8) was applied. ^{14}C or ^{15}N were used less often. In some instances chemical analogues ('chem. analog.') were used.

6.8.10 Application of mass spectrometry in forensic chemistry

In forensic science gc–ms is used very effectively [331]. Marijuana extracts are difficult to analyse. A reliable solution of this problem is offered by gc–ms [332]. The high resolution power of capillary columns in combination with the specific detection by ms is essential for developing complex chromatographic profiles that are unique for a given marijuana sample. In this way it is possible to differentiate samples according to their origin.

Etorphine is an analgesic of high potency. When given subcutaneously it is about $1\text{-}8 \times 10^4$ times more potent than morphine. Its ability to cause catatonia, even at very low dose levels, has resulted in its use for the immobilization of game animals. The fact that it works sublingually and has abuse potential raises the possibility that it may become a future threat in the drug abuse field. A quantitative gc–ms assay has been developed [333] which can measure the drug at a concentration of 2 ng/ml in urine, with about 5% relative standard deviation.

A rapid technique for forensic analysis is suggested [334] in which chemical ionization is used for obtaining 'clear' spectra. It is possible to inject stomach contents or vomits directly into the gc–ms combination for detection of about 48 drugs.

Impurities in illicit amphetamine, e.g. some pyrimidines, are detected by ms after tlc separations, thus providing some indication of the drug production source [335].

Sad to relate, analysis for drugs of abuse becomes more and more a routine problem for the laboratories involved. This calls for simple, reliable procedures requiring little effort. Direct analysis of urine by ms (CI mode) can be helpful [336]. With isobutane as reagent gas the normal components of urine do not interfere with the drug detection. In cases of interference, tlc or gc can be used as preseparation steps.

Table 6.8 – Examples of the use of gc–ms for the determination of traces of naturally occurring substances in biological materials. (All compounds were submitted to gc after derivative formation).

Substance(s)	Matrix	Level	Separation before gc	Internal standard	Reference
cyclic adenosine-3′,5′-monophosphate	maize	0.1 μmole	extrn./partn./tlc	^{14}C	304
5α-androstane-3α,17β-diol	sperm, plasma	0.03–2 ng/ml	partn./chromatog.	D	305
biogenic amines (8)	heart, plasma, urine	pg	extrn.	D	306
eicosapolyenoic acids	serum	0.05–1 μg	extrn./partn.	D	307
eicosapolyenoic acids	frog tissue		extrn./partn.		308
histamine, *N*-methylhistamine	plasma	0.5–10 ng/ml	partn./chromatog.	D, ^{15}N	309
homovanillic acid	cerebrospinal fluid	0.5–500 ng	partn./tlc	D	310
homovanillic acid, vanillylmandelic acid	cerebrospinal fluid	>0.5 ng	chromatog.	D	311
hydroxy fatty acids	biological samples	>1 fg			312
6-hydroxymelatonine	urine	35–75 pg	partn./chromatog.	D	313
4-hydroxy-3-methoxyphenylethanediol	plasma	3 pmole/ml	partn.	D	314
β-hydroxyphenethylamines	tissues	200 pg	extrn./partn./tlc	D	315
N-(indol-3-ylacetyl)aspartic acid	pine shoots	1 ng	extrn./chromatog./hplc		316
normetanephrine, metanephrine	urine	10–2000 μg/l.	chromatog.	D	317
oestrone, 17β-oestradiol, oestriol	serum		partn.	D	318
organic acids (50)	urine, serum		partn.		319
prostaglandins	plasma, urine		chromatog.	T, D	320
prostaglandins	synthetic mixture	1–100 ppm		D	321
prostaglandins, thromboxane B_2	pleural fluid, plasma	10–100 pg			322
prostaglandin E_3	renal medullary tissue		extrn./chromatog.	D	323
6-ketoprostaglandin $F_{1\alpha}$	plasma, exudates	40–400 pg	partn./tlc	T	324
6-ketoprostaglandin $F_{1\alpha}$	release from aorta	10–100 ng	partn./chromatog.	chem. analog.	325
trimethylprostaglandin, E_2 analogue	plasma	100 pg/ml	partn.	D	326
serotonin	cerebrospinal fluid	1 ng	partn.		327
testosterone, dehydroepiandrosterone	saliva	30 pg	partn.		328
tuberculostearic acid	sputum	1 pg	partn.		329
tyramine	brain	1 ng/g	extrn.	chem. analog.	330

Morphine at plasma levels of 5 ng/ml can be determined by gc–ms after derivative formation [337]. Δ^9-Tetrahydrocannabinol has been determined at 10^{-18} g level by negative-ion chemical ionization mass spectrometry [338, 339], which must surely be the current record for trace analysis. LSD has been determined in amounts of about 15 μg by mass spectral studies of ultraviolet-irradiated extracts from illicit preparations [340].

In cases of suspected arson, fire residues are investigated [341]. Distillation and steam-distillation of such residues is used to separate volatile components for gc–ms.

6.8.11 Application of mass spectrometry in pharmaceutical trace analysis

The broadest application of ms, mainly as gc–ms, is in the field of pharmaceutical analysis. About 2% of all publications on organic analysis arise from this topic. Very often, compounds under investigation for possible medical application are studied, to satisfy legal or administrative requirements for new drugs.

New effective drugs are usually administered in low concentrations. Any metabolites formed must consequently be determined at very low concentrations – either in urine or in plasma samples (which are not available in large amounts). Mass spectrometry with its specificity and low working ranges is therefore the method of choice, but it demands a very good sample preparation procedure, with special emphasis on separation of the parent compound from its metabolites. Column chromatography, liquid–liquid partition and thin-layer chromatography are the most often used separation methods.

Metabolites are often conjugated to polar molecules of biological origin (e.g. glucuronic acid, glycine, taurine or sulphuric acid) to form highly polar conjugates. To separate these from non-conjugated compounds dialysis is used as the initial step, followed by partition and chromatography [342].

Table 6.9 gives a few selected examples of the use of gc–ms for the determination of therapeutically used drugs in biological material. In all the cases quoted, plasma or urine was the sample material, except for one case [373] in which embalmed tissue was analysed. As can be seen from Tables 6.9, 6.13 and 6.17 a very frequently used technique in drug analysis is extraction of the analyte into an organic solvent, which is then analysed directly. Internal standards are very often used. The application of deuterium-labelled compounds is of special advantage when a mass spectrometric determination follows.

Recoveries are of the usual levels for such work. Examples are: 92–95% at 2 μg/ml [347], 105% at 4 ng/ml [350], 75–90% at pg/ml levels of benzbromarone and its metabolites [347], chlorpheniramine [350] and ethynyl steroids [356]. Analogously, relative standard deviations are of the same order of magnitude as for other procedures, e.g. 1–5% for 2 ng of atropine per ml [345], 3% for 2 μg of benzbromarone and its metabolites per ml [347], 5% for 10 ng of chlorambucil per ml [348], 5% for chlorpheniramine at 4 μg/ml [350], 11%

Table 6.9 – Examples of the use of gc–ms for determination of traces of drugs in biological samples.

Substance(s)	Level	Separation	Intern. stand.	Reference
anabolic steroids	25 ng/ml	partition/tlc/deriv.		343
apomorphine	0.05–4 μg/ml	partition/deriv.	chem. analogue	344
atropine	1–40 ng/ml	partition/deriv.	D	345
azelastine	1–500 ng/ml	partition	D	346
benzbromarone, bromobenzarone	0.01–2 μg/ml	partition/deriv.	chem. analogue	347
chlorambucil	10 ng/ml	partition/deriv.	D	348
2-chloroprocaine	5–100 ng/ml	partition	chem. analogue	349
chlorpheniramine	4 ng/ml	partitions	D	350
clebopride	2–80 ng/ml	partition/deriv.	chem. analogue	351
clonidine	0.1–1 ng/ml	partitions/deriv.	D	352
cocaine + metabolites	8–1000 ng/ml	partition/deriv.	D	353
codeine	5–50 ng/ml	deprotein./partition	chem. analogue	354
ethambutol	> 36 ng/ml	partition/deriv.	chem. analogue	355
ethynyl steroids	> 5 pg/ml	enrichm. by chromatog./chromatog./deriv.		356
fentanyl	0.5–100 ng/ml	partition	D	357
fluoruracil	> 10 ng/ml	partitions/deriv.	chem. analogue	358
glyceryl trinitrate	50–1600 pg/ml	partition	chem. analogue	359
glyceryl trinitrate	20 ng/ml	chromatog./enrichm./chromatog.	D, ^{15}N	360
p-hydroxyphenobarbitone	0.2–1.5 μg/ml	partition/deriv.		361
medroxyprogesterone	3–22 ng/ml	partition/deriv.	D	362
methadone + metabolites	20 ng/0.5 ml	partition	chem. analogue	363
methylphenidate	5–25 ng/ml	partition/deriv.	chem. analogue	364
metyrapone + reduced metabolite	250 pg	partition/deriv.	chem. analogue	365
nitrourea antitumour agents (lomustine, semustine)	1.0–1000 ng/ml	partition/deriv.		366
19-norandrosterone	20 ng/ml	partition/deriv./tlc	^{14}C	367
pemoline, fenozolone, thozalinone	15 ng	partition/deriv.		368
pethidine	50–150 ng/ml	partition	D	369
phenytoin + *p*-hydroxy-metabolite		partition/deriv.	$^{13}C^{15}N$, D	370
procarbazine + metabolites	10–500 ng/ml	partition/deriv.	chem. analogue	371
psychotropic drugs (e.g. diazepam, methaqualone, amitriptyline)	ng/l.	partition deriv.		372
succinyl choline	25–5000 ng/g	extraction/partition	D	373
trimetoquinol	2–50 ng/ml	partition/deriv.	D	374
tricyclic antidepressants (7)	20–240 ng/ml	partition	D	375
trifluoperazine	0.5–200 ng/ml	partitions	D	376
trifluoperazine	80 pg/ml	partition	D	377
quinine, quinidine	50 pg	partition		378

for clonidine at 0.25 ng/ml and 5% at 0.5 ng/ml [352], 6% for 1 ng of fentanyl per ml [357], 4% for 100 pg of glyceryl trinitrate per ml [359], 7% for pethidine at 70 ng/ml [369] and 3% for trifluoperazine at 2 ng/ml [376].

6.8.12 Application of mass spectrometry to the analysis of samples of ecological interest to man and of food samples

In food analysis many examples for the application of ms or gc–ms are found [379]. Volatile nitrosamines are determined at the picogram level [380]. An aliquot of the sample is irradiated with uv light (366 nm), which decomposes *N*-nitrosamines. After 60 min the photolysis is complete (CH_2Cl_2 as solvent, melting point capillaries as vessels). Mass spectrometry is applied to the irradiated and non-irradiated sample and differences between the two spectra are attributed to nitrosamines.

N-Nitrosamines are often determined with a thermal energy detector after gx (see p, 269). Owing to the hygienic and toxicological significance of these carcinogens and the economic consequences of their detection in, for example, smoked food, an alternative method should be used in addition to the thermal energy method. Thus gc–ms is suggested for the determination of 1–10 ppb of *N*-nitroso-3-hydroxypyrrolidine in cured meat products [381]. Analytical methods for the determination of *N*-nitrosamines in foods at levels below 1 μg/kg have been reviewed (183 references) [382]. For volatile nitrosamines only gc–ms gives reliable data. So far there is a lack of adequate methodology for non-volatile *N*-nitrosamines.

Pesticide residues, e.g. 50 ppb of chlorinated norbornene in fish, can be determined by gc–ms [383]. Many examples of confirmation of drug residues in animal tissue by means of gc–ms are described in the literature [384]. The legal, analytical and practical requirements in developing and submitting these procedures must be considered against the background of current requirements and past experience. Flavour components have been determined in bread [385], coffee [386] and cooked chicken [387], and this has led either to identification of new compounds (20 in bread, 30 in chicken) or improvement of the evaluation methods.

Chloroform has been searched for in tissues of poultry carcases which had been exposed to water that contained chlorine or chlorine dioxide. Actually, $CHCl_3$ was formed in the treatment with water and could be detected after extraction of the tissue with light petroleum and ms with SIM [388].

Chlorinated dibenzo-*p*-dioxins have been measured in 58 different samples of 2,4-D products [389] by extraction/partition/chromatography/gc–ms. No dioxins (limit of detection 1 ppb) were found in samples of 2,4-D, whereas 20 of 21 samples of 2,4-D esters contained from 5 ppb up to 24 ppm(!) of dioxins.

Further examples of the determination of traces in samples of food or that are of ecological relevance are given in Table 6.10.

Table 6.10 – Examples of the use of gc–ms for the analysis of samples of food or with other relevance to the human environment.

Substance	Matrix	Level	Separation before gc–ms	Reference
aldicarb	water	>0.3 ppb	partn./hplc	390
aliphatic amines	water	1–3 ppb	deriv./partn./chromatog.	391
anabolic steroid drugs	meat	1 ppb	extrn./partn./chromatog./deriv.	392
anabolic steroid drugs	veal	>0.02 ppb	extrn./deriv.	393
chlorinated alkanes, other hazardous chemicals	fish		extrn./chromatog.	394
chlorinated hydrocarbons	human adipose tissue	ppb	extrn./chromatog.	395
chloromethane	air	20 ng/l.	trapping–heat desorption	396
chlorophenoxyalkanoic acids (e.g. 2,4-D)	water, soil	>0.2 ppb	extrn./partn./deriv.	397
hexachlorobenzene	milk, drinking water	>0.8 ng/ml	partition	398
hexachlorocyclopentadiene	drinking water	>50 ng/l.	partition	399
histamine	tuna	ng	extrn./deriv.	400
methyl 4-chloroindol-3-ylacetate	canned peas	0.2 ppm	AOAC multiple pesticide-residue clean-up procedure	401
nitroaromatics	diesel exhaust particulates		extrn. (of filter)/chromatog.	402
N-nitrosodimethylamine	beer	>0.4 ppb	distn./partn.	403
phenolic compounds	water	>0.5 μg/l.	partition	404
phenolic acids	plants	<2 μg	extrn./partn./deriv.	405
phenoxyalkyl acids	fruit, vegetables	0.5 ppm	extrn./partn./chromatog./partn./deriv.	406
pirimiphos methyl	dried legumes	0.02–1 ppm	extraction	407
polychlorinated biphenyls	soils	1 ng	extraction	408
polycyclic aromatic hydrocarbons	crabs		digestion (NaOH)/extrn./chromatog.	409

polycyclic aromatic hydrocarbons (47)	fish tissue	0.2 ppb	digestion (KOH)/partn./ chromatog.	410
polycyclic aromatic hydrocarbons (150)	used oil from engines	>1.5 ppm	partition	411
styrene	tissue, blood of mice	>0.5 ng	head-space	412
2,3,7,8-tetrachlorobenzo-*p*-dioxin	fish	4–700 pg/g	digestion (KOH)/partn./ chromatog.	413
2,3,7,8-tetrachlorobenzo-*p*-dioxin	rabbits (Seveso)	30 ng/g	digestion (KOH)/partn./ chromatog.	414
2,3,7,8-tetrachlorobenzo-*p*-dioxin	fish		6 clean-up methods compared	415
2,3,7,8-tetrachlorobenzo-*p*-dioxin	fish, animal tissue	>10 pg	digestion/partn./chromatog./ hplc	416
2,3,7,8-tetrachlorobenzo-*p*-dioxin	goat milk, tissue	>1 ppt	digestion/partn./chromatog./ hplc	417
tetra-, hexa-, hepta-, octa-chlorodibenzo-*p*-dioxins	human milk	ppt	partn./chromatog.	418
polychlorodibenzo-*p*-dioxins	soil	ppt	extrn./chromatog.	419
polychlorodibenzo-*p*-dioxins (115)	fly ash (municipal incinerator)	<170 ppb	extraction	420
tetrachlorodibenzodioxins	air-filter	0.5 pg	extrn./chromatog.	421
T-2-toxin	maize	>5 ppb	extrn./partn./chromatog./ deriv.	422
xyloidine (a moulting agent)	egg yolk	>2 ppb	extrn./partn.	423

6.8.13 Some comments on mass spectrometry

Mass spectrometry is a powerful tool of organic trace analysis. The development of gc–ms (together with hplc) has proved to be "the most helpful innovation for pharmaceutical analysis and clinical chemistry" [429] and in food analysis [425] during recent years. There are some limitations to ms, though, which necessitate either additional techniques or substitution.

The amount of 'overinformation' can even become a burden. In the detection of acids in urine (after several steps, e.g. ion-exchange, tlc, derivative formation with diazomethane, and gc) about 500 compounds could be monitored by ms but only about 40% identified [426].

By 1975 more than 360 compounds had been identified in European potable waters, with many of them (and some others) found in American drinking water [427]. These data were largely obtained by ms. Since only about 10–20% of the organic matter in drinking water is volatile or otherwise amenable to gc–ms, little information exists on the bulk of the organic matter present in potable supplies. This was emphasized in a report to WHO [428] which gave high priority to the development of analytical techniques for organics not detectable by gc–ms.

Mass spectrometry is a very reliable technique. It can be used as a very specific detector in gc, thus providing information in those rare cases when two compounds appear not to be separable by gc (e.g. diphenylhydantoin and probenecid behave similarly on columns of SE 30, SP 2100, Dexsil 300 and OV 17 [429]).

On the other hand, but very rarely, even the gc–ms combination fails. The determination of the highly toxic carcinogen dichloromethyl ether in air is done by gc (on SE) and SIM-ms, but 1-chloro-2-propanol interferes even in high-resolution capillary gc, and unfortunately both compounds give the same peaks in the ms (CI mode) [430].

Polychlorinated methoxybiphenyls can be mistaken for chlorinated dibenzo-*p*-dioxins in gc–ms, as the molecular ions of these compounds show close similarity in selected-ion monitoring. It is suggested, therefore, that reported values of dioxin residues found by this (often used) method should be reappraised [431].

In coupling gc with ms and ir it is advantageous to use the additional information from ir in analysis for isomers which can be identified only inadequately by ms [432].

On the other hand it is necessary to confirm by ms the results obtained by gc–ecd* [433] when polychlorinated biphenyls are to be determined.

---

*Electron-capture detection.

6.9 DETECTORS USED IN COMBINATION WITH CHROMATOGRAPHY

6.9.1 Detectors for liquid chromatography

Two recent reviews have described the possibilities in this area [434, 435].

Very often a flow-through cell is used for monitoring of optical properties (absorbance, fluorescence, refractive index) or electrochemical properties (conductivity, voltammetric behaviour etc.) of the effluent. Such methods are described in the corresponding chapters of this book and in a monograph [436]. Problems and their solutions, in the coupling of liquid chromatography with mass spectrometry, have been critically reviewed [437–439].

Instead of passage of the eluate directly through a cell, the column effluent can be transported to the measuring device by means of a moving belt or wire. The solvent is usually evaporated during this transportation process. Such systems have also been reviewed [440, 441]. For examples see Table 6.11.

Table 6.11 – Examples of applications of combinations of chromatography with other detection systems by means of moving wires or chains.

Compound(s)	Level	Detection	Comment	Reference
steroids, drugs, coumarins, flavours	50 pg	ms	yield of transfer 25–40%	442
methyl stearate	1–100 ng	ms	yields 25–40%	443
carbohydrates, lipids, cholesterol	ng	fid	1.4% relative std. devn.	444
fluoro-organics		F-sensitive flame detector		445
radioactive compounds				446
fats, fatty acids		fid		447
petroporphyrins	μg	ms		448
phospho-organics		thermionic detection		449, 450
pesticides		N,P,Cl-sensitive detectors		451
nitrogen compounds		N-sensitive detectors		452
cytosine	1m*M*	ms	secondary ion excitation	453
coumarins, alkaloids, drugs		ms		454
ion pairs		ms		455
non-volatiles, thermally unstable compounds		ms	secondary ion excitation	456
phenolic compounds		ms		457
organic compounds (8)		ms		458
carbamate pesticides	2–30 ng	ms		459
kepone, kelevan, Mirex	1 ppm	ms		460

A slightly different principle for combining an electron-capture detector (ecd) with liquid chromatography is based on passage of the eluate into a stainless-steel tube heated to 350°C. Vapours of the solvent and the trace constituent are passed into the ecd by means of a purging gas [461–465].

Examples of use of direct coupling of the outlet of a chromatographic column with a mass spectrometer are the determination of fairly non-volatile components, e.g. tryptophan, adenosine, mono- or disaccharides [466], nucleic acid components, amino-acids, peptides, sugars [467], phenylalanine, arginine, AMP [468], steroids [469, 470], PAHs [471], polar drugs [472] and aromatic amino-acids [473]. The limit of detection is a few nanograms [467, 469, 470, 472].

A nepholometric detector for analysis of lipids after hplc has been described [474]. The chromatography is done with a non-polar solvent; at the end of the column an aqueous buffer solution is added and the opacity (or light-scattering ability) of the emulsion formed is determined. Microgram amounts of lipids can be determined.

Other new detectors have been developed, e.g. the flame aerosol detector [475]. The effluent from the column is nebulized into a turbulent flame. On elution, an organic or inorganic compound will produce ionized species in the flame. The resulting charge-flow is measured electronically. Water can be used as solvent. The limit of detection is comparable to those of other detection systems.

Other detection systems are a micro polarimeter for detection of optical activity with a detection limit of 0.5 μg [476], a 'streaming current detector' [477], which measures changes of the streaming potential with a limit of detection of 7 ng for heptanoic acid in methanol, and a photoconductivity detector, which measures changes in electrical conductivity produced by irradiation of the eluate with uv light [478]. Similarly, uv-photoinduced currents are measured [479]. A flame photometric detector [480] or a nitrogen-selective detector [481] can be coupled with hplc columns, and give detection limits in the low nanogram range.

Occasionally, several detectors are coupled, such fluorescence, uv detection and electrochemical measurement [482].

Several voltammetric detectors have been compared [483]. The lowest limit of detection was given by a graphite-paste wall-jet system, with adrenaline as test substance.

6.9.2 Detector systems for gas chromatography

About 39% of all publications on organic trace analysis describe an application of gas chromatography. Of the detector systems, mass spectrometry is most often mentioned (in about 10% of all publications), closely rivalled by the electron-capture detector (9%) and the flame-ionization detector (9%). The N,P-selective flame-ionization detector and the flame-photometric detector for S or P are each mentioned in about 3% of papers. Some monographs and surveys on gc detectors have appeared [484–486]. Some properties of these detection systems are given in Table 6.12.

Table 6.12 – Some properties of the most often used gc detectors (after [487], by permission of the copyright holder, National Bureau of Standards, Washington, D.C.).

Detector	Specificity	Limit of detection (ng)	Linear range
thermal conductivity	none	10	10^5
flame ionization	none	0.1	10^7
electron capture	selective	0.001	10^2
alkali metal flame ionization	P, S, N, halogens	0.01 (P)	10^3
flame photometric	P, S	0.1 (P)	none
coulometric	halogens, S, N	1	10^3

As can be seen from Table 6.12 the limit of detection is sufficiently low for trace work, while specificity is often low or zero when material is detected by use of very general principles. Consequently, a lot of work has been done to increase the specificity of gc detection, mainly by development of new methods.

Electron capture detection

Examples of use of this popular detection system are given in Table 6.13. Obviously, in most cases a derivative-formation reaction must be included in the analytical scheme.

A new strategy in detection of bacteria and other micro-organisms is based on the search for specific chemical components in these biological systems and the determination of such indicator components. For example, 3-hydroxydodecanoic acid is an indicator for *Neiserria gonorrhoeae* ($>10^5$ cells, which corresponds to low pg amounts of this acid). It is isolated from the biological material, converted into a derivative and determined by gc–ecd [552].

Flame-ionization detection

Some examples are given in Table 6.14.

Coulometric and conductivity detectors ('Coulson-detector', 'Hall-detector')

The coulometric detector – often called the Coulson-detector after its inventor [571] – is based on the principle that column effluents are mixed with oxygen and burnt to form HCl from any organochlorine compound. The HCl formed is titrated by coulometric generation of silver ions. Later, the principle was slightly changed. The column effluent is reacted with oxygen or hydrogen for oxidation or reduction respectively, and the reaction products are passed into water, the conductivity of which is changed by the presence of acids (formed from organohalogen compounds on oxidation) or ammonia (formed from N-compounds on reduction with hydrogen). The conductivity detector is used mostly in the form of the Hall conductivity detector [572, 573]. Examples of use of these detectors are given in Table 6.15.

Table 6.13 – Examples of the application of gas chromatography with electron-capture detection.

Substance(s)	Matrix	Level	Separation	Derivative formation	Recovery (%)	Relative standard devn. (%)	Reference
alkylbenzyldimethylammonium compounds	synthetic solution	>15–20 ng		+			488
alphadolone	plasma	10–600 ng/ml	partition	+			489
4-aminobutyric acid	cerebrospinal fluid	>0.1 mole	chromatography	+			490
amphetamine, methylamphetamine	urine	>10 ng/ml	partition	+			491
biogenic amines	rat brain	>20–30 pg		+			492
bretylium	urine, plasma	4–100 ng/ml	partition	+		1–11	493
cadaverine, putrescine	foods	>1 ppm	extraction	+			494
camazepam, temazepam	plasma	>1 ng/ml	partition		89	5	495
cannabidiol	plasma	>50 ng/ml	partition	+			496
carboxylic acids, phenols	water	>1–10 μg/l.	steam distillation	+			497
carbendazim	walnuts, fruit	0.01 ppm	extraction/partition	+			498
catecholamines	rat brain	0.2–2 ng	extn./adsorp. (Al_2O_3)	+			499
chlorinated hydrocarbons	bone marrow		extrn./partn./chromatog.				500
4-chlorophenoxyacetic acids	bean sprouts	>50 ppb	extraction/partition	+			501
4-chlorophenoxyacetic acids	surface waters	>20 pg	partition	+			502
chlorpyrifos	pea-vine	0.01 ppm	extraction/chromatog.	+			503
clonazepam	plasma	40 ng/ml	partition			4–7	504
clopidol	chicken tissue	2–20 ppb	extraction/chromatog.	+	80–87		505
cytokinins	plant extracts	1 pg		+			506
p,p'-DDT, γ-HCH	camomile, senna	<0.5 ppm	extraction/chromatog.				507
1,2-dibromoethane	grapefruit	>0.4 μg/kg	steam distilln. chromatog.				508
dichlorobenzilic acid	urine	0.05 ppm	partition		97	9	509
diclofop, diclofop-methyl	soil	>0.05 ppm	extraction/partition	+			510
exaprolol	plasma	>1 ng/ml	partition	+			511
explosives	hand-swab extracts	ng	extn./XAD 7 clean up				512
explosives		100–500 pg				3–13	513
flunitrazepam	plasma	0.5–5 ng/ml	partition				514

N-(fluoro-2-enyl) acetamide	microsomal prep.	7–40 ng/ml	partition	+			515
flurazepam	plasma	> 1–2 ng/ml	enrichment on XAD 2	+			516
guaiphenesin	plasma	> 15 ng/ml	partition	+			517
halogenated hydrocarbons	drinking water	< 1 μg/l.					518
3-methoxy-4-hydroxyphenyl ethanediol	urine		partition	+	93–97		519
methyl bromide	grapefruit	> 2 μ̂g/kg	head-space				520
metoprolol	plasma	0–400 ng/ml	partition	+		3–8	521
mexiletine	plasma	20–1000 ng	partition	+			522
morphine	plasma, urine	pg	partition	+			523
naltrexone	plasma, urine	1–10 ng/ml	partition	+			524
naloxone	plasma	5–200 ng	partition	+		4	525
nitroglycerine	plasma	0.1–100 ng/ml	partition		89	5	526
norflurazon	fruit, cotton, soil	> 10 ppb	extrn./partn./tlc				527
pentachlorophenol	water	> 0.1 ppb	partition		80		528
pentachlorophenol	urine	> 10 pg	partition	+			529
pentachlorophenol	mushrooms, wood	1–200 μg/kg	extraction/partition	+			530
pentachlorophenol	milk	> 4 ppb	partn./H_2SO_4 clean up		> 90		531
pentachlorophenol	plasma	> 1–2 ppb	partition	+			532
pesticide residues	food of plant origin	1 ppb	extraction/chromatog.				533
pesticides (33)	soil, cabbage leaves		extrn./partn./chromatog.				534
pesticides	seal tissue		extraction/chromatog.		73–100		535
pesticides (chlorinated)	penguin eggs	1–1000 ppb	extrn./partn./chromatog.				536
pesticides (chlorinated)	tobacco	5–200 ng/g	extrn./partn./chromatog.				537
phenols	alcoholic beverages	0.08–0.9 mg/l.		+		5	538
phenols	water	1–1000 μg/l.	partition	+		3–6	539
pindolol	plasma, urine	2–30 ng/ml	partition	+		6–12	540
pindolol	plasma	1–30 ng/ml	partition	+		2–9	541
PCBs, polybrominated biphenyls	serum	> 10 ng	partition/chromatog.				542
primaquine	blood	10–40 ng/ml	partition	+		6	543
probenecid	plasma	1–10 μg/ml	partition			8	544
pyrimethamine	plasma	5–400 ng/ml	partition				545
roniltan, rovral	grapes	> 10 ppb	extraction/chromatog.		94–98		546
sulphamethazine	swine feed	> 0.2 ppm	extraction/chromatog.	+			547
sulphamethazine	urine, waste water	1.0 ppm	partition	+			548
tetrachlorvinphos	plant products	> 0.5 ppm	extraction			5	549
toxaphene	fish	ppb	extraction/chromatog.				550
triazolam	plasma	> 0.2 ng/ml	partition		95	5–6	551

Table 6.14 – A few examples of the application of gc with detection by fid.

Substance(s)	Matrix	Level	Separation	Recovery (%)	Relative standard devn. (%)	Reference
acrylonitrile	ambient air	>0.5 ppm	trapping (carbon)/elution			553
amantadine	plasma, urine	500 ng/ml	partition	98	0.3	554
apovincaminic acid ester	plasma, urine	20–200 ng	partition			555
benzo(*a*)pyrene	petroleum products	2 ppb	partition/chromatog.	87–90		556
butylated hydroxyanisole, tert.-butylquinol, 2,6-di-tert.-butyl-*p*-cresol	cottonseed oil	100 ppm	partition/deriv.	84–108	8–10	557
dibenzazepine derivatives	blood	0.5–5 μg/ml	partition	79	3	558
1,2-dichloroethane	blood	>25 ng/ml	head-space			559
dimethindene	serum, urine	4–200 ng/ml	partition			560
gluthetimide, phenacetin	post-mortem tissue	2–10 μg/g	extraction/partition	47–89		561
fentanyl + analogues	plasma	0.1–3 ng/ml	partition	84		562
morpholine-4-carboxaldehyde	hydrocarbons	10–100 ppm				563
naltrexone, 6β-naltrexol	plasma, saliva, urine	1–20 ng/ml	partition/deriv.			564
tocainide, lignocaine	plasma	>0.2 μg/ml	partition			565
tranquillizer	slaughtered pigs	2–20 ng/g	extraction/chromatog.			566
polycyclic aromatic hydrocarbons (13)	smoked fish	1 μg/kg	saponification/several partitions/chromatog.			567
propadiene, propyne	propene	>1 ppm			10	568
valproic acid, ethosuximide	serum	1 ng/ml	adsorption (charcoal)/elution		3	569
valproic acid, ethosuximide	serum	1 ng–200 μg/ml	partition		5–7	570

Table 6.15 – Examples of the application of the coulometric and conductivity detectors.

Substance(s)	Matrix	Level	Reference
N-nitrosamines		>50 pg	574
N-containing herbicides		0.01–0.1 ppm	575
carbamate insecticides	food	>1 ng	576
urea herbicides	food	0.01 ppm	577
aldicarb, aldicarb sulphone	soil, water	1 ppm	578
diazepam	plasma	>1 ng	579
vinyl chloride	water	0.1 μg/1.	580
chloroanilines, chloronitroanilines	water		581

Flame photometric detector (FPD)

This detector was introduced in 1966 [582]. It is based on burning phosphorus or sulphur compounds in a hydrogen flame and observing the emission of the flame, usually with use of filters (394 nm for S, 562 nm for P). Examples of its application are listed in Table 6.16.

A review has been made of the limitations of the FPD in atmospheric analysis [604]. The response mechanism has been debated [605]. Some suggestions have been made for solving the 'flame-out problem', which results from extinction of the flame by a separated compound [606]. Dual flame photometric detectors have also been introduced [607–609].

Thermionic detector (TID)

This detector was originally described in 1964 [610]. Today, a version is often used which is based on developments in 1974 [611]. In it a rubidium silicate bead is electrically heated, and a hot plasma is formed around the bead. The gas from the column effluent – together with the separated trace compound – is fed into a flame placed under the rubidium silicate bead. A collector electrode above the flame-bead combination measures changes in the flame ionization characteristics, which are sensitive to the presence of nitrogen or phosphorus atoms in the separated trace compounds. Kolb explained the mechanism of his detector [612] as reaction of Rb atoms, vaporized from the silicate source, with the radicals of the flame to form Rb^+ and e^-. The electrons are transferred to the compounds to be detected, and the Rb^+ returns to the silicate bead, which explains the long-term stability of this detector. Other authors explain the ionization process as a catalytic one in which Rb atoms are only surface catalysts for the electron transfer, and do not leave the bead surface. The inexhaustibility of the silicate bead can be better understood by this explanation. Examples of the application of this detector are presented in Table 6.17.

Table 6.16 – Examples of the application of the flame photometric detector.

Substance(s)	Matrix	Separation before gas chromatography	Level	Reference
acephate	cabbage, bush beans, squash	partition	0.1–0.7 ppm	583
alcohols, phenols (after derivative formation with chlorophosphate)			>10 ng	584
aldicarb, malathion			15 ng	585
secondary amines (after derivative formation)	fish, sausage, ham, spinach	partition	10 ppm	586
carbamates (S-containing)				587
ethylene thiourea	potatoes, spinach		60–300 ppb	588
famphur	animal tissues	partition/column chromatog.	25 ppb	589
fenamiphos, fensulfthion	tobacco	partition/chromatog.	>50 ppb	590
fenamiphos	plants, soil	partition/chromatog.	0.01–1 ppm	591
flavours (S-containing)	beer	distillation/XAD 2	ppb	592
insect chemosterilizants (19)			ng	593
methiocarb	plants, animals, soil	partition/chromatog.	0.02–1 ppm	594
methanesulphonate of *m*-aminobenzoic acid ethyl ester	fish	partition	1–19 μg/g	595
methylthiouracil	meat	partition/chromatog.	0.5 ng	596
3-(methylthio)propionaldehyde	air		1–30 ng	597
parathion, malathion, diazinon		partition		598
parathion methyl	honey, bees	partition	0.1–10 ppm	599
phenothiazine tranquillizer	meat of slaughtered animals	partition	>1 ng	600
pirimiphos methyl	water, fish, snails	partition	0.2–61 ng	601
tebuthiuron	grass, sugar cane	partition/chromatog.	0.1 ppm	602
trichlorphon (dipterex)	eggs, milk		0.4 ppm	603

Table 6.17 –Examples of the application of a thermionic N-sensitive or P-sensitive detector.

Substance(s)	Matrix	Separation before chromatography	Level	Reference
acephate	asparagus, trout, sediments		0.01 ppm	613
acrylonitrile	air, water, plastic packaging		mg/kg	614
	monomers			615–617
amines, amino-acids			500 fg	618
amitryptiline	plasma	partition	1 ng/ml	619
amitryptiline, nortryptiline	plasma	partition	10 ng/ml	620
antidepressant drugs (28)	plasma	partition	10 ng/ml	621
antimalaria drugs (4)	urine	partition	0.2 ng	622
apovincamic acid	plasma	partition	2 ng/ml	623
aromatic amines	air		2–200 ng	624
aryl phosphates	fish	partition	0.04–1 ppm	625
bromacil, lenacil	soil	XAD 2	0.2–5 mg/kg	626
bromphenvinphos	milk, animal tissue	XAD 2	2 ppb	627
butanilicaine	urine	partition	5–500 ng/ml	628
carbamate insecticides	foliage, soil, fish	partition	0.5–5 ppm	629
carbamate insecticides	water	XAD 2	0.5 ppm	630
carbamazepine	plasma	partition		631
chloroquine	blood	partition	20–300 ng/ml	632
chlorpheniramine	plasma	partition	1 ng/ml	633
chlorthalidone	plasma, urine		10 ng	634
clofexamide	plasma	partition	50–400 ng	635
clonidine	rat brain	partition	10 ng/g	636
cocaine	plasma	partition	20 ng/ml	637
cocaine, benzoylecgonine	plasma, urine	partition	10 ng/ml	638
cyclobenzaprine	plasma	partition	5 nmole/ml	639
cyclophosphamid	plasma	partition	10 ng/ml	640, 641
cytokinins	grapevines	partition	> 10 pg	642
dextramethorpan	plasma	partition	1 ng/ml	643
diltiazem	plasma	partition	10 ng/ml	644
doxylamine succinate	animal feed, urine	partition, chromatog.	0.1–1 ppb	645
fenitrothion	wheat	partition	0.5–12 ng	646
fluphenazine	plasma	partition	10 ng/ml	647

Table 6.17 – *continued*

Substance(s)	Matrix	Separation before chromatography	Level	Reference
hexamethylmelamine	plasma	partition	800 ppb	648
indoles	saliva	partition	>1 ppb	649
lignocaine	plasma	partition	0.5–20 μg/ml	650
lignocaine	plasma	partition	0.1–9 μg/ml	651
maprotiline	blood	partition	20–400 ng/ml	652
metapramine	plasma	partition	15 ng/ml	653
methadone + metabolite	plasma, urine	partition	5 ng/ml	654
methatrimeprazine	plasma	partition		655
methomyl	plants	XAD 2	ng	656
N-methylcarbamate insecticides			1 ng–5 μg	657
N-methylimidazolylacetic acid	urine	chromatography	1–40 μmole/1.	658
morphine	urine	partition	10 ng/ml	659
N-compounds	oil spill	XAD 2		660
nefopam	plasma, saliva	partition	10–70 ng/ml	661
nicotine	plasma	partition	ng/ml	662
nicotine, cotinine	plasma	partition	1–100 μg/l.	663
nicotine, cotinine	plasma	partition	2–50 ng/ml	664
N-nitrosamines	water	adsorption on charcoal, extraction	1 ppt	665
N-nitrosamines	food, beer		0.1–100 ng	666
N-nitrosamines	cigarette smoke condensate	partition		667
nomifensine	plasma	partition	2 ng/ml	668
odorous volatiles	air	charcoal adsorption		669
organophosphorus insecticides	water	XAD 2	10–100 ng/l.	670
organophosphorus insecticides	food		0.5 mg/kg	671
organophosphorus insecticides			4–40 pg	672
orphenadrine	plasma	partition	1 ng/ml	673
oxyphenbutazone	plasma	partition	0.5 μg/ml	674
pesticides	urine	partition		675
pethidine, norpethidine	plasma, urine	partition	50–1000 ng/ml	676
pethidine, norpethidine	blood	partition	>5 ng/ml	677

pemoline	urine	partition		678
phenobarbital primidone	plasma	partition	0.5 mg/l.	679
phenothiazines	plasma	partition	>4 ng/ml	680
phospho-organic pesticides	plants		0.04 mg/kg	681
prajmalium bitartrate	plasma	partition	10–200 ng/ml	682
procyclidine	plasma	partition	2–160 ng	683
propranolol	plasma	partition	12 ng/ml	684
pyrazon	sugar beet	partition	>20 ppb	685
tranylcypromine	plasma, urine	partition	2–125 ng/ml	686
theophylline	plasma	partition	50 μmole/l.	687
triazine herbicides	soil, plants	partition	500 ppb	688–691
triazine herbicides	crops	chromatography	10–40 pg	692
triazine herbicides, metabolites	urine	partition	0.1 ppm	693
tricyclic antidepressants	plasma	partition	<20 ng/ml	694
tricyclic antidepressants	plasma	partition	25–175 μg/l.	695, 696
trifluoperazine	plasma	partition	0.5–15 ng/ml	697
xylazine	meat, milk	partition	>90 ppb	698

Chemiluminescence detectors

Chemiluminescence detectors are used mainly for the determination of nitrosamines [699]. An initial pyrolysis reaction cleaves *N*-nitrosamines at the N–NO bond to produce nitric oxide or the nitrosyl radical NO· [700], which react with ozone to give a chemiluminescence reaction. Such detectors have considerable selectivity for nitrosamines. The light emitted from the $NO–O_3$ reaction is in the near ir, while other known chemiluminescence reactions with ozone emit light in the visible or near uv region. Selectivity can be further improved by placing a cold trap between the pyrolysis chamber and the $NO–O_3$ reaction chamber. Direct injection of the samples into the pyrolysis chamber or combination of the detector with gas chromatography are both possible. The use of the detector, which is also called a 'thermal energy analyser', has been investigated [710, 702]. Interference of ethanol in determination of *N*-nitrosodimethylamine has been reported [703]. *N*-Nitrosodimethylamine has been determined at a level of 0.1–4 ppm in dimethylamine, by gc–ms, gc–TID and gc–TEA (thermal energy analyser). The results from these three independent methods agreed [704]. Other examples of use of the TEA-detector are given in Table 6.18.

The use of chemiluminescence reactions for the determination of other compounds, mainly reactive hydrocarbons, has been described [722]. Ozone reacts with olefins at room temperature and with alkanes, alcohols, aromatics, vinyl chloride, acetylene, NO and H_2S at above 150°C. The limit of detection is very often a few ppb. The reaction with vinyl chloride can be used for detection of about 10 ppm after gc [723]. Polycyclic aromatics, N-heterocyclics, and S-containing compounds give an ozone-induced chemiluminescence reaction when deposited on silica gel [724].

Hydrocarbons, both saturated and unsaturated, and with or without halogen constituents, can be detected at a level of a few tenths of a nanogram (for vinyl fluoride) if the sample is introduced into active nitrogen. A chemiluminescence reaction takes place [725].

Molecular emission cavity analysis

This method, which is useful for inorganic compounds [726], can be modified so that sulphur or phosphorus compounds can be detected at the ng-level [727], after gc separation.

Microwave plasma detector (MPD)

The principle of this detector was published in 1965 [728]. It is based on passage of the gc eluate into a strong microwave field which excites emission of the characteristic spectra of the atoms in the separated trace components. An argon atmosphere is often used. The detector can be set to detect C, H, D(!), P, O, N, S, halogens or metals. The limit of detection is of the order of 0.1–1 ng. The technique has been used for the determination of pesticides [729] and of

Table 6.18 – Examples of the use of a thermal analyser (TEA) as a detector in gc.

Substance(s)	Matrix	Level	Separation	Reference
N-nitrosodimethylamine	malt beverages	125 ng/g	adsorption/elution	705
N-nitrosodimethylamine	malted barley	3 ppb	distillation/partition	706
N-nitrosodimethylamine	diesel engine emissions	>0.1 μg/m³	trapping/partition	707
N-nitrosodimethylamine	beer, wort, malt		distillation/partition	708
N-nitrosodimethylamine	beer	>0.1 ppb	distillation/partition	709
N-nitrosodimethylamine	beer	0.1 ppb	distillation/partition	710
N-nitrosodimethylamine, *N*-nitrosopyrrolidine	dried foods	1–67 ppb		711
N-nitrosodimethylamine	meat products	2–7 μg/kg		712
N-nitrosopyrrolidine	meat products	0.5 μg/kg		
N-nitrosopiperidine	meat products	0.5 μg/kg		
N-nitrosamines	amines	30–40 pg		713
N-nitrosodiethanolamine	cosmetics	5 ppb	partition/chromatography	714
N-nitrosodiethanolamine	cosmetics	10 ppb	chromatography	715
N-nitrosodiethanolamine	cosmetics	20–30 ppb	extraction/chromatography	716
N-nitrosoproline	food	5 μg/kg	extraction/partition/derivative formation	717
nitrosamines, volatiles	pickled herring	0.3 μg/kg	distillation/partition	718
nitrosamines, volatiles	biological material	3 ng	extraction	719
nitrosamines, volatiles	cooked-out bacon fat	1 ppb	extraction/partition	720
N-nitrosothiazolidine	fried bacon	32 ppb	distillation/partition/ chromatography	721

trihalomethanes (at below 1 ppb) in water [730]. Recent improvements to the method have been reported [731-734]. Trichloroacetic acid at the ppb level in water [735] and polybrominated biphenyls at the level of more than 1 ng [736] have been determined.

Photoionization detector (PID)
In this detector, ions are produced when the column effluent is exposed to intense radiation. The excitation energy can be controlled so that materials with a high ionization potential do not give a response. Very good results have been obtained for toluene, cycloheptene and n-decane [737]. A PID has been used for determining ppt concentrations of benzene, toluene and *m*-xylene in air [738]. Likewise, chlorobenzenes (5 ppb–15 ppm) in air, urine and blood samples have been measured [739]. Other applications are the determination of CF_4, C_3F_8 and C_4F_{10} (with better responses than by ecd) at the 8 ppb–0.3 ppm level [740], of thiols at the $\mu g/m^3$-level in air [741], and of oil in air (1 ppb) [742]. Quenching or enhancement of the PID signal has been investigated [743].

Piezo quartz detector
A piezoelectric quartz crystal coated with a gc stationary phase can take up organic compounds from a carrier gas stream passing over it. The consequent change in weight of the coating causes the oscillation frequency to change. In this way 1 ppb of nitrotoluene [744] or 30-300 ppm of toluene in air [745], gaseous pollutants [746] and 0.02 ppm of di-isocyanatotoluene [747, 748] have been detected or measured.

6.10 ELECTROCHEMICAL METHODS

About 4% of all publications on organic trace analysis mention an electrochemical method for the determination of trace constituents.

6.10.1 Potentiometric methods

In organic trace analysis potentiometric titrations are considerably less used than in inorganic analysis, as only a rather limited number of redox systems can be examined potentiometrically. Only those organic reactions that exhibit electrochemical reversibility can be studied directly.

In organic trace analysis a specialized type of potentiometry is more important, viz. establishment of a membrane potential. Membrane electrodes include glass electrodes (for measurement of H^+, K^+, Na^+, Ca^{2+}, NH_3 etc.), liquid-membrane electrodes and solid-state electrodes.

The use of ion-selective electrodes as biosensors has been reviewed [749]. Their principle is demonstrated in Fig. 6.11. The field of such enzyme electrodes will expand in the near future. Miniaturization of the electrodes will also enlarge the field of application. Enzyme electrodes exhibit the very high specificity of enzyme reactions. The sensitivity depends on the enzyme-substrate relations, construction of the membrane, and other parameters. Twenty-five enzyme-electrode probes for about 40 substrates have been described [750]. Insecticides which inhibit cholinesterase (organophosphorus compounds, carbamates) can be determined in the ng/ml range. Control of membrane enzyme activity is constantly necessary, however, as bacterial growth or other effects may interfere.

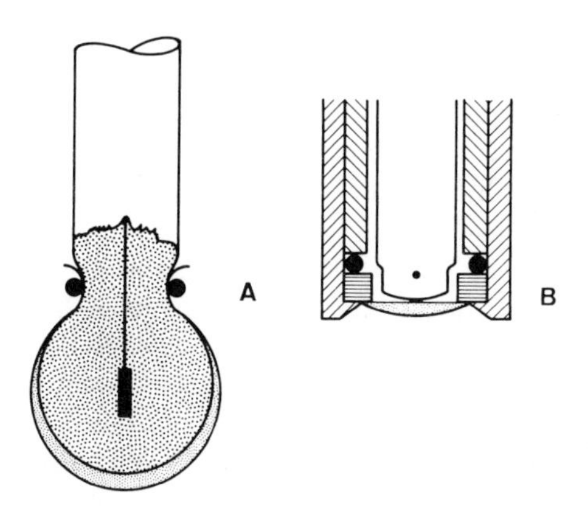

Fig. 6.11 – Enzyme electrodes. (A) A cross-linked, enzyme-coated glass electrode. (B) A triple-membrane electrode; *outer membrane* = cellulose; *middle membrane* = gas-permeable Teflon; *inner membrane* = ion-selective membrane of the indicator electrode (in the case of CO_2 or NH_3 sensors, a pH glass electrode). (Reproduced from [749] by permission of the copyright holders, Springer Verlag).

Glass membrane micro electrodes [751, 752] can be used for the direct measurement of pH in tissues [753]. Such small electrodes can also be helpful in organic trace analysis, for pH control in tiny droplets of solutions.

An example of application of potentiometric methods is the determination of traces of penicillamine (by reacting it with excess of Hg^{2+} and titrating the surplus) [754]. Traces of phenols (10^{-7}–10^{-4} mole) can be reacted with tyrosinase, immobilized on the surface of a Pt electrode; benzoquinone is formed, which is reduced by $Fe(CN)_6^{4-}$ and the $Fe(CN)_6^{3-}$ formed is determined potentiometrically [755]. In another example, specificity is obtained by passing the volatile *N*-nitrosamines to be determined (10^{-8} mole) into an aqueous solution of NaOH. This solution is then irradiated with light from a xenon lamp, and nitrite is formed by photodecomposition. The NO_2^- is separated on an ion-exchange column and determined potentiometrically [756].

6.10.2 Coulometric methods

Coulometric methods are based on the direct proportionality between the quantity of electricity and the amount of oxidation or reduction that occurs upon passage of a current through an electrochemical cell. By measurement of the quantity of electricity the amount of chemical change brought about is determined. A coulometric cell can be used to generate very small amounts of a reagent, which can be measured quite simply in that way.

The end-point of the reaction can be indicated in several ways. Very often a potentiometric or voltammetric method is used, in which an additional electrode couple is introduced for indication purposes. The change of current or potential is measured. Photometry or almost any other convenient solution method can also be used [757].

Reagents are generated either directly within the coulometric cell or are added to it after generation elsewhere [758, 759]. Several applications in organic trace analysis have been described. Coulometry has been used in combination with gc, especially for the detection of traces of halogen-containing compounds.

Amines are titrated in the same way as ammonia [760]. A borate buffer of pH 8.5 is used as supporting electrolyte, and contains KBr from which hypobromite is generated. The hypobromite reacts rapidly with the amines or ammonia, which can thus be titrated in submicrogram amounts [761].

Examples of applications are the use of coulometric titration in pollution analysis [762], for determination of chlorine-containing pollutants in water [763] and determination of 1 ppm of vinyl chloride in air [764]. Combination with hplc has been suggested for determination of perphenazine or fluphenazine in serum by partition into hexane, followed by the hplc [765].

6.10.3 Voltammetry

Voltammetry is based on the potential–current behaviour of a polarizable electrode in the solution to be analysed. In contrast to potentiometry (which has zero current flow) in voltammetry a net current flow is measured. Mass transport to the electrode surface is necessary to ensure constant flow of current as reduction or oxidation takes place. There are three types of mass-transport mechanisms:

(1) convection, which is simply brought about by stirring or other mechanical agitation of the solution;
(2) diffusion in a layer very close to the surface area of the electrode;
(3) migration of ions within the solution.

There are several types of voltammetric method. Polarography uses a dropping mercury electrode, the surface of which is constantly renewed, thus excluding certain side-effects of other voltammetric procedures. The process is diffusion-controlled. Components of the solution become electrochemically active, depending on the potential applied to the electrode. The 'half-wave potential' is a means for qualitative identification of a substance. The specificity of voltammetric procedures is increased by use of the proper potentials. In certain variants of this method solid electrodes are used.

Specialized forms – such as oscillopolarography, pulse polarography and chronopotentiometry – are also useful in organic trace analysis, as they offer high sensitivity in many cases. Differential methods can be used if the sensitivity is to be enhanced.

The following reversibly reacting systems exhibit better analytical features than irreversible reactions.

(1) Quinones-hydroquinones

$$O{=}C_6H_4{=}O \xrightleftharpoons{2e^-,\ 2H^+} HO{-}C_6H_4{-}OH$$

(2) Azo compounds

$$R\text{-}N{=}N\text{-}R \xrightarrow{2e^-,\ 2H^+} R\text{-}NH\text{-}NH\text{-}R$$

(3) *N*-nitroso compounds

$$R\text{-}NO \xrightarrow{2e^-,\ 2H^+} R\text{-}NHOH \xrightarrow{2e^-,\ 2H^+} R\text{-}NH_2$$

The second reaction proceeds only in acid solution.

Irreversible reactions are observed with the following systems.

(1) Halogen-containing compounds, e.g.

$$R\text{-}CH_2\text{-}Br \xrightarrow{2e^-,\ H^+} R\text{-}CH_3 + Br^-$$

(2) Nitro compounds

$$R\text{-}NO_2 \xrightarrow{4e^-,\ 4H^+} R\text{-}NHOH \xrightarrow{2e^-,\ 2H^+} R\text{-}NH_2$$

The second step proceeds with aromatic compounds only in acidic solution.

(3) Disulphides

$$R\text{-}S\text{-}S\text{-}R' \xrightarrow{2e^-,2H^+} RSH + R'SH$$

(4) Oximes, hydrazones, N-heterocyclics, compounds carrying the C=N– group

$$=C{=}N\text{-} \xrightarrow{2e^-,\ 2H^+} {=}CH{-}NH{-}$$

(5) Hydrazo compounds

$$R\text{-}CO\text{-}NH\text{-}NH_2 \xrightarrow{2e^-,\ 2H^+} R\text{-}CO\text{-}NH_2 + NH_3$$

(6) *N*-Oxides

$$\equiv N{-}O \xrightarrow{2e^-,\ 2H^+} \equiv N + H_2O$$

As can be seen from the chemical equations, many reactions are highly dependent on the pH of the solution.

Obviously, not all functional groups are electrochemically active. On the other hand, those few groups which react can be determined rather specifically. There are several examples of application of electrochemical determinations with little or even no separation before the final determination.

Electrochemical methods provide greater flexibility and specificity than uv absorbance measurements for the determination of theophylline after hplc separation [766]. Analogously, 2-phenylphenol residues in orange rind can be determined at the 1 ppm level after extraction with CH_2Cl_2 and Florisil column clean-up [767]. The differential pulse determination of 1,4-benzodiazepines is simpler and more flexible than fluorimetry, gc (with fid), spectrophotometry and immunology [768]. In the determination of catecholamines after hplc separation the electrochemical separation is about 10 times more sensitive than uv or fluorescence detection [769].

A few articles describe general aspects of the applicability of voltammetric methods for the determination of organic compounds with biological significance [770] and of pesticides [771].

Equipment for voltammetric analysis – even for highly sophisticated techniques – has been commercially available for several years. Recent interest focuses on the combination with chromatography, mainly hplc. Equipment with flow-through cells of about 1 μl volume is offered by several firms. Solid electrodes made of vitreous carbon, carbon paste, amalgamated gold or platinum are used. Carbon paste – a mixture of graphite powder and silicone oil or paraffin wax – is the most popular electrode material for organic chemical reactions. Quantities of a few ng – and often less – are detectable with such micro cells.

General aspects of electrochemical detection in hplc have been presented [772-776], and special detector systems evaluated [777-780]. Application to electrolytically non-conductive column effluents is possible by addition of conductive electrolytes to the eluate [781].

Examples of the use of electrochemical methods in organic trace analysis are given in Table 6.19.

Several reagents, containing one or two nitro groups, have been suggested for formation of electrochemically active derivatives from compounds which are not themselves active [841]. These reagents are used in precolumn techniques, i.e. the trace compounds to be separated and determined are converted into derivatives before the chromatographic separation. Excess of reagent can be separated at the same time as the trace compounds. The column effluent is monitored electrochemically. Table 6.20 gives examples of such reagents. As the nitro group can be reduced in a reversible, highly sensitive, 6-electron reaction, good sensitivity is attainable.

Table 6.19 – Examples of electrochemical determination of organic traces.

Compound(s)	Matrix	Level	Separation	Determination	Reference
catecholamines, metabolites	biological fluids		rev. phase lc	amperometric	782
epinephrine, norepinephrine		pg	hplc	amperometric	783
DOPA, dopamine, epinephrine	tissue	0.4–1 ppm	rev. phase hplc	amperometric	784
normetanephrine, 3-methoxytyramine	urine	20 μg/l.	ion-exch./partn./lc	amperometric	785
catecholamines		50–200 ng	rev. phase hplc	amperometric	786
norepinephrine, serotonin	brain tissue	subpicomole	ion-exch. hplc	amperometric	787
catecholamines	plasma	>0.2 pg	lc	amperometric	788
epinephrine, DOPA		ng		amperometric	789
catecholamines	brain		*in situ* (!)	amperometric	790
serotonin, dopamine	brain	2–900 ng/g	hplc	amperometric	791
vanillylmandelic acid	urine		partn./lc	amperometric	792
serotonin	biological materials	25 pg	ion-exch./hplc	amperometric	793
N-nitrosodiethanolamine	grinding fluid	μmole		diff. pulse polarog.	794
N-nitrosamines		10^{-7} mole		diff. pulse polarog.	795
N-nitrosamines		10^{-8} *M*	partn.	diff. pulse polarog.	796
N-nitroso compounds	metal-working coolant fluids	>0.5 μg		diff. pulse polarog.	797
nitrophenol, nitrobenzene		>4 ng	hplc	dropping Hg electr.	798
nitropyridines			lc	dropping Hg electr.	799
nitration of contaminants	water			dropping Hg electr.	800
pyridine-*N*-oxides		0.5 μg/ml		pulse polarog.	801
imidazo-1,4-benzodiazepines	blood	>50 ng/ml	partn.	diff. pulse polarog.	802
1,4-benzodiazepines	blood	>10 ng/ml	partn.	diff. pulse polarog.	803
1,4-benzodiazepines	biological fluids			diff. pulse polarog.	768
1,4-benzodiazepines		3 ng	hplc	amperometry	804
phenothiazine sulphoxides		100 nmole	tlc	pulse polarog.	805
triamterene		>1 μg/ml		polarog.	806
paraquat	urine, serum	30 ng/ml		diff. pulse polarog.	807
tryptophan, metabolites		ng	hplc	amperometry	808
uric acid	cereal products	2 μg/g	partn./hplc	amperometry	809
theophylline		10–20 ppm	hplc	amperometry	810
phenolic compounds	water	>1 ppb	rev. phase hplc	amperometry	811

Table 6.19 –*continued*

Compound(s)	Matrix	Level	Separation	Determination	Reference
2-phenylphenol	oranges	1 ppm	partn./lc	amperometry	812
pentachlorophenols	water	>0.3 ppm		diff. pulse polarog.	813
halogenated anilines		pg	hplc		814
2-mercaptobenzimidazole	rubber	0.1–3 μg	extrn. (acetone)	a.c. polarog.	815
dimethylsulphide, dimethyldisulphide	water	>3 μg/l.	partn.	oscillopolarog.	816
disulphiram		>0.2 mg/l.		diff. pulse polarog.	817
thiourea		10^{-8} mole		polarog.	818
cystine		>9 ng	hplc	amperometry	819
tetramethylthiuramdisulphide	plants	0.1 ppm		polarog.	820
dithiocarbamates	fruits	0.5 μg		oscillopolarog.	821
metafos	water	1 μg/l.		diff. oscillopolarog.	822
phosmet	apples	0.1 ppm		polarog.	823
aflatoxins	foodstuffs		extrn./lc	diff. pulse polarog.	824
cephalosporins		0.1 μg/ml		diff. pulse polarog.	825
tensides		30 ppb	hplc	polarog. (tensamm.)	826
detergents	effluents	0.5 ppm		polarog. (tensamm.)	827
nitriloacetic acid	water	24 ppm		deriv. pulse polarog.	828
emulsifiers	kitchen goods			polarog. (tensamm.)	829
PAHs		$10^{-8}M$		diff. pulse polarog. (non-aq. solvents)	830
phosmet, cyclohexane	water	μg/l.	partn	diff. oscillopolarog.	831
acrylamide monomer	polyacrylamide	1 ppm	extrn./lc	diff. pulse polarog.	832
morphine, morphine-3-glucuronide	blood	1 ng	hplc	amperometry	833
saccharine	beverages	0.5 ppm	partn.	diff. pulse polarog.	834
adriamycin	plasma	400 ng/ml	partn.	diff. pulse polarog.	835
robenidine	chicken tissue	0.1 ppm	partn./ion-exch./lc	cathode-ray polarog.	836
robenidine	animal food	5 mg/l.	extrn.	diff. pulse polarog.	837
acetaminophen	plasma	>0.2 μg/ml	partn./hplc	amperometry	838
N-phenyl-*N*-isopropylacetamide	soil	50 ng/kg	extrn./lc	polarog.	839
4-formylbenzoic acid	terephthalic acid	0.1 mg/l.		polarog.	840

Table 6.20 – Reagents for introduction of an electroactive group.

Reagent	Compounds from which derivatives are to be made	Reference
$(O_2N)_2C_6H_3$–C(=O)–Cl	alcohols, amines	842
O_2N–$C_6H_3(NO_2)$–F	amines, amino-acids	843
O_2N–$C_6H_3(NO_2)$–O–SO_2–OH	amines, amino-acids	844
O_2N–C_6H_4–CH_2–O–C(=N–$CH(CH_3)_2$)(NH–$CH(CH_3)_2$)	carboxylic acids	845
O_2N–C_6H_4–CH_2–Br	carboxylic acids	846
O_2N–$C_6H_3(NO_2)$–NH–NH_2	aldehydes, ketones	847

Another technique is 'post-column derivative formation', often by oxidation or reduction. In this way catecholamines can be determined in ng amounts. The oxidation or reduction reagent is added in solution in a homogeneous reaction or is applied bound to a solid electrode surface (often graphite or an oxide semiconductor) [849].

6.11 CHEMICAL DETECTION OR DETERMINATION

Besides the many methods of instrumental analysis there are useful micro procedures which are based on chemical reactions. Combination with gc is suggested [850]. The effluent from the column is either directed into a vessel containing the reagent or onto a filter paper or tlc plate impregnated with the reagent [851]. Table 6.21 gives some test reagents for functional group analysis (after [850]). Minimum detection levels are 0.4 $\mu g/\mu l$.

Table 6.21 – Functional group classification tests (after [850], by permission of the copyright holders, Wiley-Interscience, Inc.).

Compound type	Reagent	Type of positive test
Alcohols	$K_2Cr_2O_7$-HNO_3	Blue colour
	Ceric nitrate	Amber colour
Aldehydes	2,4-DNP	Yellow ppte.
	Schiff's	Pink colour
Ketones	2,4-DNP	Yellow ppte.
Esters	Ferric hydroxamate	Red colour
Mercaptans	Sodium nitroprusside	Red colour
	Isatin	Green colour
	$Pb(OAc)_2$	Yellow ppte.
Sulphides	Sodium nitroprusside	Red colour
Disulphides	Sodium nitroprusside	Red colour
	Isatin	Green colour
Amines	Hinsberg	Orange colour
	Sodium nitroprusside	Red colour 1°
		Blue colour 2°
Nitriles	Ferric hydroxamate-propylene glycol	Red colour
Aromatics	HCHO-H_2SO_4	Red-wine colour
Aliphatic unsaturation	HCHO-H_2SO_4	Red-wine colour
Alkyl halide	Alc. $AgNO_3$	White ppte.

Specific reagents that can be used in gc, for detection of functional groups, have been surveyed [852].

A few micrograms of pure (i.e. single peak) compounds from gc separations are condensed in the centre of a glass 100-μl glass capillary from the gc effluent [853]. The capillary contains a small quantity of a catalyst, and an inert or reactive gas (hydrogen, nitrogen, ozone) is passed into the capillary, which is then sealed and heated for a short time (1–10 min at 160–300°C). After completion of the reaction, the capillary is introduced into the gas chromatograph and crushed. The technique has been applied successfully to the analysis of food-flavour volatiles [854–858].

After trapping of ng amounts of analyte at liquid-nitrogen temperature [859], micromethods for chemical identification have been used, e.g. reduction with $LiAlH_4$ [860]. Ethers, alcohols, phenols and carbonyl compounds have been tested. After conversion into the corresponding hydrocarbons these were analysed by gc.

6.12 ENZYMATIC REACTIONS

Enzymatic reactions are highly specific [861]. Naturally, the determination of 'physiological' substrates is the major area of application of enzymatic analysis. The limit of detection is low. By means of highly developed special methods, e.g. 'enzymatic cycling', a few selected compounds can be determined even in amounts of 10^{-16} mole. A promising method is 'radioenzymatic labelling', in

Table 6.22 – Examples of the use of enzymatic reactions in organic trace analysis.

Trace	Level	Enzyme used	Determination	Reference
aldehyde sex pheromones	20 pg	luciferase	luminescence	865
medroxyprogesterone	30 fmole	20β-hydroxysteroid dehydrogenase	fluorimetry	866
pyridoxal-5′-phosphate	10 pg	tyrosine decarboxylase	scintillation	867
acetaldehyde	120 ppm	aldehyde dehydrogenase	photometry	868
enitrothion	1 ng	choline esterase	tlc densitometry	869
paraoxon	0.3 ng	choline esterase	tlc densitometry	870
catecholamines	1–10 pg	catechol *O*-methyltransferase	radiochemical	871–876
normetanephrine	5 pg	phenylethanolamine methyltransferase	radiochemical	877
catecholamine sulphate conjugates	20 pg/ml	catechol methyltransferase	radiochemical	878
DOPA	50 pg/ml	catechol methyltransferase	radiochemical	879
histamine	1 μg/l.	histamine methyltransferase	radiochemical	880

which a radioactively marked group is transferred to the substrate. After separation of the labelled product formed, its mass is determined radiochemically.

Man-made molecules of pharmaceutical, toxicological or ecological interest are determinable in some cases. If these substances act *in vivo* by inhibiting enzyme systems, a determination is possible. Thus, some insecticides inhibit cholinesterase at ng/ml concentrations [863]. Some groups of herbicides decouple oxidative phosphorylation [864]. Further examples of enzymatic determinations can be found in Table 6.22.

6.13 IMMUNOLOGICAL REACTIONS

After application of a non-physiological compound – the 'antigen' (AG) – to an animal, 'antibody' molecules (AB), active against the added antigen, are formed. Antibodies are globulins. They react very specifically with the corresponding antigen: AG + AB = (AB–AG). Very often a precipitate is formed at higher concentration. The method depends on the availability of the antibody AB, which acts as a specific reagent.

Usually, this method is applied in the form of a substoichiometric reaction. In this, a subequivalent amount of reagent is added to a solution containing a compound which reacts with it, and is present in two forms – labelled and unlabelled. Both forms react absolutely identically with the reagent and thus compete for the substoichiometric amount added. The ratio of labelled to unlabelled form bound to the reagent will be the same as the ratio in the solution. Thus from an analysis of the reaction product conclusions can be drawn about the composition of the solution and thus the sample.

In immunology, labelling is achieved by use of a radioactively marked antigen, and this technique is called radioimmunoassay (ria). The method depends on the availability of this labelled antigen and, of course, the antibody. The antigen from the sample and a small amount of radioactively labelled antigen added simultaneously compete for reaction with the antibody. If there is a large quantity of non-labelled antigen (originating from the sample), correspondingly less of the labelled antigen will be bound to the antibody. The radioactivity found after isolation of the antigen–antibody complex is a measure of the antigen concentration of the sample. High radioactivity corresponds to a low concentration of the (inactive) trace which is to be determined.

In routine work the use of radioactive materials is generally avoided because of the hazards and handling problems. Consequently, other labelling methods have been searched for.

Labelling with fluorescent markers has been suggested. To prepare the marker a fluorescent group – usually fluorescein [881, 882] – is bound covalently to the trace molecule (M_{tr}), giving the marker (M_{tr}–Flu_{lab}). In the assay (M_{tr}) from the sample and (M_{tr}–Flu_{lab}) compete for reaction with the antigen. The fluorescence intensity of the antigen–antibody complex is measured. Alternatively, the fluorescence of the solution is measured after separation of the complex. Of course, it has to be established that the antigen-binding properties of (M_{tr}) and (M_{tr}–Flu_{lab}) are identical.

In luminescence immunoassay (lia) a chemiluminescent label, e.g. isoluminol, is linked to the analyte, analogously to the fluorescence labelling system. The fraction of labelled material in the antibody–antigen complex is determined by reaction with H_2O_2 and peroxidase.

Analogously, marking by means of an electrochemically active group which can be determined by highly sensitive electrochemical methods has been suggested [883, 884].

Enzymes have also been used as labels. A test system has been developed on this basis and is sold commercially (EMIT® = enzyme multiplied immunoassay technique). In it, the drug to be determined is labelled with an enzyme (see Fig. 6.12). The free drug and the enzyme-labelled drug can both react when the enzyme-labelled drug binds to the antibody, its enzyme activity is inhibited (or sometimes enhanced). The analysis is done by measuring the activity of the enzyme in terms of the amount of a suitable substrate S that reacts with it. The greater the number of unlabelled drug molecules present, the less the amount of enzyme bound to AB and the greater the amount of substrate S that reacts.

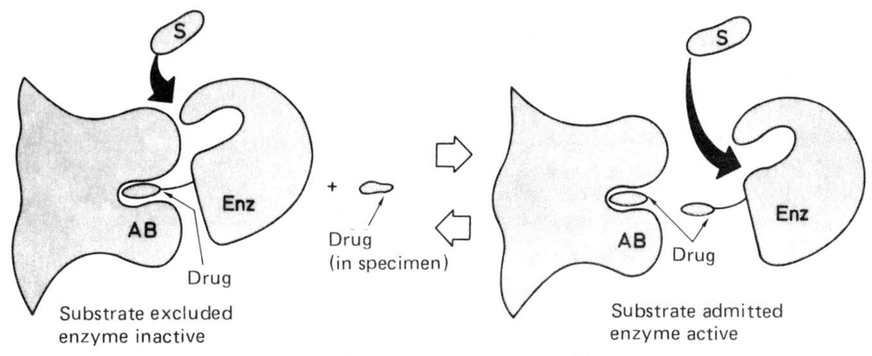

Fig. 6.12 – Schematic representation of the EMIT® homogeneous enzyme immunoassay for drug determination. (Reproduced from [896] by permission of the copyright holders, National Bureau of Standards, Washington, D.C.).

The AG–AB reaction is performed in either a homogeneous system or in a heterogeneous system in which the AB is fixed to the wall of a container. Such vessels can be prepared in advance and are sold commercially. The heterogeneous system is called 'ELISA' (enzyme-linked immunosorbent assay). The homogeneous mode is used more often for low molecular-weight antigens, while antigens with greater molecular weight are mostly determined by the two-phase system.

Several reviews have appeared on fluorescence immunoassay [885–887] and enzyme immunoassay [888–895]. Other reviews concern recent developments in homogeneous immunoassay techniques [896, 897] and the analytical reliability of immunoassay techniques [898]. Other surveys describe the application of ria to determination of antibiotics and chemotherapeutic drugs [899], drugs [900], and to analysis of cereals [901].

To produce the reagent, viz. the antibody, the analytical chemist has to use techniques which are usually unfamiliar to him. To prepare antibodies for low molecular-weight compounds, these compounds have to be fixed to some carrier proteins, as very often they will not give rise to antibody formation by themselves. Bovine albumin is frequently used for binding [902]. In the coupling it is necessary that the coupling reaction does not influence those parts of the antigen molecule which participate in the AB–AG reaction. These portions of the molecule are necessary to induce the production of the specific ('fitting') globulin antibody after injection of the AG–albumin complex into the animal to be used for production of the antibody. (The coupling to the albumin is often done by using a water-soluble carbodi-imide which fixes the antigen to free NH_2-groups of the albumin.)

In the analytical procedure the AG–AB complex has to be separated from the non-reacted label. For this purpose gel filtration, precipitation by addition of poly(ethylene glycol) (m.w. 6000) and adsorption on the walls of specially prepared containers can be used. Another method is fixation to dextran-coated charcoal or to a magnetizable solid phase consisting of cellulose-covered iron oxide (collection of which is achieved and improved by means of a magnet).

An ingenious method (the 'second antibody' or 'double antibody' method) uses a second antibody to precipitate the globulin AB (and AG–AB). Most of the sensitization reactions involve the use of rabbits. A rabbit globulin is formed as a specific reagent for the immunoassay. When horse or goat antibodies are used against rabbit immunoglobulin, precipitation of this AG, together with the AG–AB complex, is achieved.

Examples of recently developed immunoassays are given in Table 6.23. The data demonstrate the very low limits of detection which are typical of immunoassay, especially with radioactive indication. Further, the increasing popularity and importance of these methods can be seen from the great number of papers published withint the 3 years 1980–1982, and listed in Table 6.23. The number of publications in that period was about triple that for the years 1975–1979.

Table 6.23 – Examples of the application of immunoassay techniques in organic trace analysis.

Substance	Level	Method	Reference
3′-iodothyronine		ria	903
3,5-di-iodothyronine	12 fmole	ria	904
thyroxine, tri-iodothyronine		enzyme	905
thyroxine	20 ng/l.	enzyme	906
tri-iodothyronine, thyroxine		ria	907–913
tri-iodothyronine		ria	914
thyroxine		ria	915, 916

Table 6.23 – *continued*

Substance	Level	Method	Reference
thyroxine	2 μg/l.	enzyme	917
thyroxine	50 μg/l.	fluoresc.	918
aldosterone		ria	919–924
androstendione		ria	925–927
cortisol		ria	928–935
cortisol		enzyme	936–938
cortisol	20 pg	luminescence	939, 940
11-deoxyhydrocortisone		ria	941, 942
deoxycorticosterone	10 pg	ria	943
dehydroepiandrosterone	25 pg	enzyme	944
equilin	5 pg	ria	945, 946
17-epimethyltestosterone	6 pg	ria	947
2-hydroxyoestradiol	10 pg/ml	ria	948
17-hydroxyprogesterone	40 pg/ml	ria	949
4-hydroxyoestrone	5 ng/ml	ria	950
oestriol		ria	951
17β-oestradiol	2 pg	luminescence	952
17β-oestradiol	8 pg/ml	ria	953
oestrone	10 pg	ria	954, 955
oestrone 3-glucuronide	25 pg	enzyme	956
progesterone	8 pg	ria	957, 958
progesterone	5 pg	enzyme	959, 960
progesterone	15 pg	luminescence	961, 962
prednisolone		ria	963
several steroids		ria	964, 965
trenbolone	24 pg/ml of urine	ria	966
prostaglandin $F_{2\alpha}$	20 pg	ria	967–969
prostaglandin $F_{2\alpha}$	pmole	enzyme	970, 971
prostaglandin E_2		ria	972
prostaglandin E_1-derivatives		ria	973
prostaglandin E_2-derivatives		ria	974, 975
derivatives of prostaglandins $F_{1\alpha}$ and $F_{2\alpha}$			976–979
thromboxane B_2	10 pg/ml	ria	980, 981
12-hydroxy-L-eicosa-5,8,10,14-tetraenoic acid		ria	982
epinephrine, norepinephrine		ria	983
normetanephrine	5 ng	ria	984
insect juvenile hormones	15 fmole	ria	985
digoxin		ria	986–989
digoxin		ria or enzyme	990
digoxin		enzyme	991
digoxin		fluorescence	992
digitoxin		ria, enzyme	993
dihydrodigoxin		ria	994
amitriptyline, clomipramine		ria	995
amphetamine		ria	996
methylamphetamine	ng	ria	997
acebutolol	10 pg	ria	998
benzodiazepines		enzyme	999
benzodiazepines		ria	1000
carbamazepine-10,11-epoxide		enzyme	1001
carteolol	0.4 ng/ml	ria	1002
cocaine		ria	1003

Table 6.23 – *continued*

Substance	Level	Method	Reference
colchicine	70 pg	ria	1004
chlordiazepoxide	2 ng/ml	ria	1005
fluphenazine	50 pg	ria	1006
flunisolide	20 pg/ml	ria	1007
haloperidol	2 ng/ml	ria	1008
hydromorphone, hydrocodone		ria	1009
methylergometrine	0.5 ng/ml	ria	1010
methadone	3 ng/ml	ria	1011
levorphanol	1 ng/ml	ria	1012
perphenazine	50 pg	ria	1013
propylthiouracil		ria	1014
phencyclidine	2 ng/ml	ria	1015, 1016
propranolol		fluorescence	1017
trifluperazine	0.2 ng/ml	ria	1018
tobramycin		fluorescence, enzyme	1019, 1020
stilboestrol	50 pg	enzyme	1021
antiepileptic and narcotic substances		enzyme	1022
several drugs		fluorescence	1023
theophylline, phenytoin and other drugs			1024, 1025
drugs of abuse (DAU)		DAU-enzyme test	1026
aflatoxin B_1		ria	1027
aflatoxin B_1		enzyme	1028, 1029
T_2 toxin	0.5 ppb	ria	1030
ochratoxin A	25 pg	enzyme	1031
fusicoccin	1 ng	ria	1032
vitamin A	1 ng	ria	1033
vitamin B_6	pmole	ria	1034
1,25-dihydroxycholecalciferol	5 pg	ria	1035, 1036
4-acetamidobiphenyl	66 pg	ria	1037
diacetylbenzidine	6 pg	ria	1038
doxorubicin	10 μg/l.	ria	1039
gentamicin	50 μg/l.	fluorescence	1040
penicillin	10 ng/ml	ria	1041
penicillin	10 ng/ml	enzyme	1042
gentamicin		fluorescence	1043
gentamicin		enzyme	1044
pepleomycin	50 pg	enzyme	1045
daunomycin, adriamycin	1 pmole	enzyme	1046
vancomycin	40 pg/ml	ria	1047
mitomycin C	0.2 ng	enzyme	1048
serotonin	2 nmole	ria	1049, 1050
5-methyltetrahydrofolate	0.1 ng	ria	1051
cholylglycine	25 μg/l.	ria	1052
ATP	10 fmole	ria	1053
cannabinoids	ng/ml	enzyme	1054–1056
nicotine	10 ng/ml	ria	1057
cytokinins	pg	ria	1058
abscisic acid	25 pg	enzyme	1059
5-hydroxyindol-3-ylacetic acid	100 pg	ria	1060, 1061
parathion	10 ng	ria	1062
PCBs		ria	1063
2,3,7,8-tetrachlorodibenzofuran	20 pg	ria	1064
non-ionic detergents	15 ng/ml	enzyme	1065

Immunoassay kits are commercially available for substances which are to be routinely determined. Several of them have been compared with each other or otherwise evaluated. Examples are the investigation of kits for thyroxine [908, 910, 911], aldosterone [924], cortisol [928, 932, 933, 935], β-oestradiol [986], digoxin [987, 989-991, 993], amphetamine [996], benzodiazepines [999, 1000], tobramycin [1019], drugs of abuse [1026], doxorubicin [1039], gentamycin [1044], cannabinoids [1054, 1055].

Immunassay methods are offered with the reagents fixed to paper strips, giving rapid and convenient procedures [1066, 1076].

Occasionally, 'cross-reactions' (i.e. non-specific reactions with compounds other than the analyte) are reported. Examples (analyte/interfering substance) are: 3,5-di-iodothyronine/3,3′,5-tri-iodothyronine [904]; thyroxine/tri-iodothyronine [911]; cortisol/other steroids [928]; cortisol/dexamethasone [933]; cortisol/different steroids [936]; equilin/oestrone-3-sulphate [946]; prednisolone/20-dihydroprednisolene [963]; 6-oxoprostaglandin E_1/prostaglandin E_1 [973]; digitoxin/digoxin [993]; carbamazepin-10,11-epoxide/carbamazepine [1001]; colchicine/lumicolchicine [1104]; phencyclidine/several drugs [1026]; 1,25-dihydroxycholecalciferol/other cholecalciferol metabolites [1035, 1036]; serotonin/tryptamine [1049]; 2,3,7,8-tetrachlorodibenzofuran/compounds similar in structure [1064]. On the other hand, in some cases no cross-reactions are clearly stated. By adequate immunization using only an enantiomer as antigen it can even be possible to build up a stereospecific assay for the enantiomer.

Cross-reactions can lead to falsely positive reactions. In screening methods, e.g. drug screening, this effect is not too disastrous, as positively reacting samples will be checked by other methods anyway. Of course, for economic and organizational reasons there should not be too much cross-reaction. From the list above it is clear that only very closely chemically related substances interfere. This can become a problem, though, in metabolic studies, where very often such substances are precursors or metabolic products of the substance under test.

Although the specificity of immunoassay methods is very high, separation methods are more and more often mentioned in the literature, for prevention of cross-reactions or other interferences. Examples of such preseparations are given in Table 6.24.

Table 6.24 – Examples of separation steps before immunoassays.

Substance	Matrix	Separation	Reference
aldosterone	serum	partition	921, 922
androstendione	serum	partition or chromatog.	923, 926, 927
cortisol	serum	partition	928, 933, 938 940
cortisol	urine	hplc	929
11-deoxyhydrocortisone	serum	partn./gel chromatog.	941, 942
11-deoxyhydrocortisone	plasma	partn./chromatog.	943
dehydroepiandrosterone	plasma	partn./chromatog.	944
equilin	plasma	chromatography	945

Table 6.24 – *continued*

Substance	Matrix	Separation	Reference
2-hydroxyoestradiol	plasma	partn./chromatog.	948
17-hydroxyprogesterone	plasma	partition	949
4-hydroxyoestrone	urine	XAD 2 chromatography	950
oestrone	plasma	SepPak C_{18} chromatog.	954
progesterone	saliva	partition	958
steroids	prostatic tissue	extraction/partition	964
corticosteroids	plasma	partn./chromatog.	965
prostaglandins	plasma	partition	972
prostaglandin derivatives	plasma	SepPak C_{18} chromatog.	979
thromboxane B_2	plasma	partition	981
digoxin	serum	partition	988
clomipramine, amitriptyline	plasma	partition	995
amphetamine	stains from blood, semen	partition	996
benzodiazepines	biological fluids, tissues	partition	999, 1000
methylergometrine	milk, plasma	partition	1010
T_2-toxin	serum, milk	partition	1030
1,25-dihydroxycholecalciferol	serum	partn./chromatog.	1035
cannabinoids	body fluids	hplc	1056
PCBs	milk	partn./chromatog.	1063

The results of immunoassay assays have been compared with those obtained by other methods. Some examples are given in Table 6.25.

6.14 PROTEIN BINDING

In some cases a protein is used that binds a component specifically. The procedure is similar to immunoassay except that no production of antibodies is stimulated, to obtain the reagent. Instead, naturally occurring proteins are used. The substoichiometric technique is employed. The limits of detection are comparable to those in immunoassay methods. Examples are the determination of vitamin D [1100, 1101], 25-hydroxycalciferol [1102, 1103], 1,25-dihydroxyvitamin D [1104, 1105], retinoic acid [1106], 5-hydroxytryptamine [1107], cortisol [1108], progesterone [1109], corticosterone [1110], prednisolone [1111], nucleic acids [1112], and nitrazepam and triazolam [1113].

6.15 BIOLOGICAL METHODS

By means of biological methods surprisingly low concentrations of certain substances can be determined, very often with amazing selectivity. In principle, biological systems (cells, whole animals, plants etc.) are used, and some reaction is monitored which appears after exposure to the trace substance.

Riboflavin, nicotinic acid, pantothenic acid, thiamine, pyridoxine, and *p*-aminobenzoic acid have been determined in ng-amounts, and biotin, folic acid

Table 6.25 – Comparison of immunoassay methods with other methods.

Trace	Immunomethod	Method(s) of comparison	Results	Reference
cortisol	ria	hplc or ms (reference method)	hplc more accurate than ria	1068
cortisol	ria	ms	ria results 25% higher than ms	1069
6β-hydroxycortisol	ria	hplc	good correlation, ria simple, rapid	1070
derivative of prostaglandin E_2	ria	gc–ms, tlc	ria least specific	1071
benzodiazepines	EMIT	tlc	EMIT faster than tlc	1072
benzodiazepines	EMIT	gc	EMIT frequently gives low values for elevated diazepine concentrations	1073
diazepam	ria	gc	ria consistently higher by about 60 ng/ml	1074
diazepam	EMIT	gc	good agreement	1075
bupropion	ria	hplc	excellent agreement	1076
doxepin	ria	hplc	good correlation ($r = 0.99$)	1077
haloperidol	ria	gc	good agreement only after extraction before ria determination	1078
imipramine	ria	gc, gc–ms	results of all 3 methods agree	1079
nomifensine	ria	gc	results agree	1080
procainamide	EMIT	hplc	results agree ($r = 0.98$)	1081
procainamide	ria	gc	good correlation	1082
phenycyclidine	ria	gc, EMIT	correlation between 3 methods acceptable	1083
phenycyclidine	EMIT	gc	correlation good ($r = 0.98$)	1084
phenobarbitone, phenytoin	EMIT	hplc	correlation ($r = 0.99$ and 0.97)	1085
methotrexate	ria	hplc, EMIT	agreement	1086
gentamycin	ria	hplc, fluoresc. immunoassay	fluorescence immunoassay gives best results	1087
gentamycin	ria	EMIT, microbiological assay, adenylation	ria gives higher results, other methods agree	1088
indol-3-ylacetic acid	ria	gc–ms	agreement after chromatography before ria	1089
cannabinoids	EMIT	ria, gc–ms	EMIT preferable to ria or gc	1090
Δ^9-tetrahydrocannabinol	ria	gc–ms	results agree	1091
tobramycin	EMIT	microbiol. assay	results agree	1092
tobramycin	ria	EMIT, microbiol. assay	EMIT more precise	1093
drugs of abuse	EMIT	gc–ms	excellent agreement in most cases	1094
morphine	ria	gc	ria gives higher results	1095
morphine	EMIT	gc	gc more precise	1096
quinidine	EMIT	fluorimetry	EMIT more specific	1097
quinidine	EMIT	hplc	EMIT gives higher results, hplc more specific	1098
quinidine	EMIT	hplc	hplc more specific, correlation $r = 0.95$	1099

and vitamin B_{12} in pg amounts, through their action as growth factors for certain micro-organisms [1114].

Fungicides can be detected (after their tlc separation) by their inhibition of the growth of microbial organisms, such as fungi etc. The tlc plate is brought into contact with a medium containing a culture of the test fungi and incubated for an appropriate period, then the culture is examined for signs of inhibition of growth.

Smooth muscle contracts on addition of ng amounts of histamine. The presence of antihistamines prevents the effect. Both components can be determined in that way, by use of guinea pig ileum. Analogously, small amounts of oxytocin are measured from contraction of a pregnant uterus. Prostaglandins act on smooth muscle [1115].

If male cotton leafworm moths are placed in a small cage at the exit of a gas chromatograph in which an extract of abdominal segments from female moths is being separated, the elution of an active peak is characterized by the "lifting of antennae, vibrating of the wings and trying to copulate with one another" [1116]. A highly specific search for natural or synthetic sex attractants is greatly simplified by this method.

The human nose is a very sensitive detection 'instrument' with high sensitivity for odorous substances. About 10^6 molecules of ionone, ethyldisulphide, skatole, vanillin, coumarin or butyric acid per cm^3 can be detected by smell [1117]. Such substances can thus be detected after gas chromatography (see Figs. 6.13 and 6.14). Odorants have been determined in raw and potable water [1118], wine [1119] and manure [1120]. Sensory evaluation was used in

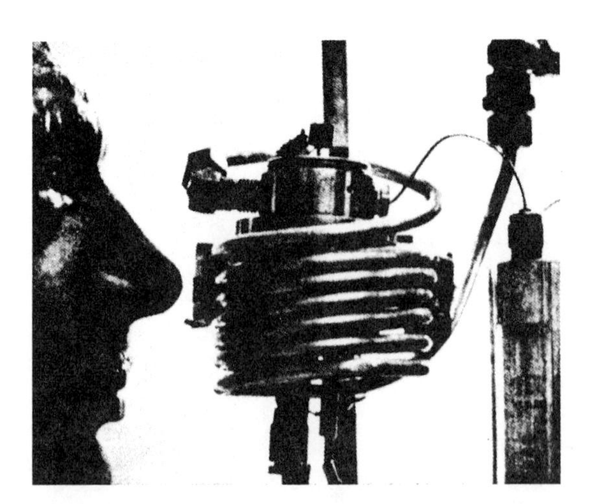

Fig. 6.13 – Simplified nasal detector. [Reproduced from L. S. Ettre and W. H. McFadden (eds.), *Ancillary Techniques of Gas Chromatography*, p. 367, by permission of the copyright holders, Wiley–Interscience, Inc.].

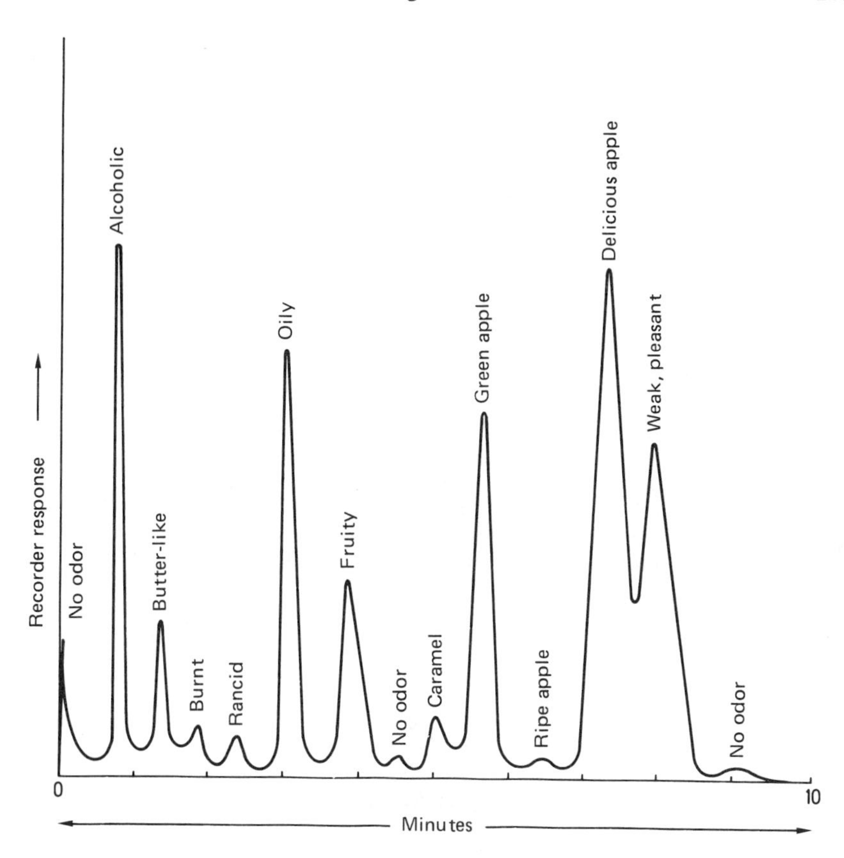

Fig. 6.14 – Odour evaluation of a chromatogram from a 'Delicious apple extract'. [Reproduced by permission from R. A. Flath, D. R. Black, D. G. Guadogni, W. H. McFadden and T. H. Schultz, *J. Agr. Food Chem.*, **15**, 29 (1967). Copyright 1967 American Chemical Society].

conjunction with gas chromatography. Threshold values are given in the publications cited.

Quantitative determination of odorous substances can be done by means of 'sniffle boxes' which hold the gas sample containing the trace to be determined. At an outlet of the box the analyst tests for odour. The mixture in the box is diluted with clean gas until no odour is detected. The quantity of diluting gas is related to the original concentration of the odoriferous trace(s). The analysts do not need to be highly trained. "Reasonably intelligent and careful young girls make excellent testers" [1121].

REFERENCES

[1] D. R. Baker, R. C. Williams and J. C. Steichen, *J. Chromatog. Sci.*, **12,** 499 (1974).
[2] T. J. Jupille, *J. Chromatog. Sci.*, **170,** 160 (1979).
[3] K. Beyermann, *Fortschritte Chem. Forschung,* **11,** 473 (1969).
[4] R. Geick, *Z. Anal. Chem.*, **288,** 1 (1977).
[5] J. R. Ferraro and L. J. Basile, *Fourier Transform IT Spectroscopy,* 2 Vols., Academic Press, New York (1979).
[6] R. Cournoyer, J. C. Shearer and D. H. Anderson, *Anal. Chem.*, **60,** 2275 (1977).
[7] R. S. McDonald, *Anal. Chem.*, **54,** 1250 (1982).
[8] H. J. Sloan and R. J. Obremski, *Appl. Spectrosc.*, **31,** 506 (1977).
[9] C. R. Kopec, W. D. Washington and C. R. Midkiff, *J. Forensic Sci.*, **23,** 57 (1978).
[10] R. G. J. Miller (ed.), *Laboratory Methods in Infrared Spectroscopy,* Heyden, London (1965).
[11] J. P. Berthelemy, A. Chopin, R. Deleu and J. L. Closset, *Talanta,* **26,** 885 (1979).
[12] H. Maulhardt and D. Kunath, *Talanta,* **29,** 237 (1982).
[13] K. Fuwa (ed.), *Recent Advances in Analytical Spectroscopy,* pp. 285–289. Pergamon Press, Oxford (1982).
[14] M. Mamiya, E. Ishii and T. Murakami, *Bunseki Kagaku,* **30,** 385 (1981).
[15] M. P. Fuller and P. R. Griffiths, *Appl. Spectrosc.*, **34,** 533 (1980).
[16] K. Beyermann, *Z. Anal. Chem.*, **226,** 16 (1967).
[17] K. Beyermann, *Clin. Chim. Acta,* **18,** 143 (1967).
[18] K. Beyermann and E. Röder, *Z. Anal. Chem.*, **230,** 347 (1967).
[19] K. Beyermann, *Z. Anal. Chem.*, **230,** 414 (1967).
[20] K. Beyermann and J. Dietz, *Z. Anal. Chem.*, **268,** 192 (1974).
[21] J. E. Katon and D. B. Phillips, *Appl. Spectrosc. Rev.*, **7,** 1 (1974).
[22] W. S. Brickell, *Proc. Anal. Div. Chem. Soc.*, **15,** 343 (1978).
[23] R. Freymann, A. Bullier, G. Capderroque and M. Selim, *Analusis,* **6,** 258 (1976).
[24] P. F. Lynch, S. Tang and C. W. Brown, *Anal. Chem.*, **47,** 1696 (1975).
[25] H. E. Hallam, *Vibrational Spectroscopy of Trapped Species,* Wiley, New York (1973).
[26] B. Meyer, *Low Temperature Spectroscopy,* Elsevier. Amsterdam (1971).
[27] G. Mamantov, E. L. Wehry, R. R. Kemmerer and E. R. Hinton, *Anal. Chem.*, **49,** 86 (1977).
[28] R. C. Stroupe, P. Tokousbalides, R. B. Dickinson, E. L. Wehry and G. Mamantov, *Anal. Chem.*, **49,** 701 (1977).
[29] E. L. Wehry and G. Mamantov, *Anal. Chem.*, **51,** 643A (1979).
[30] G. T. Reedy, S. Bourne and P. T. Cunningham, *Anal. Chem.*, **51,** 1535 (1979).
[31] S. Bourne, G. T. Reedy and P. T. Cunningham, *J. Chromatog. Sci.*, **17,** 460 (1979).
[32] E. V. Shpol'skii, *Usp. Fiz. Nauk,* **68,** 51 (1959).
[33] G. F. Kirkbright and G. C. de Lima, *Analyst,* **99,** 338 (1974).
[34] D. L. Wall and A. W. Mantz, *Appl. Spectrosc.*, **31,** 525 (1977).
[35] P. R. Griffiths, *Appl. Spectrosc.*, **31,** 284 (1977).
[36] K. Krishnan, R. Curbelo, P. Chiha and R. C. Noonan, *J. Chromatog. Sci.*, **17,** 413 (1979).
[37] D. Hanna, G. Hangac, B. A. Hohne, G. W. Small, R. C. Wieboldt and T. L. Isenhour, *J. Chromatog. Sci.*, **17,** 423 (1979).
[38] M. J. D. Low and S. K. Freeman, *Anal. Chem.*, **39,** 194 (1967).
[39] K. Freeman, in L. S. Ettre and W. H. McFadden (eds.), *Ancillary Techniques of Gas Chromatography,* p. 227. Wiley, New York (1969).
[40] D. Lephardt and B. J. Bulin, *Anal. Chem.*, **45,** 706 (1973).
[41] C. W. Low and J. F. Richards, *Appl. Spectrosc.*, **29,** 15 (1975).
[42] B. Katlafsky and M. W. Dietrich, *Appl. Spectrosc.*, **29,** 24 (1975).
[43] R. F. Brady, *Anal. Chem.*, **47,** 1425 (1975).
[44] V. Rossiter, *Intern. Lab.*, **12,** No. 2, 42 (1982).
[45] K. Krishnan, R. H. Brown, S. L. Hill, S. C. Simonoff, M. L. Olson and D. Kuehl, *Intern. Lab.*, **11,** No. 4, 66 (1981).

[46] S. S. Williams, R. B. Lam, D. T. Sparks, T. L. Isenhour and J. R. Hass, *Anal. Chim. Acta,* **138,** 1 (1982).
[47] S. R. Lowry and D. Huppler, *Anal. Chem.,* **53,** 889 (1981).
[48] K. L. Kizer, *Am. Lab.,* **5,** No. 6, 40 (1973).
[49] L. V. Azarraga and A. C. McCall, *FTIR-Spectrometry of GC Effluents,* Environmental Protection Technology, EPA Series 660/2-73-034 (1974).
[50] D. R. Mattson and R. L. Julian, *J. Chromatog. Sci.,* **17,** 416 (1979).
[51] M. M. Gomez-Taylor and P. R. Griffiths, *Anal. Chem.,* **50,** 422 (1978).
[52] P. R. Griffiths, *Appl. Spectrosc.,* **31,** 284 (1977).
[53] K. Krishnan, R. Curbelo, P. Chiha and R. C. Noonan, *J. Chromatog. Sci.,* **17,** 413 (1979).
[54] K. H. Shafer, S. V. Lucas and R. J. Jakobsen, *J. Chromatog. Sci.,* **17,** 464 (1979).
[55] M. M. Gomez-Taylor, D. Kvehl and R. Griffiths, *Intern. J. Environ. Anal. Chem.,* **51,** 103 (1978).
[56] P. C. Uden, A. P. Carpenter, H. M. Hackett, D. E. Henderson and S. Siggia, *Anal. Chem.,* **51,** 38 (1979).
[57] H. Malissa, *Z. Anal. Chem.,* **311,** 123 (1982).
[58] D. M. Hembree, A. A. Garrison, R. A. Crocombe, R. A. Yokley, E. L. Wehry and G. Mamantov, *Anal. Chem.,* **53,** 1783 (1981).
[59] K. H. Shafer, M. Cooke, F. DeRoos, R. J. Jakobsen, O. Rosario and J. D. Mulik, *Appl. Spectrosc.,* **35,** 469 (1981).
[60] V. Rossiter, *Intern. Lab.,* **12,** No. 2, 42 (1982).
[61] D. F. Gurka, P. R. Laska and R. Titus, *J. Chromatog. Sci.,* **20,** 145 (1982).
[62] G. E. Podolak, R. M. McKenzie, D. S. Rinehart and J. F. Mazur, *Am. Ind. Hyg. Assoc. J.,* **42,** 734 (1981).
[63] M. M. Gomez-Taylor, D. Kuehl and P. R. Griffiths, *Appl. Spectrosc.,* **30,** 447 (1976).
[64] M. Lemar, P. Versaud and M. Porthault, *J. Chromatog.,* **132,** 295 (1977).
[65] N. A. Parris, *J. Chromatog. Sci.,* **17,** 541 (1979).
[66] R. M. Cassidy and C. M. Niro, *J. Chromatog.,* **126,** 787 (1976).
[67] D. Kuehl and P. R. Griffiths, *J. Chromatog. Sci.,* **17,** 471 (1979).
[68] K. L. Kizer, A. W. Mantz and L. C. Bonar, *Am. Lab.,* **7,** No. 5, 85 (1975).
[69] M. P. Fuller and P. R. Griffiths, *Am. Lab.,* **10,** No. 10, 69, 1978; *Anal. Chem.,* **50,** 1906 (1978).
[70] K. Oba, K. Miura, H. Sekiguchi, R. Yagi and A. Mori, *Water Res.,* **10,** 149 (1976).
[71] E. Kunkel, G. Peitscher and K. Espeter, *Tenside,* **14,** 199 (1977).
[72] H. Hellmann, *Z. Anal. Chem.,* **295,** 393 (1979).
[73] H. Hellmann, *Z. Anal. Chem.,* **293,** 359 (1978).
[74] H. Hellmann, *Z. Anal. Chem.,* **294,** 379 (1979).
[75] J. A. Gag and N. F. Merck, *J. Forens. Sci.,* **22,** 358 (1977).
[76] W. L. Brannon, W. R. Benson and C. Schwartzman, *J. Assoc. Off. Anal. Chem.,* **59,** 1404 (1976).
[77] G. Vilcins, *Beitr. Tabaksforsch.,* **8,** 181 (1979).
[78] L. J. Luskus, H. J. Kilian, W. W. Lackey and J. D. Biggs, *J. Forens. Sci.,* **22,** 500 (1977).
[79] J. Diaz-Rueda, H. J. Sloane and R. J. Obremski, *Appl. Spectrosc.,* **31,** 298 (1977).
[80] B. B. Baker, *Am. Ind. Hyg. Assoc. J.,* **43,** 98 (1982).
[81] S. M. Freund, W. B. Maier, R. F. Holland and W. H. Beattie, *Anal. Chem.,* **50,** 1260 (1978).
[82] A. A. Karnishin, T. S. Fedorova, Y. A. Tsarfin and N. I. Zorina, *Zh. Analit. Khim.,* **32,** 2250 (1977).
[83] S. R. Lowry and J. D. Banzer, *Anal. Chem.,* **50,** 1187 (1978).
[84] D. R. Wilkinson, F. Pavlikowski and P. Jenson, *J. Forens. Sci.,* **21,** 564 (1976).
[85] A. C. Moffat, *Proc. Anal. Chem. Soc.,* **13,** 355 (1976).
[86] V. Blackpowders, W. D. Washington, R. J. Kopec and C. R. Midkiff, *J. Assoc. Off. Anal. Chem.,* **60,** 1331 (1977).

[87] L. Kahn, B. F. Dudenbostel, D. N. Speis and G. Karras, *Intern. Lab.*, May/June, 74 (1977).
[88] D. E. Caddy, *Proc. Anal. Div. Chem. Soc.*, **13**, 45 (1976).
[89] H. Brajerová and J. Knizek, *Chem. Prum.*, **30**, 601 (1980).
[90] R. G. Riley and R. G. Beyn, in *Trace Organic Analysis*, pp. 33–40. NBS, Washington (1979).
[91] F. Zürcher and M. Thüer, *Environ. Sci. Technol.*, **12**, 838 (1978).
[92] J. W. Worley, M. L. Rueppel and F. L. Rupel, *Anal. Chem.*, **52**, 1845 (1980).
[93] S. K. Ganguly, K. P. Bhattacharya and P. Kar., *Sci. Cult. (Calcutta)*, **43**, 301 (1977).
[94] A. Coutselenis, P. Kentarchou and D. Boukis, *Forens. Sci.*, **8**, 251 (1976).
[95] J. O. DeBeer, C. H. Van Peteghem and A. M. Hendrickx, *J. Assoc. Off. Anal. Chem.*, **61**, 1140 (1978).
[96] M. Cotugno, G. Sansone, U. Gallone, E. Barone, L. Rossi and A. Biondi, *Boll. Soc. Ital. Biol. Sper.*, **58**, 295 (1982).
[97] M. D. Erickson, D. L. Newton, E. D. Pellizzari, K. B. Tomer and D. D. Dropkin, *J. Chromatog. Sci.*, **17**, 449 (1979).
[98] J. J. Manura, J. M. Chao and R. Saferstein, *J. Forens. Sci.*, **23**, 44 (1978).
[99] W. A. Trinler and D. J. Reuland, *J. Forens. Sci.*, **23**, 37 (1978).
[100] K. H. Shafer, S. V. Lucas and R. L. Jakobsen, *J. Chromatog. Sci.*, **17**, 464 (1979).
[101] K. L. Gallaher and J. G. Grasselli, *Appl. Spectrosc.*, **31**, 456 (1977).
[102] J. Loader, *Basic Laser Raman Spectroscopy*, Heyden–Sadtler, New York (1970).
[103] M. E. Anderson and R. Z. Muggli, *Anal. Chem.*, **53**, 1772 (1981).
[104] L. B. Rogers, J. D. Stuart, L. P. Gross, T. B. Mallory and L. A. Carreira, *Anal. Chem.*, **49**, 959 (1977).
[105] L. Van Haverbeke and M. A. Herman, *Anal. Chem.*, **51**, 932 (1979).
[106] E. S. Etz, S. A. Wise and K. F. J. Heinrich, in *Trace Organic Analysis*, pp. 723–729. NBS, Washington (1979).
[107] H. Nakamura, K. Yoshimori, J. Itoh, M. Takagi and K. Ueno, *Mikrochim Acta*, 131 (1981 II).
[108] K. Ikeda, S. Higuchi and S. Tanaka, *Bunseki Kagaku*, **30**, 701 (1981).
[109] L. Van Haverbeke, J. F. Janssens and M. A. Herman, *Intern. J. Environ. Anal. Chem.*, **10**, 205 (1981).
[110] L. C. Hoskins and V. Alexander, *Anal. Chem.*, **49**, 695 (1977).
[111] M. D. Morris, *Anal. Lett.*, **9**, 469 (1976).
[112] S. Udenfriend, *Fluorescence Assay*, 2 Vols. Academic Press, New York (1969).
[113] G. G. Guilbault, *Fluorescence*, Dekker, New York (1967).
[114] M. Zander, *Phosphorimetry*, Academic Press, New York (1968).
[115] E. L. Wehry, *Modern Fluorescence Spectroscopy*, 4 Vols. Plenum Press, New York (1976, 1981).
[116] L. W. Hershberger, J. B. Callis and G. D. Christian, Jr., *Anal. Chem.*, **51**, 1444 (1979).
[117] T. S. Herman and R. S. Harvey, *Appl. Spectrosc.*, **23**, 435, 451, 461, 473 (1969).
[118] B. Mathiasch, *Anal. Lett.*, **4**, 519 (1971).
[119] A. Colmsjö and L. Stenberg, *Anal. Chem.*, **51**, 145 1979; *J. Chromatog.*, **169**, 205 (1978).
[120] J. N. Miller and J. W. Bridges, in E. Reid (ed.), *Blood Drugs and Other Analytical Challenges*, Horwood, Chichester (1978).
[121] R. T. Parker, R. S. Freelander and R. B. Dunlap, *Anal. Chim. Acta*, **120**, 1 (1980).
[122] R. T. Parker, R. S. Freelander and R. B. Dunlap, *Anal. Chim. Acta*, **119**, 189 (1980).
[123] J. N. Miller, *TrAC*, **1**, 31 (1981).
[124] J. L. Ward, G. L. Walden and J. D. Winefordner, *Talanta*, **28**, 201 (1981).
[125] J. J. Donkerbroek, N. J. R. van Eikema Hommes, C. Gooijer, N. H. Velthorst and R. W. Frei, *Chromatographia*, **15**, 218 (1982).
[126] L. A. Gifford, J. N. Miller, D. T. Burns and J. W. Bridges, *J. Chromatog.*, **103**, 15 (1975).

[127] H. Dekirmenjian, J. I. Javaid, U. Liskevych and J. M. Davis, *Anal. Biochem.*, **105**, 6 (1980).
[128] H. Tokunaga, T. Kimura and J. Kawamura, *Chem. Pharm. Bull.*, **30**, 228 (1982).
[129] K. A. Muhiddin and A. Johnston, *Br. J. Clin. Pharmacol.*, **12**, 283 (1981).
[130] S. Folestad, L. Johnson, B. Josefsson and B. Galle, *Anal. Chem.*, **54**, 925 (1982).
[131] J. B. Luten, *J. Food Sci.*, **46**, 958 (1981).
[132] S. J. Lewis and M. R. Fennessy, *Agents Action*, **11**, 228 (1981).
[133] G. Myers, M. Donlon and M. Kaliner, *J. Allergy Clin. Immunol.*, **67**, 305 (1981).
[134] M. Iino, R. S. T. Yu and D. J. Carr, *Plant Physiol.*, **66**, 1099 (1980).
[135] M. Furusawa, M. Tachibana and T. Tsuchiya, *Bunseki Kagaku*, **29**, 588 (1980).
[136] J. C. Brown, J. A. Duncanson and G. J. Small, *Anal. Chem.*, **52**, 1711 (1980).
[137] J. H. Richardson, K. M. Larson, G. R. Haugen, D. C. Johnson and J. E. Clarkson, *Anal. Chim. Acta*, **116**, 407 (1980).
[138] A. S. Attiyat, G. D. Christian, Jr. and J. B. Callis, *Microchem. J.*, **26**, 344 (1981).
[139] E. P. Lai, E. L. Inman and J. D. Winefordner, *Talanta*, **29**, 601 (1982).
[140] N. Strojny and J. A. F. de Silva, *Anal. Chem.*, **52**, 1554 (1980).
[141] J. J. Aaron, J. L. Ward and J. D. Winefordner, *Analusis*, **10**, 98 (1982).
[142] W. Kitzrow, L. Franke, R. Uebelhack and K. Seidel, *Acta Biol. Med. Ger.*, **39**, 489 (1980).
[143] S. M. Ramasamy and R. J. Hurtubise, *Anal. Chem.*, **54**, 1642 (1982).
[144] K. Hirauchi and A. Fujishita, *Chem. Pharm. Bull.*, **28**, 1986 (1980).
[145] T. Vo-Dinh, R. B. Gamage and P. R. Martinez, *Anal. Chim. Acta*, **118**, 313 (1980).
[146] E. L. Inman, A. Jurgensen and J. D. Winefordner, *Talanta*, **29**, 423 (1982).
[147] Y. H. Pao, *Optoacoustic Spectroscopy and Detection*, Academic Press, New York (1977).
[148] A. Rosencwaig, *Photoacoustics and Photoacoustic Spectroscopy*, Wiley, Chichester (1980).
[149] B. C. Beadle and B. R. Donoghue, *U.K. At. Energy Res. Establ. Rept.*, AERE-R 9568 (1980).
[150] L. B. Lloyd, R. Yeates and E. M. Eyring, *Anal. Chem.*, **54**, 549 (1982).
[151] S. L. Castleden, C. M. Elliott, G. F. Kirkbright and D. E. M. Spillane, *Anal. Chem.*, **51**, 2152 (1979).
[152] C. M. Ashworth, S. L. Castleden, G. F. Kirkbright and D. E. M. Spillane, *J. Photoacoust.*, **1**, 151 (1982).
[153] S. Schneider, U. Moeller, H. P. Koest and H. Coufal, *J. Photoacoust.*, **1**, 309 (1982).
[154] E. Voigtman, A. Jurgensen and J. D. Winefordner, *Analyst*, **107**, 408 (1982).
[155] M. J. Adams, B. C. Beadle and G. F. Kirkbright, *Anal. Chem.*, **50**, 1371 (1978).
[156] L. B. Kreuzer, *Anal. Chem.*, **50**, 597A (1978).
[157] J. M. Chalmers, B. J. Stay, G. F. Kirkbright, D. E. M. Spillane and B. C. Beadle, *Analyst*, **106**, 1179 (1981).
[158] K. Krishnan, *Appl. Spectrosc.*, **35**, 549 (1981).
[159] K. Feser and W. Kögler, *J. Chromatog. Sci.*, **17**, 57 (1979).
[160] R. M. Milberg and J. C. Cook, *J. Chromatog. Sci.*, **17**, 17 (1979).
[161] M. C. Ten Noever de Brauw, *J. Chromatog.*, **165**, 207 (1979).
[162] R. P. W. Scott, in *Trace Organic Analysis*, pp. 637–645. NBS, Washington (1979).
[163] A. L. Burlingame, C. H. L. Shackleton, I. Howe and O. S. Chizhov, *Anal. Chem.*, **50**, 346R (1978).
[164] J. Seibl, *Ann. Chim. (Rome)*, **67**, 599 (1977).
[165] W. D. Reynolds, *Anal. Chem.*, **51**, 283A (1979).
[166] P. Leinster, R. Perry and R. J. Young, *Talanta*, **24**, 205 (1977).
[167] L. H. Keith, *J. Chromatog. Sci.*, **17**, 48 (1979).
[168] S. K. Pierce, H. L. Gearhart and D. Payne-Bose, *Talanta*, **24**, 473 (1977).
[169] G. Vander Velde and J. F. Ryan, *J. Chromatog. Sci.*, **13**, 322 (1975).
[170] A. Alford, *Biomed. Mass. Spectrom.*, **5**, 259 (1978).
[171] D. Jahr and P. Binnemann, *Z. Anal. Chem.*, **297**, 341 (1979); **298**, 337 (1979).
[172] V. Kubelka, J. Mitera, J. Novák and J. Mostecký, *Sci. Papers Prague Inst. Chem. Technol.*, F **22**, 115 (1978).

[173] T. A. Gough, Analyst, **103,** 785 (1978).
[174] C. F. Poole and A. Zlatkis, *J. Chromatog. Sci.,* **17,** 115 (1979).
[175] F. Cattabeni, *Ann. Chim. (Rome),* **71,** 77 (1981).
[176] Anon., *Anal. Proc.,* **17,** 320 (1980).
[177] S. J. Gaskell, *TrAC,* **1,** 110 (1982).
[178] J. Gilbert and R. Self, *Chem. Soc. Rev.,* **10,** 255 (1980).
[179] R. C. Dougherty, *Biomed. Mass Spectrom.,* **8,** 283 (1981).
[180] N. H. Mahle and L. A. Shadoff, *Biomed. Mass Spectrom.,* **9,** 45 (1982).
[181] H. Remane and R. Herzschuh, *Massenspektrometrie in der organische Chemie,* Vieweg, Braunschweig (1977).
[182] A. P. de Leenheer and R. R. Roncucci (eds.), *Quantitative Mass Spectrometry in Life Sciences,* Elsevier, Amsterdam (1977).
[183] A. Frigerio and N. Castagnoli (eds.), *Advances in Mass Spectrometry in Biochemistry and Medicine,* Vol. I, Wiley, New York (1976).
[184] A. Frigerio (ed.), *Advances in Mass Spectrometry in Biochemistry and Medicine,* Vol. II, Wiley, New York (1977).
[185] K. Metzner, *Gaschromatographische Spurenanalyse,* Akad. Verlag, Leipzig (1976).
[186] R. P. W. Scott, *Liquid Chromatography Detectors,* Elsevier, Amsterdam (1977).
[187] S. Facchetti (ed.), *Application of Mass Spectrometry to Trace Analysis,* Elsevier, Amsterdam (1982).
[188] A. Frigerio (ed.), *Recent Developments in Mass Spectrometry in Biochemistry, Medicine and Environmental Research,* Elsevier, Amsterdam (1981).
[189] B. S. Middleditch, S. R. Missler and H. B. Hines, *Mass Spectrometry of Priority Pollutants,* Plenum Press, New York (1981).
[190] B. S. Middleditch, *Practical Mass Spectrometry,* Plenum Press, New York (1979).
[191] C. J. Porter, J. H. Beynon and T. Ast, *Org. Mass Spectrom.,* **16,** 101 (1981).
[192] J. A. Page, *Anal. Proc.,* **19,** 307 (1982).
[193] R. C. Dougherty, *Anal. Chem.,* **53,** 625A (1981).
[194] P. J. Wood, *Biomed. Mass Spectrom.,* **9,** 302 (1982).
[195] J. Roboz, J. Greaves, J. F. Holland and G. J. Bekesi, *Anal. Chem.,* **54,** 1104 (1982).
[196] R. C. Dougherty and J. D. Wander, *Biomed. Mass Spectrom.,* **7,** 401 (1980).
[197] H. R. Schulten and W. D. Lehmann, *Mikrochim. Acta,* 113 (1978 **II**).
[198] B. Soltmann, C. C. Sweeley and J. F. Holland, *Anal. Chem.,* **49,** 1164 (1977).
[199] G. L. Peele and D. A. Brent, *Biomed. Mass Spectrom.,* **5,** 180 (1978).
[200] T. Matsuo, H. Matsuda and I. Katakuse, *Anal. Chem.,* **51,** 69 (1979).
[201] K. L. Olson, J. C. Cook and K. L. Rinehart, *Biomed. Mass. Spectrom.,* **1,** 358 (1975).
[202] G. W. Wood, P. Y. Lau and G. N. S. Rao, *Biomed. Mass. Spectrom.,* **2,** 172 (1976).
[203] D. F. Hundt, J. Shabanowitz, F. K. Botz and D. A. Brent, *Anal. Chem.,* **49,** 1160 (1977).
[204] H. R. Schulten, *Z. Anal. Chem.,* **293,** 273 (1978).
[205] J. A. Sphon, P. A. Dreifuss and H. R. Schulten, *J. Assoc. Off. Anal. Chem.,* **60,** 73 (1977).
[206] J. C. Prome and G. Puzo, *Organ. Mass Spectrom.,* **12,** 28 (1977).
[207] D. I. Carroll, I. Dzidic, R. N. Stillwell and E. C. Horning, in *Organic Trace Analysis,* pp. 655–671. NBS, Washington (1979).
[208] R. K. Mitchum, J. R. Althaus, W. A. Korfmacher, K. L. Rowland, K. Nam and J. F. Young, *Biomed. Mass Spectrom.,* **8,** 539 (1981).
[209] H. Vogt, H. J. Heinen, S. Meier and R. Wechsung, *Z. Anal. Chem.,* **308,** 195 (1981).
[210] H. J. Heinen, S. Meier, H. Vogt and R. Wechsung, *Z. Anal. Chem.,* **308,** 290 (1981).
[211] R. J. Day, J. Zimmermann and D. M. Hercules, *Spectroscop. Lett.,* **14,** 773 (1981).
[212] K. L. Busch, S. E. Unger, A. Vincze, R. G. Cooks and T. Keough, *J. Am. Chem. Soc.,* **104,** 1507 (1982).
[213] R. Ryhage, *Anal. Chem.,* **48,** 1829 (1976).
[214] R. M. Milberg and J. C. Cook, *J. Chromatog. Sci.,* **17,** 17 (1979).
[215] K. Feser and W. Kögler, *J. Chromatog. Sci.,* **17,** 57 (1979).
[216] F. W. McLafferty, *Acc. Chem. Res.,* **13,** 33 (1980).

[217] F. Maquin, D. Fraisse and J. C. Tabet, *Spectra 2000,* **9,** 25 (1981).
[218] R. G. Cooks and G. L. Glish, *Chem. Eng. News,* **59,** 40 (1981).
[219] F. W. McLafferty, *Biomed. Mass Spectrom.,* **8,** 446 (1981).
[220] F. W. McLafferty and E. R. Lory, *J. Chromatog.,* **203,** 109 (1981).
[221] D. V. Davis and R. G. Cooks, *J. Agr. Food Chem.,* **30,** 495 (1982).
[222] F. W. Karasak and F. I. Onuska, *Anal. Chem.,* **54,** 309A (1982).
[223] M. J. Lacey and C. G. Macdonald, *Anal. Chem.,* **54,** 135 (1982).
[224] D. Zakett, J. D. Ciupek and R. G. Crooks, *Anal. Chem.,* **53,** 723 (1981).
[225] M. H. Bozorgzadeh, R. P. Morgan and J. H. Beynon, *Analyst,* **103,** 613 (1978).
[226] F. W. McLafferty and F. M. Bockhoff, *Anal. Chem.,* **50,** 69 (1978).
[227] R. G. Cooks, *Intern. Lab.,* No. 1, 79 (1979).
[228] R. K. Boyd and J. H. Beynon, *Org. Mass Spectrom.,* **12,** 163 (1977).
[229] D. S. Millington and J. A. Smith, *Org. Mass Spectrom.,* **12,** 264 (1977).
[230] A. M. Greenway and C. F. Simpson, *J. Phys.,* **13,** 1131 (1980).
[231] M. C. ten Noever de Brauw, *J. Chromatog.,* **165,** 207 (1979).
[232] F. Etzweiler, *J. Chromatog.,* **167,** 133 (1978).
[233] D. Henneberg, U. Henrichs, H. Husmann and G. Schomburg, *J. Chromatog.,* **167,** 139 (1978).
[234] K. Grob, *Chromatographia,* **9,** 509 (1976).
[235] W. H. Dürbeck, I. Büker and W. Leymann, *Chromatographia,* **11,** 295, 372 (1978).
[236] W. H. McFadden, *J. Chromatog. Sci.,* **17,** 2 (1979).
[237] P. J. Arpino and G. Guiochon, *Anal. Chem.,* **51,** 683A (1979).
[238] M. A. Baldwin and F. W. McLafferty, *Org. Mass Spectrom.,* **7,** 1111 (1973).
[239] J. D. Henion, *Anal. Chem.,* **50,** 1687 (1978).
[240] P. J. Arpino, G. Guiochon, P. Krien and G. Devant, *J. Chromatog.,* **185,** 529 (1979).
[241] S. Imabayashi, T. Nakamuro, Y. Sawa, J. Hasegawa, K. Sakaguchi, T. Fujita, Y. Mori and K. Kawabe, *Anal. Chem.,* **51,** 534 (1979).
[242] C. R. Blakley, M. J. McAdams and M. L. Vestal, *J. Chromatog.,* **158,** 261 (1978).
[243] T. Takeuchi, Y. Hirata and Y. Ohumura, *Anal. Chem.,* **50,** 660 (1978).
[244] R. P. W. Scott, C. G. Scott, M. Munroe and J. Hess, *J. Chromatog.,* **99,** 395 (1974).
[245] W. H. McFadden, H. L. Schwartz and S. Evans, *J. Chromatog.,* **122,** 389 (1976).
[246] W. H. McFadden, D. C. Bradford, G. Eglinton and H. Nicolaides, *J. Chromatog. Sci.,* **17,** 518 (1979).
[247] D. I. Carroll, I. Dzidic, R. N. Stillwell, K. D. Haegele and E. C. Horning, *Anal. Chem.,* **47,** 2369 (1975).
[248] P. R. Jones and S. K. Yang, *Anal. Chem.,* **47,** 1000 (1975).
[249] E. Haahti and T. Nikari, *Acta Chem. Scand.,* **17,** 2565 (1963).
[250] C. R. Blakley, M. J. McAdams and M. L. Vestal, *J. Chromatog.,* **158,** 261 (1978).
[251] M. L. Vestal, in *Trace Organic Analysis,* pp. 647–654. NBS, Washington (1979).
[252] M. Chaigneau and M. Chastagnier, *Bull. Soc. Chim. France,* 40 (1976).
[253] S. N. Chesler, B. H. Gump, H. S. Hertz, W. E. May and S. A. Wise, *Anal. Chem.,* **50,** 805 (1978).
[254] R. Shinohara, T. Hori and M. Koga, *Bunseki Kagaku,* **27,** 400 (1978).
[255] H. Matsushima and T. Hanya, *Bunseki Kagaku,* **24,** 505 (1975).
[256] J. Albaiges and P. Albrecht, *Intern. J. Environ. Anal. Chem.,* **6,** 171 (1979).
[257] B. S. Middleditch, B. Basile and E. S. Chang, *J. Chromatog.,* **142,** 777 (1977).
[258] R. Shinohara, M. Koga, J. Shinohara and T. Hori, *Bunseki Kagaku,* **26,** 856 (1977).
[259] G. Grimmer, H. Böhnke and H. Borwitzky, *Z. Anal. Chem.,* **289,** 91 (1978).
[260] F. M. Benoit, G. L. Lebel and D. T. Williams, *Intern. J. Environ. Anal. Chem.,* **6,** 277 (1979).
[261] T. A. Bellar, J. J. Lichtenberg and J. W. Eichelberger, *Environ. Sci. Technol.,* **10,** 926 (1976).
[262] P. L. Sherman, A. M. Kemmer, L. Metcalfe, H. D. Toy and G. E. Parris, in *Trace Organic Analysis,* pp. 205–212. NBS, Washington (1979).
[263] T. Fujii, *Anal. Chim. Acta.,* **92,** 117 (1977).
[264] L. L. Ingram, G. D. McGinnis and S. V. Parikh, *Anal. Chem.,* **51,** 1077 (1979).

[265] T. Fujii, *J. Chromatog.*, **139**, 297 (1977).
[266] A. J. Burgasser and J. F. Colaruotolo, *Anal. Chem.*, **51**, 1588 (1979).
[267] J. Freudenthal, *Intern. J. Environ. Anal. Chem.*, **5**, 311 (1978).
[268] J. Rivera, M. R. Cuberes and J. Albaiges, *Bull. Environ. Contam. Toxicol.*, **18**, 624 (1977).
[269] T. Fujii, *Anal. Chem.*, **49**, 1985 (1977).
[270] M. D. Erickson, L. C. Michael, R. A. Zweidinger and E. D. Pellizzari, *J. Assoc. Off. Anal. Chem.*, **61**, 1335 (1978).
[271] E. E. McNeill, P. Otson, W. F. Miles and F. J. M. Rajabalee, *J. Chromatog.*, **132**, 277 (1977).
[272] G. D. Veith, D. W. Kuehl, F. A. Puglisi, G. E. Glass and J. G. Eaton, *Arch. Environ. Contam. Toxicol.*, **5**, 487 (1977).
[273] R. L. Harless, D. E. Harris, G. W. Sovocool, R. D. Zehr, N. K. Wilson and E. O. Oswald, *Biomed. Mass Spectrom.*, **5**, 232 (1978).
[274] H. Nakamura, *Suido Kyokai Zasshi*, 37 (1977).
[275] H. Hon-Nami and T. Hanya, *J. Chromatog.*, **161**, 205 (1978).
[276] R. T. Coutts, G. R. Jones and S. F. Liu, *J. Chromatog. Sci.*, **17**, 551 (1979).
[277] R. Tatsukawa, J. Itoh and T. Wakimoto, *J. Agr. Chem. Soc. Japan*, **51**, 315 (1977).
[278] A. Yasuhara and K. Fuwa, *J. Chromatog.*, **172**, 453 (1979).
[279] K. H. Bergert and V. Betz, *Chromatographia*, **7**, 681 (1974).
[280] D. Schuetzle, *Biomed. Mass. Spectrom.*, **2**, 288 (1975).
[281] M. Dong, D. C. Locke and E. Ferrand, *Anal. Chem.*, **48**, 368 (1976).
[282] E. B. Rietz, *Anal. Lett.*, **12**, 143 (1979).
[283] D. Hoffmann, K. D. Brunnemann, I. Schmeltz and E. L. Wynder, in *Trace Organic Analysis*, pp. 131–141. NBS, Washington (1979).
[284] C. D. Briggs and S. J. A. Hawthorne, *Proc. Anal. Div. Chem.*, **15**, 181 (1978).
[285] K. D. Bartle, M. L. Lee and M. Novotny, *Proc. Anal. Div. Chem. Soc.*, **13**, 304 (1976).
[286] R. C. Lao, R. S. Thomas and J. L. Monkman, *J. Chromatog.*, **112**, 681 (1975).
[287] M. L. Lee, D. L. Vassilaros, W. S. Pipkin and W. L. Sorensen, in *Trace Organic Analysis*, pp. 731–738. NBS, Washington (1979).
[288] M. L. Lee, M. Novotny and K. D. Bartle, *Anal. Chem.*, **48**, 1566 (1976).
[289] G. Grimmer, H. Böhmke and A. Glasser, *Erdöl, Kohle, Erdgas, Petrochemie*, **30**, 411 (1977).
[290] D. B. Walters, W. J. Chamberlain, F. J. Akin, M. E. Snook and O. T. Chortyk, *Anal. Chim. Acta*, **99**, 143 (1978).
[291] M. L. Lee, M. Novotny and K. D. Bartle, *Anal. Chem.*, **48**, 405 (1976).
[292] R. F. Severson, M. E. Snook, O. T. Chortyk and R. F. Arrendale, *Beitr. Tabaksforsch.*, **8**, 273 (1976).
[293] I. Schmeltz, J. Tosk and D. Hoffmann, *Anal. Chem.*, **48**, 645 (1976).
[294] K. Kuwata, *Bunseki Kagaku*, **27**, 650 (1978).
[295] E. D. Pellizari, J. E. Bunch, R. E. Berkley and J. McRae, *Anal. Chem.*, **48**, 803 (1976).
[296] D. E. Harsch and R. A. Rasmussen, *Anal. Lett.*, **10**, 1041 (1977).
[297] R. Tatsukawa, T. Okamoto and T. Wakimoto, *Bunseki Kagaku*, **27**, 164 (1978).
[298] R. Ciupe, E. Fagarasan and L. Prop, *Rev. Chim. (Bucharest)*, **28**, 575 (1977).
[299] B. J. Tyson, *Anal. Lett.*, **8**, 807 (1975).
[300] D. R. Cronn and D. E. Harsch, *Anal. Lett.*, **9**, 1015 (1976).
[301] E. P.Grimsrud and R. A. Rasmussen, *Atmos. Environ.*, **9**, 1010 (1975).
[302] E. Jellum, *J. Chromatog.*, **143**, 427 (1977).
[303] M. Vanhaelen, R. Vanhaelen-Fastre and J. Geeraerts, *J. Chromatog.*, **144**, 108 (1977).
[304] B. Janistyn, *Z. Naturforsch. C, Biosci.*, **36**, 193 (1981).
[305] G. Moneti, M. Pazzagli, G. Fiorelli and M. Serio, *J. Steroid Biochem.*, **13**, 623 (1980).
[306] R. Kobelt, G. Wiesener and R. Kuehn, *J. High Resolut. Chromatog., Chromatog. Commun.*, **4**, 520 (1981).

[307] M. Suzuki, M. Nishizawa, T. Miyatake and Y. Kagawa, *J. Lipid Res.*, **23**, 363 (1982).
[308] G. A. Eiceman, V. A. Fuavao, K. D. Doolittle and C. A. Herman, *J. Chromatog.*, **236**, 97 (1982).
[309] H. Mita, H. Yasueda and T. Shida, *J. Chromatog.*, **221**, *Biomed. Appl.*, **10**, 1 (1980).
[310] W. Vogt, K. Jacob, A. B. Ohnesorge and G. Schwertfeger, *J. Chromatog.*, **199**, 191 (1980).
[311] D. Louis, F. Belleville, P. Nabet and P. Paysant, *Ann. Med. Nancy Est*, **20**, 1097, 1100 (1981).
[312] H. J. Stan and M. Scheutwinkel-Reich, *Lipids*, **15**, 1044 (1980).
[313] M. Tetsuo, S. P. Markey, R. W. Colburn and I. J. Kopin, *Anal. Biochem.*, **110**, 208 (1981).
[314] D. C. Jimerson, S. P. Markey, J. A. Oliver and I. J. Kopin, *Biomed. Mass Spectrom.*, **8**, 256 (1981).
[315] D. A. Durden, A. V. Juorio and B. A. Davis, *Anal. Chem.*, **52**, 1815 (1980).
[316] B. Andersson and G. Sandberg, *J. Chromatog.*, **238**, 151 (1982).
[317] C. Canfell, S. R. Binder and H. Khayam-Bashi, *Clin. Chem.*, **28**, 25 (1982).
[318] M. Iwai, H. Kanno, M. Hashino, J. Suzuki, T. Yanaihara and T. Nakayama, *J. Chromatog.*, **225**, *Biomed. Appl.*, **14**, 274 (1981).
[319] T. Niwa, K. Maeda, T. Ohki, A. Saito and I. Tsuchida, *J. Chromatog.*, **225**, *Biomed. Appl.*, **14**, 1 (1981).
[320] H. Mueller, R. Mongrovius and H. J. Seyberth, *J. Chromatog.*, **226**, *Biomed. Appl.*, **15**, 450 (1981).
[321] R. G. Megargle, L. E. Slivon, J. E. Graas and A. H. Andrist, *Anal. Chim. Acta*, **120**, 193 (1980).
[322] H. Miyazaki, M. Ishibashi, K. Yamashita, Y. Nishikawa and K. Katori, *Biomed. Mass Spectr.*, **8**, 521 (1981).
[323] A. Ferretti, N. W. Schoene and V. P. Flanagan, *Lipids*, **16**, 800 (1981).
[324] M. Suzuki, I. Morita, M. Kawamura, S. Murato, M. Nishizawa, T. Miyatake, H. Nagase, K. Ohno and H. Shimizu, *J. Chromatog.*, **221**, *Biomed. Appl.*, **10**, 361 (1980).
[325] C. Chiabrando, A. Noseda, M. A. Noe and R. Fanelli, *Prostaglandins*, **20**, 747 (1980).
[326] B. H. Min, J. Pao, W. A. Garland, J. A. F. de Silva and M. Parsonnet, *J. Chromatog.*, **183**, *Biomed. Appl.*, **9**, 411 (1980).
[327] S. P. Markey, R. W. Colburn and J. N. Johannessen, *Biomed. Mass Spectrom.*, **8**, 301 (1981).
[328] S. J. Gaskell, A. W. Pike and K. Griffiths, *Steroids*, **36**, 219 (1980).
[329] L. Larsson, P. A. Mardh, G. Odham and G. Westerdahl, *J. Chromatog.*, **182**, *Biomed. Appl.*, **8**, 402 (1980).
[330] G. B. Baker, D. F. LeGatt and R. T. Coutts, *J. Neurosci. Methods*, **5**, 181 (1982).
[331] R. M. Smith, *Intern. Lab.*, May/June, 115 (1978).
[332] M. Novotny, M. L. Lee, C. E. Low and A. Raymond, *Anal. Chem.*, **48**, 24 (1976).
[333] S. P. Jindal, T. Lutz and P. Vestergaard, *Anal. Chem.*, **51**, 269 (1979).
[334] D. P. Beggs and A. G. Day, *J. Forens. Sci.*, **4**, 891 (1974).
[335] A. M. Van der Ark, A. B. E. Theeuwen and A. M. A. Verweij, *Pharm. Weekbl.*, **112**, 977 (1977).
[336] R. Saferstein, J. J. Manura and P. K. De, *J. Forens. Sci.*, **23**, 29 (1978).
[337] W. J. Cole, J. Parkhouse and Y. Y. Yousef, *J. Chromatog.*, **136**, 409 (1977).
[338] D. F. Hunt and F. W. Crow, *Anal. Chem.*, **50**, 1781 (1978).
[339] D. F. Hunt and F. W. Crow, in *Trace Organic Analysis*, pp. 601–607. NBS, Washington (1979).
[340] R. W. Ulrich, D. L. Bowerman, P. H. Wittenberg, B. L. McGaha, D. K. Schisler, J. A. Anderson, J. A. Levisky and J. L. Pflug, *Anal. Chem.*, **47**, 581 (1975).
[341] I. C. Stone, J. N. Lomonte, L. A. Fletcher and W. T. Lowry, *J. Forens. Sci.*, **23**, 78 (1978).

[342] D. Arndts and K. L. Rominger, *Arzneim. Forsch.*, **28,** 1951 (1978).
[343] E. Houghton and P. Teale, *Biomed. Mass Spectrom.*, **8,** 358 (1981).
[344] H. Watanabe, S. Nakano, N. Ogawa and T. Suzuki, *Biomed. Mass Spectrom.*, **7,** 160 (1980).
[345] M. Eckert and P. H. Hinderling, *Agents Action,* **11,** 520 (1981).
[346] J. Hasegawa, Y. Tomono, M. Tanaka, T. Fujita, K. Sugiyama and N. Hirose, *Arzneim. Forsch.*, **31, II,** 1215 (1981).
[347] W. H. Stueber and W. H. Moeller, *J. Chromatog.*, **224,** *Biomed. Appl.*, **13,** 327 (1981).
[348] H. Ehrsson, S. Eksborg, I. Wallin, Y. Marde and B. Joansson, *J. Pharm. Sci.*, **69,** 710 (1980).
[349] B. R. Kuhnert, P. M. Kuhnert and A. L. P. Reese, *J. Chromatog.*, **224,** *Biomed. Appl.*, **13,** 488 (1981).
[350] J. A. Thompson and F. H. Leffert, *J. Pharm. Sci.*, **69,** 707 (1980).
[351] Y. Hayasaka, S. Murata and K. Umemura, *Chem. Pharm. Bull.*, **29,** 1478 (1981).
[352] S. Murray, K. A. Waddell and D. S. Davies, *Biomed. Mass Spectrom.*, **8,** 500 (1981).
[353] D. M. Chinn, D. J. Crouch, A. M. Peat, B. S. Finkle and T. A. Jennison, *J. Anal. Toxicol.*, **4,** 37 (1980).
[354] G. Lhoest and M. Mercier, *Pharm. Helv. Acta,* **55,** 316 (1980).
[355] M. R. Holdiness, Z. H. Israili and J. B. Justice, *J. Chromatog.*, **224,** *Biomed. Appl.*, **13,** 415 (1981).
[356] M. Tetsuo, M. Axelson and J. Sjovall, *J. Steroid Biochem.*, **13,** 847 (1980).
[357] S. N. Lin, T. F. Wang, R. M. Caprioli and P. N. B. Mo, *J. Pharm. Sci.*, **70,** 1276 (1981).
[358] H. Isomura, S. Higuchi and S. Kawamura, *J. Chromatog.*, **224,** *Biomed. Appl.*, **13,** 423 (1981).
[359] P. Ottoila, J. Taskinen and A. Sothmann, *Biomed. Mass Spectrom.*, **9,** 108 (1982).
[360] G. Idzu, M. Ishibashi, H. Miyazaki and Y. Yamamoto, *J. Chromatog.*, **229,** *Biomed. Appl.*, **18,** 327 (1982).
[361] I. M. Kapetanovic and H. J. Kupferberg, *J. Pharm. Sci.*, **70,** 1218 (1981).
[362] G. Phillipou and R. G. Fritz, *Clin. Chim. Acta,* **103,** 129 (1980).
[363] G. I. Kang and F. S. Abbott, *J. Chromatog.*, **231,** *Biomed. Appl.*, **20,** 311 (1982).
[364] Y. M. Chan, S. J. Soldin, J. M. Swanson, C. M. Deber, J. J. Thiessen and S. MacLeod, *Clin. Biochem.*, **13,** 266 (1980).
[365] K. Ichimura, Y. Nozaki, K. Kato, T. Harada and K. Ishii, *Koenshu-Iyo Masu Kenkyukai,* **5,** 181 (1980).
[366] R. G. Smith and L. K. Cheung, *J. Chromatog.*, **229,** *Biomed. Appl.*, **18,** 464 (1982).
[367] I. Bjorkhem and H. Ek, *J. Steroid Biochem.*, **17,** 447 (1982).
[368] W. Gielsdorf, *J. Clin. Chem. Clin. Biochem.*, **20,** 65 (1982).
[369] W. Ihn, W. Schade, B. Mueller and G. Peiker, *Pharmazie,* **35,** 598 (1980).
[370] A. Van Langenhove, C. E. Costello, J. E. Biller, K. Biemann and T. R. Browne, *Clin. Chim. Acta,* **115,** 263 (1981).
[371] R. M. Gorsen, A. J. Weiss and R. W. Manthei, *J. Chromatog.*, **221,** *Biomed. Appl.*, **10,** 309 (1980).
[372] K. K. Midha, C. Charette, I. J. McGilveray, D. Webb and M. C. McLean, *Clin. Toxicol.*, **18,** 713 (1982).
[373] R. B. Forney, F. T. Carroll, I. K. Nordgren, B. M. Pettersson and B. Holmstedt, *J. Anal. Toxicol.*, **6,** 115 (1982).
[374] T. Suzuki, K. Tsuzurahara, T. Murata and S. Takeyama, *Biomed. Mass Spectrom.*, **9,** 94 (1982).
[375] A. Lapin, *Eur. J. Mass Spectrom. Biochem. Med. Environ. Res.*, **1,** 121 (1980).
[376] R. Whelpton, S. H. Curry and G. M. Watkins, *J. Chromatog.*, **228,** *Biomed. Appl.*, **17,** 321 (1982).
[377] K. K. Midha, R. M. H. Roscoe, K. Hall, E. M. Hawes, J. K. Cooper, G. McKay and H. U. Shetty, *Biomed. Mass Spectrom.*, **9,** 186 (1982).
[378] R. L. Furner, G. B. Brown and J. W. Scott, *J. Anal. Toxicol.*, **5,** 275 (1981).
[379] R. Self, *Biomed. Mass Spectrom.*, **6,** 361 (1979).

[380] R. C. Doerr and W. Fiddler, *J. Chromatog.*, **140**, 284 (1977).
[381] C. Janzowski, G. Eisenbrand and R. Preussmann, *J. Chromatog.*, **150**, 216 (1978).
[382] J. H. Hotchkiss, *J. Assoc. Off. Anal. Chem.*, **64**, 1037 (1981).
[383] M. P. Yurawecz and J. A. Roach, *J. Assoc. Off. Anal. Chem.*, **61**, 26 (1978).
[384] J. Sphon, *J. Assoc. Off. Anal. Chem.*, **61**, 1247 (1978).
[385] T. Obretenov and P. Hadjeva, *Z. Lebensm. Unters. Forsch.*, **165**, 195 (1977).
[386] P. Hubert, H. Kwasny, P. Werkhoff and U. Turner, *Z. Anal. Chem.*, **285**, 242 (1977).
[387] R. J. Horvat, *J. Agr. Food Chem.*, **24**, 953 (1976).
[388] D. Robinson, G. C. Mead and K. A. Barnes, *Bull. Environ. Contam. Toxicol.*, **27**, 145 (1981).
[389] W. P. Cochrane, J. Singh, W. Miles and B. Wakeford, *J. Chromatog.*, **217**, 289 (1981).
[390] L. H. Wright, M. D. Jackson and R. G. Lewis, *Bull. Environ. Contam. Toxicol.*, **28**, 740 (1982).
[391] M. Koga, T. Akiyama and R. Shinohara, *Bunseki Kagaku*, **30**, 745 (1981).
[392] H. J. Stan and B. Abraham, *J. Chromatog.*, **195**, 231 (1980).
[393] B. Bergner-Lang and M. Kaechele, *Dtsch. Lebensm. Rundsch.*, **77**, 305 (1980).
[394] D. W. Kuehl, E. N. Leonard, K. J. Welch and G. D. Veith, *J. Assoc. Off. Anal. Chem.*, **63**, 1238 (1980).
[395] E. Schulte and L. Acker, *Nahrung*, **24**, 577 (1980).
[396] B. E. Wilkes, L. J. Priestley and L. K. Scholl, *Microchem. J.*, **27**, 420 (1982).
[397] V. D. Chmil, *Zh. Analit. Khim.*, **36**, 1121 (1981).
[398] H. Sekita, M. Takeda, Y. Saito and M. Uchiyama, *Eisei Shikenso Hokoku*, **138** (1980).
[399] F. M. Benoit and D. T. Williams, *Bull. Environ. Contam. Toxicol.*, **27**, 303 (1981).
[400] J. D. Henion, J. S. Nosanchuk and B. M. Bilder, *J. Chromatog.*, **213**, 475 (1981).
[401] D. L. Heikes, *J. Assoc. Off. Anal. Chem.*, **63**, 1224 (1980).
[402] Newton, D. L., M. D. Erickson, K. B. Tomer, E. D. Pellizari, P. Gentry and R. B. Zweidinger, *Environ. Sci. Technol.*, **16**, 206 (1982).
[403] D. Andrzejewski, D. C. Havery and T. Fazio, *J. Assoc. Off. Anal. Chem.*, **64**, 1457 (1981).
[404] K. Kusunoki, K. Mitarai and T. Asano, *Bunseki Kagaku*, **30**, 646 (1981).
[405] R. J. Horvat and S. D. Senter, *J. Agr. Food Chem.*, **28**, 1292 (1980).
[406] W. Specht and M. Tillkes, *Z. Anal. Chem.*, **307**, 257 (1981).
[407] T. L. Barry, G. Petzinger and F. M. Gretch, *Bull. Environ. Contam. Toxicol.*, **27**, 524 (1981).
[408] K. Jiang, Z. Kang and Y. Bian, *Huanjiing Kexue Xuebao*, **2**, 85 (1982).
[409] L. S. Ramos and P. G. Prohaska, *J. Chromatog.*, **211**, 284 (1981).
[410] D. L. Vassilaros, P. W. Stoker, G. M. Booth and M. L. Lee, *Anal. Chem.*, **54**, 106 (1982).
[411] G. Grimmer, J. Jacob, K. W. Naujack and G. Dettbarn, *Z. Anal. Chem.*, **309**, 13 (1981).
[412] C. Pantarotto, R. Fanelli, I. Belletti and F. Bidoli, *Anal. Biochem.*, **105**, 340 (1980).
[413] R. L. Harless, E. O. Oswald, R. G. Lewis, A. E. Dupuy, D. D. McDaniel and H. Tai, *Chemosphere*, **11**, 193 (1982).
[414] R. Fanelli, M. P. Bertoni, M. Bonfanti, M. G. Castelli, C. Chiabrando, G. P. Martelli, M. A. Noe, A. Noseda and C. Sbarra, *Bull. Environ. Contam. Toxicol.*, **24**, 818 (1980).
[415] W. C. Brumley, J. A. G. Roach, J. A. Sphon, P. A. Dreifuss, D. Andrzejewski, R. A. Niemann and D. A. Firestone, *J. Agr. Food Chem.*, **29**, 1040 (1981).
[416] R. K. Mitchum, G. F. Moler and W. A. Korfmacher, *Anal. Chem.*, **52**, 2278 (1980).
[417] S. Raisanen, R. Hiltunen, A. U. Arstila and T. Sipilainen, *J. Chromatog.*, **208**, 323 (1981).
[418] M. L. Langhorst and L. A. Shadoff, *Anal. Chem.*, **52**, 2037 (1980).
[419] A. Liberti, P. Ciccioli, E. Brancaleoni and A. Cecinato, *J. Chromatog.*, **242**, 111 (1982).

[420] G. A. Eiceman, R. E. Clement and F. W. Karasek, *Anal. Chem.*, **53**, 955 (1981).
[421] D. J. Harvan, J. R. Hass, J. L. Schroeder and B. J. Corbett, *Anal. Chem.*, **53**, 1755 (1981).
[422] J. P. Chaytor and M. J. Saxby, *J. Chromatog.*, **237**, 107 (1982).
[423] J. E. Taylor, A. van Iderstine, R. M. Weppelman, G. Olson, R. W. Walker, H. E. Mertel and W. J. A. vandenHeuvel, *J. Agr. Food Chem.*, **30**, 858 (1982).
[424] M. A. Evenson
[425] J. A. Yerensian, K. G. Sloman and A. K. Foltz, *Anal. Chem.*, **51**, 105R (1979).
[426] M. Spiteller and G. Spiteller, *J. Chromatog.*, **164**, 253 (1979).
[427] B. Crathorne, C. D. Watts and M. Fielding, *J. Chromatog.*, **185**, 671 (1979).
[428] A. W. Garrison, *Technical Paper No. 9*, WHO Intern. Reference Centre for Community Water Supply, The Hague (1976).
[429] J. M. Steyn and H. K. L. Hundt, *J. Chromatog.*, **143**, 207 (1977).
[430] F. L. Schulting and E. R. J. Wils, *Anal. Chem.*, **49**, 2365 (1977).
[431] D. W. Philipson and B. Puma, *Anal. Chem.*, **52**, 2328 (1980).
[432] C. L. Wilkins, G. N. Giss, G. Brissey and S. Steiner, *Anal. Chem.*, **53**, 113 (1981).
[433] T. Cairns, *Anal. Chem.*, **53**, 1599 (1981).
[434] R. M. Smith, *Anal. Proc.*, **19**, 361 (1982).
[435] R. B. Meiris, *Dev. Chromatog.*, **2**, 1 (1980).
[436] R. P. W. Scott, *Liquid Chromatography Detectors*, Elsevier, Amsterdam (1977).
[437] W. H. McFadden, *Anal. Proc.*, **19**, 258 (1982).
[438] R. J. Chapman, *Anal. Proc.*, **19**, 235 (1982).
[439] P. J. Arpino, *TrAC*, **1**, 154 (1982).
[440] K. Aitzetmueller, *J. Chromatog. Sci.*, **13**, 454 (1975).
[441] R. P. W. Scott, in *Trace Organic Analysis*, pp. 637–645. NBS, Washington (1979).
[442] D. E. Games, J. L. Gower, M. G. Lee, I. A. S. Lewis, M. E. Pugh and M. Rossiter, *Proc. Anal. Div. Chem. Soc.*, **15**, 101 (1978).
[443] W. H. McFadden, H. L. Schwartz and S. Evans, *J. Chromatog.*, **122**, 389 (1976).
[444] S. Privett and W. H
[445] K. Rüdiger, M. Höhne and R. Herrmann, *Z. Anal. Chem.*, **293**, 353 (1978).
[446] N. S. Radin and D. del Vecchio, *Anal. Chem.*, **50**, 824 (1978).
[447] K. Aitzetmüller, *J. Chromatog. Sci.*, **13**, 454 (1975).
[448] W. H. McFadden, D. C. Bradford, G. Eglinton, S. K. Hajbrahim and N. Nicolaides, *J. Chromatog. Sci.*, **17**, 518 (1979).
[449] B. J. Compton and W. C. Purdy, *J. Chromatog.*, **169**, 39 (1979).
[450] K. Slais and M. Krejči, *J. Chromatog.*, **91**, 181 (1974).
[451] M. E. Getz, G. W. Hanes and K. R. Hill, in *Trace Organic Analysis*, pp. 345–353. NBS, Washington (1979).
[452] K. R. Hill and H. L. Christ, *J. Chromatog. Sci.*, **17**, 395 (1979).
[453] A. Benninghoven, A. Eicke, M. Junack, W. Sichtermann, J. Krizek and H. Peters, *Org. Mass Spectrom.*, **15**, 459 (1980).
[454] D. E. Games, P. Hirter, W. Kuhnz, E. Lewis, N. C. A. Weerasinghe and S. A. Westwood, *J. Chromatog.*, **203**, 131 (1981).
[455] D. P. Kirby, P. Vouros, B. L. Karger, B. Hidy and B. Petersen, *J. Chromatog.*, **203** 139 (1981).
[456] R. D. Smith and J. E. Burger, *U.S. Dept. Energy, Rept.*, PNL-SA-8274 (1980).
[457] D. E. Games, M. S. Lant, S. A. Westwood, M. J. Cocksedge, N. Evans, J. Williamson and B. J. Woodhall, *Biomed. Mass Spectrom.*, **9**, 215 (1982).
[458] R. D. Smith and A. L. Johnson, *Anal. Chem.*, **53**, 1120 (1981).
[459] L. H. Wright, *J. Chromatog. Sci.*, **20**, 1 (1982).
[460] T. Cairns, E. G. Siegmund and G. M. Doose, *Anal. Chem.*, **54**, 953 (1982).
[461] U. A. Th. Brinkman, P. M. Onel and G. de Vries, *J. Chromatog.*, **171**, 424 (1979).
[462] F. W. Willmott and R. J. Dolphin, *J. Chromatog. Sci.*, **12**, 695 (1974).
[463] A. F. Chamberlain and J. S. Marlow, *J. Chromatog. Sci.*, **15**, (1977).
[464] R. J. Dolphin and F. W. Wilmott, *J. Chromatog.*, **149**, 161 (1978).
[465] R. J. Dolphin, F. W. Wilmott, A. D. Mills and L. P. J. Hoogeveen, *J. Chromatog.*, **122**, 259 (1976).

[466] H. Yoshida, K. Matsumoto, K. Itoh, S. Tsuge, Y. Hirata, K. Mochizuki, N. Kokubun and Y. Yoshida, *Z. Anal. Chem.*, **311**, 674 (1982).
[467] C. R. Blakley, J. C. Carmody and M. L. Vestal, *Clin. Chem.*, **26**, 1467 (1980).
[468] C. R. Blakley, J. J. Carmody and M. L. Vestal, *Anal. Chem.*, **52**, 1636 (1980).
[469] J. D. Henion and T. Wachs, *Anal. Chem.*, **53**, 1963 (1981).
[470] E. Yamauchi, T. Mizuno and K. Azuma, *Shitsuryo Bunseki*, **28**, 227 (1980).
[471] T. Takeuchi, D. Ishii, A. Saito and T. Ohki, *J. High Resolut. Chromatog. Chromatog. Commun.*, **5**, 91 (1982).
[472] J. D. Henion and G. A. Haylin, *Spectra 2000*, **8**, 53 (1980).
[473] R. D. Voyksner, J. R. Hass and M. M. Bursey, *Anal. Lett.*, **15**, 1 (1982).
[474] S. L. Smith, J. W. Jorgenson and M. Novotny, *J. Chromatog.*, **187**, 111 (1980).
[475] S. A. Wise, R. A. Mowery and R. S. Juvet, *J. Chromatog. Sci.*, **17**, 601 (1979).
[476] E. S. Yeung, L. E. Steenhoek, E. Larry, S. D. Woodruff and J. C. Kuo, *Anal. Chem.*, **52**, 1399 (1980).
[477] S. Terabe, K. Yamamoto and T. Ando, *Spectra-Phys., Chromatog. Rev.*, **7**, 9 (1981).
[478] W. A. McKinley, *J. Anal. Toxicol.*, **5**, 209 (1981).
[479] D. C. Locke, B. S. Dhingra and A. D. Baker, *Anal. Chem.*, **54**, 447 (1982).
[480] V. L. McGuffin and M. Novotny, *J. Chromatog.*, **218**, 179 (1981).
[481] R. J. Lloyd, *J. Chromatog.*, **216**, 127 (1981).
[482] A. N. Masoud and Y. N. Chan, *J. High Resolut. Chromatog., Chromatog. Commun.*, **5**, 299 (1982).
[483] K. Štulík and V. Pacáková, *J. Chromatog.*, **208**, 269 (1981).
[484] D. J. David, *Gas Chromatographic Detectors*, Wiley-Interscience, New York (1974).
[485] J. Ševcík, *Detectors in Gas Chromatography*, Elsevier, Amsterdam (1976).
[486] L. S. Ettre, in *Trace Organic Analysis*, pp. 547–585. NBS, Washington (1976).
[487] R. G. Lewis, in *Accuracy in Trace Organic Analysis*, p. 11. NBS, Washington (1976).
[488] S. L. Abidi, *J. Chromatog.*, **200**, 216 (1980).
[489] A. J. Pateman, *J. Chromatog.*, **226**, *Biomed. Appl.*, **15**, 213 (1981).
[490] T. Asakura and M. Matsuda, *Jekeikai Med. J.*, **28**, 167 (1981).
[491] M. Terada, T. Yamamoto, T. Yoshida, Y. Kuroiwa and S. Yoshimura, *J. Chromatog.*, **237**, 285 (1982).
[492] U. E. G. Bock and P. G. Waser, *J. Chromatog.*, **213**, 413 (1981).
[493] C. M. Lai, B. L. Kamath, M. Zee and Y. Avraham, *J. Pharm. Sci.*, **69**, 681 (1980).
[494] W. F. Staruszkiewicz and J. F. Bond, *J. Assoc. Off. Anal. Chem.*, **64**, 584 (1981).
[495] R. Riva, F. Albani and A. Baruzzi, *Farmaco, Ed. Prat.*, **37**, 15 (1982).
[496] A. B. Jones, M. A. Elsohly, J. A. Bedford and C. E. Turner, *J. Chromatog.*, **226**, *Biomed. Appl.*, **15**, 99 (1981).
[497] E. Fogelqvist, B. Josefson and C. Roos, *J. High Resolut. Chromatog., Chromatog. Commun.*, **3**, 568 (1980).
[498] S. Cline, A. Felsot and L. Wei, *J. Agr. Food Chem.*, **29**, 1087 (1981).
[499] P. S. Doshi and D. J. Edwards, *J. Chromatog.*, **210**, 505 (1981).
[500] S. Dmochewitz and K. Ballschmiter, *Z. Anal. Chem.*, **310**, 6 (1982).
[501] Y. S. Wong, *J. Assoc. Off. Anal. Chem.*, **65**, 1118 (1982).
[502] M. Baldi, A. Bovolenta, L. Penazzi and L. Zanoni, *Inquinamento*, **23**, 59 (1981).
[503] R. A. Guinivan, N. P. Thompson and P. C. Bardalaye, *J. Assoc. Off. Anal. Chem.*, **64**, 1201 (1981).
[504] P. Larking, *J. Chromatog.*, **221**, *Biomed. Appl.*, **10**, 399 (1980).
[505] E. Suzuki, M. Matsuda, A. Momose and M. Namekata, *J. Assoc. Off. Anal. Chem.*, **63**, 1211 (1980).
[506] M. Ludewig, K. Doerfling and W. A. Koenig, *J. Chromatog.*, **243**, 93 (1982).
[507] D. Fehr and S. L. Ali, *Pharm. Ztg.*, **125**, 558, 561–562 (1980).
[508] J. R. King, D. L. von Windeguth and A. K. Burditt, *J. Agr. Food Chem.*, **28**, 1049 (1980).
[509] S. S. Brady, H. F. Enos and K. A. Levy, *Bull. Environ. Contam. Toxicol.*, **24**, 813 (1980).

[510] J. D. Gaynor and D. C. MacTavish, *Analyst,* **107,** 700 (1982).
[511] A. Schultzová, O. Gattnar, D. Kovacová and Z. Mahrla, *Cesk. Farm.,* **31,** 64 (1982).
[512] J. M. F. Douse, *J. Chromatog.,* **234,** 415 (1982).
[513] J. M. F. Douse, *J. Chromatog.,* **208,** 83 (1981).
[514] Y. C. Sumirtapura, C. Aubert and J. P. Cano, *Arzneim. Forsch.,* **32, I,** 252 (1982).
[515] M. Batardy-Gregoire, C. Razzouk, E. Evrard and M. Roberfroid, *Anal. Biochem.,* **122,** 199 (1982).
[516] R. Aderjan, P. Fritz and R. Mattern, *Arzneim. Forsch.,* **30, II,** 1944 (1980).
[517] S. Singhawangcha, C. F. Poole and A. Zlatkis, *J. Chromatog.,* **183,** *Biomed. Appl.,* **9,** 433 (1980).
[518] L. V. McCarthy, E. B. Overton, C. K. Raschke and J. L. Laseter, *Anal. Lett.,* **13,** 1417 (1980).
[519] C. Kim and G. R. van Loon, *Clin. Chem.,* **27,** 1284 (1981).
[520] J. R. King, C. A. Benschoter and A. K. Burditt, *J. Agr. Food Chem.,* **29,** 1003 (1981).
[521] C. D. Kinney, *J. Chromatog.,* **225,** *Biomed. Appl.,* **14,** 213 (1981).
[522] A. Pachecus, Y. Santoni, M. Fornaris, S. Magnan, C. Aubert, A. Ragon and J. P. Cano, *Arzneim. Forsch.,* **32, I,** 688 (1982).
[523] M. Hirst, R. G. Herne and J. R. Robinson, *Subst. Alcohol. Actions/Misuse,* **1,** 361 (1980).
[524] R. H. Reuning, S. B. Ashcraft and B. E. Morrison, *NIDA Res. Monogr.,* 25 (1980).
[525] P. J. Meffin and K. J. Smith, *J. Chromatog.,* **183,** *Biomed. Appl.,* **9** 352 (1980).
[526] C. C. Wu, T. D. Sokoloski, A. M. Burkman, M. F. Blanford and L. S. Wu, *J. Chromatog.,* **17,** 333 (1982).
[527] V. W. Winkler, J. R. Patel, M. Januszanis and M. Colarusso, *J. Assoc. Off. Anal. Chem.,* **64,** 1309 (1981).
[528] C. S. Giam, D. A. Trujillo, S. Kira and Y. Hrung, *Bull. Environ. Contam. Toxicol.,* **25,** 824 (1980).
[529] M. E. P. B. Siqueira and N. A. G. G. Fernicola, *Bull. Environ. Contam. Toxicol.,* **27,** 380 (1981).
[530] H. A. Meemken, P. Fuerst and K. Habersaat, *Dtsch. Lebensm. Rundsch.,* **78,** 282 (1982).
[531] A. P. Borsetti, *J. Agr. Food Chem.,* **28,** 710 (1980).
[532] L. L. Needham, R. E. Cline, S. L. Head and J. A. Liddle, *J. Anal. Toxicol.,* **5,** 283 (1981).
[533] J. Delventhal, *Lebensmittelchem. Gerichtl. Chem.,* **34,** 122 (1980).
[534] M. A. Saleh, *J. Environ. Sci. Health,* **B, 17** 35 (1982).
[535] G. H. Takei and G. H. Leong, *Bull. Environ. Contam. Toxicol.,* **27,** 489 (1981).
[536] K. Ballschmiter, C. Scholz, H. Buchert, M. Zell, K. Figge, K. Polzhofer and H. Hoerschelmann, *Z. Anal. Chem.,* **309,** 1 (1981).
[537] British Standards Institution, *BS 5202:Part 10:1982* (ISO 4389–1981).
[538] M. Lehtonen, *J. Chromatog.,* **202,** 413 (1980).
[539] L. Renberg, *Chemosphere,* **10,** 767 (1981).
[540] M. Guerret, D. Lavene and J. R. Kiechel, *J. Pharm. Sci.,* **69,** 1191 (1980).
[541] M. Guerret, *J. Chromatog.,* **221,** *Biomed. Appl.,* **10,** 387 (1980).
[542] L. L. Needham, V. W. Burse and H. A. Price, *J. Assoc. Off. Anal. Chem.,* **64,** 1131 (1981).
[543] T. G. Rajagopalan, B. Anjaneyulu, V. D. Shanbag and R. S. Grewal, *J. Chromatog.,* **224,** *Biomed. Appl.,* **13,** 265 (1981).
[544] P. Hartvig, C. Fagerlund and B. M. Emanuelsson, *J. Chromatog.,* **228,** *Biomed. Appl.* **17,** 340 (1982).
[545] R. C. Jones, P. R. Ryle and B. C. Weatherley, *J. Chromatog.,* **224,** *Biomed. Appl.,* **13,** 492 (1981).
[546] E. Lemperle, N. Emmanoulidis and E. Kerner, *Dtsch. Lebensm. Rundsch.,* **78,** 6 (1982).
[547] R. K. Munns and J. E. Roybal, *J. Assoc. Off. Anal. Chem.,* **65,** 1048 (1982).

[548] C. L. Holder, H. C. Thompson and M. C. Bowman, *J. Chromatog. Sci.*, **19**, 625 (1981).
[549] P. Vasileva-Aleksandrova, A. Neicheva, E. Kovacheva and G. Marudov, *J. Chromatog.*, **241**, 404 (1982).
[550] M. A. Ribick, G. R. Dubay, J. D. Petty, D. L. Stalling and C. J. Schmitt, *Environ. Sci. Techn.*, **16**, 310 (1982).
[551] D. J. Greenblatt, M. Divoll, L. J. Moschitto and R. I. Shader, *J. Chromatog.*, **225**, *Biomed. Appl.*, **14**, 202 (1981).
[552] I. J. Sud and D. S. Feingold, *J. Invest. Dermatol.*, **73**, 521 (1979).
[553] R. Soma, C. Confortini and G. Trinci, *Boll. Chim. Unione Ital. Lab. Prov., Parte Sci.*, **6**, 161 (1980).
[554] P. M. Belanger and O. Grech-Belanger, *J. Chromatog.*, **228**, *Biomed. Appl.*, **17**, 327 (1982).
[555] A. Marzo, G. Quadro, R. Stradi, M. Saccarello, L. Molteni and G. Bruno, *Arzneim. Forsch.*, **31, II**, 1681 (1981).
[556] A. Radecki, H. Lamparczyk, J. Grzybowski and J. Halkiewicz, *Z. Anal. Chem.*, **303**, 397 (1980).
[557] D. M. Wyatt, *J. Am. Oil Chem. Soc.*, **58**, 917 (1981).
[558] T. Kaniewska, *Acta Pol. Pharm.*, **38**, 217 (1981).
[559] E. Zuccato, F. Marcucci and E. Mussini, *Anal. Lett.*, **13**, 363 (1980).
[560] M. M. Wermeille and G. A. Huber, *J. Chromatog.*, **228**, *Biomed. Appl.*, **17**, 187 (1982).
[561] G. R. Nakamura, Y. Liu and T. T. Noguchi, *J. Anal. Toxicol.*, **5**, 162 (1981).
[562] T. J. Gillespie, A. J. Gandolfi, R. M. Maiorino and R. W. Vaughan, *J. Anal. Toxicol.*, **5**, 133 (1981).
[563] A. H. Tameesh, T. M. Sarkisian and K. R. Hameed, *Chromatographia*, **13**, 617 (1980).
[564] K. Vereby, A. De Pace, D. Jukofsky, J. V. Volavka and S. J. Mule, *J. Anal. Toxicol.*, **4**, 33 (1980).
[565] S. D. Gettings, R. J. Flanagan and D. W. Holt, *J. Chromatog.*, **225**, *Biomed. Appl.*, **14**, 469 (1981).
[566] M. Olling, R. W. Stephany and A. G. Rauws, *Dev. Anim. Vet. Sci.*, **6**, 335 (1980).
[567] B. K. Larsson, *Z. Lebensm. Unters. Forsch.*, **174**, 101 (1982).
[568] Y. Ju, X. Sun and J. Kong, *Fenxi Huaxue*, **10**, 21 (1982).
[569] C. Manfredi and L. Zinterhofer, *Clin. Chem.*, **28**, 246 (1982).
[570] A. Kumps and Y. Mardens, *Clin. Chem.*, **26**, 1759 (1980).
[571] D. M. Coulson and L. A. Cavanagh, *Anal. Chem.*, **32**, 1245 (1960).
[572] R. C. Hall, *J. Chromatog. Sci.*, **12**, 152 (1974).
[573] B. E. Pape, D. H. Rodgers and T. C. Flynn, *J. Chromatog.*, **134**, 1 (1977).
[574] E. von Rappard, G. Eisenbrand and R. Preussmann, *J. Chromatog.*, **124**, 247 (1976).
[575] J. F. Lawrence, *Intern. J. Environ. Anal. Chem.*, **5**, 95 (1978).
[576] J. F. Lawrence, *J. Chromatog.*, **123**, 287 (1976).
[577] J. F. Lawrence, *J. Assoc. Off. Anal. Chem.*, **59**, 1061 (1976).
[578] M. Galoux, J. C. Van-Damme, A. Bernes and J. Potvin, *J. Chromatog.*, **177**, 245 (1979).
[579] B. E. Pape and M. A. Ribick, *J. Chromatog.*, **136**, 127 (1977).
[580] R. C. Dressman and E. F. McFarren, *J. Chromatog. Sci.*, **15**, 69 (1977).
[581] V. Lopez-Avila and R. Northcutt, *J. High Resolut. Chromatog., Chromatog. Commun.*, **5**, 67 (1982).
[582] S. S. Brody and J. E. Chaney, *J. Gas Chromatog.*, **4**, 42 (1966).
[583] N. K. Narain, M. Hanif, M. A. Latheef and C. C. Lewis, *Anal. Lett.*, **13**, 213 (1980).
[584] R. G. Deo and P. H. Howard, *J. Assoc. Off. Anal. Chem.*, **61**, 210 (1978).
[585] J. M. Zehner and R. A. Simonaitis, *J. Chromatog. Sci.*, **14**, 348 (1976).
[586] T. Hamano, Y. Mitsuhashi and Y. Matsuki, *Agr. Biol. Chem.*, **45**, 2237 (1981).
[587] R. H. Bromilow and K. A. Lord, *J. Chromatog.*, **125**, 495 (1976).
[588] J. H. Onley, *J. Assoc. Off. Anal. Chem.*, **60**, 1111 (1977).
[589] M. C. Ivey, *J. Assoc. Off. Anal. Chem.*, **59**, 261 (1976).

[590] A. N. Sagredos and W. R. Eckert, *Beitr. Tabaksforsch.*, **9**, 107 (1977).
[591] M. J. Brown, *J. Agr. Food Chem.*, **29**, 1129 (1981).
[592] T. L. Peppard and J. M. F. Douse, *J. Chromatog.*, **176**, 444 (1979).
[593] M. C. Bowman, *J. Chromatog. Sci.*, **13**, 307 (1975).
[594] K. J. Strankowski and C. W. Stanley, *J. Agr. Food Chem.*, **29**, 1034 (1981).
[595] J. B. Sills and C. W. Luhning, *J. Assoc. Off. Anal. Chem.*, **60**, 961 (1977).
[596] L. Laitem and P. Gaspar, *J. Chromatog.*, **140**, 266 (1977).
[597] Y. Hoshika, *J. Chromatog.*, **237**, 439 (1982).
[598] H. Bargnoux, D. Pepin, J. L. Chabard, F. Vedrine, J. Petit and J. A. Berger, *Analusis*, **6**, 107 (1978).
[599] B. Ross and A. J. Harvey, *J. Agr. Food Chem.*, **29**, 1095 (1981).
[600] L. Laitem, I. Bello and P. Gaspar, *J. Chromatog.*, **156**, 327 (1978).
[601] L. H. Zakitis and E. M. McCray, *Bull. Environ. Contam. Toxicol.*, **28**, 334 (1982).
[602] A. Loh, S. D. West and T. D. Macy, *J. Agr. Food Chem.*, **26**, 410 (1978).
[603] M. Imanaka, J. Matsunaga and T. Ishida, *Shokuhin Eiseigaku Zasshi.*, **22**, 472 (1981).
[604] S. O. Farwell and R. A. Rasmussen, *J. Chromatog. Sci.*, **14**, 224 (1976).
[605] M. Maryama and M. Kakemoto, *J. Chromatog. Sci.*, **16**, 1 (1978).
[606] S. Hasinski, *J. Chromatog.*, **119**, 207 (1976).
[607] P. L. Patterson, R. L. Howe and A. Abu-Shumays, *Anal. Chem.*, **50**, 339 (1978).
[608] S. Kapila and C. R. Vogt, *J. Chromatog. Sci.*, **17**, 327 (1979).
[609] C. R. Vogt and S. Kapila, *J. Chromatog. Sci.*, **17**, 546 (1979).
[610] A. Karmen and L. Giuffrida, *Nature*, **201**, 1204 (1964).
[611] B. Kolb and J. Bischoff, *J. Chromatog. Sci.*, **12**, 625 (1974).
[612] B. Kolb, M. Auer and P. Rospisil, *J. Chromatog. Sci.*, **15**, 53 (1977).
[613] S. Y. Szeto, J. Yee, M. J. Brown and P. C. Oloffs, *J. Chromatog.*, **240**, 526 (1982).
[614] G. B. M. Gawell, *Analyst*, **104**, 106 (1979).
[615] R. S. Marano, S. P. Levine and T. M. Harvey, *Anal. Chem.*, **50**, 1948 (1978).
[616] M. E. Brown, C. V. Breder and T. P. McNeal, *J. Assoc. Off. Anal. Chem.*, **61**, 1383 (1978).
[617] G. B. M. Gawell, *Analyst*, **104**, 106 (1979).
[618] K. Jacob, C. Falkner and W. Vogt., *J. Chromatog.*, **167**, 67 (1978).
[619] J. Vasiliades and K. C. Bush, *Anal. Chem.*, **48**, 1708 (1976).
[620] H. B. Hucker and S. C. Stauffer, *J. Chromatog.*, **138**, 437 (1977).
[621] J. Kristinson, *Acta Pharmacol. Toxicol.*, **49**, 390 (1981).
[622] M. Bonino, F. Mokofio and S. Barazi, *J. Chromatog.*, **224**, *Biomed. Appl.*, **13**, 332 (1981).
[623] M. Polgar and L. Vereczky, *J. Chromatog.*, **241**, 29 (1982).
[624] G. Becher, *J. Chromatog.*, **211**, 103 (1981).
[625] P. Lombardo and I. J. Egry, *J. Assoc. Off. Anal. Chem.*, **62**, 47 (1979).
[626] D. J. Caverly and R. C. Denney, *Analyst*, **102**, 576 (1977).
[627] M. Drygas, *Chem. Anal. (Warsaw)*, **22**, 517 (1977).
[628] E. T. Delbeke and M. Debackere, *J. Chromatog.*, **237**, 344 (1982).
[629] S. Y. Szeto and K. M. S. Sundaram, *J. Chromatog.*, **200**, 179 (1980).
[630] K. M. Sundaram, S. Y. Szeto and R. Hindle, *J. Chromatog.*, **177**, 29 (1979).
[631] R. R. Chambers, *J. Chromatog.*, **154**, 272 (1978).
[632] A. R. Viala, E. Deturmeny, M. Estadieu, A. Durand and J. P. Cano, *J. Chromatog.*, **224**, *Biomed. Appl.*, **13**, 503 (1981).
[633] J. E. O'Brien, O. Hinsvark, W. Bryant, L. Amsel and F. E. Leaders, *Anal. Lett.*, **10**, 1163 (1977).
[634] H. L. J. M. Fleuren and J. M. van Rossum, *J. Chromatog.*, **152**, 41 (1978).
[635] K. Kigasawa, M. Tanaka, H. Shimizu and M. Saito, *Yakugaku Zasshi*, **102**, 343 (1982).
[636] P. B. M. W. M. Timmermans, A. Brands and P. A. van Zwieten, *J. Chromatog.*, **144**, 215 (1977).
[637] B. H. Dvorchik, S. H. Miller and W. P. Graham, *J. Chromatog.*, **135**, 141 (1977).

[638] M. J. Kogan, K. G. Verebey, A. C. dePace, R. B. Resnick and S. J. Mule, *Anal. Chem.*, **49**, 1965 (1977).
[639] H. B. Hucker and S. C. Stauffer, *J. Chromatog.*, **124**, 164 (1976).
[640] T. Facchinetti, M. d'Incalci, G. Martelli, L. Cantoni, G. Belvedere and M. Salmona, *J. Chromatog.*, **145**, 315 (1978).
[641] N. Van den Bosch, O. Driessen, A. Emonds, A. T. van Oosterom, P. J. A. Timmermans, D. de Vos and P. H. T. J. Slee, *Methods Find. Exp. Clin. Pharmacol.*, **3**, 377 (1981).
[642] A. Zelleke, G. C. Martin and J. M. Labavitch, *J. Am. Soc. Hortic. Sci.*, **105**, 50 (1980).
[643] J. E. O'Brien, O. N. Hinsvark, W. R. Newman, L. P. Amsel, J. E. Giering and F. E. Leader, in *Trace Organic Analysis*, pp. 481–485. NBS, Washington (1979).
[644] V. Rovei, M. Mitchard and P. L. Morselli, *J. Chromatog.*, **138**, 391 (1977).
[645] H. C. Thompson, C. L. Holder and M. C. Bowman, *J. Chromatog. Sci.*, **20**, 373 (1982).
[646] M. H. K. Abdel-Kader and G. R. B. Webster, *Intern. J. Environ. Anal. Chem.*, **11**, 153 (1982).
[647] J. I. Javaid, H. Dekirmenjian, U. Liskevych, R. L. Lin and J. M. Davis, *J. Chromatog. Sci.*, **19**, 439 (1981).
[648] M. M. Ames and G. Powis, *J. Chromatog.*, **174**, 245 (1979).
[649] Y. Hoshika, *J. Chromatog. Sci.*, **19**, 444 (1981).
[650] J. Gal, M. D. Freedman, E. Kumar and C. R. Freed, *Ther. Drug Monit.*, **3**, 177 (1981).
[651] M. E. Kruczek, *J. Pharmacol. Methods*, **5**, 137 (1981).
[652] A. Sioufi and A. Richard, *J. Chromatog.*, **221**, *Biomed. Appl.*, **10**, 393 (1980).
[653] A. R. Viala, J. P. Cano, A. G. Durand, T. Erlenmaier and R. M. Garreau, *Anal. Chem.*, **49**, 2354 (1977).
[654] P. Jacob, J. F. Rigod, S. M. Pond and N. L. Benowitz, *J. Anal. Toxicol.*, **5**, 292 (1981).
[655] S. G. Dahl, T. Bratlid and O. Lingjaerde, *Ther. Drug Monit.*, **4**, 81 (1982).
[656] J. N. Ogata, K. H. Yanagihara, J. W. Hylin and A. Bevenue, *J. Chromatog.*, **157**, 401 (1978).
[657] T. A. Wehner and J. N. Seiber, *J. High Resolut. Chromatog. Chromatog. Commun.*, **4**, 348 (1981).
[658] J. J. Keyzer, B. G. Wolthers, H. Breukelman, H. F. Kaufman and J. G. R. De-Monchy, *Clin. Chim. Acta*, **121**, 379 (1982).
[659] H. V. Street, W. Vycudilik and G. Machata, *J. Chromatog.*, **168**, 117 (1978).
[660] G. M. Frame, G. A. Flanigan and D. C. Carmody, *J. Chromatog*, **168**, 365 (1978).
[661] S. F. Chang, C. S. Hansen, J. M. Fox and R. E. Ober, *J. Chromatog.*, **226**, *Biomed. Appl.*, **15**, 79 (1981).
[662] C. Feyerabend, T. Levitt and M. A. H. Russell, *J. Pharm. Pharmacol.*, **27**, 434 (1975).
[663] N. Hengen and M. Hengen, *Clin. Chem.*, **24**, 50 (1978).
[664] M. J. Kogan, K. Vereby, J. H. Jaffee and S. J. Mule, *J. Forens. Sci.*, **26**, 6 (1981).
[665] Z. H. Chen, *Fenxi Huaxue*, **9**, 631 (1981).
[666] F. Yin, J. H. Ding and S. L. Liu, *Anal. Lett.*, **14**, 977 (1981).
[667] W. J. Chamberlain and R. F. Arrendale, *J. Chromatog.*, **234**, 478 (1982).
[668] E. Bailey, M. Fenoughty and L. Richardson, *J. Chromatog.*, **131**, 347 (1977).
[669] A. Zeman and K. Koch, *J. Chromatog.*, **216**, 199 (1981).
[670] G. L. LeBel, D. T. Williams, G. Griffiths and F. M. Benoit, *J. Assoc. Off. Anal. Chem.*, **62**, 241 (1979).
[671] J. Hild, E. Schulte and H. P. Thier, *Chromatographia*, **11**, 397 (1978).
[672] M. Wolf, R. Deleu and A. Copin, *J. High Resolut. Chromatog. Chromatog. Commun.*, **4**, 346 (1981).
[673] J. J. M. Labout, C. T. Thijssen and W. Hespe, *J. Chromatog.*, **144**, 201 (1977).
[674] M. Bertrand, C. Dupuis, M. A. Gagnon and R. Dugal, *J. Chromatog.*, **171**, 377 (1979).

[675] D. E. Bradway, E. M. Lores and T. R. Edgerton, *Residue Rev.*, **75**, 51 (1980).
[676] J. Tse and K. Chan, *Methods Find. Exp. Clin. Pharmacol.*, **3**, 99 (1981).
[677] P. Jacob, J. F. Rigod, S. M. Pond and N. L. Benowitz, *J. Pharm. Sci.*, **71**, 166 (1982).
[678] N. P. E. Vermeulen, D. de Roode and I. D. D. Breimer, *J. Chromatog.*, **137**, 333 (1977).
[679] F. L. Vandermark and R. F. Adams, *Clin. Chem.*, **22**, 1062 (1976).
[680] V. Radulescu, S. Budrugeac and R. Dragan, *Rev. Chim. (Bucharest)*, **31**, 492 (1980).
[681] I. P. Nesterova, *Zh. Analit. Khim.*, **32**, 1790 (1977).
[682] M. Thoma, L. Farinello and A. Mueller, *Arzneim. Forsch.*, **31**, 1020 (1981).
[683] K. Dean, G. Land and A. Bye, *J. Chromatog.*, **221**, *Biomed. Appl.*, **10**, 408 (1980).
[684] A. G. DeBoer, D. D. Breimer and J. M. Gubbens-Stibbe, *Pharm. Weekbl., Sci. Ed.*, **2**, 1105 (1980).
[685] F. Kuhlmann, *Z. Lebensm. Unters. Forsch.*, **173**, 35 (1981).
[686] E. Bailey and E. J. Barron, *J. Chromatog.*, **138**, *Biomed. Appl.*, **9**, 25 (1980).
[687] R. E. Chambers, *J. Chromatog.*, **171**, 473 (1979).
[688] N. Totir, S. Marchidan, C. Volanschi, N. Cimpoeru and R. Andrei, *Rev. Chim. (Bucharest)*, **27**, 523 (1976).
[689] A. Colas, J. Royer and R. Simon, *Analusis*, **3**, 355 (1975).
[690] E. Matisová, J. Krupčik and O. Liška, *J. Chromatog.*, **173**, 139 (1979).
[691] T. Gabrio and D. Ennet, *Pharmazie*, **37**, 375 (1982).
[692] H. Roseboom and H. A. Herbold, *J. Chromatog.*, **202**, 431 (1980).
[693] D. E. Bradway and R. F. Moseman, *J. Agr. Food Chem.*, **30**, 244 (1982).
[694] D. R. Abernethy, D. J. Greenblatt and R. I. Shader, *Pharmacology*, **95**, 57 (1981).
[695] P. Volin, *Clin. Chem.*, **27**, 1785 (1981).
[696] V. Rovei, M. Sanjuan and P. D. Hrdina, *J. Chromatog.*, **182**, *Biomed. Appl.*, **8**, 349 (1980).
[697] R. M. H. Roscoe, J. K. Cooper, E. M. Hawes and K. K. Midha, *J. Pharm. Sci.*, **71**, 625 (1982).
[698] A. Rogstad and M. Yndestad, *J. Chromatog.*, **216**, 350 (1981).
[699] T. J. Hansen, M. C. Archer and S. R. Tannenbaum, *Anal. Chem.*, **51**, 1526 (1979).
[700] I. S. Krüll, E. U. Goff, G. G. Hoffmann and D. H. Fine, *Anal. Chem.*, **51**, 1706 (1979).
[701] T. Y. Fan, R. Ross, D. H. Fine, L. Keith and A. W. Garrison, *Environ. Sci. Technol.*, **12**, 692 (1978).
[702] J. H. Hotchkiss, J. F. Barbour, L. M. Libbey and R. A. Scanlan, *J. Agr. Food Chem.* **26**, 884 (1978).
[703] G. V. Alliston, K. S. Webb and T. A. Gough, *J. Chromatog.*, **175**, 194 (1979).
[704] D. M. Parees, *Anal. Chem.*, **51**, 1675 (1979).
[705] J. H. Hotchkiss, D. C. Havery and T. Fazio, *J. Assoc. Off. Anal. Chem.*, **64**, 929 (1981).
[706] Institute of Brewing Analysis Committee, *J. Inst. Brew.*, **88**, 266 (1982).
[707] E. U. Goff, J. R. Coombs, D. H. Fine and T. M. Baines, *Anal. Chem.*, **52**, 1833 (1980).
[708] K. S. Markl and R. A. Latimer, *J. Am. Soc. Brew. Chem.*, **39**, 59 (1981).
[709] N. P. Sen, S. Seaman and M. Bickis, *J. Assoc. Off. Anal. Chem.*, **65**, 720 (1982).
[710] N. P. Sen and S. Seaman, *J. Assoc. Off. Anal. Chem.*, **64**, 933 (1981).
[711] N. P. Sen and S. Seaman, *J. Assoc. Off. Anal. Chem.*, **64**, 1238 (1981).
[712] D. Kuehne and A. Mirna, *Fleischwirtsch.*, **61**, 111 (1981).
[713] D. M. Parees and S. R. Prescott, *J. Chromatog.*, **205**, 429 (1981).
[714] D. B. Black, R. C. Lawrence, E. G. Lovering and J. R. Watson, *J. Assoc. Off. Anal. Chem.*, **64**, 1474 (1981).
[715] D. Klein, A. M. Girard, J. De Smedt, Y. Fellion and G. Debry, *Food Cosmet. Toxicol.*, **19**, 233 (1981).
[716] J. L. Ho, H. H. Wisneski and R. L. Yates, *J. Assoc. Off. Anal. Chem.*, **64**, 800 (1981).

[717] J. Kann, O. Tauts and A. Lettner, *Vopr. Pitan.*, 65 (1981).
[718] E. Pedersen and I. Meyland, *Z. Lebensm. Unters. Forsch.* **173**, 359 (1981).
[719] T. Maki, *Bull. Environ. Contam. Toxicol.*, **25**, 751 (1980).
[720] N. P. Sen and S. Seaman, *J. Agr. Food Chem.*, **30**, 364 (1982).
[721] W. I. Kimoto, J. W. Pensabene and W. Fiddler, *J. Agr. Food Chem.*, **30**, 757 (1982).
[722] S. v. Heusden and L. P. J. Hoogeveen, *Z. Anal. Chem.*, **282**, 307 (1976).
[723] W. A. McClenny, B. E. Martin, R. W. Baumgardner, R. K. Stevens and A. E. O'Keeffe, *Environ. Sci. Technol.*, **10**, 810 (1976).
[724] E. A. Hill, J. K. Nelson and J. W. Birks, *Anal. Chem.*, **54**, 541 (1982).
[725] D. G. Sutton, K. R. Westberg and J. E. Melzer, *Anal. Chem.*, **51**, 1399 (1979).
[726] R. Belcher, S. L. Bogdanski and A. Townshend, *Anal. Chim. Acta*, **67**, 1 (1973).
[727] R. Belcher, S. L. Bogdanski, M. Burguera, E. Henden and A. Townshend, *Anal. Chim. Acta*, **100**, 515 (1978).
[728] A. L. McCormack, S. C. Tong and W. D. Cook, *Anal. Chem.*, **37**, 1470 (1965).
[729] C. A. Bache and D. I. Lisk, *Anal. Chem.*, **39**, 786 (1967).
[730] B. D. Quimby, M. F. Delaney, P. C. Uden and R. M. Barnes, *Anal. Chem.*, **51**, 875 (1979).
[731] G. Chevrier, T. Hanai, K. C. Tran and J. Hubert, *Can. J. Chem.*, **60**, 898 (1982).
[732] H. A. Dingjan and H. J. De Jong, *Spectrochim. Acta*, **36B**, 325 (1981).
[733] I. S. Krull and S. Jordan, *Intern. Lab.*, **10**, No. 8, 13 (1980).
[734] K. Tanabe, H. Haraguchi and K. Fuwa, *Spectrochim. Acta*, **36B**, 633 (1981).
[735] J. W. Miller, P. C. Uden and R. M. Barnes, *Anal. Chem.*, **54**, 485 (1982).
[736] K. J. Mulligan, J. A. Caruso and F. L. Fricke, *Analyst*, **105**, 1060 (1980).
[737] D. A. Leathard and B. C. Shurlock, *Identification Techniques in Gas Chromatography*, Wiley, New York (1970).
[738] N. E. Hester and R. A. Meyer, *J. Air Poll. Control. Assoc.*, **29**, 107 (1979).
[739] M. L. Langhorst and T. J. Nestrick, *Anal. Chem.*, **51**, 2018 (1979).
[740] F. F. Andrawes, E. K. Gibson, Jr. and D. A. Bafus, *Anal. Chem.*, **52**, 1377 (1980).
[741] V. B. Stein and R. S. Narang, *Anal. Chem.*, **54**, 991 (1982).
[742] A. N. Freedman and B. D. H. Worker, *Intern. Environ. Saf.*, 10 (1980).
[743] G. I. Senum, *J. Chromatog.*, **205**, 413 (1981).
[744] Y. Tomita, M. H. Ho and G. G. Guilbault, *Anal. Chem.*, **51**, 1475 (1979).
[745] M. H. Ho, G. G. Guilbault and B. Rietz, *Anal. Chem.*, **52**, 1489 (1980).
[746] T. E. Edmonds and T. S. West, *Anal. Chim. Acta*, **117**, 147 (1980).
[747] J. F. Alder and C. A. Isaac, *Anal. Chim. Acta*, **129**, 163 (1981).
[748] J. F. Alder and C. A. Isaac, *Anal. Chim. Acta*, **129**, 175 (1981).
[749] K. Camman, *Z. Anal. Chem.*, **287**, 1 (1977).
[750] G. G. Guilbault, *Enzyme Microbiol. Technol.*, **2**, 258 (1980).
[751] M. Lavallee, O. F. Schanne and N. C. Herbert, *Glass Microelectrodes*, Wiley, New York (1969).
[752] J. L. Walker, *Anal. Chem.*, **43**, 69A (1971).
[753] L. R. Pucacco and N. W. Carter, *Anal. Biochem.*, **89**, 151 (1978).
[754] S. M. Donahe, G. E. Janauer and T. D. Zucconi, *Anal. Lett.*, **11**, 721 (1978).
[755] J. G. Schiller, A. K. Chen and C. C. Liu, *Anal. Biochem.*, **85**, 25 (1978).
[756] B. G. Snider and D. C. Johnson, *Anal. Chim. Acta*, **106**, 1 (1979).
[757] E. Bishop, *Coulometric Analysis*, Elsevier, Amsterdam (1975).
[758] W. C. Purdy, *Electroanalytical Methods in Biochemistry*, McGraw-Hill, New York (1965).
[759] G. Dryhurst, *Electrochemistry of Biological Molecules*, Academic Press, New York (1977).
[760] C. M. Arcand and E. H. Swift, *Anal. Chem.*, **28**, 440 (1965).
[761] K. Beyermann, unpublished work.
[762] K. Tóth, G. Nagy, Z. Fehér and E. Pungor, *Z. Anal. Chem.*, **282**, 379 (1976).
[763] U. Fritschi, G. Fritschi and H. Kussmaul, *Z. Wasser und Abwasser-Forsch.*, **11**, 165 (1978).
[764] A. Cedergren and S. A. Frederikson, *Talanta*, **23**, 217 (1976).

[765] U. R. Tjaden, J. Lankelma, H. Poppe and R. G. Muusze, *J. Chromatog.*, **125,** 275 (1976).
[766] M. S. Greenberg and W. J. Mayer, *J. Chromatog.*, **169,** 321 (1979).
[767] D. E. Ott, *J. Assoc. Off. Anal. Chem.*, **61,** 1465 (1978).
[768] M. A. Brooks and J. A. F. de Silva, *Talanta,* **22,** 849 (1975).
[769] G. A. Scratchley, A. N. Masoud, S. J. Stohs and D. W. Wingard, *J. Chromatog.*, **169,** 313 (1979).
[770] M. R. Smyth and W. F. Smyth, *Analyst,* **103,** 529 (1978).
[771] R. R. Rowe, *Proc. Anal. Div. Chem. Soc.*, **13,** 232 (1976).
[772] P. T. Kissinger, C. S. Bruntlett, K. Bratin and J. R. Rice, in *Trace Organic Analysis,* pp. 705–712. NBS, Washington (1979).
[773] C. Bollet, M. Caude and R. Rosset, *Analusis,* **6,** 54 (1978).
[774] W. R. Heineman and P. T. Kissinger, *Anal. Chem.*, **50,** 166R (1978).
[775] P. T. Kissinger, *Anal. Chem.*, **49,** 447A (1977).
[776] K. Brunt, *Pharm. Weekbl.*, **113,** 689 (1978).
[777] B. Fleet and C. J. Little, *J. Chromatog. Sci.*, **12,** 747 (1974).
[778] H. B. Hanekamp, W. H. Voogt, P. Bos and R. W. Frei, *Anal. Lett.*, **12,** 175 (1979).
[779] H. B. Hanekamp, P. Bos, U. A. Th. Brinkman and R. W. Frei, *Z. Anal. Chem.*, **297,** 404 (1979).
[780] W. Lund, M. Hannisdal and T. Greibrokk, *J. Chromatog.*, **173,** 249 (1979).
[781] M. Lemar and M. Porthault, *J. Chromatog.*, **130,** 372 (1977).
[782] L. J. Felice, C. S. Bruntlett, R. E. Shoup and P. T. Kissinger, *in Trace Organic Analysis,* pp. 391–397. NBS, Washington (1979).
[783] Y. Yui, M. Kimura, Y. Itokawa and C. Kawai, *J. Chromatog.*, **177,** 376 (1979).
[784] J. Wagner, M. Palfreyman and M. Zraika, *J. Chromatog.*, **164,** 41 (1979).
[785] R. E. Shoup and P. T. Kissinger, *Clin. Chem.*, **23,** 1268 (1977).
[786] G. A. Scratchley, A. N. Masoud, S. J. Stohs and D. W. Wingard, *J. Chromatog.*, **169,** 313 (1979).
[787] S. Sasa and C. L. Blank, *Anal. Chim. Acta,* **104,** 29 (1979).
[788] R. J. Fenn, S. Siggia and D. J. Curran, *Anal. Chem.*, **50,** 1067 (1978).
[789] A. N. Strohl and D. J. Curran, *Anal. Chem.*, **51,** 1045 (1979).
[790] J. L. Ponchon, R. Cespuglio, F. Gonon, M. Jouvet and J. F. Pujol, *Anal. Chem.*, **51,** 1483 (1979).
[791] L. Sasa and C. L. Blank, *Anal. Chem.*, **49,** 354 (1977).
[792] L. J. Felice and P. T. Kissinger, *Clin. Chim. Acta,* **76,** 317 (1977).
[793] C. Hansson and E. Rosengren, *Anal. Lett.*, **11,** 901 (1978).
[794] R. Samuelsson, *Anal. Chim. Acta,* **102,** 133 (1978).
[795] K. Hasebe and J. Osteryoung, *Anal. Chem.*, **47,** 2412 (1975).
[796] R. Samuelson, *Anal. Chim. Acta,* **108,** 213 (1979).
[797] M. R. Smyth, J. G. Osteryoung, P. G. Rowley and S. J. Weininger, *Z. Anal. Chem.*, **298,** 17 (1979).
[798] L. Michel and A. Zatka, *Anal. Chim. Acta,* **105,** 109 (1979).
[799] T. Wasa and S. Musha, *Bull. Chem. Soc. Japan,* **48,** 2176 (1975).
[800] J. P. Hart, W. F. Smyth and B. J. Birch, *Proc. Anal. Div. Chem. Soc.*, **13,** 336 (1976).
[801] T. S. Ma, M. R. Hackman and M. A. Brooks, *Mikrochim. Acta,* 617 (1975 **II**).
[802] C. V. Puglisi, J. C. Meyer, L. D'Arconte. M. A. Brooks and J. A. F. de Silva, *J. Chromatog.*, **145,** 81 (1978).
[803] M. A. Brooks and M. R. Hackman, *Anal. Chem.*, **47,** 2059 (1975).
[804] W. Lund, M. Hannisdal and T. Greibrokk, *J. Chromatog.*, **173,** 249 (1979).
[805] W. J. M. Underberg, A,. J. F. Ebskamp and J. M. Pillen, *Z. Anal. Chem.*, **287,** 296 (1977).
[806] F. Pellérin, J. F. Letavernier and N. Chama, *Analusis,* **5,** 19 (1977).
[807] G. Franke, W. Pietrulla and K. Preussner, *Z. Anal. Chem.*, **298,** 38 (1979).
[808] D. D. Koch and P. T. Kissinger, *J. Chromatog.*, **164,** 441 (1979).
[809] L. A. Pachla and P. T. Kissinger, *Anal. Chim. Acta,* **88,** 385 (1977).
[810] M. S. Greenberg and W. J. Mayer, *J. Chromatog.*, **169,** 321 (1979).

[811] D. N. Armentrout, J. D. McLean and M. W. Long, *Anal. Chem.*, **51**, 1039 (1979).
[812] D. E. Ott, *J. Assoc. Off. Anal. Chem.*, **61**, 1465 (1978).
[813] A. L. Wade, F. M. Hawkridge and H. P. Williams, *Anal. Chim. Acta*, **105**, 91 (1979).
[814] E. M. Lores, D. W. Bristol and R. F. Moseman, *J. Chromatog. Sci.*, **16**, 358 (1978).
[815] F. Kusano, H. Kawasaki, K. Itadani and K. Hagiwara, *Bunseki Kagaku*, **27**, 687 (1978).
[816] L. A. Zatuchnaya, V. A. Kremer, G. F. Tsvindina, V. A. Kaminskaya and N. A. Benedis, *Zh. Analit. Khim.*, **32**, 1431 (1977).
[817] C. A. Mairesse-Ducarmois, G. J. Patriarche and J. L. Vandenbalck, *Anal. Chim. Acta*, **84**, 47 (1976).
[818] H. Sohr and K. Wienhold, *J. Electroanal. Chem.*, **68**, 113 (1976).
[819] R. Eggli and R. Asper, *Anal. Chim. Acta*, **101**, 253 (1978).
[820] P. Nangniot, L. Zenon-Roland and M. Berlemont-Frennet, *Analusis*, **6**, 273 (1978).
[821] G. K. Budnikov, G. S. Supin, N. A. Ulachovic and N. K. Sakurova, *Zh. Analit. Khim.*, **30**, 2275 (1975).
[822] I. Ya. Kheifets, N. A. Sobina and N. A. Romanov, *Zh. Analit. Khim.*, **32**, 1984 (1977).
[823] J. Davidek, M. Neméthová and J. Seifert, *Z. Anal. Chem.*, **287**, 286 (1977).
[824] M. R. Smyth, D. W. Lawellin and J. G. Osteryoung, *Analyst*, **104**, 73 (1979).
[825] A. G. Fogg, N. M. Fayad, C. Burgess and A. McGlynn, *Anal. Chim. Acta*, **108**, 205 (1979).
[826] J. Lankelma and H. Poppe, *J. Chromatog. Sci.*, **14**, 310 (1976).
[827] Z. Kozarac, V. Žutic and B. Cosovic, *Tenside*, **13**, 260 (1976).
[828] B. J. A. Haring and W. v. Delft, *Anal. Chim. Acta*, **94**, 201 (1977).
[829] K. Linhart, *Dtsch. Lebensmittel Rundsch.*, **71**, 417 (1975).
[830] J. F. Coetzee, G. H. Kazi and J. C. Spurgeon, *Anal. Chem.*, **48**, 2170 (1976).
[831] N. A. Sobina, L. Y. Kheifets and N. A. Romanov, *Zh. Analit. Khim.*, **33**, 137 (1978).
[832] S. R. Betso and J. D. McLean, *Anal. Chem.*, **48**, 766 (1976).
[833] M. W. White, *J. Chromatog.*, **178**, 229 (1979).
[834] G. Sontag and K. Kral, *Z. Anal. Chem.*, **294**, 278 (1979).
[835] L. A. Sternson and G. Thomas, *Anal. Lett.*, **10**, 99 (1977).
[836] J. E. Smith, N. R. Passarela and J. C. Wyckoff, *J. Assoc. Off. Anal. Chem.*, **60**, 1310 (1977).
[837] P. Hocquellet and C. Lespagne, *Analusis*, **6**, 215 (1978).
[838] J. W. Munson, R. Weierstall and H. B. Kostenbauder, *J. Chromatog.*, **145**, 328 ((1978).
[839] M. M. Filimonova, V. E. Gortunova and B. F. Filimonov, *Zh. Analit. Khim.*, **32**, 140 (1977).
[840] Yu. S. Ignat'ev, M. P. Strukova and V. V. Dolbanova, *Zh. Analit. Khim.*, **29**, 1434 (1974).
[841] P. T. Kissinger, K. Bratin, G. C. Davis and L. A. Pachla, *J. Chromatog. Sci.*, **17**, 137 (1979).
[842] M. A. Carey and H. E. Persinger, *J. Chromatog. Sci.*, **10**, 537 (1972).
[843] A. D. Smith and J. B. Jepson, *Anal. Biochem.*, **18**, 36 (1976).
[844] C. Penaino and A. E. Harper, *Anal. Chem.*, **33**, 1863 (1961).
[845] E. Grushka, H. D. Durst and E. J. Kitka, *J. Chromatog.*, **112**, 673 (1975).
[846] D. R. Knapp and S. Krueger, *Anal. Lett.*, **8**, 603 (1975).
[847] F. A. Fitzpatrick, S. Siggia and J. Dingman, *Anal. Chem.*, **44**, 221 (1972).
[848] L. J. Papa and L. P. Turner, *J. Chromatog. Sci.*, **10**, 747 (1972).
[849] W. R. Heinemann and P. T. Kissinger, *Anal. Chem.*, **50**, 166R (1978).
[850] C. Merritt, Jr., in L. S. Ettre and W. H. McFadden (eds.), *Ancillary Techniques of Gas Chromatography*, p. 337. Wiley, New York (1969).
[851] B. Casu and L. Cavalotti, *Anal. Chem.*, **34**, 1514 (1962).
[852] T. Jupille, *J. Chromatog. Sci.*, **17**, 160 (1979).
[853] G. Stanley, *J. Chromatog.*, **178**, 487 (1979).
[854] G. Stanley and B. H. Kennett, *J. Chromatog.*, **75**, 304 (1973).

[855] F. B. Whitfield, G. Stanley and K. E. Murray, *Tetrahedron Lett.*, 95 (1973).
[856] D. R. Adams, S. P. Bhatnagar, R. C. Cookson. G. Stanley and F. B. Whitfield, *J. Chem. Soc., Chem. Commun.*, 469 (1974).
[857] G. D. Prestwich, F. B. Whitfield and G. Stanley, *Tetrahedron*, **32**, 2945 (1976).
[858] B. S. Goodrich, E. R. Hesterman, K. E. Murray. R. Mykotowycz, G. Stanley and G. Sugowdz, *J. Chem. Ecol.*, **4**, 581 (1978).
[859] I. Klimes, W. Stünzi and D. Lamparsky, *J. Chromatog.*, **136**, 23 (1977).
[860] I. Klimes, W. Stünzi and D. Lamparsky, *J. Chromatog.*, **166**, 469 (1978).
[861] H. U. Bergmeyer, *Methoden der enzymatischen Analyse*, 2 Vols. Verlag Chemie, Weinheim (1974).
[862] O. H. Lowry, J. V. Passoneau, D. W. Schulz and K. M. Rock, *J. Biol. Chem.*, **236**, 2746 (1962).
[863] G. Guilbault, *Enzymatic Methods of Analysis*, Pergamon Press, Oxford (1970).
[864] K. H. Büchel (ed.), *Pflanzenschutz und Schädlingsbekämpfung*, Thieme, Stuttgart (1977).
[865] E. A. Meighen, N. Slessor and G. G. Grant, *Experientia*, **37**, 555 (1981).
[866] H. Adlercreutz and M. Harkonen, *J. Steroid Biochem.*, **13**, 507 (1980).
[867] Y. S. Shin-Buehring, R. Rasshofer and W. Endres, *J. Inherited Metab. Diseases*, **4**, 123 (1981).
[868] L. P. McCloskey and P. Mahaney, *Am. J. Enol. Vitic.*, **32**, 159 (1981).
[869] S. U. Bhaskar, *Talanta*, **29**, 133 (1982).
[870] H. Breuer, *J. Chromatog.*, **243**, 183 (1982).
[871] L. Bauce, J. A. Thornhill, K. E. Cooper and W. L. Veale, *Life Sci.*, **27**, 1921 (1980).
[872] S. Suzaki, K. Morita, T. Ohuchi and M. Oka, *Igaku no Ayumi*, **113**, 85 (1980).
[873] M. J. Brown and D. A. Jenner, *Clin. Sci.*, **61**, 591 (1981).
[874] S. Demassieux, L. Corneille, S. Lachance and S. Carriere, *Clin. Chim. Acta*, **115**, 377 (1981).
[875] J. Bosak, E. Knoll, D. Ratge and H. Wisser, *J. Clin. Chem. Clin. Biochem.*, **18**, 413 (1980).
[876] G. Koch, U. Johansson and E. Arvidson, *J. Clin. Chem. Clin. Biochem.*, **18**, 367 (1980).
[877] K. Kobayashi, A. Foti, V. DeQuattro, R. Kolloch and L. Miano, *Clin. Chim. Acta*, **107**, 163 (1980).
[878] G. A. Johnson, C. A. Baker and R. T. Smith, *Life Sci.*, **26**, 1591 (1980).
[879] M. J. Brown and C. T. Dollery, *Br. J. Clin. Pharmacol.*, **11**, 79 (1981).
[880] L. Guilloux, D. Hartmann and G. Ville, *Clin. Chim. Acta*, **116**, 269 (1981).
[881] H. T. Kernes, *Clin. Chem.*, **27**, 249 (1981).
[882] B. L. Allman, *Clin. Chem.*, **27**, 1176 (1981).
[883] W. Heineman, C. W. Anderson and H. B. Halsall, *Science*, **204**, 865 (1979).
[884] S. G. Weber and W. C. Purdy, *Anal. Lett.*, **12**, 1 (1979).
[885] P. S. C. Van der Plas, F. A. Huf and H. J. DeJong, *Pharm. Weekbl.*, **116**, 1341 (1981).
[886] D. S. Smith, M. H. Al-Hakiem and J. Landon, *Ann. Clin. Biochem.*, **18**, 253 (1981).
[887] R. M. Nakamura and W. R. Dito, *Lab. Med.*, **11**, 807 (1980).
[888] G. Gunzer and E. Rieke, *Pharm. Heute*, **89**, 37 (1980).
[889] M. C. Garcia, *Rev. Asoc. Bioquim. Argent.*, **44**, 32 (1980).
[890] A. H. M. W. Schuurs and B. K. van Weemen, *J. Immunoassay*, **1**, 229 (1980).
[891] M. J. O'Sullivan, J. W. Bridges and V. Marks, *Ann. Clin. Biochem.*, **16**, 221 (1979).
[892] E. T. Maggio, *Enzyme Immunoassay*, CRC-Press, Boca Raton (1980).
[893] A. H. W. M. Schuurs and B. K. van Weemen, *Clin. Chim.*, **81**, 1 (1977).
[894] H. J. Kyrein, *Ärztl. Lab.*, **24**, 57 (1978).
[895] M. Schöneshöfer, *Ärztl. Lab.*, **24**, 94 (1978).
[896] D. S. Kabakoff, in *Trace Organic Analysis*, pp. 533–539. NBS, Washington (1979).
[897] K. Lübke and G. Nieuweboer, *Immunologische Teste für niedermolekulare Wirkstoffe*, Thieme, Stuttgart (1978).
[898] W. Vogt, *Ärztl. Lab.*, **23**, 173 (1977).
[899] A. Broughton and J. E. Strong, *Clin. Chem.*, **22**, 726 (1976).

[900] J. Landon and A. C. Moffat, *Analyst*, **101**, 225 (1976).
[901] S. Baudner, *Getreide, Brot, Mehl*, **32**, 330 (1978).
[902] B. F. Erlanger, *Pharmacol. Rev.*, **25**, 271 (1973).
[903] I. J. Chopra, *J. Clin. Endocrinol. Metab.*, **51**, 117 (1980).
[904] K. Kirkegaard, J. Faber, K. Siersbaek-Nielse and T. Fris, *Acta Endocrinol.*, **97**, 196 (1981).
[905] R. Yamamoto, S. Hattori, T. Inukai, A. Matsuura and K. Yamashita, *Clin. Chem.*, **27**, 1721 (1981).
[906] H. H. Weetall, W. Hertl, F. B. Ward and L. S. Hersh, *Clin. Chem.*, **28**, 666 (1982).
[907] T. J. Reimers, R. G. Cowan, H. P. Davidson and E. D. Colby, *Am. J. Vet. Res.*, **42**, 2016 (1981).
[908] T. J. Wilke and P. A. Turnbull, *Ann. Clin. Biochem.*, **19**, 104 (1982).
[909] E. Cooper, A. Anderson, M. J. Bennett, A. H. MacLennan, G. M. Stirrat and C. W. Burke, *Clin. Chim. Acta*, **118**, 57 (1982).
[910] A. Iio, M. Ata, N. Sumoto, M. Shimooka, M. Kawamura and K. Hamamoto, *Kaku Igaku*, **18**, 345 (1981).
[911] J. Konishi, T. Kousaka, Y. Iida, K. Kasagi, K. Ikekubo and K. Torizuka, *Kaku Igaku*, **18**, 371 (1981).
[912] F. Pennisi, P. B. Romelli, L. Vancheri, C. Multinu and P. Cornale, *Nuklearmedizin, Suppl.*, 787 (1980).
[913] E. Cooper, A. Anderson, M. J. Bennett, A. H. MacLennan, G. M. Stirrat and C. W. Burke, *Clin. Chim. Acta*, **118**, 57 (1982).
[914] H. Buckmueller, S. Thiemann and H. Schmidt-Gayk, *Ärztl. Lab.*, **27**, 155 (1981).
[915] J. M. Obregon, *Ann. Clin. Biochem.*, **19**, 29 (1982).
[916] D. Inbar, Y. Tamir, S. Derfler, J. Feingers and D. Wagner, *Clin. Chim. Acta*, **106**, 129 (1980).
[917] J. M. Izquierdo, P. Sotorrio and A. Quiros, *Clin. Chem.*, **28**, 123 (1982).
[918] R. D. Nargessi, J. Ackland, M. Hassan, G. C. Forrest, D. S. Smith and J. Landon, *Clin. Chem.*, **26**, 1701 (1980).
[919] T. P. Lee and C. H. Tan, *Clin. Chem.*, **27**, 2018 (1981).
[920] E. A. S. Al-Dujaili and C. R. W. Edwards, *Clin. Chim. Acta*, **116**, 277 (1981).
[921] Z. Putz, R. Hampl, J. Veleminsky and L. Starka, *Cas. Lek. Cesk.*, **120**, 724 (1981).
[922] H. Witzgall and S. Hassan-Ali, *J. Clin. Chem. Clin. Biochem.*, **19**, 387 (1981).
[923] P. R. Walsh, M. C. Wang and M. L. Gitterman, *Ann. Clin. Lab. Sci.*, **11**, 138 (1981).
[924] F. M. Stearns, *Clin. Chem.*, **27**, 1471 (1981).
[925] G. D. Nordblom, R. Webb, R. E. Counsell and B. G. England, *Steroids*, **38**, 161 (1981).
[926] J. G. Vieira, E. M. K. Russo, R. M. B. Maciel and O. A. Germek, *Arq. Brasil. Endocrinol. Metabol.*, **25**, 47 (1981).
[927] A. Montemurro, M. W. Johnson, G. Barile and E. Youssefnejadian, *J. Obstet. Gynaecol.*, **1**, 247 (1981).
[928] G. E. Nolan, J. B. Smith, V. J. Chavre and W. Jubiz, *J. Clin. Endocrinol.*, **52**, 1242 (1981).
[929] M. Schoeneshoefer, A. Fenner and H. J. Dulce, *Clin. Chim. Acta*, **101**, 125 (1980).
[930] G. K. Stalla, G. Giesemann, O. A. Mueller, W. G. Wood and P. C. Scriba, *J. Clin. Chem. Clin. Biochem.*, **19**, 427 (1981).
[931] R. Hiramatsu, *Clin. Chim. Acta*, **117**, 239 (1981).
[932] M. G. Del Chicca, A. Clerico, M. Ferdeghini and G. Mariani, *J. Nucl. Med. Allied Sci.*, **25**, 35 (1981).
[933] R. M. Gough and G. Ellis, *Clin. Biochem.*, **14**, 74 (1981).
[934] Z. Kraiem, L. Kahana, V. Elias, S. Ghersin and M. Sheinfeld, *Clin. Chem.*, **26**, 1916 (1980).
[935] O. Lantto, I. Bjorkhem, R. Blomstrand and A. Kallner, *Clin. Chem.*, **26**, 1899 (1980).
[936] H. Hosoda, N. Kobayashi, S. Miyairi and T. Nambara, *Chem. Pharm. Bull.*, **29**, 3606 (1981).
[937] T. Nako, F. Tamamura, N. Tsunoda and K. Kawata, *Steroids*, **38**, 111 (1981).

[938] G. Kominami, I. Fujisaka, A. Yamauchi and M. Kono, *Clin. Chim. Acta,* **103,** 381 (1980).
[939] M. Pazzagli, J. B. Kim, G. Messeri, F. Kohen, G. Bolleli, G. F. Tommasi, R. Salerno, G. Moneti and M. Serio, *J. Steroid Biochem.,* **14,** 1005 (1981).
[940] F. Kohen, M. Pazzagli, J. B. Kim and H. R. Lindner, *Steroids,* **36,** 421 (1980).
[941] P. R. Walsh, M. C. Wang and J. C. Su, *Clin. Biochem.,* **14,** 47 (1981).
[942] M. Matsuki, K. Kakita, A. Tenku, S. Matsumura, H. Oyama, S. Nishida and M. Horino, *Kawasaki Med. J.,* **6,** 129 (1981).
[943] S. Nishida, M. Matsuki, M. Horino, H. Oyama, A. Tenku and K. Kakita, *Kawasaki Med. J.,* **7,** 31 (1981).
[944] H. Arakawa, M. Maeda, A. Tsuji and A. Kambegawa, *Steroids,* **38,** 453 (1981).
[945] B. R. Bhavnani, I. R. Sarda and C. A. Woolever, *J. Clin. Endocrinol. Metab.,* **52,** 741 (1981).
[946] M. R. A. Morgan, P. G. Whittacker, B. P. Fuller and P. D. G. Dean, *J. Steroid Biochem.,* **13,** 551 (1980).
[947] V. Mann, A. B. Benko and L. T. Kocsar, *Steroids,* **37,** 593 (1981).
[948] S. Kono, G. R. Merriam, D. D. Brandon, D. L. Loriaux and M. B. Lipsett, *J. Clin. Endocrinol. Metabol.,* **54,** 150 (1982).
[949] M. Capelli, A. Cassio, A. Balsamo, M. Zappulla, G. Bolelli, D. Ventura and P. Picchietti, *Lab. (Milan),* **8,** 19 (1981).
[950] G. Emons, C. Mente, R. Knuppen and P. Ball, *Acta Endocrinol.,* **97,** 251 (1981).
[951] C. T. Brooks, J. B. Copas and R. W. A. Oliver, *Clin. Chem.,* **28,** 499 (1982).
[952] J. B. Kim, G. J. Barnard, W. P. Collins, F. Kohen, H. R. Lindner and Z. Eshar, *Clin. Chem.,* **28,** 1120 (1982).
[953] J. Saumande, *Steroids,* **38,** 425 (1981).
[954] T. Nambara, T. Ohkubo and K. Shimada, *Clin. Chim. Acta,* **119,** 81 (1982).
[955] M. Luisi, D. Silvestri, G. Maltinti, A. L. Catarsi and F. Franchi, *Lancet,* 542 (1980, **II**).
[956] J. P. Shah and U. M. Joshi, *J. Steroid Biochem.,* **16,** 283 (1982).
[957] K. Stevens, S. E. Long and G. C. Perry, *Br. Vet. J.,* **137,** 17 (1981).
[958] M. Luisi, F. Franchi, P. M. Kikovic, D. Silvestri, G. Cossu, A. L. Catarsi, D. Barletta and M. Gasperi, *J. Steroid Biochem.,* **14,** 1069 (1981).
[959] B. G. Joyce, A. H. Othick, G. F. Read and D. Riad-Fahmy, *Ann. Clin. Biochem.,* **18,** 42 (1981).
[960] T. Nakao, *Acta Endocrinol.,* **93,** 223 (1980).
[961] F. Kohen, J. B. Kim, H. R. Lindner and W. P. Collins, *Steroids,* **38,** 73 (1981).
[962] M. Pazzagli, J. B. Kim, G. Messeri, G. Martinazzo, F. Kohen, F. Franceschetti, G. Moneti, R. Salerno, A. Tommasi and M. Serio, *Clin. Chim. Acta,* **115,** 277 (1981).
[963] S. M. H. Al-Habet, W. A. C. McAllister, J. V. Collins and H. J. Rogers, *J. Pharmacol. Methods,* **6,** 137 (1981).
[964] J. A. Belis, *Invest. Urol.,* **17,** 332 (1980).
[965] W. G. Sippell, *Forschungsber,* BMFT-FB-T 81-036 (1981).
[966] D. Schopper, *Arch. Lebensmittelhyg.,* **32,** 203 (1981).
[967] T. P. Deheny, W. S. Murdoch, L. Boyle, W. A. W. Walters and A. L. A. Boura, *Prostaglandins,* **21,** 1003 (1981).
[968] R. Vijayakumar, S. Murdoch, T. Deheny and W. A. W. Walters, *J. Chromatog.,* **225,** *Biomed. Appl.,* **14,** 57 (1981).
[969] K. L. Klein, W. J. Scott and K. E. Clark, *Prostaglandins,* **22,** 623 (1981).
[970] T. Yano, H. Tomoko, Y. Hayashi and S. Yamamoto, *J. Biochem. (Tokyo),* **90,** 773 (1981).
[971] Y. Hayashi, T. Yano and S. Yamamoto, *Biochem. Biophys. Acta,* **663,** 661 (1981).
[972] P. J. Lijnen, L. J. Verschueren and A. K. Amery, *Bull Soc. Chim. Belg.,* **90,** 229 (1981).
[973] D. G. B. Chang and H. H. Tai, *Prostaglandins, Leukotriens Med.,* **8,** 11 (1982).
[974] S. A. Metz, M. G. Rice and R. P. Robertson, *Adv. Prostaglandins Thromboxane Research,* **6,** 183 (1980).

[975] F. A. Kimball, J. C. Cornette, G. L. Bundy and K. T. Kirton, *Prostaglandins,* **20,** 559 (1980).
[976] W. Schlegel, J. Urdinola and H. P. G. Schneider, *Acta Endocrinol.,* **100,** 98 (1982).
[977] O. Ylikorkala and L. Viinikka, *Prostaglandins Med.,* **6,** 427 (1981).
[978] H. Sinzinger, K. Silberbauer, Z. Detre, C. Leithner, K. Klein, M. Warumm and E. Csonka, *Radioakt. Isot. Klin. Forsch.,* **14,** 485 (1980).
[979] W. Siess and F. Dray, *J. Lab. Clin. Med.,* **99,** 388 (1982).
[980] D. S. McCann, J. Tokarsky and R. P. Sorkin, *Clin. Chem.,* **27,** 1417 (1981).
[981] L. Viinikka and O. Ylikorkala, *Prostaglandins,* **20,** 759 (1980).
[982] L. Levine, I. Alam, J. Gjika, T. J. Carty and E. J. Goetzl, *Prostaglandins,* **20,** 923 (1980).
[983] W. J. Raum and R. S. Swerdloff, *Life Sci.,* **28,** 2819 (1981).
[984] M. Nassel-Hiemke and H. J. Schuemann, *J. Biochem. Biophys. Methods,* **4,** 255 (1981).
[985] C. Strambi, A. Strambi, M. L. Dereggi and M. A. Delaage, *Eur. J. Biochem.,* **118,** 401 (1981).
[986] S. Vemuri and K. Sambamurthy, *Indian J. Exp. Biol.,* **18,** 669 (1980).
[987] R. H. Christenson, J. E. Hammond, J. H. Hull and J. A. Bustrack, *Clin. Chim. Acta,* **120,** 13 (1982).
[988] P. Pohle, *Ärtzl. Lab.,* **26,** 275 (1980).
[989] B. Bergdahl and L. Molin, *Clin. Biochem.,* **14,** 67 (1981).
[990] P. F. Angelino, F. Matta, G. C. Bulgarelli, A. Altieri, E. Rolando, L. Riva and D. Gastando, *G. Ital. Cardiol.,* **10,** 1031 (1980).
[991] R. Allner and H. Kruepe, *Ärtzl. Lab.,* **27,** 69 (1981).
[992] M. H. H. Al-Hakiem, R. D. Nargessi, M. Pourfarzaneh and A. J. Hodgkinson, *J. Clin. Chem. Clin. Biochem.,* **20,** 151 (1982).
[993] J. M. Scherrmann and R. Bourdon, *Clin. Chem.,* **26,** 670 (1980).
[994] V. P. Butler, D. Tse-Eng, J. Lindenbaum, S. M. Kalman, J. J. Preibisz, D. G. Rund and P. S. Wissel, *J. Pharmacol. Exp. Ther.,* **221,** 123 (1982).
[995] C. Goyot, B. Golse and J. Grenier, *Pathol. Biol.,* **29,** 330 (1981).
[996] F. P. Smith, *Forensic Sci. Intern.,* **17,** 225 (1981).
[997] S. Inayama, Y. Tokunaga, E. Hosaya, T. Nakadate, T. Niwaguchi, K. Aoki and S. Saito, *Chem. Pharm. Bull.,* **28,** 2779 (1980).
[998] B. Gourmel, J. Fiet, R. F. Collins, J. M. Vilette and C. Dreux, *Clin. Chim. Acta,* **108,** 229 (1980).
[999] E. L. Slightom, J. C. Cagle, H. H. McCurdy and F. Castagna, *J. Anal. Toxicol.,* **6,** 22 (1982).
[1000] K. Robinson, M. G. Rutherford and R. N. Smith, *J. Pharm. Pharmacol.,* **32,** 773 (1980).
[1001] F. Monaco and S. Piredda, *Epilepsia,* **21,** 475 (1980).
[1002] S. Y. Chu, S. M. Vega, A. Ali and L. T. Sennello, *J. Pharm. Sci.,* **70,** 990 (1981).
[1003] R. D. Budd, *Clin. Toxicol.,* **18,** 773 (1981).
[1004] J. M. Scherrmann, L. Boudet, R. Pontikis, Nguyen-Hoang-Nam and E. Fournier, *J. Pharm. Pharmacol.,* **32,** 800 (1980).
[1005] R. Lucek and R. Dixon, *Res. Commun. Chem. Pathol. Pharmacol.,* **27,** 397 (1980).
[1006] K. K. Midha, J. K. Cooper and J. W. Hubbard, *Commun. Psychopharmacol.,* **4,** 107 (1980).
[1007] C. Nerenberg and S. B. Matin, *J. Pharm. Sci.,* **70,** 900 (1981).
[1008] R. J. Wurzburger, R. L. Miller, E. A. Marcum, W. A. Colburn and S. Spector, *J. Pharmacol. Exp. Ther.,* **217,** 757 (1981).
[1009] I. L. Honigberg and J. T. Stewart, *J. Pharm. Sci.,* **69,** 1171 (1980).
[1010] S. Iwamura and A. Kambegawa, *J. Pharmacobio.-Dynam.,* **4,** 275 (1981).
[1011] G. S. F. Ling, J. G. Umans and C. E. Inturrisi, *J. Pharmacol. Exp. Ther.,* **217,** 147 (1981).
[1012] R. Dixon, T. Crews, E. Mohacsi, C. Inturrisi and K. Foley, *Res. Commun. Chem. Pathol. Pharmacol.,* **32,** 545 (1981).

[1013] K. K. Midha, C. Mackonka, J. K. Cooper, J. W. Hubbard and P. K. F. Yeung, *Br. J. Clin. Pharmacol.*, **11**, 85 (1981).
[1014] D. S. Cooper, V. C. Saxe, F. Maloof and E. C. Ridgway, *J. Clin. Endocrinol. Metab.*, **52**, 204 (1981).
[1015] J. E. Heveran, M. Anthony and C. Ward, *J. Forens. Sci.*, **25**, 79 (1980).
[1016] J. E. Heveran and C. Ward, *J. Forens. Sci.*, **25**, 719 (1980).
[1017] M. H. H. Al-Hakiem, G. W. White, D. S. Smith and J. Landon, *Ther. Drug Monit.*, **3**, 159 (1981).
[1018] K. K. Midaha, J. W. Hubbard, J. K. Cooper, E. M. Hawes, S. Fournier and P. Yeung, *Br. J. Clin. Pharmacol.*, **12**, 189 (1981).
[1019] B. Gerson, L. Dean and F. Bell, *Ther. Drug Monit.*, **3**, 167 (1981).
[1020] G. Kominami, M. Nakamura, S. Mori and M. Kono, *Clin. Chim. Acta*, **117**, 189 (1981).
[1021] K. I. Arnstadt, *Z. Lebensm. Unters. Forsch.*, **173**, 255 (1981).
[1022] G. Palotti, V. Fabbri, C. Mambelli, S. Ruggeri and R. Nunziati, *Lab. (Milan)*, **8**, 33 (1981).
[1023] A. J. Quattrone, C. M. O'Donnell, J. McBride, P. B. Mendershausen and R. S. Putnam, *J. Anal. Toxicol.*, **5**, 245 (1981).
[1024] M. E. Jolley, *J. Anal. Toxicol.*, **5**, 236 (1981).
[1025] D. Knight, *Clin. Biochem.*, **14**, 14 (1981).
[1026] L. V. Allen and M. L. Stiles, *Clin. Toxicol.*, **18**, 1043 (1981).
[1027] B. C. Yu, C. Chang, S. H. Kwan and P. C. Ho, *Proc. Natl. Sci. Counc. Repub. China*, **B, 5**, 404 (1981).
[1028] A. Biermann and G. Terplan, *Arch. Lebensmittelhyg.*, **31**, 51 (1980).
[1029] O. El-Nakib, J. J. Pestka and F. S. Chu, *J. Assoc. Off. Anal. Chem.*, **64**, 1077 (1981).
[1030] S. Lee and F. S. Chun, *J. Assoc. Off. Anal. Chem.*, **64**, 684 (1981).
[1031] J. J. Pestka, B. W. Steinert and F. S. Chu, *Appl. Environ. Microbiol.*, **41**, 1472 (1981).
[1032] R. Federico, P. Aducci, M. Marra and C. Pini, *Z. Pflanzenphysiol.*, **104**, 275 (1981).
[1033] S. S. Westfall and G. H. Wirtz, *Experientia*, **36**, 1351 (1980).
[1034] J. W. Thanassi and J. A. Cidlowski, *J. Immunol. Methods*, **33**, 261 (1980).
[1035] R. Bouillon, P. de Moor, E. G. Bagglioni and M. R. Uskokovic, *Clin. Chem.*, **26**, 562 (1980).
[1036] M. Peacock, G. A. Taylor and W. Brown, *Clin. Chim. Acta*, **101**, 93 (1980).
[1037] H. J. Johnson, S. F. Cernosek, R. M. Gutierrez-Cernosek and L. L. Brown, *J. Anal. Toxicol.*, **4**, 86 (1980).
[1038] H. J. Johnson, R. M. Gutierrez-Cernosek and L. L. Brown, *J. Anal. Toxicol.*, **5**, 157 (1981).
[1039] E. Piall, G. W. Aherne and V. Marks, *Clin. Chem.*, **28**, 119 (1982).
[1040] Y. G. Tsai, L. Wilson and E. Keefe, *Clin. Chem.*, **26**, 1610 (1980).
[1041] J. M. Wal and G. F. Bories, *Anal. Biochem.*, **114**, 263 (1981).
[1042] T. Miura, H. Kouna and T. Kitagawa, *J. Pharmacobio-Dynam.*, **4**, 706 (1981).
[1043] C. M. O'Donnell, J. McBride, S. Suffin and A. Broughton, *J. Immunoassay*, **1**, 375 (1980).
[1044] C. J. Voegeli and G. J. Burckart, *Clin. Chem.*, **28**, 248 (1982).
[1045] K. Fujiwara, M. Yasuno and T. Kitagawa, *Cancer Res.*, **41**, 4121 (1981).
[1046] K. Fujiwara, M. Yasuno and T. Kitigawa, *J. Immunol. Methods*, **45**, 195 (1981).
[1047] K. L. Fong, D. W. Ho, L. Bogerd, T. Pan, N. S. Brown, L. Gentry and G. P. Bodey, *Antimicrob. Agents Chemother.*, **19**, 139 (1981).
[1048] K. Fujiwara, H. Saikusa, M. Yasuno and T. Kitagawa, *Cancer Res.*, **42**, 1487 (1982).
[1049] F. Engbaek and B. Voldby, *Clin. Chem.*, **28**, 624 (1982).
[1050] M. A. Delaage and J. J. Puizillout, *J. Physiol. (Paris)*, **77**, 339 (1981).
[1051] J. L. Langone, *Anal. Biochem.*, **104**, 347 (1980).
[1052] P. Miller, S. Weiss, M. Cornell and J. Dockery, *Clin. Chem.*, **27**, 1698 (1981).
[1053] C. Gonzalez and J. Garcia-Sancho, *Anal. Biochem.*, **114**, 285 (1981).
[1054] B. Law, K. Pocock and A. C. Moffat, *J. Forens. Sci. Soc.*, **22**, 275 (1982).

[1055] B. Riesselmann, *Dtsch. Apotheker Ztg.*, **121**, 2078 (1981).
[1056] D. J. Harvey, *TrAC*, **1**, 66 (1981).
[1057] A. Castro, N. Monji, A. Hacer, M. Yi, E. R. Bowman and H. McKennis, *Eur. J. Biochem.*, **104**, 331 (1980).
[1058] E. M. S. MacDonald, D. E. Akiyoshi and R. O. Morris, *J. Chromatog.*, **214**, 101 (1981).
[1059] J. Daie and R. Wyse, *Anal. Biochem.*, **119**, 365 (1982).
[1060] J. J. Pizillout and M. A. Delaage, *J. Pharmacol. Exp. Ther.*, **217**, 791 (1981).
[1061] E. W. Weiler, *Planta*, **153**, 319 (1981).
[1062] C. D. Ercegovic, R. P. Vallejo, R. R. Gettig, L. Woods, E. R. Bogus and R. O. Mumma, *J. Agr. Food Chem.*, **29**, 559 (1981).
[1063] W. H. Newsome and J. B. Shields, *Intern. J. Envir. Anal. Chem.*, **10**, 295 (1981).
[1064] M. J. Luster, P. W. Albro, K. Chae, L. D. Lawson, J. T. Corbett and J. D. McKinney, *Anal. Chem.*, **52**, 1497 (1980).
[1065] S. I. Wie and B. D. Hammock, *Anal. Biochem.*, **125**, 168 (1982).
[1066] A. C. Greenquist, B. Walter and T. M. Li, *Clin. Chem.*, **27**, 1614 (1981).
[1067] E. C. Metcalf, M. R. Morgan and P. D. G. Dean, *J. Chromatog.*, **235**, 501 (1982).
[1068] O. Lantto, *Clin. Chem.*, **28**, 1129 (1982).
[1069] C. Lindberg, S. Jonsson, P. Hedner and A. Gustafsson, *Clin. Chem.*, **28**, 174 (1982).
[1070] E. Gerber-Taras, B. K. Park and E. E. Ohnhaus, *J. Clin. Chem. Clin. Biochem.*, **19**, 525 (1981).
[1071] K. Green, F. A. Kimball, B. A. Thornburgh and A. J. Wickramasinghe, *Prostaglandins*, **20**, 767 (1980).
[1072] A. Poklis, *J. Anal. Toxicol.*, **5**, 174 (1981).
[1073] Y. Pegon, E. Pourcher and J. J. Vallon, *J. Anal. Toxicol.*, **6**, 1 (1982).
[1074] M. G. Rutterford and R. N. Smith, *J. Pharm. Pharmacol.*, **32**, 449 (1980).
[1075] J. E. Wallace, S. C. Harris and E. L. Shimek, *Clin. Chem.*, **26**, 1905 (1980).
[1076] R. F. Butz, D. H. Schroeder, R. M. Welch, N. B. Mehta, A. P. Phillips and J. W. A. Findley, *J. Pharmacol. Exp. Ther.*, **217**, 602 (1981).
[1077] R. Virtanen, J. S. Salonen, M. Scheinin, E. Iaslo and V. Mattila, *Acta Pharmacol. Toxicol.*, **47**, 274 (1980).
[1078] R. E. Poland and R. T. Rubin, *Life Sci.*, **29**, 1837 (1981).
[1079] K. K. Midha, C. Charette, J. K. Cooper and I. J. McGilvray, *J. Anal. Toxicol.*, **4**, 237 (1980).
[1080] I. M. McIntyre, T. R. Norman, G. D. Burrows and K. P. Maguire, *Bt. J. Clin. Pharmacol.*, **12**, 691 (1981).
[1081] W. C. Griffiths, P. Dextraze, M. Hayes, J. Mitchell and I. Diamond, *Clin. Toxicol.*, **16**, 51 (1980).
[1082] P. Mojaverian and G. D. Chase, *J. Pharm. Sci.*, **69**, 721 (1980).
[1083] H. L. Weingarten and E. Trevias, *J. Anal. Toxicol.*, **6**, 88 (1982).
[1084] C. B. Walberg and R. S. Gupta, *J. Anal. Toxicol.*, **6**, 97 (1982).
[1085] D. Couri, C. Suarez de Villar and P. Toy-Manning, *J. Anal. Toxicol.*, **4**, 227 (1980).
[1086] R. G. Buice, W. E. Evans, J. Karas, C. A. Nicholas, P. Sidhu, A. B. Straughn, M. C. Meyer and W. R. Crom., *Clin. Chem.*, **26**, 1902 (1980).
[1087] W. Hospes, R. J. Boksma and J. R. J. B. Brouwers, *Pharm. Weekbl., Sci. Ed.*, **4**, 32 (1982).
[1088] R. M. Ratcliff, C. Mirelli, E. Moran, D. O'Leary and R. White, *Antimicrob. Agents Chemother.*, **19**, 508 (1981).
[1089] W. L. Pengelly, R. S. Bandurski and A. Schulze, *Plant Physiol.*, **68**, 96 (1981).
[1090] J. E. O'Connor and T. A. Rejent, *J. Anal. Toxicol.*, **5**, 168 (1981).
[1091] E. P. Yeager, U. Goebelsmann, J. R. Soares, J. D. Grant and S. J. Gross, *J. Anal. Toxicol.*, **5**, 81 (1981).
[1092] H. Fukuchi, M. Yoshida, M. Okihara, S. Tokuoka and T. Konishi, *Am. J. Hosp. Pharm.*, **38**, 1933 (1981).
[1093] W. J. Acton, O. M. Van Duyn, L. V. Allen and D. J. Flournoy, *Clin. Chem.*, **28**, 177 (1982).

[1094] J. Fenton, M. Schaffer, N. W. Chen and E. W. Bermes, *J. Forens. Sci.*, **25**, 314 (1980).
[1095] R. Donald, L. Paalzow and P. O. Edlund, *J. Pharm. Sci.*, **71**, 314 (1982).
[1096] L. von Meyer, G. Kauert and G. Drasch, *Beitr. Gerichtl. Med.*, **39**, 113 (1981).
[1097] B. E. Pape, *Ther. Drug Monit.*, **3**, 357 (1981).
[1098] H. R. Ha, G. Kewitz, M. Wenk and F. Follath, *Br. J. Clin. Pharmacol.*, **11**, 312 (1981).
[1099] P. G. Dextraze, J. Foreman, W. C. Griffiths and I. Diamond, *Clin. Toxicol.*, **18**, 291 (1981).
[1100] R. L. Horst, T. A. Reinhardt, D. C. Beitz and E. T. Littledike, *Steroids*, **37**, 581 (1981).
[1101] D. A. Seamark, D. J. H. Trafford and H. L. J. Makin, *J. Steroid Biochem.*, **14**, 111 (1981).
[1102] B. Garcia-Pascual, A. Peytremann, B. Courvoisier and D. E. M. Lawson, *Clin. Chim. Acta*, **68**, 99 (1976).
[1103] S. Dokoh, R. Morita, M. Fukunaga, I. Yamamoto, S. Itsuo, Y. Chohei, A. Yamamoto and K. Torizuka, *Nippon Naibumpi Gakkai Zasshi*, **57**, 1209 (1981).
[1104] J. P. Mallon, J. G. Hamilton, C. Nauss-karol, R. J. Karol, C. J. Ashley, D. S. Matuszewski, C. A. Tratnyek, G. G. Bryce and O. N. Miller, *Arch. Biochem. Biophys.*, **201**, 277 (1980).
[1105] M. W. France and B. Lalor, *J. Med. Sci.*, **150**, 310 (1981).
[1106] Y. Shidoji and N. Hosoya, *Anal. Biochem.*, **104**, 457 (1980).
[1107] M. C. De Felipe, J. A. Fuentes and A. H. Drummond, *Biochem. Pharmacol.*, **31**, 1661 (1982).
[1108] A. de la Pena and J. W. Goldzieher, *Clin. Chem.*, **20**, 1376 (1974).
[1109] F. Schoen, K. Hackenberg, D. Paar and D. Reinwein, *J. Clin. Chem. Clin. Biochem.*, **18**, 355 (1980).
[1110] S. J. Henning, *Steroids*, **35**, 673 (1980).
[1111] T. Z. Tan, S. P. Deng and W. B. Zhou, *SSu-Ch'uan I Hsueh Yuan Hsueh Pao*, **11**, 89 (1980).
[1112] M. Yarus, *Anal. Biochem.*, **70**, 346 (1976).
[1113] R. Jochemsen, G. J. M. J. Horbach and D. D. Breimer, *Res. Commun. Chem. Pathol. Pharmacol.*, **35**, 259 (1982).
[1114] H. F. Milton and W. A. Waters, *Methods of Quantitative Microanalysis*, Arnold, London (1955).
[1115] U. S. von Euler and R. Eliasson, *Prostaglandins*, Academic Press, New York (1967).
[1116] C. T. Malone and W. H. McFadden, in *Ancillary Techniques in Gas Chromatography*, L. S. Ettre, W. H. McFadden (eds.), p. 370. Wiley-Interscience, New York (1969).
[1117] P. Karrer, *Organische Chemie*, Thieme, Stuttgart (1950).
[1118] Department of Environment National Water Council (U.K.), *Methods of Examination of Waters and Associated Matter*, London (1982).
[1119] A. A. Williams, *J. Inst. Brew.*, **88**, 43 (1982).
[1120] Y. Hoshika, Y. Nihei and G. Muto, *Analyst*, **106**, 1187 (1981).
[1121] R. W. Moncrieff, *Odours*, Heinemann, London (1970).

7

Special topics in organic trace analysis

7.1 TOPICS RELATED TO GROUPS OF SUBSTANCES

So far, the methods available for the separation and determination of organic traces have been described. This chapter will deal with the substances (as groups or individually) to which the methods can be applied, and with the general analytical aspects of trace work. Many reviews of interest have appeared on applications of organic trace analysis, such as the determination of organics in water [1] (242 references), [2] (94 references), with special regard to 114 priority pollutants [3] and to continuous monitoring [4]. Aliphatic amines form a group of special interest to the environmental chemist, as can be seen from a review with 211 references [5]. Several assessments have been made of the analytical methods for chlorinated dibenzodioxins, marking the current significance of these compounds in organic trace analysis [6-9]. PCBs are similarly a focus for analytical interest [10]. Analytical methodology for pesticides has been reviewed several times (e.g. [11, 12]. *N*-Nitroso compounds in foods, work-places and pesticides are another group of materials of ecological interest deserving review [13]. Many compounds exhibit mutagenic properties, and as this class of substances is chemically heterogeneous many analytical procedures must be made available and described [14]. Plant growth substances (183 references) are complex chemicals, and have a resulting complicated analytical chemistry [15]. More and more substances in work-places are recognized as hazards. At present, no official methods exist for the determination of some 130 such compounds. Existing procedures have been summarized [16]. Drugs of abuse are another increasing burden and analytical task [17] (362 references). Cocaine is one of the most dangerous compounds in this respect, and requires all too frequent analysis [18] (80 references). Doping has also become a problem in sport [19]. Therapeutically used drugs and their metabolites often have to be determined, and quantitative selected-ion mass spectrometry can be useful for this purpose [20] (613 references).

7.2 DETERMINATION OF TRACE COMPOUNDS WITHOUT SEPARATION

It is obviously of great interest to use or develop methods which avoid tedious separations before the final determination, but such methods must be examined in the context of analytical strategy in general. This is well illustrated by the new developments in mass spectrometry [21]. It is possible to use extensive sample preparation and clean-up followed by mass spectrometric investigation with a low-resolution instrument or use of only a few separation and clean-up steps, but subsequent high-resolution mass spectrometry. The selection of the method is influenced by the task itself. Determinations in the ppm or ppb region are much more simply solved by direct working procedures, whereas determinations in the ppt or ppqd region require good separation techniques before the application of mass spectrometry, even with high-resolution instruments. "Many parameters influence the choice of methods – like cost of man power (which is different at high level and at low level training of personnel), the availability of the required instrument and reagent, the chemical nature of the trace(s) and the matrix etc. There is not one answer to a given problem. Often several schools of thought exist. Probably, for organic trace analysis the increase of specificity of the mass spectrometric equipment is the method of choice" [21].

So far, only 2% of all papers published on organic trace analysis offer methods with very little or no separation or sample preparation work before direct measurement. Examples are given in Table 7.1.

Table 7.1 – Examples of determinations of organic traces without separation steps or with a very simple sample preparation method.

Substance	Matrix	Method	Reference
	μg-range or ppm-range		
acetylene	oxygen	photometry	22
urea, formaldehyde	effluents	photometry	23
methyl ethyl ketone	air	photometry	24
acrolein	air, tobacco smoke	photometry	25
trichlorethanol, trichloroacetic acid	urine	photometry	26
methyl parathion	technical formulations	photometry	27
vinyl chloride	liquid air	ir	28
chloromethyl methyl ether	chloroform	Fourier-ir	29
acrolein	air	fluorescence	30
2,3-benzofluorene	air	fluorescence	31
humic acids	water	fluorescence	32
substituted benzenes	aerosols from air	ms (high-resoln.)	33
triamterene	serum	polarography	34
nitrovin	animal feed	polarography	35
acrolein	water	polarography	36
N-nitrosodiethylamine	grinding fluid	pulse polarog.	37
phosphoethanolamine	biological material	enzymatic	38
physostigmine	blood	enzymatic	39
guanine ribonucleotides	biological material	enzymatic	40
thyroxine	blood	ria	41
2-ethylhexylphthalate	plasma	ria	42

Table 7.1 – *continued*

Substance	Matrix	Method	Reference
gibberellic acid	malt	ria	43
diphenylhydantoin	blood	ria/EMIT	44
antibiotics	milk	bioassay	45
chlortetracycline	feed	bioassay	46
bacitracin	feed	bioassay	47
streptomycin	eggs	bioassay	48
thiopeptine	feeds	bioassay	49
	ng-range or ppb range		
tryptophan	plasma	fluorescence	50
porphyrins	urine	spectroscopy	51
polycyclic aromatics	mixture of PAHs	Fourier-ir	52
arginine	proteins	fluorescence	53
morphine, LSD	biological material	fluorescence	54
PAHs	other PAHs	luminescence	55
p-aminobenzoic acid	vitamin tablets	phosphorescence	56
adriamycin	plasma	diff. pulse polarog.	57
nitrilotriacetic acid	water	diff. pulse polarog.	58
P-containing insecticides	milk	enzymatic	59
ATP	ocean water	enzymatic	60
glucose-1-phosphate	wheat germs	enzym. + radiom.	61
glutamine		enzym. + radiom.	62
S-adenosylmethionine	rat brain	enzym. + radiom.	63
UDP-glucuronic acid	tissues	enzym. + radiom.	64
catecholamines	plasma	enzym. + radiom.	65
uric acid	plasma	enzym. + fluoresc.	66
oestriol	urine	ria	67
3,3′,5-tri-iodothyronine	serum	ria	68, 69
digoxin	blood	ria	70
metanephrine	urine	ria	71
lithocholates	serum	ria	72
lysergic acid, LSD	serum	ria	73
thromboxane	blood(?)	ria	74
testosterone		ria	75
conjugated oestriol	urine	ria	76
digoxin	biological material	ria	77, 78
glibenclamide	serum	ria	79
calcitonin	urine	ria	80
isoniazid	serum	ria	81
morphine, barbiturates	urine	ria	82
phallotoxins	mushrooms	protein binding	83
penicillins	milk	bioassay	84
	pg-range (or less) and ppt range		
PAHs	mixture of PAHs	fluorescence	85
amino compounds	protein solutions	fluorescence	86
clomipramine	plasma	ria	87
oestrogens, testosterone	edible animal products	ria	88
vitamin B_{12}	food	ria	89
tri-iodothyronine	plasma	ria	90
oestriol	serum	ria	91
aminobenzimidazole	biological material	ria	92
25-hydroxycalciferol	plasma	protein binding	93

7.3 METHODS TO IMPROVE THE LIMIT OF DETECTION OF ORGANIC TRACE ANALYTICAL METHODS

The working ranges most widely investigated can be identified from a statistical evaluation of the literature, as given in Table 7.2.

Table 7.2 – Concentration or mass ranges that have been examined; frequency according to the analytical literature (data calculated from *Analytical Abstracts* 1978–1980).

Level	% of all articles
μg	8
ng	26
pg (or less)	8
ppm	20
ppb	32
ppt	6

Some methods with a very low limit of detection or working range have been published (see Table 7.3).

Organic trace analysis will have to develop more and more methods with extremely low limits of detection. There are several approaches to the solution of this problem.

One that is often used is 'miniaturization', i.e. replacement of large reaction vessels by micro or ultramicro equipment. Though smaller absolute quantities can thus be handled, the techniques become more complicated with such tiny glassware. Performance of the reaction in a flow-system often helps with sample transfer. Reagent solutions are pumped through micro vessels, cuvettes etc. which are otherwise difficult to fill.

The limit of detection can be improved considerably when a reliable blank system is available and measurement is made against this blank. In spectrophotometry this method is commonplace (double-beam systems). In electrochemistry it is less often used.

Another way for improvement is subtraction of the spectra of solvents in order to obtain the spectra of only the compounds of interest. Because the Fourier technique allows the measurement of very accurate photometric data at extremely precisely defined wavenumbers, this computer technique is suggested as especially useful [121, 122].

Dual-beam FT-infrared spectrometry improves the limit of detection of FT-infrared measurements by about a factor of four (relative to single-beam measurements) [123–125].

Electronic differentiation of a signal often permits better resolution of curves, plots etc. (see Fig. 7.1). Examples of the application of derivative spectroscopy are the rapid determination of 50 μg of paraquat per litre of serum in emergency cases of poisoning [126], the identification of 4-hydroxycoumarin

Table 7.3 – Examples of methods with very low limits of detection or working ranges.

Compound(s)	Matrix	Level	Separation	Determination	Reference
aniline (derivatives)		90 ag	gc	ecd	94
alcohols		25 fg	gc	ecd	95
biotinyloestrone		300 fg		protein binding	96
PAHs		600 fg	hplc	fluorescence	97
impurities	hydrogen	1 ppqd	gc		98
phthalate esters	air	0.02 ppt	gc	ecd	99
chlorinated pesticides	water	0.1 ppt	gc	ms	100
dimethyl sulphide	air, water	0.1 ppt	gc	fpd	101
PAHs	water	0.1 ppt	tlc	densitometry	102
benzene	sea-water	0.2 ppt	gc	fid	103
methyl bromides	air	>0.5 ppt	gc	ms	104
TCDD	soil	>0.2 ppt	gc	ms	105
riboflavin		0.5 ppt		fluorimetry	106
tri-iodothyronine	serum	0.2 pg	gc	ecd	107
thyroid hormones	blood	>0.5 pg	gc	ms	108
chlorinated insecticides		0.5 pg	gc	ecd	109
phthalate	biological material	>0.1 pg	gc	ecd	110
dibutyl phthalate	air	3 ppt	gc	ms	111
TCDD	vegetables	1 ppt	chromatog/gc	ms	112, 113
aflatoxins	milk	5 ppt	chromatog./tlc	densitometry	114
giberellic acid	plant extracts	2 pg		immunoassay	115
chlorinated compounds	arctic air	1 pg	gc	ms	116
iodothyronine		0.5 pg	gc	ecd	117
aromatic diamines	urine	30 ppt	partition/gc	ms	118
TCDD	fish	50 ppt	chromatog./gc	ms	119
PAHs	air	0.1 ng/10 m^3	chromatog./gc	fid	120

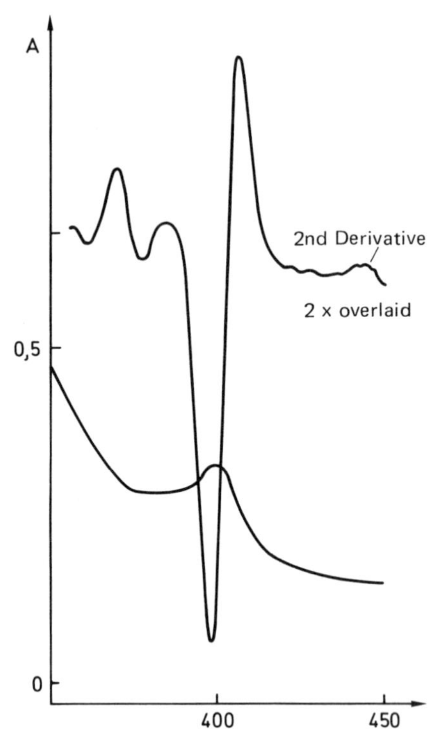

Fig. 7.1 – Absorption spectrum of porphyrins in urine and second derivative of original curve [130]. (Reproduced by permission of the copyright holders, Verlag Ernst Giebeler, Darmstadt).

derivatives in organ and blood homogenates [127] and the determination of ng–μg amounts of anilines and phenols [128]. Electronic differentiation has been reviewed [129].

Electronic averaging improves the signal to noise ratio. Two ways of averaging signals are used.

(1) The measurement is taken over a sufficient period, at a given point of a spectrum. The noise averages out to a certain extent.

(2) Spectra are averaged digitally. The analogue spectrum is digitized at regular, predetermined intervals along the wavelength axis. A second spectrum is then produced, independently of the first, and digitized at exactly the same points. In this way n spectra are obtained and collected. The corresponding points of each are summed in the computer memory. The noise tends to average out and the signal increases in size. Theoretically the signal to noise ratio is increased by $\sqrt{n}$. Thus, a tenfold increase in sensitivity requires the recording of 100 spectra. Consequently, procedures in which spectra are produced very rapidly are of most advantage.

Figure 7.2 shows a single recording of the ir spectrum of 1 μg of tyrosine. Figure 7.3 shows the effect of signal summation of the spectrum of 25 ng of tyrosine after one registration (bottom), after 53 registrations (middle) and after 612 accumulations (top). Figure 7.4 demonstrates the spectrum of the same 25 ng of tyrosine after 2448 accumulations and 30 "smoothing" operations.

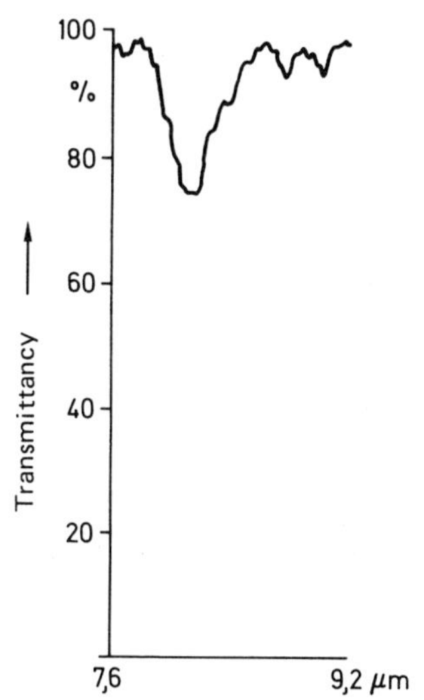

Fig. 7.2 – Infrared micro-reflectance spectrum of 1 μg of tyrosine. (Reproduced from [131], by permission of the copyright holders, Springer Verlag).

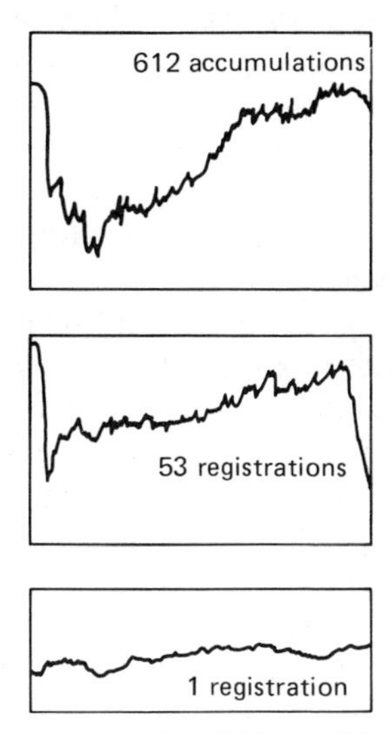

Fig. 7.3 – Infrared reflectance spectra of 25 ng of tyrosine after several repetitions of time-averaging registrations. (Reproduced from [131], by permission of the copyright holders, Springer Verlag).

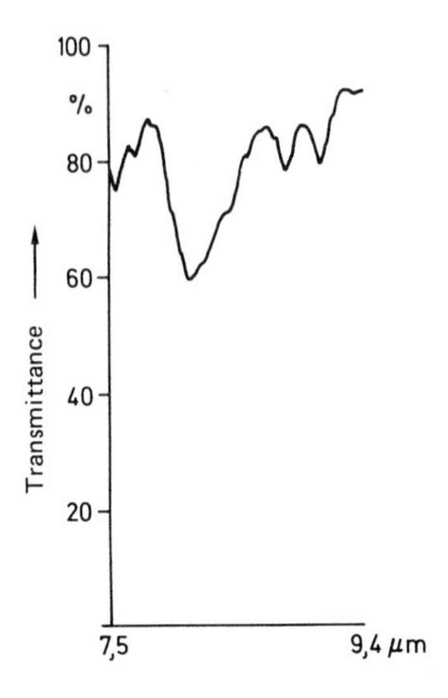

Fig. 7.4 – Infrared microreflectance spectrum of 25 ng of tyrosine after 2448 accumulations and 30 smoothing operations. (Compare with Fig. 7.2). (Reproduced from [131], by permission of the copyright holders, Springer Verlag).

The limit of detection can also be influenced by chemical means. An 'active' group can be attached to an otherwise inactive molecule. For example, by addition of a uv-absorbing group to a non-absorbing molecule the sensitivity for uv detection is created or increased. Analogously, fluorescent, radioactive or electroactive groups can be linked to an inactive trace. Such derivatives can be made before or after any separation steps. Ketosteroids have been reacted with phenylhydrazine to attach a uv-absorbing group to them [132]. *p*-Nitrobenzoylchloride serves a corresponding purpose for hydroxysteroids and digitalis glucosides [133] and for carbohydrates [134].

Sensitivity and selectivity cannot be regarded independently. In gas chromatography more selective detectors are now available than the once universally used fid. They often reduce the background of the chromatogram, decrease the noise and record only the peaks of interest. Because the selectivity is greater, the resolution and hence the limit of detection will be improved. One example of this strategy is the 'chopping technique' in which a current is divided (chopped) into time-resolved segments. Electronic amplification of the resulting chopped signal is achieved more easily than that of the non-modulated signal.

The principle may be illustrated by the performance of a solute-switched electron-capture detector for pesticides [135], which is based on the concept of solute switching [136–138] (see also Fig. 7.5). Secondary electrons (produced by β-particle irradiation of the nitrogen used as carrier gas) will ionize

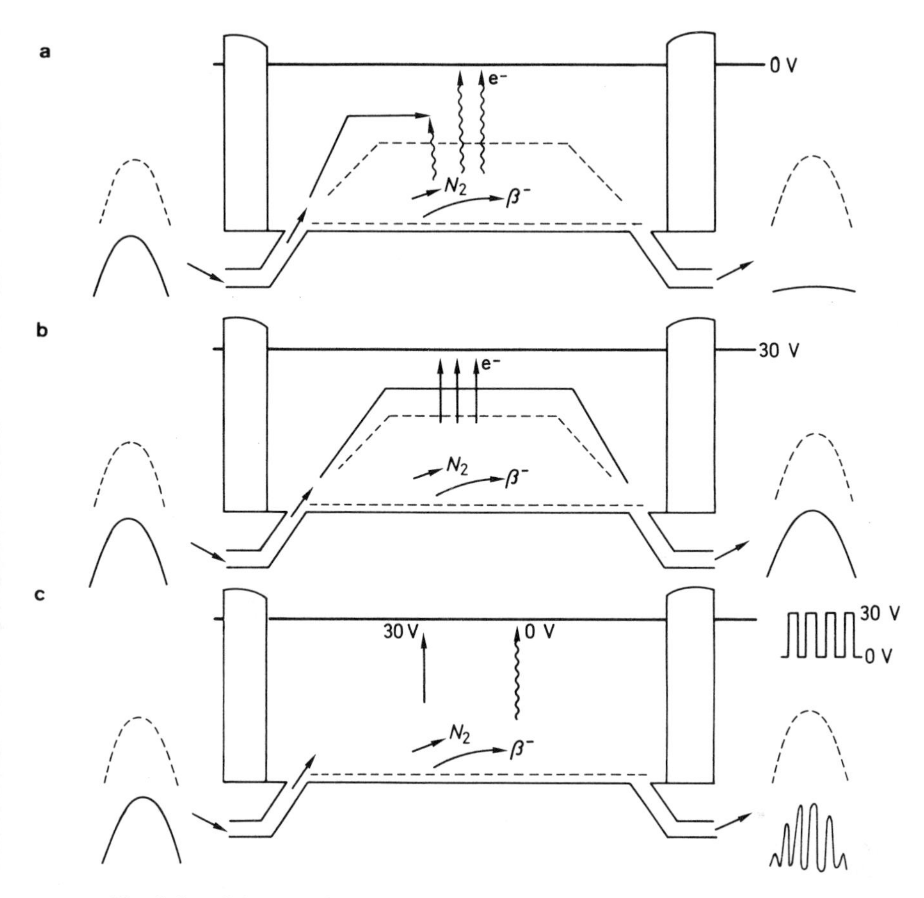

Fig. 7.5 – Schematic diagram of the processes involved in the solute switching. Solid lines represent peaks due to solutes of high electron-affinity, broken lines refer to low electron-affinity solutes (after [135]). (Reproduced from *Journal of Chromatographic Science*, by permission of the copyright holders, Preston Publications Inc.).

components of high electron-affinity, and thereby destroy them, while those with low affinity will pass through the cell unaffected (Fig. 7.5a). If a high enough potential field (e.g. 30 V) is applied to the cell the velocity of the electrons will be too high for them to be captured and all compounds entering the cell will pass through it unaffected (Fig. 7.5b). By alternate switching of the cell between the two states for the field (0 V and 30 V) the peaks for substances with high electron-affinity (e.g. chlorinated pesticides) will be chopped while those for substances with low electron-affinity will remain unaffected (Fig. 7.5c). The resulting modulated signals from the ecd are isolated electronically by a high-pass filter. Only compounds which are capable of being chopped (switched) are thus recorded in the chromatogram (see Fig. 7.6).

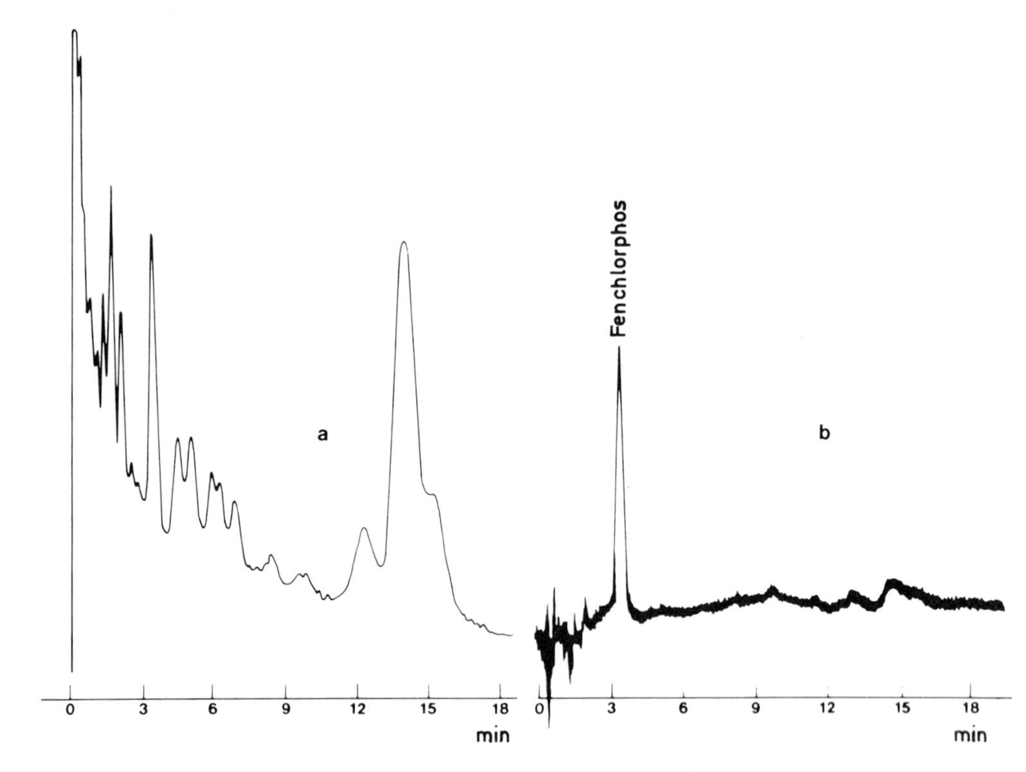

Fig. 7.6 – Normal ecd (a) and the solute-switched ecd (b) chromatograms of a vegetable extract in which no pesticide was originally present. Fenchlorphos was added as a reference compound. (From [135]). (Reproduced from *Journal of Chromatographic Science*, by permission of the copyright holders, Preston Publications Inc.).

Another way of enhancing sensitivity and improving the detection limit is by 'amplification' or 'multiplication'. A well-known example in inorganic trace analysis is the reaction for determination of iodide. It is oxidized to give iodate which can then be reacted with iodide to give iodine.

$$I^- \longrightarrow IO_3^-$$

$$IO_3^- + 5I^- \longrightarrow 3I_2$$

The original quantity of iodide is thus multiplied by a factor of 6. However, any interference in the reaction is also multiplied. Examples of amplification reactions in organic trace analysis are enzymatic cycling reactions.

A further example is 'catalytic amplification' [139, 140]. In a thin-layer amperometric transducer two working electrodes (w) are placed opposite each other (Fig. 7.7). If a chemically reversible electroactive component is passed through the device it will be oxidized at one electrode, diffuse to the other side,

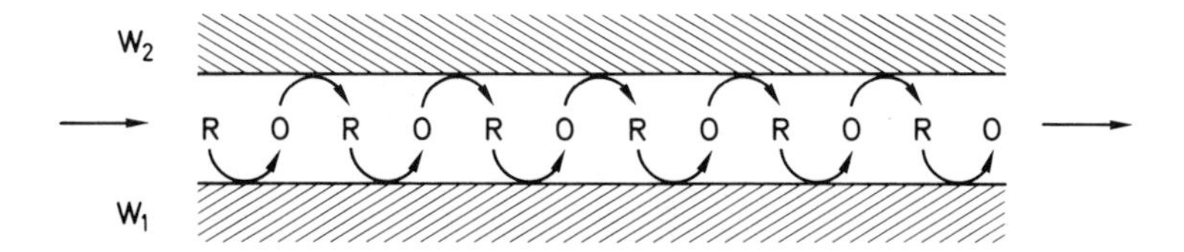

Fig. 7.7 – Twin-electrode thin-layer detector cell for electrochemical recycling of reversible redox-couples (from [139]). (Reproduced from *Journal of Chromatographic Science,* by permission of the copyright holders, Preston Publications Inc.).

and be reduced. In order to greatly enhance the sensitivity the number of diffusion and reduction-oxidation steps must be large. The cell must be thin and the solute flow slow.

Occasionally, an effect is discovered which leads towards a strong enhancement of the analytical signal. The response of an electron-capture detector to polycyclic aromatic hydrocarbons is selectively increased by about two orders of magnitude by addition of 0.2% of oxygen to the carrier gas [141]. Similarly sub-ng amounts of bis(chloromethyl) ether become detectable by a ^{63}Ni ecd, which corresponds to a 40-fold improvement over the detection in oxygen-free carrier gas [142]. The sensitivity of a ^{63}Ni ecd for vinyl chloride is improved about 1000-fold by doping the nitrogen carrier gas with 10–50 ppm of nitrous oxide [143]. In laser-induced infrared fluorescence the gas-phase spectra of ethyl ether, propyl ether and ethyl methyl ketone show considerably enhanced intensity after addition of an excess of propene to the gas sample [144]. The chemiluminescence reaction for detection of nerve gases or insecticides is improved by a factor of 1000 by addition of chloride; the detection limit is shifted from 1 μg to 1 ng [145].

7.4 SOME WAYS TO INCREASE THE SPECIFICITY OF METHODS

Very often specificity is increased by combination of two methods of lower specificity. This can be done by

(1) tandem methods with a sequential combination in which one procedure follows the first, which must be non-consumptive;
(2) stream-splitting and distribution of the isolated trace so that several methods of determination can be used (see Fig. 7.8).

The second variant is less sensitive than the first. Examples of such a combination are stream splitting in gc and simultaneous detection of the effluent by fid and ecd (see Fig. 7.9). Examples of the first variant are a micro flow-cuvette for uv-absorption measurements after hplc, followed by electrochemical detection. The coupling of chromatography with mass spectrometry is probably the

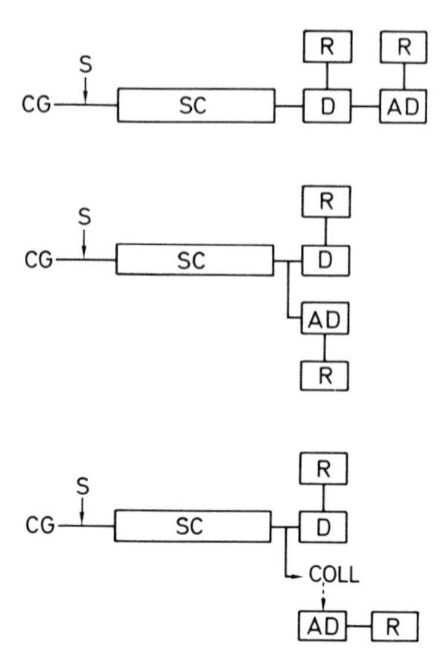

Fig. 7.8 – Block diagram of the various possibilities for combining ancillary methods with the standard gc system. Symbols are: CG carrier gas; S sample; SC separation column; D detector; R recorder (read-out); AD ancillary analytical method; COLL fraction collection (after [146]). [Reproduced from L. S. Ettre, in *Ancillary Techniques of Gas Chromatography*, L. S. Ettre and W. H. McFadden (eds.), p. 3, by permission of the copyright holders, Wiley–Interscience Inc.).

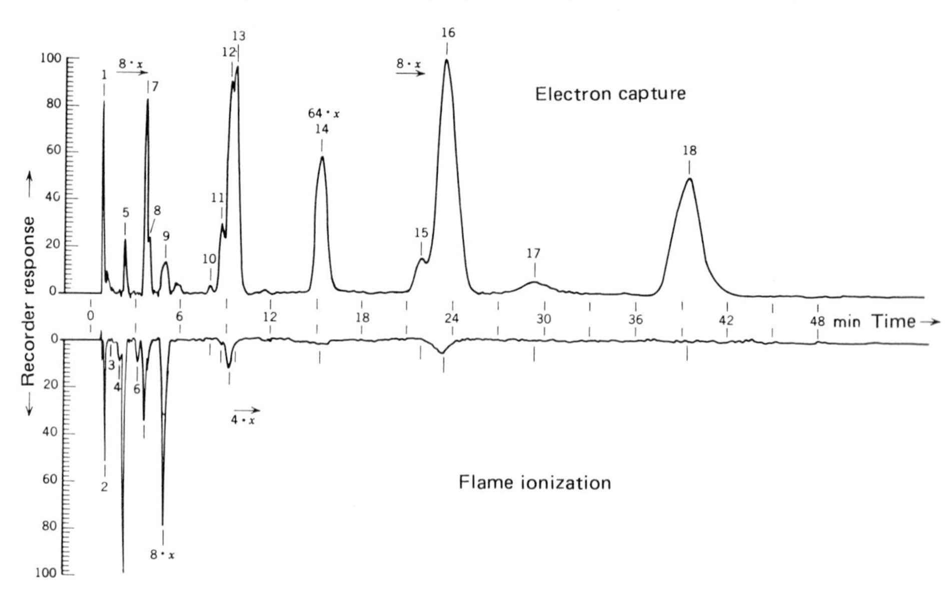

Fig. 7.9 – Chromatogram from head-space analysis of garlic cloves. [Reproduced by permission from D. E. Oaks, H. Hartmann and K. P. Dimick, *Anal. Chem.*, **36**, 1563 (1964). Copyright 1964 American Chemical Society].

best combination so far, in which the fid or ecd is used for highly sensitive but non-specific quantitative determination, and identification is done by the mass spectrometer.

The specificity of a given reaction can be influenced and often improved by suitable treatment of the sample before the determination. In gas chromatography precolumn reactions are often used [146]. The principle is shown diagramatically in Fig. 7.10. Abstraction is mainly used, in which the trace compound is converted into some non-volatile or non-detectable product. Molecular sieves abstract straight chain alkanes or olefins from hydrocarbon mixtures. Other compounds are also abstracted, however, so the selectivity of the method has to be checked (see Table 7.4).

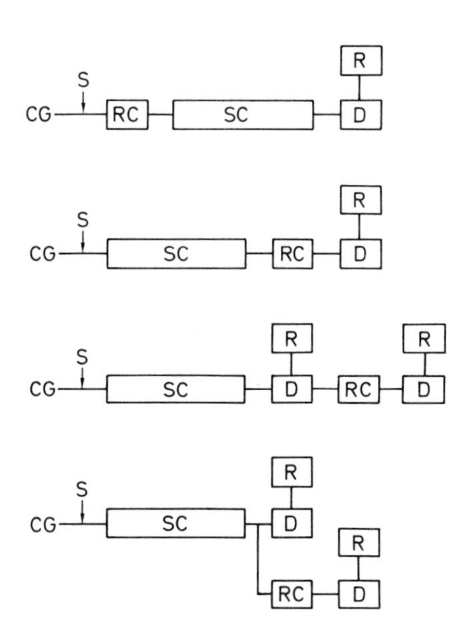

Fig. 7.10 – Block diagrams of the various principal possibilities for reaction gas chromatography. The basic parts of the standard gas chromatographic system are indicated with heavy lines of symbols in italics. *CG*, carrier gas; *S*, sample; *SC*, separation column; *D*, detector; *R*, recorder (read-out); RC, reaction chamber (reactor); D and R, detector and recorder after the reactor. [Reproduced from L. S. Ettre, in *Ancillary Techniques of Gas Chromatography*, L. S. Ettre and W. H. McFadden (eds.), p. 8, by permission of the copyright holders, Wiley–Interscience, Inc.

For abstraction of aldehydes, ketones (and alcohols) a precolumn reactor filled with boric acid, benzidine or dianisidine [148], 2,4-dinitro- or 4-nitrophenylhydrazine [149, 150] or *N,N′*-disubstituted *p*-phenylendiamines [151] has been proposed.

Table 7.4 – Precolumn reactions of various classes of substances (from [147], by permission of the copyright holders, Wiley–Interscience, Inc.).

Type of compound	Reaction used
Alcohols	Conversion into nitrites Dehydration to olefins Reduction to hydrocarbons
Acids	Esterification by flash-exchange with potassium ethyl sulphate Esterification by decomposition of tetramethylammonium salts in injection port
Malonic acids	Decarboxylation to monocarboxylic acid in injection port
α-Hydroxycarboxylic acids	Periodic acid oxidation to carbonyl compounds in injection port
Amino-acids	Oxidation to aldehyde by ninhydrin
Phenols	Formation of methyl ethers by decomposition of tetramethylammonium phenolates in injection port
Purines, pyrimidines, barbiturates	Methylation in injection port by decomposition of tetramethylammonium salt
Amides	Conversion into nitriles in phosphoric acid column
Esters	Saponification in a wet KOH column for analysis of the free alcohols
Salts of acids, mercaptans, and amines	Conversion into the free acid, mercaptan, or amine in an acidic or basic precolumn
Tertiary amines	Conversion into olefins by Hofmann exhaustive methylation
Carbamates	Hydrolysis to phenols on short phosphoric acid precolumn

Another precolumn method is ozonolysis, which permits location of the positions of double bonds, from the fragments produced. The carbon skeleton technique has already been mentioned (see p. 177).

In precolumn partition the trace compound is distributed between two immiscible liquids. An aliquot of one phase is then injected, and a reduction in peak height is observed which is in mathematical relation to the peak height of the untreated sample [152]. This method has the disadvantage that the signal is always reduced in intensity, and that the detector must give linear response.

Users of such combined techniques have been warned [153]: ". . . results may be obtained which were not expected; similarly, results which have been observed with the existing methods when used alone may be found incorrect. These facts may be very irritating and may require additional work. The author . . . wants to caution the readers not to be discouraged".

7.5 SCREENING METHODS IN ORGANIC TRACE ANALYSIS

The trace analyst is very often confronted with the problem of analysis of a large number of samples, e.g. in the analysis of drugs of abuse, in clinical chemistry, industrial hygiene etc. This calls for rapid and simple methods of investigation. Screening methods are used. They are designed so that though falsely positive results may occur, falsely negative results can not. All 'positive' samples can then be re-analysed by a more specific method, whereas all negative results from the screening test can be accepted with confidence, which greatly reduces the load on the more specific system. It is mandatory, of course, that the screening method gives a positive result when the substance to be detected is present. In a typical example, from 419 samples of milk 19% gave a positive result with a screening method [154] for aflatoxins. From 197 samples of cheese 69% seemed to contain the mycotoxins. Reinvestigation by mass spectrometry confirmed the results only for the milk samples, demonstrating the need for a confirmatory test [155]. In another case, for confirmation of positive results of a screening technique for lyncomycin at the 100-μg/kg level in animal tissues (after extraction, clean-up by solvent extraction, and detection by microbiological plate assay) an additional method was used (clean-up column chromatography, then thin-layer chromatography and detection by bioautoradiography [156]).

LSD can be detected in amounts of about 0.5 ng by an ria test [157]. Any samples (blood, urine, stomach fluid etc.) giving negative results are excluded from further investigation, which considerably reduces the number of samples to be further tested. Any positively reacting sample is analysed by reversed phase hplc. The eluted fraction containing the LSD is tested by fluorescence or another ria test [158].

The reliability of results is improved by use of two independent screening methods, e.g. by immunoassay and a tlc system [159]. If the results from these two methods diverge, gas chromatography is used as a third method. Other examples of confirmation of the result of the first screening procedure by a second investigation are the determination of cannabinoids in blood by using EMIT and additional more specific procedures [16], of Δ^9-tetrahydrocannabinol by ria and gc-ms [161], of phencyclidine by tlc with gc or gc-mas as second confirmatory method [162], of phencyclidine by ria and tlc [163], of phencyclidine, morphine, codeine and benzoylecgonine by tlc and EMIT [164], of cocaine by tlc and confirmation through gc [165] and of barbiturates, diazepam and methaqualone by tlc and gc [166].

'Screening analytical chemistry' uses all methods of highly developed qualitative or semiquantitative analytical chemistry [167]. Spot-tests, simple colour reactions in test-tubes and reactions in gas-test tubes (e.g. Dräger tubes) are applied. For these last, a solid support (silica or similar material) is impregnated with a reagent and the gas sample is passed over the reagent layer. Specially prepared reagents fixed onto carriers (paper, polymer films, foams) are also very useful [1968]. Such test-strips are used extensively in clinical chemistry [169, 170] but few examples are known of their use outside this field. There is definitely a great need, though, for development of more such systems, as they are needed for control of the working atmosphere in factories, or on farms after insecticide spraying. For example, microgram amounts of DDT, aldrin, chlordan or endosulphan give coloured spots on a chromogenic paper produced by spraying filter paper with a fresh solution of 1% *o*-tolidine in acetone, drying, and storing in the dark until needed. The pesticides can even be detected by pressing the moist, cut end of a sprayed vegetable against the impregnated paper for 30 sec [171]. In clinical chemistry and forensic chemistry, more recent techniques employ enzyme immunoassay test-reactions on carriers, e.g. for detecting drugs in biological systems [172].

'Wipe tests' are applied when samples are to be taken from outdoor surfaces. In field tests at Seveso, 1 ng of TCDD per m^2 on soil or buildings has been detected by a combination of cotton-swab wiping with gc–ms [173]. There is definitely a great need for more of such simple tests in screening analytical chemistry.

Data on the behaviour of 570 commonly used drugs and organophosphorus pesticides in tlc, gc (with OV-1 and OV-17) and hplc (reversed phase C_{18}) have been published [174]. Analogously, more than 60 narcotic and addictive drugs have been investigated with respect to their liquid-liquid distribution and subsequent identification by tlc [175]. Use of hplc has been suggested for the determination of more than 40 drugs [176]. A system has been described which uses twelve gas chromatographs, together with computer-assisted data-handling, to screen about 500 samples daily for central nervous system stimulants, sympathomimetic amines and narcotic analgesics [177]. All these investigations have been undertaken in order to build up rapid and reliable screening systems.

Examples of screening procedures are given in Table 7.5.

7.6 LOCAL ANALYSIS AND ANALYSIS OF SURFACES

Although the analysis of restricted small areas is more a problem of microanalysis, the knowledge of a few principles can be helpful in solving some problems of trace analysis.

Surface areas are analysed – to a depth of less than 1.6 μm – by infrared attenuated reflection (ir-ATR) spectroscopy [201]. By combination with Fourier techniques and electronic time-averaging the sensitivity for surface investigations can be greatly improved [202].

In 'laser desorption mass spectrometry' [203] the surface area is exposed to a submicrosecond laser pulse producing cationic species from the molecules of the surface, and these are then analysed by mass spectrometry. About 1 ng of ethanol, acetone or sodium stearate per cm^2 of a solid surface can be determined by ms after excitation and evaporation of the traces by laser [204].

A carbon fibre electrode implanted in the neostriatum of a rat's brain has been used for *in vivo* measurement of extracellular concentration of dopamine [205]. The authors compared the results obtained by *in vivo* and *in vitro* measurements [206].

Table 7.5 – Examples of screening methods in organic trace analysis.

Substance(s) screened	Level	Matrix	Method	Reference
common mycotoxins (14)	100 mg/kg	food	extrn./partn./tlc	178
mycotoxins (12)	3–1000 μg/kg	animal feed	membrane clean-up, tlc	179
aflatoxins, ochratoxin	4–16 μg/kg	corn, peanuts, food	extrn./partn./lc, fluorescence	180
aflatoxin	5 μg/kg	corn	extrn./partn./tlc	181
mycotoxins	5–35 μg/kg	cereals	extrn./gel filtn./tlc	182
paraquat		marihuana	partn./tlc	183
paraquat		marihuana	gc or hplc/photometry	184
barbiturates, benzodiazepines, amphetamines, several opiates	0.2–10 mg/l.	urine	EMIT/tlc (‘Drug screen’)	185
cocaine, benzoylecgonine	0.1 mg/l.	urine	partn./tlc	186
drugs of abuse	0.5 μmole/ml	urine	partn./ms (CI-mode)	187
LSD	0.5 ng	body fluids	ria	188
drugs of abuse	1 mg/l.	urine	partn./tlc	189
street drugs	3 μg	drug preparations	extrn./tlc	190
street drugs		drug preparations	extrn./tlc	191
diazepam, chlordiazepoxide		urine	hydrolysis/partn./tlc	192
barbiturates, hydantoins	0.5 μg	drug preparations	extrn./spot-test	193
several drugs		biological material	several partns./tlc	194
isoniazid	50 μg/l.	serum	ria	195
triglycerides	1.6 g/l.	serum	direct ria	196
antibiotics	1 mg/l.	milk	microbiological	197
lincomycin	0.1 mg/kg	animal tissue	extrn./partn./microbiol. test	198
histamine		tuna	extrn./tlc	199
butylhydroxyanisole	25 μg/kg	food	water-vap. distn./fluorescence	200

7.7 TELEMETRY IN TRACE ANALYSIS

In some cases the concentration of trace constituents in very remote places is to be determined. This problem can be attacked by

(1) sampling at the remote place and transportation of the sample to the laboratory;
(2) measurement of the trace constituent locally and 'transportation' of a measured signal to the laboratory;
(3) investigation over a long distance (very long path-length).

An example of remote sampling and transportation is the measurement of 0.2 ppb of CF_2Cl_2 and 0.1 ppb of $CFCl_3$ at an altitude of 4000 m (North-west Pacific, Nov. 1974). Air was collected in special steel containers, the surface of which had been electropolished by a special method to prevent adsorption. The actual analysis of the gas sample was done by mass spectrometry [207]. A rather similar approach has been used [208] for high-altitude air samples, which were collected by a U-2 aircraft. A freeze-out loop was used for concentration of the components of the sample, and gas chromatography with ecd was performed afterwards in the laboratory.

An automatic water sampler has been devised [209]. It is capable of unattended operation for up to 7 days in a remote location. The sampler collects and seals up to 26 grab samples according to preset sampling frequencies, sequences, flow-rates, and collection periods. The equipment concentrates up to 14 samples on various accumulators, such as XAD resins. After the collection period the samples are taken to the laboratory. A battery-operated sampler for gas chromatographic determination of traces of pesticides in air has been described [210]. Portable gas chromatographs have been invented for *in-situ* balloon-borne application to measure stratospheric halocarbons [211], or from 0.1 to 100 ppm of air contaminants by photoionization [212], or for an on-the-spot investigation of arson [213]. A portable microwave multi-gas analyser has detection limits in the ppm region [214]. A mobile miniature infrared analyser has been designed for the detection of gases and vapours at the ppb level [215]. A small and portable quadrupole ms system can be applied to measure atmospheric pollution [216]. An interesting micro gas chromatograph has been described under the title 'Let's get small' [217]. A spiral grove (0.5 m long) is etched into a silicon wafer and capped with a glass plate. A channel, about 200 μm wide and 20 μm deep, results. A microbead thermistor is implanted at the outlet of this little gas chromatograph, which has a diameter of about 5 cm. Injection of sample volumes of about 1 nl is possible. A detection limit of 10–50 ppm has been reported. Intended uses are application in future space missions of NASA or as a pocket-size contaminant monitor. With expanding space technology, specially designed equipment is transported into space or the stratosphere. Special column fillings for gc analysis of the atmosphere of Jupiter and Venus have been prepared [218].

Measurement of the concentration of a trace constituent and transmission of an analogue electrical signal to the surveillance laboratory is a technique which is often used in technical chemistry. Monitoring of water systems or of process streams are such examples. Remote testing of body fluids during monitoring of hospital patients, e.g. by means of ion-selective electrodes, is still a field undergoing development [219].

Examples of measurement over long distances are reviewed in some articles [220-222] and a book [223]. Investigation of the optical properties of a volume of air containing traces is done by absorption (see Figs. 7.11 and 7.12), or by measurement of back-scattered light, or by measurement of the radiant energy emitted from the pollution source (Fig. 7.13).

Obviously, lasers are very useful for "doing analysis at a place you are *not*. For example, when your sample happens to be a chemical warfare agent, local analysis method would tell you, 'You are dead'." [224]. The potential of lasers for the detection of atmospheric pollutants in very low concentrations, at a

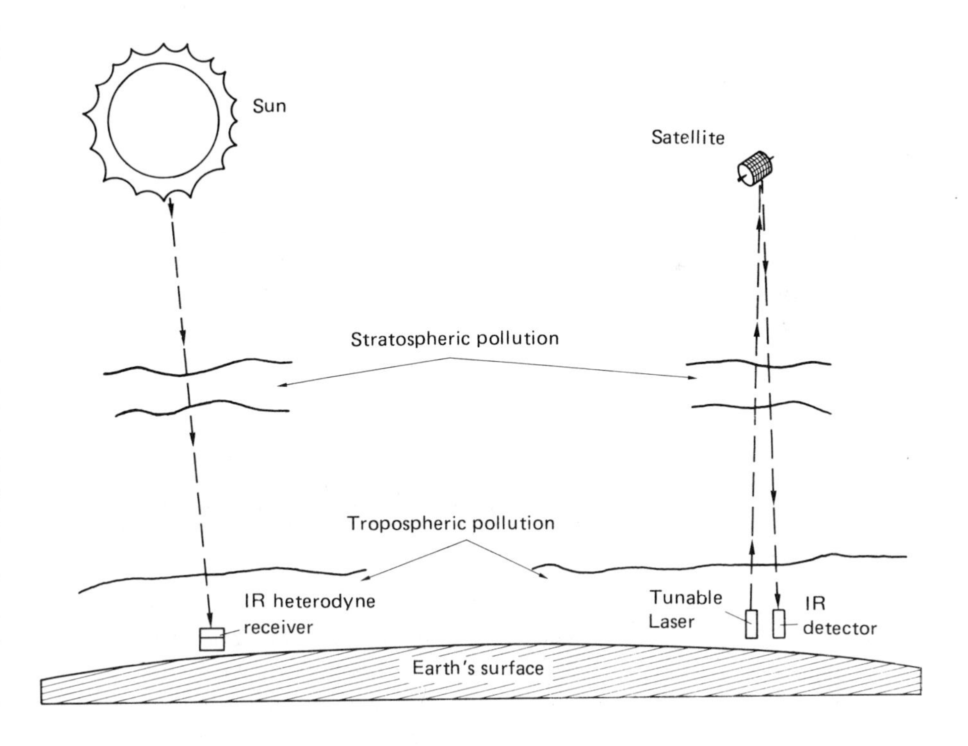

Fig. 7.11 – Long-path vertical profile monitoring of the atmosphere, using (a) passive system with sun as radiation source and tunable laser detection, and (b) active, ground-based system employing co-operative satellite reflector (after [221], reproduced by permission of the copyright holders, Academic Press Inc.).

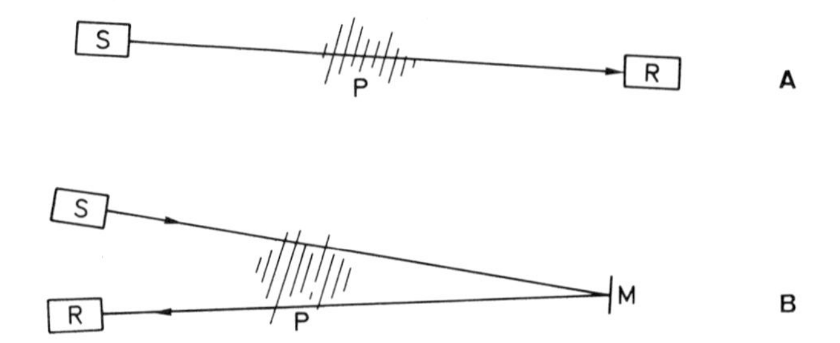

Fig. 7.12 – Measurement of the optical properties of air by optical absorption over a long path (after [220]). Method (A) direct absorption; Method (B) measurement by reflection at a mirror M, S source, P pollutant, R receiver. (Reproduced by permission of the copyright holders, Springer Verlag).

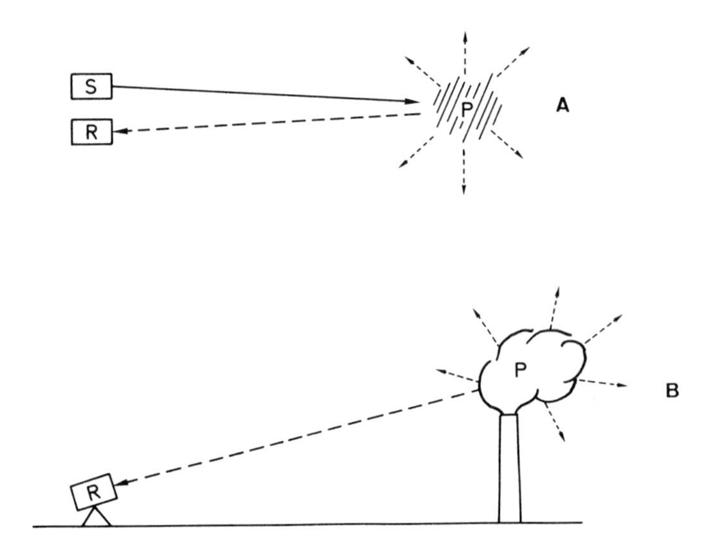

Fig. 7.13 – (A) Measuring a cloud of pollutant by the light that it scatters back to the source. S luminous source; P pollutant; R receiver. (B) Measuring a cloud of pollutant that is emitting radiant energy (after [220]). (Reproduced by permission of the copyright holders, Springer Verlag).

range of many kilometres, is mentioned in another review on the current potential applications of lasers [225]. Several types of lasers have been described [221] and the potentials of lasers in inorganic and organic trace analysis have been outlined [226]. From 0.02 to 45 ppm of ethylene has been determined by using a computer-controlled carbon-dioxide laser and a long-path absorption system [227]. Analogously, 3 ppm of ethylene in automobile exhaust can be measured [228].

The light emitted from planets or the sun can be used as a source for absorption measurements (see Fig. 7.11). In this way terrestrial and planetary atmospheres are analysed by using FT-ir methods [229].

Within the scientific area of 'analysis over great distances' and 'extraterrestrial analysis' only a few applications to organic trace analysis have been made, so far. Most applications aim at inorganic compounds (NO_2, CO_2 etc.) which are determined more easily. There will be an increasing demand for detection of organic molecules – an interesting field for the future of organic trace analysis.

7.8 APPLICATION OF TRACE ANALYSIS TO RECOGNITION OF DISEASES BY ANALYSIS PROFILES OF CONSTITUENTS OF BODY FLUIDS

A complete profile of all the constituents of a body fluid is impossible, at present [23]. The goal, therefore, is the development of a complete profile of a

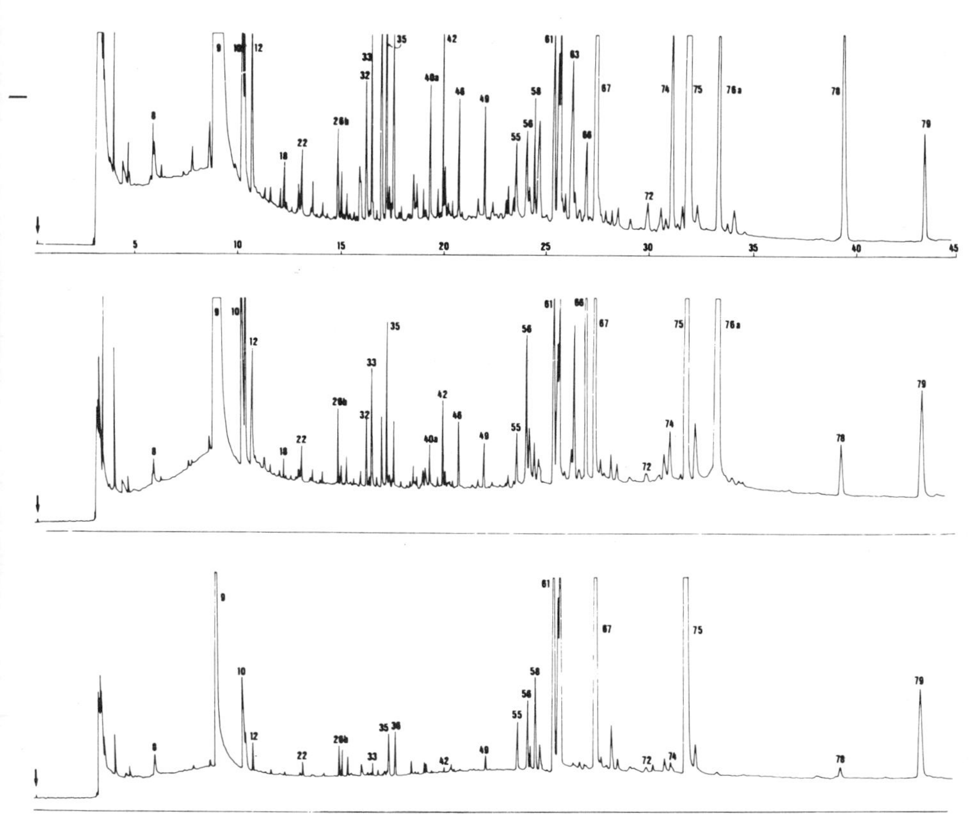

Fig. 7.14 – Gas chromatographic profiles of ultrafiltered serum from a uremic patient, before (top) and after (middle) hemodialysis treatment. For comparison (bottom): pool of non-uremic sera (after [251]). (Reproduced by permission of the copyright holders, Elsevier Science Publishing Co., Amsterdam).

selected group of substances (e.g. steroids, amino-acids etc). 'Normal' samples are compared with 'pathological' samples. Any differences might become of value for diagnostic purposes or lead to a possible biochemical explanation for the disease. Large screening programs must be run and as many substances as possible should be included in the profile analysis. This leads to analytical problems and the necessity for the application of methods with the highest resolution, such as capillary gc. The use of pattern-recognition techniques and computer sorting is mandatory for handling the vast amount of information obtained [231-250]. Metabolic profiles have been determined by gc-ms (CI-mode) in investigation of uremia (see also Fig. 7.14). The same disease was the object of studies based on use of gel chromatography [252-255], reversed-phase chromatography [256] and isotachophoresis [257]. Organic acids are the most often investigated group of components [258]. High-resolution gas chromatography [259, 260] and hplc have been used for their study [261].

7.9 THE FUTURE

Clearly, the enormous field of organic trace analysis has been only partly explored, and as new techniques become available, or even more ingenious exploitation of the existing ones is made, more and more of the problems in this area will be solved. It is hoped that this volume will help to stimulate interest and inventiveness in this vast and largely untapped field of scientific research.

REFERENCES

[1] F. B. DeWalle, C. Lo, J. Sung, D. Kalman, E. S. K. Chian, M. Giabbai and M. Ghosal, *J. Water Pollut. Control Fed.*, **54**, 555 (1982).
[2] D. Grange and P. Clement, *Rapp. Rech. LPC*, No. 103 (1981).
[3] Q. V. Thomas, J. R. Stork and S. L. Lammert, *J. Chromatog. Sci.*, **18**, 583 (1980).
[4] R. Briggs, *Public Health Engineer*, **10**, 73 (1982).
[5] H. Heberer and G. Bittersohl, *Z. Chem.*, **20**, 361 (1980).
[6] O. Hutzinger, R. W. Frei, E. Merian and F. Pocchiari (eds.), *Chlorinated Dioxins and Related Compounds*, Pergamon Press, Oxford (1982).
[7] National Research Council of Canada, *Natl. Res. Counc. Can. Rept.*, No. 18576 (1981).
[8] T. Cairns, L. Fishbein and R. K. Mitchum, *Biomed. Mass Spectrom.*, **7**, 484 (1980).
[9] P. G. Baker, R. A. Hoodless and J. F. C. Tyler, *Pestic. Sci.*, **12**, 297 (1981).
[10] T. Cairns and E. G. Siegmund, *Anal. Chem.*, **53**, 1183A (1981).
[11] J. Harvey and G. Zweig (eds.), *Pesticide Analytical Methodology*, American Chemical Society, Washington, D.C. (1980).
[12] T. A. Pressley and J. A. Longbottom, *US Environ. Prot. Agency, Off. Res. Dev.* (Rept.), EPA-600/4-82-006 (1982).
[13] R. A. Scanlan and S. R. Tannenbaum (eds.), *N-Nitroso Compounds*, American Chemical Society, Washington, D.C. (1981).
[14] R. W. Frei and U. A. T. Brinkman (eds.), *Mutagenicity Testing and Related Analytical Techniques*, Gordon and Breach, New York (1981).
[15] M. L. Brenner, *Ann. Rev. Plant Physiol.*, **32**, 511 (1981).
[16] E. C. Gunderson, C. C. Anderson, R. H. Smith and L. J. Doemeney, *DHHS (NIOSH) Publ. (US)*, No. 80-133 (1980).

[17] T. A. Gough and P. B. Baker, *J. Chromatog. Sci.*, **20**, 289 (1982).
[18] J. E. Lindgren, *J. Ethnopharmacol.*, **3**, 337 (1981).
[19] Anon., *Anal. Proc.*, **18**, 114 (1981).
[20] W. A. Garland and M. L. Powell, *J. Chromatog. Sci.*, **19**, 392 (1981).
[21] S. Facchetti (ed.), *Application of Mass Spectrometry to Trace Analysis*, p. 313. Elsevier, Amsterdam (1982).
[22] M. N. Capelloni, L. G. Frederick, D. R. Latshaw and J. B. Wallace, *Anal. Chem.*, **49**, 1218 (1977).
[23] M. Mauzac, F. Guerard, J. Matthieu and J. Laroche, *Analusis*, **4**, 236 (1976).
[24] H. Tyras and A. Blochowicz, *Chem. Anal. (Warsaw)*, **21**, 647 (1976).
[25] B. Nagorski and M. Boguslawska, *Chem. Anal. (Warsaw)*, **22**, 127 (1977).
[26] M. Mantel and R. Nothmann, *Analyst*, **102**, 672 (1977).
[27] N. Ramakrishna and B. Ramachandran, *J. Indian Chem. Soc.*, **55**, 185 (1978).
[28] S. M. Freund, W. B. Maier, R. F. Holland and W. H. Beattie, *Anal. Chem.*, **50**, 1260 (1978).
[29] S. R. Lowry and J. D. Banzer, *Anal. Chem.*, **50**, 187 (1978).
[30] Y. Suzuki, S. Imai and A. Hamaguchi, *Taiki Osen Kenkyu*, **12**, 255 (1978).
[31] I. Hornyak, *Acta Chim. Acad. Sci. Hung*, **90**, 103 (1976).
[32] G. L. Brun and D. L. D. Milburn, *Anal. Lett.*, **10**, 1209 (1977).
[33] D. Schuetzle, *Biomed. Mass Spectrom.*, **2**, 288 (1975).
[34] F. Pellerin and J. F. Letavernier, *Analusis*, **5**, 19 (1977).
[35] A. Rogstad and K. Hogberg, *Anal. Chim. Acta*, **94**, 461 (1977).
[36] L. H. Howe, *Anal. Chem.*, **48**, 2167 (1976).
[37] R. Samuelsson, *Anal. Chim. Acta*, **102**, 133 (1978).
[38] D. E. Leelavathi and R. W. Guynn, *Anal. Biochem.*, **72**, 261 (1976).
[39] W. A. Groff, R. I. Ellin and R. L. Skalsky, *J. Pharm. Sci.*, **66**, 389 (1977).
[40] D. M. Karl, *Anal. Biochem.*, **89**, 581 (1978).
[41] R. Stedry and B. Ehrlichova, *Radiochem. Radioanal. Lett.*, **34**, 209 (1978).
[42] M. I. Luster, P. W. Albro, K. Chae, G. Clark and J. D. McKinney, *Clin. Chem.*, **24**, 429 (1978).
[43] S. Donhauser, *Brauwiss.*, **30**, 325 (1977).
[44] S. Stavcharsky, T. Ludden, J. P. Allen and P. Wu, *Anal. Chim. Acta*, **92**, 213 (1977).
[45] L. A. Ouderkirk, *J. Assoc. Off. Anal. Chem.*, **59**, 1122 (1976).
[46] H. S. Ragheb, *J. Assoc. Off. Anal. Chem.*, **60**, 1119 (1977).
[47] C. A. Fassbender and S. E. Katz, *J. Assoc. Off. Anal. Chem.*, **59**, 1113 (1976).
[48] J. M. Inglis and S. E. Katz, *J. Assoc. Off. Anal. Chem.*, **61**, 1098 (1978).
[49] N. Takano, K. Nomura and A. Arakawa, *J. Assoc. Off. Anal. Chem.*, **60**, 206 (1977).
[50] H. Steinhart and J. Sandmann, *Anal. Chem.*, **49**, 950 (1977).
[51] A. Schmitt, *J. Clin. Chem. Clin. Biochem.*, **15**, 303 (1977).
[52] G. Mamantov, E. L. Wehry, R. R. Kemmerer and E. R. Hinton, *Anal. Chem.*, **49**, 86 (1977).
[53] R. E. Smith and R. MacQuarrie, *Anal. Biochem.*, **90**, 246 (1978).
[54] J. C. Andre, P. Baudot and M. Niclause, *Clin. Chim. Acta*, **76**, 55 (1977).
[55] R. Farooq and G. F. Kirkbright, *Analyst*, **101**, 566 (1976).
[56] R. M. A. von Wandruszka and R. J. Hurtubise, *Anal. Chem.*, **48**, 1784 (1976).
[57] L. A. Sternson and G. Thomas, *Anal. Lett.*, **10**, 99 (1977).
[58] B. J. A. Haring and W. von Delft, *Anal. Chim. Acta*, **94**, 201 (1977).
[59] H. Konrad and T. Gabrio, *Nahrung*, **20**, 395 (1976).
[60] L. Hofer-Siegrist, *Schweiz. Z. Hydrologie*, **38**, 49 (1976).
[61] C. P. Cheung and R. J. Suhadolnik, *Anal. Biochem.*, **83**, 57 (1977).
[62] V. F. Kalb, T. J. Donohue, M. G. Corrigan and R. W. Bernlohr, *Anal. Biochem.*, **90**, 47 (1978).
[63] P. H. Yu, *Anal. Biochem.*, **86**, 498 (1978).
[64] K. P. Wong, *Anal. Biochem.*, **82**, 559 (1977).
[65] E. J. Moerman, M. G. Bogaert and A. F. de Schaepdryver, *Clin. Chim. Acta*, **72**, 89 (1976).
[66] P. Kamoun, G. Lafourcade and H. Jerome, *Clin. Chem.*, **22**, 964 (1976).

[67] D. W. Anderson and U. Goebelsmann, *Clin. Chem.*, **22**, 611 (1976).
[68] M. Hüfner and M. Grussendorf, *Clin. Chim. Acta*, **69**, 497 (1976).
[69] K. Horn, I. Marschner and P. C. Scriba, *J. Clin. Chem. Clin. Biochem.*, **14**, 353 (1976).
[70] W. J. F. van der Vijgh, *Pharm. Weekbl.*, **113**, 585 (1978).
[71] R. W. Lam, R. Artal and D. A. Fisher, *Clin. Chem.*, **23**, 1264 (1977).
[72] A. E. Cowan, M. G. Korman, A. F. Hoffmann, J. Turcotte and J. A. Carter, *J. Lipid Res.*, **18**, 692 (1977).
[73] W. A. Ratcliffe, S. W. Fletcher, A. C. Moffat, J. G. Ratcliffe, W. A. Harland and T. E. Levin, *Clin. Chem.*, **23**, 169 (1977).
[74] F. A. Fitzpatrick, R. R. Germann, J. M. McGuire, R. C. Kelly, M. A. Wynalda and F. F. Sun, *Anal. Biochem.*, **82**, 1 (1977).
[75] M. Janoušková, H. Fingerová and M. Talaš, *Radiochem. Radioanal. Lett.*, **34**, 289 (1978).
[76] H. Fingerová, *Radiochem. Radioanal. Lett.*, **34**, 261 (1978).
[77] D. A. Brent, *J. Assoc. Off. Anal. Chem.*, **59**, 1006 (1976).
[78] M. T. Carlton, T. R. Witty, M. J. Hasler, R. E. Bjomsen and K. H. Painter, *J. Radioanal. Chem.*, **43**, 389 (1978).
[79] P. Glogner, N. Henri and L. Nissen, *Arzneim. Forsch./Drug Res.*, **27**, 1703 (1977).
[80] R. H. Snider, C. F. Moore, O. L. Silva and K. L. Bekker, *Anal. Chem.*, **50**, 449 (1978).
[81] R. Schwenk, K. Kelly, K. S. Tse and A. H. Sehon, *Clin. Chem.*, **21**, 1059 (1975).
[82] M. Usategui-Gomez, J. E. Heveran, R. Cleeland, B. McGhee, Z. Telischak, T. Awdziej and E. Grunberg, *Clin. Chem.*, **21**, 1378 (1975).
[83] A. J. Schäfer and H. Faulstich, *Anal. Biochem.*, **83**, 720 (1977).
[84] L. A. Ouderkirk, *J. Assoc. Off. Anal. Chem.*, **60**, 1116 (1977).
[85] R. C. Stroupe, P. Tokousbalides, R. B. Dickinson, E. L. Wehry and G. Mamantov, *Anal. Chem.*, **49**, 701 (1977).
[86] D. J. Reeder, L. T. Sniegoski and R. Schaffer, *Anal. Biochem.*, **86**, 490 (1978).
[87] G. F. Read and D. Riad-Fahmy, *Clin. Chem.*, **24**, 36 (1978).
[88] B. Hoffmann, *J. Assoc. Off. Anal. Chem.*, **61**, 1263 (1978).
[89] P. J. Richardson, K. J. Favell, G. C. Gidley and G. H. Jones, *Analyst*, **103**, 865 (1978).
[90] L. Quihillalt, E. Rolleri and R. Malvano, *Clin. Chim. Acta*, **72**, 187 (1976).
[91] C. A. Miller and M. C. Fetter, *J. Lab. Clin. Med.*, **89**, 1125 (1977).
[92] H. R. Lukens, C. B. Williams, S. A. Levinson, W. B. Dandliker, D. Murayama and R. L. Baron, *Environ. Sci. Technol.*, **11**, 293 (1977).
[93] B. Gracia-Pascual, A. Peytremann, B. Courvoisier and D. E. M. Lawson, *Clin. Chim. Acta*, **68**, 99 (1976).
[94] J. A. Corkill, M. Joppich, S. H. Kuttab and R. W. Giese, *Anal. Chem.*, **54**, 481 (1982).
[95] P. M. Burkinshaw, E. D. Morgan and C. F. Poole, *J. Chromatog.*, **132**, 548 (1978).
[96] D. Exley, B. Woodhams and K. Shrimanker, *J. Steroid Biochem.*, **14**, 1177 (1981).
[97] B. S. Das and G. H. Thomas, *Anal. Chem.*, **50**, 967 (1978).
[98] D. J. A. Dear, A. G. Dillon and A. N. Freedman, *J. Chromatog.*, **137**, 315 (1977).
[99] H. Yamasaki and K. Kuwata, *Bunseki Kagaku*, **26**, 1 (1977).
[100] E. E. McNeil, R. Otson, W. F. Miles and F. J. M. Rajabalee, *J. Chromatog.*, **132**, 277 (1977).
[101] A. Gaudry, B. Bonsang, B. C. Nguyen and P. Nadaud, *Chemosphere*, **10**, 731 (1981).
[102] D. K. Basu and J. Saxena, *Environ. Sci. Technol.*, **12**, 791 (1978).
[103] R. L. Petty, *Mar. Chem.*, **10**, 409 (1981).
[104] D. E. Harsch and R. A. Rasmussen, *Anal. Lett.*, **10**, 1041 (1976).
[105] A. Di Domenico, V. Silano, G. Viviano and G. Zapponi, *Ecotoxicol. Environ. Saf.*, **4**, 283 (1980).
[106] L. W. Hrubesh, *Anal. Chim. Acta*, **86**, 263 (1976).
[107] B. A. Petersen, R. N. Hanson, R. W. Giese and B. L. Karger, *J. Chromatog.*, **126**, 503 (1976).
[108] B. A. Petersen and P. Vouros, *Anal. Chem.*, **49**, 1304 (1977).

[109] V. Rus, I. Fundue, A. Crainiceanu and S. Trestianu, *Anal. Chem.*, **49**, 2133 (1977).
[110] R. Takeshita, E. Takabatake, K. Minagawe and Y. Takizawa, *J. Chromatog.*, **133**, 303 (1977).
[111] J. L. Bove, B. Dalven and V. P. Krukeja, *Intern. J. Environ. Anal. Chem.*, **5**, 189 (1978).
[112] H. G. Nowicki, C. A. Kieda, V. Current and T. H. Schaefers, *J. High Resolut. Chromatog. Chromatog. Commun.*, **4**, 178 (1981).
[113] A. Cavallaro, G. Bartolozzi, D. Carreri, G. Bandi, L. Luciani, G. Villa, A. Gorni and G. Invernizzi, *Chemosphere*, **9**, 623 (1980).
[114] K. Leuenberger and E. Baumgartner, *J. Chromatog.*, **178**, 543 (1979).
[115] E. W. Weiler and U. Wieczorek, *Planta*, **152**, 159 (1981).
[116] M. Oehme, *TrAC*, **1**, 321 (1982).
[117] J. A. Corkill, M. Joppich, A. Nazareth and R. W. Giese, *J. Chromatog.*, **240**, 415 (1982).
[118] R. E. Hurst and R. L. Settine, *Anal. Chem.*, **53**, 2175 (1981).
[119] L. L. Lamparski, T. J. Nestrick and R. H. Stehl, *Anal. Chem.*, **51**, 1453 (1979).
[120] G. Grimmer, K. W. Naujack and D. Schneider, *Z. Anal. Chem.*, **311**, 475 (1982).
[121] J. L. Koenig, *Appl. Spectrosc.*, **29**, 293 (1975).
[122] D. W. Vidrine, *J. Chromatog. Sci.*, **17**, 477 (1979).
[123] D. Kuehl and P. R. Griffiths, *Anal. Chem.*, **50**, 418 (1978).
[124] M. M. Gomez-Taylor and P. R. Griffiths, *Anal. Chem.*, **50**, 422 (1978).
[125] M. M. Gomez-Taylor, D. Kuehl and P. R. Griffiths, *Intern. J. Environ. Anal. Chem.*, **5**, 103 (1978).
[126] D. R. Jarvie, A. F. Fell and M. J. Stewart, *Clin. Chim. Acta*, **117**, 153 (1981).
[127] P. Wiegart, *Z. Rechtsmed.*, **86**, 221 (1981).
[128] G. Talsky, S. Götz-Meyer and H. Betz, *Mikrochim. Acta*, 1 (1981 **II**).
[129] T. C. O'Haver, *Anal. Chem.*, **51**, 91A (1979).
[130] F. Spreitzhofer, *GIT*, 117 (1978).
[131] K. Beyermann and J. Dietz, *Z. Anal. Chem.*, **268**, 192 (1974).
[132] S. Siggia and R. A. Dishman, *Anal. Chem.*, **42**, 253 (1974).
[133] F. Nachtmann, H. Spitzy and R. W. Frei, *Anal. Chem.*, **48**, 1576 (1976).
[134] F. Nachtmann, *Z. Anal. Chem.*, **282**, 209 (1976).
[135] P. R. Boshoff and B. J. Hopkins, *J. Chromatog. Sci.*, **17**, 588 (1979).
[136] J. E. Lovelock, *J. Chromatog.*, **99**, 3 (1974).
[137] J. E. Lovelock, *J. Chromatog.*, **112**, 29 (1975).
[138] P. G. Simmonds, A. J. Lovelock and J. E. Lovelock, *J. Chromatog.*, **126**, 3 (1976).
[139] P. T. Kissinger and K. Brat, *J. Chromatog. Sci.*, **17**, 142 (1979).
[140] D. A. Roston and P. T. Kissinger, *Anal. Chem.*, **54**, 429 (1982).
[141] D. A. Miller, K. Skogerboe and E. P. Grimsrud, *Anal.Chem.*, **53**, 464 (1981).
[142] G. J. Kallos, *Anal. Chem.*, **53**, 963 (1981).
[143] P. D. Goldan, F. C. Fehsenfeld, W. C. Kuster, M. P. Phillips and R. E. Sievers, *Anal. Chem.*, **52**, 1751 (1980).
[144] S. Higuchi, S. Tanaka, N. Kihara and H. Waki, *Chem. Lett.*, 609 (1981).
[145] U. Fritsche, *Anal. Chim. Acta*, **118**, 179 (1980).
[146] M. Beroza and M. N. Inscoe, in *Ancillary Techniques of Gas Chromatography*, L. S. Ettre and W. H. McFadden (eds.) pp. 89–144, Wiley-Interscience, New York (1969).
[147] M. Beroza and M. N. Inscoe, in *Ancillary Techniques of Gas Chromatography*, L. S. Ettre and W. H. McFadden (eds.), p. 136. Wiley–Interscience, New York (1969).
[148] T. B. Pang, *Yu Chia Hua Hsueh*, **1**, 39 (1981).
[149] K. Komárek, J. Nováková, K. Ventura and J. Churácek, *Collection Czech. Chem. Commun.*, **47**, 2121 (1982).
[150] P. Kalo, *J. Chromatog.*, **205**, 39 (1981).
[151] M. B. Evans, *Chromatographia*, **13**, 555 (1980).
[152] D. A. Leathard and B. C. Shurlock, *Identification Techniques in Gas Chromatography*, Wiley–Interscience, New York (1970).
[153] R. Kaiser, in *Ancillary Techniques of Gas Chromatography*, L. S. Ettre and W. H. McFadden (eds.), p. 301. Wiley-Interscience, New York (1969).

[154] F. Kiermeyer and G. Weiss, *Z. Lebensmittel Unters. Forsch.*, **160**, 337 (1976).
[155] G. Weiss, M. Miller and G. Behringer, *Milchwiss.*, **33**, 409 (1978).
[156] A. R. Barbiers and A. W. Neff, *J. Assoc. Off. Anal. Chem.*, **59**, 849 (1976).
[157] W. A. Ratcliff, *Clin. Chem.*, **23**, 169 (1977).
[158] P. J. Twitchett, S. M. Fletcher, A. T. Sullivan and A. C. Moffat, *J. Chromatog.*, **150**, 73 (1978).
[159] M. Oellerich and R. Haeckel, *Lab. Med.*, **3**, 65 (1979).
[160] H. W. Peel and B. J. Perrigo, *J. Anal. Toxicol.*, **5**, 165 (1981).
[161] R. A. Bergmann, T. Lukaszewski and S. Y. S. Wang, *J. Anal. Toxicol.*, **5**, 85 (1981).
[162] B. Kaul and B. Davidov, *Clin. Toxicol.*, **16**, 7 (1980).
[163] R. D. Budd and W. J. Leung, *Clin. Toxicol.*, **18**, 85 (1981).
[164] R. W. Michalek and T. A. Rejent, *J. Anal. Toxicol.*, **4**, 215 (1980).
[165] L. S. F. Hsu, J. I. Sharrard, C. Love and T. C. Marrs, *Ann. Clin. Biochem.*, **18**, 368 (1981).
[166] H. H. McCurdy, L. J. Lewellen, J. C. Cagle and E. T. Solomon, *J. Anal. Toxicol.*, **5**, 253 (1981).
[167] M. Geldmacher von Mallinckrodt, *Einfache Untersuchungen auf Gifte im klinisch-chemischen Laboratorium,* Thieme, Stuttgart (1976).
[168] V. M. Ostrovskaya, *Zh. Analit. Khim.*, **32**, 1820 (1977).
[169] J. Greyson, *J. Autom. Chem.*, **3**, 66 (1981).
[170] A. Zipp, *J. Autom. Chem.*, **3**, 71 (1981).
[171] N. G. K. Karanth, M. S. Srimathi and S. K. Majumder, *Bull. Environ. Contam. Toxicol.*, **28**, 221 (1982).
[172] H. J. Kyrein, *Ärtzl. Lab.*, **24**, 57 (1978).
[173] A. di Domenico, F. Merli, L. Boniforti, I. Camoni, A. Di Muccio, F. Taggi, L. Vergori, G. Colli, G. Elli, A. Gorni, P. Grassi, G. Invernizzi, A. Jemma, L. Luciani, F. Cattabeni, L. De Angelis, G. Galli, C. Chiabrando and R. Fanelli, *Anal. Chem.*, **51**, 735 (1979).
[174] T. Daldrup, F. Susanto and P. Michalke, *Z. Anal. Chem.*, **308**, 413 (1981).
[175] K.-A. Kovar, M. Noy and R. Pieper, *Dtsch. Apotheker Ztg.*, **122**, 3 (1982).
[176] P. M. Kabra, B. E. Stafford and L. J. Marton, *J. Anal. Toxicol.*, **5**, 177 (1981).
[177] R. Dugal, R. Masse, G. Sanchiez and M. J. Bertrand, *J. Anal. Toxicol.*, **4**, 1 (1980).
[178] C. P. Gorst-Allman and P. S. Steyn, *J. Chromatog.*, **175**, 325 (1979).
[179] B. A. Roberts and D. S. P. Patterson, *J. Assoc. Off. Anal. Chem.*, **58**, 1178 (1975).
[180] C. E. Holaday, *J. Am. Oil Chem. Soc.*, **53**, 602 (1976).
[181] L. M. Seitz and H. E. Mohr, *Cereal Chem.*, **51**, 487 (1974).
[182] B. G.-E. Josefsson and T. E. Möller, *J. Assoc. Off. Anal. Chem.*, **60**, 1369 (1977).
[183] N. H. Choulis, *J. Chromatog.*, **168**, 562 (1979).
[184] P. F. Lott, J. W. Lott and D. J. Dennis, *J. Chromatog. Sci.*, **16**, 390 (1978).
[185] M. Oellerich, W. R. Külpmann and R. Haeckel, *J. Clin. Chem. Clin. Biochem.*, **15**, 275 (1977).
[186] J. E. Wallace, H. E. Hamilton, H. Schwertner, D. E. King, J. L. McNay and K. Blum, *J. Chromatog.*, **114**, 433 (1975).
[187] R. Saferstein, J. J. Manura and P. K. De, *J. Forens. Sci.*, **23**, 29 (1978).
[188] P. J. Twitchett, S. M. Fletcher, A. T. Sullivan and A. C. Moffat, *J. Chromatog.*, **150**, 73 (1978).
[189] N. C. Jain, W. J. Leung, R. D. Budd and T. C. Sneath, *J. Chromatog.*, **115**, 519 (1975).
[190] J. A. Vinson, J. E. Hooyman and C. E. Ward, *J. Forens. Sci.*, **20**, 552 (1975).
[191] W. J. Woodford, *J. Chromatog.*, **115**, 678 (1975).
[192] K. K. Kaistha and R. Tadrus, *J. Chromatog.*, **154**, 211 (1978).
[193] M. J. de Faubert, *Analyst,* **100**, 878 (1975).
[194] W. J. Serfontein, D. Botha and L. S. de Villiers, *J. Chromatog.*, **115**, 507 (1975).
[195] R. Schwenk, K. Kelly, K. S. Tse and A. H. Sehon, *Clin. Chem.*, **21**, 1059 (1975).
[196] H. Frye, R. Kletsch and G. J. Miille, *J. Lab. Clin. Med.*, **90**, 109 (1977).
[197] J. Hamann, A. Tolle, A. Blüthgen and W. Heeschen, *Milchwiss.*, **31**, 18 (1976).
[198] A. R. Barbiers and A. W. Neff, *J. Assoc. Off. Anal. Chem.*, **59**, 849 (1976).

[199] E. R. Lieber and S. L. Taylor, *J. Chromatog.*, **153**, 143 (1978).
[200] S. Dilli and K. Robards, *Analyst*, **102**, 201 (1977).
[201] R. Kellner and G. Gidaly, *Mikrochim. Acta*, 415 (1978 **II**).
[202] H. Ishida and J. L. Koenig, *Intern. Lab.*, May/June, 49 (1978).
[203] M. A. Posthumus, P. G. Kistemaker, H. L. C. Meuzelaar and M. C. T. N. de Brauw, *Anal. Chem.*, **50**, 985 (1978).
[204] V. I. Faerman and I. L. Agafonov, *Zavodsk. Lab.*, **44**, 715 (1978).
[205] F. Gonon, R. Cespuglio, J.-L. Ponchon, M. Buda, M. Jouvet, R. N. Adams and J.-F. Pujol, *Compt. Rend.*, **286D**, 1203 (1978).
[206] J.-L. Ponchon, R. Cespuglio, F. Gonon, M. Jouvet and J.-F. Pujol, *Anal. Chem.*, **51**, 1483 (1979).
[207] E. P. Grimsrud and R. A. Rasmussen, *Atmos. Environ.*, **9**, 1010 (1975).
[208] D. E. Harsch and D. R. Cronn, *J. Chromatog. Sci.*, **16**, 363 (1978).
[209] C. K. Doll, in *Trace Organic Analysis*, pp. 65–78. NBS, Washington (1979).
[210] H. L. Gearhart, R. L. Cook and R. W. Whitney, *Anal. Chem.*, **52**, 2223 (1980).
[211] W. C. Kuster, P. D. Goldan and F. C. Fehsenfeld, *J. Chromatog.*, **205**, 271 (1981).
[212] N. J. Barker and R. C. Leveson, *Intern. Lab.*, **11**, No. 5, 65 (1981).
[213] L. A. Presley, *Arson Anal. Newsl.*, **3**, 18 (1979).
[214] L. W. Hrubesh, A. S. Maddux, D. C. Johnson, R. L. Morrison, J. N. Nielson and M. Malachosky, *U.S. Dept. Energy Rept.*, UCID-17867 (1978).
[215] R. Syrjala, *Intern. Environ. Saf.*, **23** (1978).
[216] T. W. Ottley, *Dyn. Mass Spectrom.*, **6**, 212 (1981).
[217] Anon, *Anal. Chem.*, **51**, 1066A (1979).
[218] F. H. Woeller and G. E. Pollock, *J. Chromatog. Sci.*, **16**, 137 (1978).
[219] R. P. Buck, *Proc. Anal. Div. Chem. Soc.*, **14**, 332 (1977).
[220] P. Chovin, *Z. Anal. Chem.*, **282**, 281 (1976).
[221] E. D. Hinkley and R. T. Ku, in *Environmental Analysis*, G. W. Ewing (ed.), p. 161. Academic Press, New York (1977).
[222] E. C. Tuazon, A. M. Winer, R. A. Graham and J. N. Pitts, *Adv. Environ. Sci. Technol.*, **10**, 259 (1980).
[223] E. D. Hinkley, *Laser Monitoring of the Atmosphere, Topics in Appl. Physics*, Vol. 14, Springer, Heidelberg (1976).
[224] R. Hirschfeld, C. Howard and F. Lytle, *Anal. Chem.*, **51**, 1064A (1979).
[225] B. L. Sharp, *Proc. Anal. Div. Chem. Soc.*, **13**, 104 (1976).
[226] J. D. Winefordner, *Anal. Proc.*, **18**, 284 (1981).
[227] B. Marthinson, J. Johansson and S. T. Eng, *Opt. Quantum Electron*, **12**, 327 (1980).
[228] T. Hamza, T. Kobayashi and H. Inaba, *Opt. Quantum Electron.*, **13**, 187 (1981).
[229] J. R. Ferraro and J. L. Basile (eds.), *Fourier Transform Infrared Spectroscopy, Application to Chemical Systems*, Vol. 2, Academic Press, New York (1979).
[230] A. Zlatkis, K. Y. Lee, C. F. Poole and G. Holzer, *J. Chromatog.*, **163**, 125 (1979).
[231] A. Zlatkis, W. Bertsch, H. A. Lichtenstein, A. Tishbee, F. Shunbo, H. M. Liebich, A. M. Coscia and N. Fleischer, *Anal. Chem.*, **45**, 763 (1973).
[232] A. Zlatkis, H. A. Lichtenstein and A. Tishbee, *Chromatographia*, **6**, 67 (1973).
[233] H. M. Liebich, O. Al-Babbili, A. Zlatkis and K. Kim, *Clin. Chem.*, **21**, 1204 (1975).
[234] P. A. G. Malya, J. R. Wright and W. R. Nes, *J. Chromatog. Sci.*, **9**, 700 (1971).
[235] B. Krotoszynski, G. Gabriel, H. O'Neill and M. P. A. Claudio, *J. Chromatog. Sci.*, **15**, 239 (1977).
[236] R. Teranishi, T. R. Mon, A. B. Robinson, P. Gray and L. Pauling, *Anal. Chem.*, **44**, 18 (1972).
[237] E. Stoner, D. Cowburn and L. C. Craig, *Anal. Chem.*, **47**, 344 (1975).
[238] A. B. Robinson and L. Pauling, *Clin. Chem.*, **20**, 962 (1974).
[239] H. Dirren, A. B. Robinson and L. Pauling, *Clin. Chem.*, **21**, 1970 (1975).
[240] A. B. Robinson, H. Dirren, A. Sheets, J. Miguel and P. R. Lundgren, *Exp. Gerontol.*, **11**, 11 (1976).
[241] M. Anbar, R. L. Dyer and M. E. Scolnik, *Clin. Chem.*, **22**, 1503 (1976).
[242] R. A. Chalmers, M. J. R. Healy, A. M. Lawson, J. T. Hart and R. W. E. Watts, *Clin. Chem.*, **22**, 1292 (1976).

[243] R. D. Malcolm and R. Leonards, *Clin. Chem.*, **22**, 623 (1976).
[244] T. Kitagawa, B. A. Smith and E. S. Brown, *Clin. Chem.*, **21**, 735 (1975).
[245] L. Eldjarn, E. Jellum and O. Stokke, *Clin. Chem.*, **21**, 63 (1975).
[246] F. W. Bultitude and S. J. Newham, *Clin. Chem.*, **21**, 1329 (1975).
[247] S. J. Goodman, P. Helland, O. Stokke, A. Flatmark and E. Jellum, *J. Chromatog.*, **142**, 497 (1977).
[248] E. Jellum, *J. Chromatog.*, **143**, 427 (1977).
[249] H. M. Liebich and O. Al-Babbili, *J. Chromatog.*, **112**, 539 (1975).
[250] H. M. Liebich and J. Wöll, *J. Chromatog.*, **142**, 539 (1975).
[251] A. C. Schoots, F. E. P. Mikkers and C. A. M. G. Cramers, *J. Chromatog.*, **164**, 1 (1979).
[252] R. Dzurik, P. Bozek, J. Reznicek and A. Obornikova, *Proc. Eur. Dial. Transplant Assoc.*, **10**, 263 (1973).
[253] T. M. S. Chang, M. Migchelsen, J. F. Coffey and A. Stark, *Trans. Am. Soc. Artif. Intern. Organs*, **20**, 364 (1974).
[254] G. Cueille, *J. Chromatog.*, **146**, 55 (1978).
[255] A. Gordon, J. Bergström, P. Fürst and L. Zimmerman, *Kidney Intern.*, **7**, 45 (1975).
[256] F. C. Senftleber, A. G. Halline, H. Veening and D. A. Dayton, *Clin. Chem.*, **22**, 1522 (1976).
[257] F. Mikkers, S. Ringoir and R. DeSmet, *J. Chromatog.*, **162**, 341 (1979).
[258] R. A. Chalmers and A. M. Lawson, *Organic Acids in Man*, Chapman & Hall, London (1982).
[259] D. Pinkston, G. Spiteller, H. von Henning and D. Matthaei, *J. Chromatog.*, **223**, *Biomed. Appl.*, **12**, 1 (1981).
[260] K. Tanaka, D. G. Hine, A. West-Dull and T. B. Lynn, *Clin. Chem.*, **26**, 1838, 1847 (1980).
[261] D. N. Buchanan and J. G. Thoene, *J. Liq. Chromatog.*, **4**, 1587 (1981).

Index

A

abscisic acid, 26, 137, 162, 169, 284
absorption methods, 222
absorption solution, 93, 176
abstraction, 329
accuracy, 33, 61, 92, 98
  control, 82
acebutolol, 59, 283
acenaphthene, 117
acenocoumarin, 137
acephate, 264, 265
acetaldehyde, 94, 107, 279
acetamidobiphenyl, 284
acetaminophen, 276
acetate, 184
acetic acid, 120
acetic anhydride, 96, 190
acetominophen, 60
acetone, 96,115
acetonitrile, 120, 125
acetoxy-4-bromomethylcoumarin, 190
acetylcysteine, 165, 166
acetyldopamine, 166
acetylene, 95, 268, 318
acetylglycine, 120
acetylphenothiazine, 161
acids, 192, 330
  in urine, 256
acrolein, 55, 95, 96, 318
acrylamide, 135, 247, 276
acrylonitrile, 55, 94, 95, 106, 120, 141, 142, 262, 265
actinomycetes, 248
activated carbon, 100
activated charcoal, 93, 94
addictive drugs, 332
addicts, 15
additives to food, 27
adenosylmethionine, 319
adrenaline, 258
adriamycin, 157, 276, 284, 319
adsorbent-packed traps, 106
adsorbents, 119
adsorption, 54, 102, 115, 116, 126, 171
  chromatography, 149
  on charcoal, 266
  on plastic materials, 99
adsubble methods, 183
aerosols, 93, 97
affinity chromatography, 149
aflatoxin, 19, 105, 162, 284
aflatoxins, 21, 22, 23, 24, 63, 132, 134, 168, 276, 331, 333
Aflatoxin-Verordnung, 21
ag, 18
agarose, 155
agricultural sample, 20
air, 20, 92, 238, 248, 334, 336
air-segmented reactor, 189
alcalase, 123
alcohols, 94, 190, 191, 264, 268, 277, 278, 330
  aliphatic, 94
aldehyde sex pheromones, 279
aldehydes, 96, 162, 181, 277, 278, 329
aldicarb, 254, 263, 264
  sulphone, 263
aldosterone, 283, 285
aldrin, 115, 140, 146, 179, 331
aliphatic alcohols, 94
  amines, 193, 317
  biogenic amines, 190
  unsaturation, 278
alkali metal flame ionization detector, 259
alkaloids, 195, 257
alkanes, 232, 247, 268
alkenes, 232
alkyl benzo(*c*)cinnolines, 233
  halide, 278
  quaternary ammonium ions, 152
  sulphonates, 152
alkyl-9-fluorenones, 233
alkylating reagent, 196
alkylation techniques, 189
alkylbenzenesulphonates, 162, 247
alkylbenzyldimethylammonium compounds, 260

alkylpyridines, 142
alphadolone, 260
alternate acid–base partitioning, 144
alternative hypothesis, 40
alumina, 152
aluminium oxide, 95
amantadine, 262
Amberlite, 95
Amberlyst, 101
American Society for Testing Materials, 92
amezinium, 57
amides, 330
amines, 26, 103, 120, 137, 142, 165, 181, 190, 192, 193, 235, 237, 254, 260, 264, 265, 272, 277, 278, 330
amino-acids, 20, 26, 57, 104, 118, 119, 152, 165, 166, 190, 191, 193, 258, 265, 277, 330, 338
aminobenzimidazole, 319
aminobenzoic acid, 286, 319
  ethyl ester, 125, 264
aminobutyric acid, 193, 260
aminocarb, 132
aminochlorobenzophenones, 193
amino compounds, 319
aminophenol, 96
aminopterin, 156
aminopyridine, 156, 157
amino-sugars, 118
aminotriazole, 162
amiodarone, 157
amitryptiline, 60, 156, 157, 252, 265, 283, 286
ammonium sulphate, 125
AMP, 258
amphetamine, 156, 157, 175, 191, 249, 260, 283, 285, 286
amphetamines, 192, 333
amplification, 326
anabolic steroid drugs, 254
anabolic steroids, 252
anabolics, 157
analysis, 102
  of surfaces, 332
  of variance, 51
  over great distances, 337
analytical function, 29
  regions, 45
  strategy, 318
  techniques, 20
ancillary methods, 328
ancymidol, 192
5α-androstane-3α-, 17β-diol, 250
androstendione, 283, 285
androsterone, 191
anilazine, 162
aniline, 156, 169
anilines, 276, 322
animal oils, 233
anionic surfactants, 195, 232
anions, 195
anthracene, 181
anthryldiazomethane, 165
antibiotics, 22, 27, 282, 319, 333
antibody, 280
anticonvulsants, 104, 157
anticonvulsives, 194
antidepressant drugs, 265
antigen, 280
antihistamines, 288
antimalaria drugs, 265
antioxidants, 27
API, 241
Apiezon, L., 174
apomorphine, 252
apovincamic acid, 265
apovincaminic acid, 195
  ester, 262
applicator, 168
aprotic solvent, 194
archives, 80
area for delicate and sensitive equipment, 70
  extraction and clean-up procedures, 69
  sample preparation, 69
  sample storage, 70
arginine, 258, 319
arithmetic mean, 31
Arochlor, 178, 179
aroma components, 103
aromatic acids, 194
aromatic aldehydes, 94
aromatic amines, 162, 265
aromatic amino-acids, 166
aromatic carboxylic acids, 193
aromatic hydrocarbons, 96, 103, 196
aromatics, 268, 278
arprinocid, 63
arson, 251, 334
artefact formation, 126
arterial blood, 102
aryl phosphates, 265
asarone, 162
asparagin, 185
aspartic acid, 185
aspirin, 157
assigned values, 51
ASTM, 92
asulam, 133
atenolol, 59
atmosphere of Jupiter, 334
atmospheric pressure ionization, 241
ATP, 117, 284, 319
ATR, 226
atrazine, 133
atropine, 251, 252
attenuated total reflection, 226
attogram, 18
AutoAnalyzer, 182, 187
automatic water sampler, 334
automatically operated equipment, 155
automobile exhaust, 336
azapropazon, 125
azelastine, 252
azeotrope, 139
azeotropic mixtures, 141
azinphos-methyl, 162
azo compounds, 273
Azopropazon, 159

B

bacitracin, 319
back-partitioning, 144
back-washing, 154
background contamination, 28
bacteria, 176, 259
bacterial growth, 118
  transformation, 116
band broadening, 187
barban, 133
barbiturate, 240
barbiturates, 181, 194, 236, 319, 330, 331, 333
barbituric acids, 196
basic amino-acids, 115
batch, 66
battery-operated sampler, 334
bed reactor, 188, 189
bell-shaped curve for the frequency distribution, 38
benomyl, 162
benzaldehyde, 190
benzanilide, 156
benzanthrone, 59
benzbromarone, 251, 252
benzene, 84, 85, 94, 106, 119, 270, 318
  hexachloride isomers, 100
benzidine, 96, 162
benzodiazepines, 157, 274, 275, 283, 285, 286, 287, 333
benzofluorene, 318
benzo(*a*)pyrene, 57, 96, 97, 123, 124, 126, 147, 162, 196, 233, 236, 248, 262
benzoyl chloride, 165
benzoylecgonine, 192, 265, 331, 333
benzpyrene, 59
benzyl bromide, 190
benzyloxime-perbenzoyl derivatives, 165
betamethasone, 133
bias, 52
bile acids, 57, 105, 166
billion, 18
binapacryl, 126
bioassay, 25, 319
bioavailability, 182
biochemistry, 21, 102
bioequivalence/bioavailability, 21, 65
biogenic amines, 107, 165, 250
biological methods, 286
biologists, 16
biosensors, 270
biotin, 286
biotransformation, 28
biphenyl, 162
biphenyls, 115, 136
bis(chloromethyl) ether, 327
bisantrene, 133
blank values, 32, 70, 118
bleeding, 155
  of septa, 174
blending, 123
blood, 116
  samples, 104, 122
body constituents, 22
  fluids, 337
bread, 253
breakthrough volumes, 93
bretylium, 260
bromacil, 265
bromination, 189
bromoacetanilide, 59
bromobenzarone, 252
bromo-derivatives, 103
bromomethylmethoxycoumarin, 190
bromomethylpentafluorobenzene, 190
bromo-2,3,4,5,6-pentafluorotoluene, 190
bromophenacyl esters, 165
bromophos, 196
bromouracil, 156
bromovincamine, 157
bromphenvinphos, 265
bufuralol, 192
bulk material, 91
bupirimate, 126
bupropion, 287
butane-1,4-diol, 233
butaneboric acid, 190
butanilicaine, 265
butanol, 190
butaperazine, 237
butyl iodide, 190
butyl isocyanate, 192
butylamine, 190
butylhydroxyanisole, 333
butylquinol, 262
butyric acid, 288

C

CA-MIKES, 244
cadaverine, 169, 260
caffeine, 175
  complexation, 196
calcitonin, 319
calibration function, 30
  material, 52
  standard, 32, 53
camazepam, 260
cancer policy, 65
cannabidiol, 260
cannabinoids, 105, 156, 157, 168, 284, 285, 286, 287, 331
cannabis alkaloids, 23, 104
capillary columns, 155, 172
capronium chloride, 176
captafol, 132
captan, 192
captopril, 157
carbamate herbicides, 191
  insecticides, 100, 191, 263, 265
  pesticides, 162, 257
carbamates, 194, 264, 330
carbamazepine, 104
carbamazepine, 161, 194, 265
carbamazepine-10, 11-epoxide, 283
carbaryl, 162
carbendazim, 132, 260

Carbochrome K-5/Carborafin, 94
carbofuran, 162, 191, 196
carbohydrates, 117, 257, 324
carbon fibre electrode, 332
    skeleton method, 177, 178, 330
    tetrachloride, 54, 92, 115
carbonyl compounds, 96, 190, 278
Carbopack, 98
Carbosieve B, 95
Carbowax 20 M, 173, 174
Carbowax 400, 95
carboxylic acids, 26, 190, 192, 195, 196, 260, 277
carcinogenic properties, 23
carcinogens, 253, 256
carotene, 235
carprofen enantiomers, 26
carrier gas, 327
carteolol, 283
cassettes, 154
catalysis oven, 181
catalytic amplification, 326
catalytic hydrodechlorination, 179
catecholamine sulphate conjugates, 279
catecholamines, 165, 169, 195, 235, 243, 260, 274, 275, 279, 319
catechols, 100
causal connection, 48
celiprolol, 157
cellulose, 149, 152
central nervous system stimulants, 332
centrifugation, 118
cephalosporins, 276
cereals, 282
certified reference material, 52
chain methods, 154
charcoal, 103, 176, 233
charge-transfer complexes, 196
chemical analogues, 249
    detection, 277
    hazards, 93
    ionization, 239
chemiluminescence, 327
    detector, 268
chemotherapeutic, 282
chlorambucil, 251, 252
chloramphenicol, 135, 171, 190
chlorbenoxamine, 156
chlordan, 331
chlordiazepoxide, 135, 157, 284, 333
chlordiazepoxides, 156
chlorimipramine, 195
chlorinated alkanes, 254
chlorinated aromatic pollutants, 57
chlorinated compounds, 95, 247
chlorinated dibenzodioxins, 132, 162, 238, 253, 256, 317
chlorinated diphenyl esters, 162
chlorinated hydrocarbons, 92, 94, 107, 120, 260
chlorinated norbornene, 253
chlorinated pesticides, 100, 247
chlorinated phenols, 119, 133
chlorination, 137
chlormequat, 135
chloroacetanilides, 233
chloroacetyl chloride, 95
chloroanilines, 263
chlorobenzenes, 124, 132, 270
chlorocholine, 133
chloro-derivatives, 103
chlorodesmethyldiazepam, 59
2-chloroethanol, 95
chloro-ethers, 247
chlorofluorocarbons, 248
chloroform, 84, 85, 92, 115, 253
chloro-7-iodo-8-hydroxyquinoline, 196
chloromethane, 254
chloromethyl methyl ether, 233, 318
chloronitroanilines, 263
chloronitrobenzenes, 234
chloro-4-nitrobenzo-2-oxo-1,3-diazole, 191
chlorophenols, 100, 162
chlorophenoxy acid herbicides, 233
chlorophenoxyacetic acids, 260
chlorophenoxyalkanoic acids, 254
chloroplasts, 169
chloroprocaine, 252
chloro-2-propanol, 256
chloroquine, 156, 157, 265
4-chloro-5-sulphamoylanthranilic acid, 125
8-chlorotheophylline, 59
chlorpheniramine, 195, 251, 252, 265
chlorpromazine, 156, 157, 161, 195, 236
chlorpyrifos, 95, 100, 260
chlorsulpheron, 162
chlortetracycline, 319
chlorthalidone, 196, 265
cholesterol, 104, 193, 257
choline chloride, 241
cholinesterase, 280
cholylglycine, 284
chopping technique, 324
Chromarods, 169
chromatography, see particular type
chromogenic paper, 331
chromophore, 187
chromophoric groups, 223
Chromosorb W, 170
Chromosorb 102, 101
Chromosorb 105, 101
Chromosorb 106, 101
Chromotropic acid, 95
chrysene, 59, 248
CI mode, 240, 249
CI, 239
cigarette smoke, 232
cigarettes, 23
cimetidine, 157
cinnarizine, 156, 157
citrate, 118, 184
Clean Air Act, 24
clean-up, 152, 155, 165, 318
clebopride, 252
clinical chemistry, 21, 22, 102, 155
clobazam, 133, 156, 157
clofexamide, 265
clomiphene, 157

clomipramine, 283, 286, 319
clonazepam, 57, 58, 260
clonidine, 105, 252, 253, 265
clopidol, 163, 260
clotting blood, 118
co-condensation, 176, 177, 178
co-precipitation, 181
cocaine, 104, 156, 157, 158, 175, 189, 192, 252, 265, 283, 317, 331, 333
codeine, 59, 156, 157, 175, 195, 252, 331
coefficient of correlation, 47, 50
coefficients of variation, 58, 59
coffee, 253
colchicine, 284
cold trap, 94, 102
cold-finger technique, 143
collaborative studies, 54, 61–63
collagenase, 123
collection, 178
  system, 177
collision-activation mass-analysed kinetic-energy spectrometry, 244
column back-flushing, 154
  bleeding, 173, 174
  chromatography, 232, 251
  switching, 154, 173
  switching techniques, 174
combination of columns, 154
combination of excitation modes, 241
combination of IR with liquid chromatography, 232
combination of mass spectrometry with gas chromatography, 245
combinations of chromatography with detection systems, 257
components of high electron-affinity, 325
computer input, 76
concentration plates, 168
conductivity detector, 259, 263
confidence interval, 38, 44, 49
  limits, 38
conjugated oestriol, 319
conjugates, 251
constant systematic error, 32
consumer goods, 238
consumption of reagents, 19
contamination, 119, 120, 153, 180
  control, 28
continuous monitoring, 317
continuous sampling, 98
continuous solvent extraction apparatus, 148
control of the entire procedure, 60
control substance, 64
controlled diffusion, 55
conversion of functions, 30
cooked chicken, 253
coprostanol, 193, 247
correlation, 46–50
corticosteroids, 157, 286
corticosterone, 132, 286
cortisol, 120, 132, 133, 166, 237, 283, 285, 286, 287
cost of quality control, 70
cotinine, 135, 266
coulometric detector, 259, 263
  methods, 271
Coulson-detector, 259
coumarin, 288
coumarins, 257
counter-ion, 152
coupling of ir with gc, 230
coupling of liquid chromatography with mass spectrometry, 247
coupling of ms with tlc or liquid chromatography, 245
CPSC, 65
creatine, 241
creatinine, 117
cresols, 96
critical level, 43, 44, 45
cross-reactions, 285
crude oil, 99
cryoconcentration, 99
cryogenic ir, 227
  traps, 106
cumulative frequency, 36
curcumin, 163
cyanogen chloride, 120
cyanuric acid, 163
cyclic adenosine-3′,5′-monophosphate, 250
cyclobenzaprine, 265
cyclohexane, 276
cyclohexanone, 107
cyclophosphamid, 132, 265
cyheptamid, 59
cystine, 276
cytokinins, 191, 260, 265, 284
cytosine, 257

D

2,4-D, 99, 115, 116, 190, 191, 193, 254
  products, 253
DADI, 244
damping devices, 156
dansyl chloride, 161, 190
dansylation, 194
dansylhydrazine, 165
dantrolene, 156, 157
DAU, 284
daunomycin, 284
DC 200, 173
DDE, 179
DDT, 57, 96, 98, 115, 117, 138, 141, 179, 183, 247, 260, 331
decision-making process, 29
decisions, 28
decomposition, 122
definitive method, 51, 52
  value, 52
dehalogenation, 189
dehydrocholesterol, 166
dehydroepiandrosterone, 250, 283, 285
demethylellipticine, 156
deodorants, 119
deoxyhydrocortisone, 283, 285

deproteination, 125, 126, 133, 161, 166, 167
derivative formation, 74, 161, 165, 237, 238, 249, 259
techniques, 187
derivatives of camphor, 103
desalination, 182
desipramine, 161
desmethyldiazepam, 59
desorption, 97
detection, 45, 156
limit, 44, 45
systems, 257, 258
detectors, 170, 324
for gas chromatography, 258
for liquid chromatography, 257
detergents, 276
determination, 45
limit, 33, 45
methods, 221
without separation, 318
deuterated analogue, 116
dueterium-labelled compounds, 251
Deutsche Trinkwasserverordnung, 21
deviation factor, 38
dexsil, 256, 300
dextramethorpan, 265
dextran, 155
diacetoxyscirpenol, 125
diacetylbenzidine, 284
dialkyl sulphides, 100
di-allate, 142
dialysis, 182, 251
diaminobenzoic acid, 190
diamond, 226
diamorphine, 23
diasolysis, 183
diazepam, 156, 157, 158, 252, 263, 287, 331, 333
diazinon, 264
diazoethane, 191
diazomethane, 189, 190
dibenzazepine derivatives, 262
dibromoethane, 260
1,2-dibromo-3-chloropropane, 95, 132
dibutyl phthalate, 121, 122
dibutylsulphide, 96
dichlobenil, 163
dichloroethylene, 107
dichlorobenzidine, 162
3,3′-dichlorobenzidine, 96
dichlorobenzilic acid, 260
dichloroethane, 262
1,2-dichloroethane, 92
dichloromethane, 106, 120
dichloromethyl ether, 95, 256
dichlorophenoxyacetic acid, 163
2,4-dichlorophenoxyacetic acid, 115
diclofop, 260
diclofop-methyl, 260
dichlohexyl-(7-methoxycoumarin-4-yl)-methylpseudourea, 191
dieldrin, 62, 146, 179
diesel particulate emissions, 24, 233
diesel-engined vehicles, 24
diethyl carbonate, 107
diethyl phthalate, 99, 232
diethylcarbamoyl chloride, 94
diethylhexyl phthalate, 122
diethylstilboestrol, 191, 192
difenzoquat, 163
differential oscillopolarography, 276
differential pulse polagraphy, 222, 275, 276, 319
diffuse reflectance spectroscopy, 226
diffuse reflectance spectrum, 232
diffuse reflectance techniques, 238
diflubenzuron, 126
digestion, 123, 254, 255
digital smoothing, 225
digitalis glucosides, 324
glycosides, 192
digitoxin, 283
digoxin, 57, 158, 283, 285, 286, 319
dihydrodigoxin, 283
dihydroxybenzylamine, 156
dihydroxycholecalciferol, 284, 286
dihydroxyvitamin D, 166, 286
di-iodothyronine, 282
di-isocyanates, 96
di-isocyanatohexene, 96
di-isocyanatotoluene, 270
diltiazem, 265
dimethindene, 262
dimethylamine, 97
4-dimethylaminocinnamaldehyde, 96
dimethyldisulphide, 276
dimethylhydrazine, 163
dimethylsulphate, 191
dimethylsulphide, 107, 276
dimethyl-*trans*-9-decalol, 248
diminazene, 195
dinitrobutylphenol, 138
dinitrophenylhydrazine, 95, 96, 190
2,4-dinitrotoluene, 55
dinitrotoluene isomers, 132
dioctyl phthalate, 232
dioxins, 244
diphenyl ether herbicides, 178
herbicides, 133
diphenylhydantoin, 53, 194, 256, 319
diphenylhydramine, 156
dipterex, 264
dipyridamole, 133, 156, 158
diquat, 185
direct analysis of daughter ions, 244
direct coupling, 258
direct injection, 162, 163, 164, 245
direct working procedures, 318
disaccharides, 258
diseases, 337
disopyramide, 182
dispersion, 33
disposal of used reagents and solvents, 85
dissolution, 123
distillation, 74, 132, 251, 254, 269
techniques, 136

distribution, 74
pattern, 34
disulphides, 273, 278
disulphiram, 276
di-tert.-butyl-*p*-cresol, 262
dithiocarbamate fungicides, 196
pesticides, 107
dithiocarbamates, 276
DNA bases, 143
DNBP, 138
dodecylbenzenesulphonate, 132
DOPA, 166, 192, 275, 279
dopamine, 166, 191, 192, 195, 239, 275, 332
doping, 317
dothiepin, 60
double antibody method, 282
double isotope dilution technique, 186
double labelling, 58
double reservoir evaporation vessel, 139
double-focusing spectrometer, 244
doxepin, 287
doxorubicin, 284, 285
doxylamine succinate, 265
dropping mercury electrode, 272, 275
drug analysis, 143
metabolism, 22
drugs screening, 285, 333
drugs, 22, 27, 104, 105, 115, 116, 124, 125, 157, 168, 170, 171, 172, 195, 233, 249, 251, 257, 282, 332
of abuse, 249, 284, 285, 287, 317, 330, 333
screening, 285, 333
drying of solution, 141
reagent, 141
dual flame photometric detectors, 263
dynamic sampling, 106

E

ECD, 55, 156, 222, 334
ecology, 15, 21
economy, 15
EDTA, 118, 185
effects on human health, 66
effluents, 238, 247
EI, 239
mode, 240
eicosapolyenoic acids, 250
electroactive group, 277
electrochemical detection in hplc, 274
electrochemical determination, 275, 276
electrochemical methods, 221
electrochemical properties, 156
electrochemical property, 161
electrochemical recycling, 327
electrochemically active derivatives, 274
electrolytes, 170
electron capture, 328
detection, 259, 260
detector, 15, 55, 258, 259, 327
impact, 239
electronic averaging, 322
electronic differentiation, 322
electronic subtraction spectra, 225
electrophoresis, 184, 238
elemental analysis, 221
ELISA, 281
ellipticine, 156, 158
eluate-splitting, 154
elution chromatography, 151
EMIT, 281, 287, 319, 331, 333
emitter wire, 241
emulsifiers, 276
emulsions, 147
endosulfan, 194, 233, 331
enhancement of analytical signal, 327
environmental chemist, 317
environmental protection agency, 24, 92
enzymatic cycling, 278, 326
enzymatic determination, 221, 318, 319
enzymatic hydrolysis, 123
enzymatic reactions, 278, 279
enzyme electrodes, 270, 271
enzyme immunoassay, 281, 332
enzyme-linked immunosorbent assay, 281
enzyme-multiplied immunoassay technique, 281
EPA, 24, 65, 92
guidelines for registration of pesticides, 21
ephedrine, 175
ephedrines, 191
epichlorhydrin, 95
epimethyltestosterone, 283
epinephrine, 237, 275, 283
equilin, 283, 285
equipment, 70, 118, 122
and techniques for liquid–liquid distribution, 145
ergocristine, 156
ergot alkaloids, 158
ergotamine, 156
error bar, 49
error of the first kind, 29
error of the second kind, 29
esters, 103, 107, 278, 330
ethambutol, 252
ethanediol, 96
ethanol, 107, 115, 332
ethaverine, 158
ethers, 278
ethmozine, 132, 135
ethosuximide, 262
ethyl ether, 327
iodide, 137, 190
methyl ketone, 141, 327
ethylbenzene, 26, 119
ethyldisulphide, 288
ethylene glycol dinitrate, 55
ethylene, 94, 232, 236, 336
oxide, 107, 133
ethylenethiourea, 63, 163, 264
ethylhexylphthalate, 318
ethylmorphine, 59, 175
ethynyl steroids, 251, 252
etorphine, 57, 249

eugenol, 166
eurosemid, 196
evaporation to dryness, 140
exaprolol, 260
excitation modes, 241
processes, 239
explosive residues, 233
explosives, 260
extraction, 124, 126
efficiency, 124
extractive alkylation, 194
extraterrestrial analysis, 337
extrelut, 104, 105
columns, 121

F

F-test, 41
facilities for trace analysis, 68
faecal pollution, 247
famphur, 264
fat, 126
fats, 257
fatty acids, 96, 104, 120, 142, 165, 166, 192, 257
fatty anilides, 169
FD, 240
FDA, 65
Federal Environmental Pesticide Control Act, 21
Federal Insecticide, Fungicide & Rodenticide Act, 65
Federal Water Pollution Control Act, 21
femtogram, 18
fenamiphos, 264
fenchlorphos, 326
fenfluramine, 192
fenitrothion, 100, 105, 265
fenozolone, 196, 252
fensulfthion, 264
fentanyl, 252, 253, 262
fg, 18
FI, 240
FID, 155, 222, 262
field desorption, 238, 240
field ionization, 240
filled columns, 170
filling of columns, 152
filters, 97
filtration, 74, 167
filtration-sampling, 97
final reporting of study results, 80
final reports, 68
fish, 124
flame aerosol detector, 258
flame detector, 257
flame ionization, 328
flame ionization detector, 155, 258, 259
flame photometric detector, 258, 259, 263, 264
flash-heater derivative formation, 175
flavour components, 253
flavours, 22, 27, 107, 257, 264
flecainide, 237
flophemesyl chloride, 191
florisil, 152
flow sheet, 73, 75, 76
flow-through cells, 274
flufenamic acid, 59
flunisolide, 284
flunitrazepam, 260
fluoranthene, 237
fluorescamine, 165, 189, 191
fluorescein, 236, 280
fluorescence, 235, 331
detection, 156
immunoassay, 281, 287
immunoassay, 287
intensity, 161
spectrum, 229
fluorescent markers, 187
fluorimetry, 63, 279, 287
fluoro-2-enyl acetamide, 260
5-fluoro(hexylcarbamoyl)uracil, 158
fluoro-3-nitrobenzotrifluoride, 191
fluoro-7-nitrobenzofurazan, 165
fluorouracil, 156, 158, 252
fluphenazine, 265, 272, 284
fluram, 189
flurazepam, 261
fluridone, 163, 192
herbicide, 100
focusing mass spectrometers, 242
folic acid, 286
food analysis, 155, 253
food, 20, 238, 254
dyes, 124
flavour volatiles, 278
foodstuff dyes, 124, 236
forensic analysis, 249
forensic chemistry, 249
forensic experts, 16
forensic purposes, 143
formaldehyde, 55, 94, 95, 97, 117, 118, 133, 142, 163, 318
formamide, 123
formetanate, 163
formylbenzoic acid, 276
Fourier analysis, 224
techniques, 332
Fourier–IR, 26, 318
analysis, 224
Fourier-transform gc–ir analysis, 231
infrared spectrometer, 238
FPD, 263
fraction collection, 176
freeze-out loop, 334
freezing, 176
freon, 12, 233
frequency distribution, 38
FT-ir, 337
spectra, 232
spectroscopy, 233
fucose, 166
fumarate, 184
functional connection, 30, 48
functional group analysis, 277

functional group classification tests, 278
functional groups, 25, 178
fungicides, 288
furazolidine, 163, 169
furosemide, 125, 196
fusarium toxins, 133, 168
fused silica capillary columns, 172
fusicoccin, 284

G

gas analysis, 232
cells, 224
chromatography, 170, 221, 324, 334
chromatography with flame ionization detection, 25
chromatography with Fourier transform infrared analysis, 25
chromatography-mass spectrometry, 25
cuvette, 230
mixtures, stability, 54
gas-blending procedures, 55
gas-liquid chromatography, 74
gas-permeation tubes, 55
gas-phase coulometer, 55
spectra, 327
Gaschrom Q, 170
gaseous electrophoresis, 185
gases, 54
gc, 170, 262, 287, 331, 332
detectors, 258
system, 328
(N–P), 132
gc–ecd, 190, 191, 256, 259
gc–fid, 25, 259
gc-flame photometry, 263
gc–FT–ir, 25, 233, 234
gc-mas, 331
gc–ms–coupling, 245
gc–ms, 25, 26, 234, 238, 241, 243, 245, 247, 248, 249, 250, 253, 256, 287, 331, 338
gel chromatography, 155, 170, 285
gel filtration, 149, 170, 333
gentamycin, 284, 285, 287
geosmin, 248
gestagens, 192
gibberellic acid, 319
gibberellins, 166
glass capillary columns, 170, 171
fibre filter, 96
membrane micro electrodes, 271
glassware, 122
glass-wool, 120
glibenclamide, 319
glibornuride, 191
GLP, 62 ff
inspection, 72
glucose-l-phosphate, 319
glutamic acid, 185, 247
glutamine, 185, 319
glutaraldehyde, 95
gluthethimide, 156, 262
glycerides, 232
glyceryl trinitrate, 252, 253
glyphosphate, 163
golay-type columns, 170
good clinical practices, 21, 65
good laboratory practice, 62 ff
grab-method, 98
gradient elution, 151
graphical test of normal distribution, 36
grinding, 123
griseofulvin, 125
gross signal, 42
guaiacols, 100
guaiphenesin, 261
guanidine compounds, 165, 166
guanidines, 191
guanidino compounds, 195
guanine ribonucleotides, 318
guanosine, 241
guidelines for identifying carcinogens and estimation of risks, 65
Gutermann's measure of imprecision, 34

H

H_0, 29
H_1, 29
haemolysis, 102, 104
hair, 104, 119
half-wave potential, 272
hall-detector, 259
hallucinogens, 27
haloalkanes, 163
halocarbons, 95, 103, 132, 248
halogen-containing compounds, 273
halogenated hydrocarbons, 103, 106, 261
halogenated pesticides, 15
halo-organic compounds, 100
haloperidol, 60, 284, 287
halothane metabolites, 106
handling and storage, 66
hazardous gases, 93
hazards, 105, 233
HCH, 260
head-space, 106, 249
analysis, 102, 105
gc, 247
health hazards, 19
heart-cutting, 173
heel prick, 104
hemodialysis, 337
heptachlor, 179, 194
epoxide, 146
heptachlorodibenzo-*p*-dioxins, 255
heptafluorobutyric anhydride, 191
heptanoic acid, 258
herbicides, 190, 226, 263
heroin, 23, 158, 175, 234
HETP, 150, 151
hexachlorobenzene, 63, 120, 124, 254
hexachlorocyclohexane, 26
isomers, 55
hexachlorocyclopentadiene, 254

hexachlorodibenzo-*p*-dioxin isomers, 26, 255
hexachlorohexane isomers, 132
hexachlorophene, 190
hexaflouracetylacetone, 191
hexamethylmelamine, 266
hexamine, 120
1,6-hexanediamine, 59
high performance liquid chromatography, 25
high resolution, 243
high-altitude air samples, 334
high-pressure liquid chromatography, 135, 155ff
high-resolution capillary gc, 174
  mass spectrometry, 242, 318
high-volume sampler, 96
higher alcohols, 107
higher resolution spectrometers, 242
highly toxic materials, 84
histamine, 63, 115, 116, 125, 165, 166, 167, 169, 185, 191, 192, 193, 237, 250, 254, 279, 288, 333
histidine, 191
histological degradation, 118
home-made columns, 153
homogeneity of sample, 92
  of variances, 41
homogeneous immunoassay techniques, 282
homogenization, 123
homogenizers, 123
homologous series, 25
homovanillic acid, 57, 166, 250
hormones, 22, 165
hospital patients, 335
hplc, 25, 26, 155ff, 274, 285, 287, 332, 333
  columns, 153
  detectors, 223
hplc-FT-ir, 234
hplc-gc-ms, 26
human body fluids, 248
human breath, 238
human environment, 254
human nose, 288
human skin, 119, 121
humic acid, 318
hyaluronidase, 123
hydantoins, 333
hydrazo compounds, 273
hydrazones, 273
hydrocarbons, 63, 95, 98, 103, 106, 107, 142, 170, 238, 247, 248, 268
hydrochlorothiazide, 196
hydrocodone, 284
hydrodechlorination, 181
hydrolysis, 25, 116
hydromorphone, 284
hydrophobicity, 152
hydroquinones, 273
hydroxy fatty acids, 250
hydroxy-acids, 26
hydroxyanisole, 262
hydroxycalciferol, 286, 319
hydroxycarboxylic acids, 330
hydroxycortisol, 287
hydroxydesipramine, 156, 158
hydroxydodecanoic acid, 259
hydroxy-L-eicosa-5,8,10,14-tetraenoic acid, 283
hydroxyethyltheophylline, 59
hydroxy-2-fluorenylacetamide, 137, 193
hydroxyimipramine, 156
hydroxyindoleacetic acid, 192
hydroxyindol-3-yl-acetic acid, 166, 167, 284
5-hydroxyindolylacetic acid, 57
hydroxylated steroids, 193
hydroxymelatonine, 250
hydroxy-3-methoxyphenylethanediol, 167, 250
4-hydroxy-3-methoxyphenylacetic acid, 57
hydroxy-L-nitrosophyrrolidine, 191
hydroxyoestradiol, 283, 286
hydroxyoesterone, 283, 286
hydroxyphenethylamines, 250
hydroxyphenobarbitone, 252
hydroxyphenylacetic acids, 192
hydroxyphenylamines, 191
hydroxyprogesterone, 283, 286
hydroxysteroids, 324
hydroxytryptamine, 156, 166, 286
5-hydroxytryptamine, 125
hydroxytryptophan, 166
hydroxyvitamin D, 167
hygiene, 85
hypothesis testing, 40

I

imidazo-1,4-benzodiazepines, 275
imidazolacetic acid, 193
imipramine, 287
immunoassay, 282-286, 331
  kits, 285
  methods, 287
immunological reactions, 280
imprecision, 33
impurity, 66
in vivo measurement, 332
inaccuracy, 33
inconstant blank values, 119
incorporation of traces, 54
indapamide, 158
independent methods, 61
independent screening methods, 331
indol-3-ylacetic acid, 57, 162, 163, 169, 237, 287
indol-3-ylacetyl aspartic acid, 250
indolalkylamines, 192
indoles, 95, 167, 266
indomethacin, 137, 156, 158
  dimethylamide, 156
industrial health control, 93
industrial products, 20
inflammable solvents, 84
infrared attenuated reflection spectroscopy, 332

infrared microreflectance spectrum, 323, 324
infrared microreflectance spectroscopy, 143
infrared microscopes, 197
infrared reflectance spectrometry, 168, 198
infrared spectrometry, 221, 224
infrared spectroscopy, 177
inorganic chemistry, 20
inorganic trace analysis, 15
insect chemosensory behaviour, 249
 chemosterilizants, 264
 juvenile hormones, 283
insecticides, 97, 98, 115, 280, 319, 327
 aerosol bound fraction, 98
 gaseous fraction, 98
 in air, 98
instability of organic trace constituents, 25
interdecile range, 34, 35
interface, 245
interfacing techniques, 246
interlaboratory comparisons, 61
 control, 61–63
 quality control, 82
 relative standard deviation, 62
 surveys, 82
internal standards, 56, 59, 60, 116, 136, 156, 249–251
 standardization, 56
International Organization for Standardization, 52
intralaboratory quality control, 82
investigator, 122
iodide, 326
iodofenphos, 193
iodothyronine, 282
ion-exchange, 57, 237, 247, 256
 chromatography, 135, 149, 169, 170
 membranes, 182
ion-pair formation, 194, 195
ion-pairing technique, 152
ionone, 288
ir, 318
 detectors, 224
 gas cuvette, 232
 laser, 238
 microspectrophotometer, 224
 spectrum, 323
IRLG, 65
isobutyl chloroformate, 137
 methacrylate, 231
isocitrate, 184
isocratic elution, 151
isocyanates, 96, 163, 191, 196
isoetharine, 132
isomers, 25, 233, 256
isomerization, 189
isoniazid, 319, 333
isoprene, 232
isopropylcodeine, 156
isotachophoresis, 184
isotope dilution, 56
 method, 185, 186
isotopically labelled tracers, 28

K

keeper, 138
kelevan, 257
kepone, 247, 257
ketoconazole, 105
ketoglutarate, 184
ketones, 103, 162, 277, 278, 329
ketoprostaglandin, 250
ketosteroids, 192, 294
KRS 5, 226
Kuderna-Danish evaporator, 36, 248

L

labelled internal standard, 58
labetalol, 156, 158, 161
labetol, 190
labile compound, 97
laboratory air, 122
 atmosphere, 118
 flooring, 122
 for organic trace work, 69
lactate, 184
laser desorption mass spectrometry, 241, 332
laser-induced fluorescence, 237
laser-induced infrared fluorescence, 327
laser-infrared fluorescence, 98
laser-raman microprobe, 197
lasers, 236, 335, 336
laudanosine, 161
lc, 149
legal aspects, 22
legal chemistry, 234
legislation, 15, 21, 62
legislative acts, 21
lenacil, 265
levorphanol, 57, 284
lia, 281
life sciences, 21, 238, 248
light pipes, 224, 230
light-sensitive components, 118
light-sensitive compounds, 25
lignocaine, 156, 262, 266
limit of detection, 42–46, 222, 324
 of determination, 42–46
limiting values, 39
limonene, 119
lincomycin, 333
lindane, 62, 96, 98, 138, 140
linked scan method, 244
linoleic acid, 173
linolenic acid, 173
lipids, 145, 169, 257
liquid nitrogen, 123
liquid–liquid distribution, 135, 143ff, 165, 195, 332
 partition, 251
liquid-membrane electrodes, 270
literature, 221
lithocholates, 319
local analysis, 332

lofepramine, 191
log-normal distribution, 37
lomustine, 252
lorazepam, 156, 158
Lord-test, 41
loss, 32
  in deproteination, 126
losses, 171
  improved by internal standard, 58
  of trace, 54
low resolution, 243
low-resolution instruments, 318
low-resolution ms analysis, 242
low-temperature fluorescence, 228
loxapine, 156
LSD, 23, 57, 251, 319, 331, 333
luminescence, 279, 319
  immunoassay, 281
lunar samples, 119
lutidine, 178
lyncomycin, 331
lyophilization, 99, 118
lysergic acid, 319

M

MAC-values, 92
major compound, 20
malate, 184
malathion, 264
malonic acids, 330
man-made traces, 22
maprotiline, 57, 156
marijuana, 168, 249
marine sediments, 233
marprotiline, 266
masking, 187
mass chromatogram, 245
mass spectrometer, 182
mass spectrometry, 221, 222, 238ff, 317, 318
matrix, 18, 135
matrix-isolation Fourier-transform infrared spectrum, 228
matrix-isolation mode, 229
matrix-isolation spectroscopy, 227
maximum (Walter) test, 41
maximum allowable workplace concentrations, 84
mean value, 38
measurement over long distances, 335
measures of imprecision, 33
meat, 123
  products, 253
mebendazole, 105
mechanization of liquid–liquid distribution, 147
median, 34, 37, 38
medical checks, 85
medicinal gases, 232
medicine, 21
medroxyprogesterone, 252, 279
melatonine, 57, 193
membrane, 333
  electrodes, 270
  filter, 120
membranes, 182
meperidine, 192
mepyramine, 156
mercaptans, 278, 330
mercaptobenzimidazole, 276
mercaptoethanol, 188
6-mercaptopurine, 133
mesoridazine, 156
metabolic changes, 118
metabolic disorders, 248
metabolic profiles, 338
metabolites, 161, 251, 317
metafos, 276
metal organic compound, 19
metamphetamine, 191
metanephrine, 57, 250, 319
metanephrines, 167
metapramine, 156, 158, 266
methadone, 57, 175, 252, 266, 284
  enantiomers, 26
methane, 232
methanol, 125, 142
methaqualone, 158, 252, 331
methatrimeprazine, 266
methiocarb, 264
methomyl, 163, 266
methotrexate, 156, 158, 287
methoxsalen, 158
methoxy-4-hydroxyphenylethanediol, 261
methoxy-4-hydroxyphenylglycol, 192
methoxychlorpromazine, 161
methoxypipothiazine, 161
methoxypsoralen, 156
methoxytyramine, 167, 275
methyl anthranilate, 142
  benzoquate, 63
  bromide, 54, 94, 248, 261
  chloride, 94, 248
  4-chloroindol-3-ylacetate, 254
  ethyl ketone, 96, 318
  iodide, 54, 137, 191
  parathion, 318
  phenidate, 175
  stearate, 257
  thiouracil, 191
methylamine, 188
methylaminomethyl anthracene, 96, 191
methylamphetamine, 260, 283
methylating agents, 84
6-methylbenzthiazol-2-ylphenylisocyanate, 191
methylcarbamate insecticides, 266
methylcarbamates, 187
methylene chloride, 119
methylenebis-(2-chloroaniline), 163
methylergometrine, 284, 286
methylguanidine, 195
methylhistamine, 57, 250
methylimidazolylacetic acid, 266
methylindene, 117
methyl-2-methylpyrrolidone, 100

methylnaphthalene, 117
methylphenidate, 252
methylpropranolol, 156, 161
methyltetrahydrofolate, 284
methylthiopropionaldehyde, 264
methylthiouracil, 137, 264
metoprolol, 57, 59, 156, 158, 261
metyrapone, 252
mexiletine, 261
mg, 18
mianserin, 133
micro extraction flask, 146
micro flow-cells, 153
micro gas chromatograph, 334
micro infrared spectrometry, 168, 184
micro ir-cell, 232
micro method, 19
micro polarimeter, 258
micro specular reflectance, 226
micro standard flasks, 198
micro sublimation, 143
micro syringes, 168
micro technique for derivative formation, 139
micro zone melting, 181
micro-organisms, 248, 259
microbiological assay, 287
microbiological attack, 25
microbiological test, 333
microbore columns, 155
microcuvettes, 224
microelectrodes, 271
microfiltration, 168
microgram, 18
microlitre syringe, 197
micromanipulator, 166
micropollutants, 103
microprocessors, 154
microsampling, 225
microwave multi-gas analyser, 334
microwave plasma detector, 268
mid-range (Walsh) test 41
MID, 242
midazolam, 158
milling, 123
mincing, 123, 126
minimum working concentration, 45
mirex, 194, 257
misonidazol, 158
mitomycin, 159, 284
mixed-bed columns, 174
mixed-phase columns, 174
mobile laboratory furniture, 69
mobile miniature infrared analyser, 334
mobile sampling units, 93
modular system, 154
molecular emission cavity analysis, 268
molecular filtration, 125
molecular sieve, 95
molecular weights, 25
molecular-sieve columns, 93
molecule-specific methods, 221
monoiodocetate, 118
monosaccharides, 118, 182, 193
morphine, 59, 104, 156, 159, 175, 191, 227, 251, 261, 266, 276, 287, 319, 331
morphine-3-glucuronide, 276
morpholine-4-carboxaldehyde, 262
mould fungus, 176
moving belt, 247
  chains, 257
  needle system, 174
  wires, 257
MPD, 268
ms. 156, 157, 238ff, 257, 287, 318, 333
  sources, 238
ms-ms method, 243
multiple ATR, 226
multiple selected-ion monitoring, 238
multiple-ion detection, 242
multiplication, 326
multistep distribution processes, 149
mushrooms, 249
mutagenic fractions, 25
mutagenic properties, 317
mycotoxins, 22, 182, 241, 333

N

N-heterocyclics, 268, 273
N-nitrosamines, 23, 24
N-oxides, 273
N-sensitive detectors, 257
N,P-selective flame-ionization detector, 258
N,P,Cl-sensitive detectors, 257
nalidixic acid, 59
nalorphine, 156
naloxone, 261
naltrexol, 262
naltrexone, 261, 262
nanogram, 18
naphthacene, 237
naphthalene, 163
naphthalenes, 136, 248
1-naphthol, 162
naphthylamine, 165
naphthylmethylamine, 96
narcotic analgesic, 332
  drugs, 175, 332
nasal detector, 288
National Bureau of Standards, 53
National Institute of Occupational Safety and Health, 92
naturally occurring substances, 250
naturally occurring traces, 22
NBD-chloride, 191
nefopam, 266
negative chemical ionization, 238, 239
negative-ion chemical ionization mass spectrometry, 251
Neisseria gonorrhoeae, 259
neostigmine, 59, 159
neostriatum, 332
nephelometric detector, 258
nerve gases, 327
net signal, 42

neuroleptics, 159
newborns, 102
ng, 18
nicarbazin, 196
nicotinamide, 167
nicotine, 59, 105, 115, 116, 122, 135, 266, 284
nicotinic acid, 286
ninhydrin, 165
NIOSH, 92, 93
nitrated polycyclic hydrocarbons, 163
nitrazepam, 59, 286
nitriles, 278
nitriloacetic acid, 190, 276
nitrilotriacetic acid, 319
nitro compounds, 196, 273
nitro-aromatic explosives, 163
nitroaromatics, 254
nitrobenzene, 275
nitrobenzoyl chloride, 192, 324
nitrogen-selective detector, 258
nitroglycerine, 261
nitromethaqualone, 132
nitrophenol, 275
  isomers, 26, 190
nitropyridines, 275
nitrosamines, 100, 122, 142, 155, 163, 191, 196, 253, 263, 266, 268, 269, 271, 275
nitroso compounds, 238, 273, 275, 317
nitrosoatrazine, 123
nitrosocarbaryl, 123
nitrosodiethanolamine, 133, 163, 169, 191, 269, 275
nitrosodiethylamine, 318
nitrosodimethylamine, 59, 63, 94, 254, 269
nitrosodipropylamine, 59
nitroso-3-hydroxypyrrolidine, 253
nitrosopiperidine, 269
nitrosoproline, 132, 269
nitrosopyrrolidine, 269
nitrosothiazolidine, 269
nitrotoluene, 270
nitrourea antitumour agents, 252
nitrovin, 318
nmr, 120
noludar, 193
nomenclature, 33
nomifensine, 191, 266, 287
non-ionic detergents, 100, 284
non-normally distributed, 34
non-polar substances, 143
non-electrolytes, 170
noradrenalin, 166
norandrosterone, 252
norepinephrine, 57, 191, 192, 195, 275, 283
norfenfluramine, 192
norflurazon, 261
normal distribution, 35, 38
normal distribution curve, 37
normally distributed, 34
normetanephrine, 250, 275, 279, 283
norpethidine, 266
norpsuedoephedrine, 156, 159
nortriptyline, 157, 265
not detected, 44
nucleic acid components, 258
nucleic acids, 286
nucleosides, 20, 115
null hypothesis, 29, 40
number of organic compounds, 23
number of separation steps, 134
nutmeg constituents, 243

O

OAS, 236
occlusion, 126
Occupational Safety and Health Administration, 93
ochratoxin, 333
ochratoxin A, 124, 168, 182, 284
octachlorodibenzo-*p*-dioxins, 255
ocytocin, 288
odorants, 288
odorous substances, 27
odorous volatiles, 266
odour evaluation, 289
oestradiol, 250, 283, 285
oestriol, 167, 250, 283, 319
oestrogen residues, 123
oestrogens, 57, 190, 192, 319
oestrone, 250, 283, 286
  3-glucuronide, 283
off-line analysis, 156, 230
oil, 100
olefins, 329
oleic acid, 173
oligosaccharides, 165, 167, 241
on the fly method, 231
on-column derivative-formation techniques, 174, 187
on-line analysis, 156, 230
one-sided confidence limits, 38, 39
open capillaries, 171
open columns, 170
open split coupling, 246
open tubular columns, 156
opiates, 333
optoacoustic method, 238
optoacoustic spectroscopy, 236
organic acids, 100, 248, 338
organic chemistry, 20
organic matter in drinking water, 256
organic pollutants, 94
organic vapours, 94
organochlorine compound, 260
organochlorine pesticides, 132
organochlorine residues, 100
organohalogens, 247
organophosphate pesticides, 240
organophosphorus insecticides, 266
organophosphorus pesticides, 145
organophosphorus residues, 95
orphenadrine, 266

oryzalin, 191
OSHA, 65, 93, 233
Ostion SP-1, 101
OV 1, 173
OV 17, 173, 174, 256
OV 101, 173, 174
OV 210, 173
OV 225, 173
overinformation, 256
overloading of columns, 174
oxalate, 184
oxalacetate, 184
oxalic acid, 190
oxfendazole, 163
oxidation, 25, 116
oxidative phosphorylation, 280
oximes, 273
oxocarboxylic acids, 191
oxosteroids, 165, 167
oxyphenbutazone, 266
oyster, 124
ozonolysis, 330

P

P-sensitive detector, 265-267
packed column, 171, 172
paeonol, 57
PAHs, 19, 28, 62, 99, 100, 115, 116, 121, 126, 132, 145, 147, 152, 172, 229, 231, 233, 235, 237, 238, 244, 247, 248, 258, 276, 319, see also polycyclic aromatic hydrocarbons
  standard reference material, 53
pantothenic acid, 286
papain, 123
papavarine, 159, 161
paraban, 133
paracetamol, 159
paraffin oil, 232
paraoxon, 279
paraquat, 125, 176, 185, 275, 333
parathion, 264, 284
  methyl, 168, 264
part per billion, 18
part per million, 18
part per quadrilion, 18
part per trillion, 18
particulate matter from air, 248
partition chromatography, 149
pas, 236
pattern-recognition, 338
PCBs, 63, 100, 120, 137, 155, 174, 189, 194, 247, 261, 284, 286, 317
peak broadening, 151, 154, 171
pellet techniques, 227
pemoline, 125, 196, 252, 267
penbutolol, 159
penicillamine, 271
penicillin, 284
penicillinic acid, 135
penicillins, 319
pentachloronitrobenzene, 142
pentachlorophenol, 57, 96, 124, 132, 163, 196, 247, 261
pentachlorophenols, 276
pentafluorobenzaldehyde, 192
pentafluorobenzoyl chloride, 192
pentafluorobenzyl bromide, 192
pentafluorobenzylhydroxylamine, 192
pentafluoropropionic anhydride, 192
pepleomycin, 284
peptides, 241, 258
percentile, 39
perchlorethylene, 94
perchloric acid, 125
perchlorination, 189
pericyazine, 156
permeation tube, 55
permethrin enantiomers, 26
perphenazine, 272, 284
persedon, 193
personal monitoring, 93
pervaporation, 182
perylene, 237
pesticide, 152, 238
  analysis 144, 155, 170, 172
  residues, 27, 170, 253
  standard solutions, 53
  traces 233
Pesticides, 28, 63, 95, 103, 115, 117, 118, 138, 142, 145, 172, 174, 188, 239, 241, 257, 261, 266, 268, 274, 317, 324, 332
pethidine, 57, 175, 252, 253, 266
petroleum oils, 233
petroporphyrins, 257
Pflanzenschutzgesetz, 21
pg, 18
pH-value, 74
phallotoxins, 319
pharmaceutical analysis, 102, 251
pharmaceutical industry, 22
pharmaceutical preparation, 29
pharmaceutical products, 15, 20
pharmacologically active compounds, 27
phase boundary, 147
phase-separator, 148
phaseic acid, 162
phenacetin, 59, 262
phencyclidine, 284, 331
phenmetrazine, 175
phenobarbital, 53, 60, 125, 189, 194, 267
phenobarbitone, 287
phenol, 107, 142
phenolic acids, 195
  compounds, 235, 254, 257, 275
phenols, 62, 95, 96, 100, 119, 163, 164, 190, 192, 193, 195, 196, 235, 260, 261, 264, 271, 278, 322, 330
phenothiazine, 159
  sulphoxides, 275
  tranquillizer, 264
phenothiazines, 267
phenoxyacetic acid herbicides, 100
phenoxyalkanoic acid herbicides, 192
phenoxyalkyl acids, 254

phenycyclidine, 132, 287
phenyl isothiocyanate, 192
phenylalanine, 258
phenylbutazone, 59
phenylcarbamate residues, 168
phenyldiazomethane, 192
phenylethylamine 59
phenylhydrazine, 324
phenyl-*N*-isopropylacetamide, 276
phenyl-2,4-oxazolidinedione, 190
phenylphenol, 162, 192, 274, 276
phenylpropanolamine, 159
phenylpropylamine, 192
phenyl-5-*p*-tolylhydantoin, 156
phenylurea herbicides, 172, 196
phenytoin, 194, 252, 284, 287
phosmet, 276
phosphoethanolamine, 318
phospholipids, 169, 241
phosphorescence, 221, 235, 236, 237, 319
phosphorus compounds, 263
phosphorus tribromide, 192
photoacoustic cell, 238
  spectroscopy, 236
photoconductivity, 162
  detector, 258
photodechlorination, 194
photodecomposition, 116, 194, 271
photoionization detector, 270
photoisomerization, 194
photolysis, 189, 194, 253
photometry, 195, 279, 318, 333
photomultipliers, 242
phthalaldehyde, 161, 165, 188, 189, 192, 237
phthalate esters, 95, 120, 122, 153
phthalic acids, 193
phylloquinone, 167
physostigmine, 159, 318
picogram, 18
PID, 270
piezo quartz detector, 270
pilocarpine, 191
pinazepam, 59
pindolol, 261
pipothiazine, 159, 161
pirimiphos methyl, 254, 264
planetary atmospheres, 337
planets, 337
plant growth substances, 317
  hormones, 169
plants, 124
plasma electrophoresis, 185
plasmagram, 185
plasticizers, 122
plastics, 27
pocket computers, 50
podophyllotoxines, 159
poisons, 2, 27
polar adsorbent materials, 152
  drugs, 258
  group, 187
  substances, 143
polarity, 149, 152, 187
polarography, 272
politicians, 16
politics, 15
pollutant, 336
pollutants, 63, 172, 231
pollution, 24
  control, 25
poly(ethylene glycol), 97, 170
polyacrylamide, 155
polyamines, 59, 137, 167
polybrominated biphenyls 152, 194, 240, 261, 270
polychlorinated biphenyls, 28, 95, 103, 133, 164, 254, 256, see also polychlorobiphenyls
polychlorinated dibenzo-*p*-dioxins 122, 255
polychlorinated methoxybiphenyls, 256
polychlorinated naphthalenes, 247
polychlorobiphenyls, 142, see also polychlorinated biphenyls
polychlorodibenzo-*p*-dioxins, 122, 255
polychloronaphthalenes, 194
polycyclic aromatic hydrocarbons, 19, 95, 96, 98, 196, 228, 254, 262, 327, see also PAHs
polycyclic aromatics, 268
polyenoic fatty acids, 189
polyester foam, 100
polyethylene needles, 197
polyheterocyclic compounds, 26
polymer matrices, 176
  substance, 19
polynuclear aromatic hydrocarbons, 164
polyoxyethylene alkyl ether non-ionic surfactants, 195
polyoxyethylene surfactants, 232
polyoxymethylene, 55
polyurethane, 247
  foam, 93, 95, 100
Porapak, 94, 95, 99, 100
Porapak Q, 98, 101, 121
Porapak T, 98
Porous glass beads, 155
porphyrins, 319, 322
portable battery-operated sampler, 93
portable equipment, 93
portable gas chromatographs, 334
portable ms-system, 334
post-column derivative formation, 154, 165, 187, 277
post-column fluorometric system, 188
potassium bromide pellets, 225
potentiometric methods, 270
poultry, 253
power of a test, 40, 41
ppb, 18
ppm, 18
ppqud, 18
ppt, 18
prajmalium bitartrate, 267
prazepam, 156
prazosin, 159, 161
pre-column derivative formation, 187
precipitation, 126, 181, 195

precision, 33
 control, 82
precolumn partition, 330
precolumn reactions, 329, 330
precolumn techniques, 274
precolumns, 153, 173
preconcentration, 153
prednisolone, 159, 286
prednisone, 159
preservation, 102
prifinium quaternary ammonium ion, 135
primaquine, 261
primary amines, 189
primary aromatic amines, 96
primary standards, 98
primidone, 53, 194, 267
priority pollutants, 238, 317
probability range, 37
probenecid, 159, 195, 256, 261
procainamide, 287
procarbazine, 252
prochloraz, 133
procyclidine, 267
progesterone, 59, 105, 283, 286
promethazine, 159, 161
propadiene, 262
propafenone, 159
propane-1,2-diol dinitrate, 134
propanil, 132
propan-2-ol, 118
proportional systematic error, 32
propranolol, 156, 159, 160, 161, 267, 284
 enantiomers, 26
propyl ether, 327
propylnornicotine, 59
propylthiouracil, 284
propyne, 262
prostacyclin, 57, 105
prostaglandin, 57, 283
prostaglandin E, 287
prostaglandins, 57, 167, 190, 192, 250, 286, 288
protection of the environment, 21
protein binding, 167, 286, 319
protein-bound fractions, 182
proteinases, 123
proteins, 115
protocol, 66
pseudomonas aeruginaosa, 248
 cepacia, 248
 maltophilia, 248
pseudoephedrine enantiomers, 26
psychotropic drugs, 252
pumps, 156
purging, 247
purines, 330
purity of food, 21
putrescine, 169, 190, 260
pyrazon, 164, 267
pyren-1-yl maleimide, 165
pyrethroid insecticide enantiomers, 26
pyridine-*N*-oxides, 275
pyridines, 195
pyridoxal-5′-phosphate, 279
pyridoxine, 190, 286
pyrimethamine, 261
pyrimidines, 330
pyrolysis, 74, 176
pyrostigmin, 59
pyrrolidine, 142

Q

QF 1, 173
quadrupole mass spectrometry, 238, 242, 334
qualitative analysis, 45
quality assurance programme, 66–68
quality control, 22, 82
 charts, 83
 of reagents, 70
 system, 83
quantitative analysis, 45
quartz capillaries, 171
quaternary ammonium compounds, 132, 160
quenching of fluorescence, 236
quinidine, 115, 252, 287
quinine, 237, 252
quinolines, 237
quinoline-8-sulphonic acid chloride, 190
5-quinolinol, 156
quinones, 181, 273

R

radioactive material, 85
 tracers, 54, 124
radiochemical tracer method, 185
Radioimmunoassay, 221, 222, 280, see also ria
raining out, 176, 177
raman microprobe, 235
 scattering, 234
 spectroscopy, 234
random errors, 30, 31, 38
range W, 33
ranges of yields, 134
ranitidine, 160
raw data, 66, 77
raw data sheet, 78, 79
Rayleigh scattering, 234
reaction gas chromatography, 329
reactions on carriers, 332
reactor, 173
ready-to-use columns, 152
reagent specifications, 70
 waste disposal, 19
reagents, 70, 118, 122
receipt, 66
recognition of deviations, 82
record for trace analysis, 251
recording of data, 76
recoveries, 132, 251

reference gases, 54
library, 234
material, 52
materials, 54, 71
materials from precursors, 55
method, 51, 52
method value, 51
methods, 52
solids, 54
solutions, 53
substances, 67, 70
value, 52
regression analysis, 46–50
lines, 49
relationships, causal, 47
conjectural, 47
functional, 47
stochastic, 47
relative standard deviations, 134, 251
remote sampling, 334
testing of body fluids, 335
reproducibility, 61
research hypothesis, 29
resolution, 324
retinoic acid, 26, 57, 167, 286
retinol, 125
retrieval of data, 80
reversed phase, 100, 152
chromatography, 170
hplc, 331
column, 161
review, 238
review articles, 170
ria, 280, 287, 318, 319, 331, 333, see also radioimmunoassay
riboflavine, 167, 286
robenidine, 276
rodenticides, 163
roniltan, 261
rooms for trace analysis, 68
room-temperature phosphorescence, 237
rotary evaporation, 139
vessel, 140
rovral, 261
RP, 152
rubber stoppers, 120

S

S-containing compounds, 268
S,S,S-tributylphosphorotrithioate, 96
saccharides, 192
saccharine, 176, 276
Safe Drinking Water Act, 21
safety regulations, 84
salicyclic acid, 125
saliva, 104
salsolinol, 191
sample, 91
changing with time, 91
constant with time, 91
inhomogeneity, 62
loop, 153
preparation, 251, 318
splitting, 154
sample transfer, 168, 196
sampling 91ff, 123
and enrichment procedures, 92
conditions, 102
enrichment methods, 99, 100
of biological material, 105
of gases, 94
of water, 98ff
saponification, 233, 262
sapphire, 226
satellite reflector, 335
satratoxins, 164
scale expansion, 225
science administrators, 16
screening, 330–332, 338
screening analytical chemistry, 331
screening methods, 285, 333
screw-cap sample vial, 120
SE 30, 99, 170, 173, 256
SE 52, 171
second antibody method, 282
second derivative, 322
secondary amines, 264
secondary emission methods, 222
sedatives, 125
selected-ion mode, 242
selectivity, 324
semustine, 252
sensitivity, 23, 31, 33, 222, 241
of method, 65
sensory evaluation, 288
Sep-Pak, 104
Sep-Pak C_{18} cartridges, 105
separation, 123, 285, 286
factor, 134
principle, 134
schemes, 23
septa for injection in gc, 120
serotonin, 57, 166, 167, 195, 237, 250, 275, 284
Seveso, 23, 24
sex attractants, 288
Shpol'skii glass, 237
solvents, 237
signal to noise ratio, 242, 322
Silar 5 CP, 174
silica, 152, 155
gel, 95, 96, 103, 149
silicone rubber membranes, 183
silylating reagents, 192
silylation, 137
SIM, 242
similarity of organic compounds, 25
single-ion monitoring, 242
single-step dialysis, 182
skatole, 288
skewed distributions, 37
smooth muscle, 288
sniffle boxes, 289
soap chromatography, 152
sociology, 15
sodium fluoride, 118
sulphate, 141
solid electrodes, 270, 272

solid standards, 53, 54
solubility of amorphous silica, 155
solute-switched electron-capture detector, 324
solute switching, 325
solvent evaporation techniques, 136
    sublation, 183, 184, 232
solvents, 120
sorbic acid, 142
sotalol, 160
source of contamination, 119
soxhlet extraction, 121, 124
SP 2100, 174, 256
SP 2250, 174
SP 2300, 174
speciation, 182
specificity, 33, 61, 259
specimen, 66
spectral accumulation, 224
spectral averaging, 224
spectroscopic methods, 222
spectroscopy, 319
spermidin, 190
spermine, 190
Spheron MD, 100
Spheron MD 30/70, 101
Spheron SE, 101
spiking, 56
splitless technique, 174
spot-tests, 31, 333
squalene, 121
SRM, 53
    for antiepilepsy drugs, 53
stabilizers, 102, 116
standard deviation, 31, 35, 38, 43, 45, 132
standard gas mixtures with very low concentrations, 55
standard material, 52
standard operating procedure, 66, 68, 71
standard reference material for PCBs, 53
standard reference materials, 53, 118
standardized material, 61
stationary phases, 172, 173
statistical survey, 20
steam distillation, 251
stearate, 332
steroids, 57, 104, 105, 120, 165, 257, 258, 338
stilboestrol, 164, 284
stirring, 126
stochastic connection, 48
stomach contents, 249
storage, 115ff, 123
    areas, 68
    instability, 62
    of solvents, 71
straight chain alkanes, 329
strategy of problem solution, 72
    of the organic trace analyst, 28
stratosphere ozone layer, 93
stratospheric pollution, 335
stream splitting, 174, 327
streaming current detector, 258
street drugs, 23, 333
streptomycin, 319
stripping of volatile components, 102, 103
strychnine, 133
Student's *t*-distribution, 43
study director, 66, 67
    plan, 66, 71-76
styrene, 107, 164
sublimation, 198
sub-sampling procedure, 92
substoichiometric reaction, 280
subtilisin, 124
succinate, 184
succinyl choline, 252
suction device for separation of one phase, 146
sugars, 165
sulfinalol, 160
sulphamethazine, 133, 164, 236, 261
sulphapyridine, 160
sulphasalazine, 160
sulphides, 278
sulphinpyrazone, 59
sulphonamides, 160, 196
sulphonic acids, 152, 190, 195
sulphosalicyclic acid, 125
sulphoxides, 26
sulphur compounds, 263
sun, 337
surface analysis, 236
surface-active compounds, 184
surfactants, 152, 169
survey of the literature, 16
sweat, 119
sweep co-distillation, 142
symbols, 74
    for flow sheets, 73, 74, 75
sympathomimetic amines, 332
Synachrom, 101
synthetic samples, 56
systematic errors, 61

T

2,4,5-T, 22, 99, 115
T_2-toxin, 168, 255, 284, 286
talinol, 132
tandem methods, 153, 327
tandem ms, 243
tap water, 116
tartaric acid, 96
taurine, 125, 165, 167
TCDD, 22, 24, 194, 332
TDE, 179
TEA, 269
TEA-detector, 268
tears, 104
tebuthiuron, 264
technical products, 22
teeth, 104
Teflon separator, 147
telemetry, 334
temazepam, 260
tenax, 99, 101, 232, 249

Tenax G, 98
Tenax GC, 93, 98, 121
  cartridges, 121
tensides, 276
terpenes, 103
terphenyls, 247
test, 29
test facility, 66, 68
test-facility management, 67
test for normal distribution, 36
test substance, 64, 67
test system, 66
testosterone, 59, 146, 250, 319
tetrabenazine, 160
tetrachlorobenzo-*p*-dioxin, 255
tetrachlorodibenzo-*p*-dioxin isomers, 26
tetrachlorodibenzodioxins, 24
tetrachlorodibenzofuran, 284
  isomers, 26
tetrachloroethane, 94
tetrachloroethylene, 92, 119, 120, 132, 142
tetrachlorvinphos, 261
tetracyclines, 164
tetraethylammonium hydroxide, 194
tetrahydrocannabinol, 57, 168, 251, 287, 331
  acid, 133
tetrahydrocannabinols, 193
tetramethylammonium hydroxide, 194
tetramethylthiuramdisulphide, 276
theophylline, 59, 125, 267, 274, 275, 284
therapeutically used drugs, 317
thermal analyser, 269
thermal conductivity detector, 259
thermal energy analysis, 268
thermal energy detector, 253
thermionic detection, 257
thermionic detector, 222, 263
thermionic N-sensitive detector, 265–267
thiamine, 165, 167, 286
thin-layer chromatography, 165, 168, 232, 251, see also tlc
  densitometry, 221
thiochrome, 165
thioguanine, 160
thiols, 26, 270
thiopeptine, 319
thiophene, 100
thioridazine, 160, 161
thiourea, 276
thozalinone, 196, 252
thromboxane, 250, 283, 286, 319
thymine, 143
thymol, 118
thyroid hormones, 167, 172, 191
thyroxine, 167, 191, 282, 285, 318
TID, 263
time averaging, 227, 231, 323
time of flight spectrometer, 242
time-weighted average, 99
timolol, 57
tlc, 165, 168, 236, 249, 256, 287, 331, 332
tobacco smoke, 122, 248
tobramycin, 284, 285, 287
tocainide, 262
tocopherols, 167
tofisopam, 160
toilet soaps, 119
tolbutamide, 191
tolerance interval, 38
  limits, 39
tolualdehyde, 190
toluene, 118, 119, 141, 270
  diamines, 164
tolylethylamine, 59
tosyl chloride, 193
toxaphene, 261
toxic gases, 232
  organic compounds, 94
toxic substances, 15
Toxic Substances Control Act, 21, 65
toxic trace gases, 94
toxicity classes, 85
toxicological activity, 25
trace, 18
  with unknown chemical identity, 25
tranexamic acid, 191
tranquillizer, 262
transfer pipettes, 197
tranylcypromine, 267
trapping method, 93
trenbolone, 169, 283
trends in ms, 238
tri-allate, 95, 142
tri-iodothyronine, 63, 191, 282, 319
triamterene, 160, 275, 318
triazine herbicides, 164, 168, 267
triazolam, 115, 261, 286
trichloroacetic acid, 125, 190, 270, 318
trichloroethane, 85, 106
trichloroethanol, 106, 318
trichloroethyl chloroformate, 193
trichloroethylene, 92, 106, 119
trichlorophenoxyacetic acid, 115, 190, 241
trichlorphon, 264
tricresyl phosphate, 225
tricyclic antidepressants, 105, 115, 160, 252, 267
tricyclic drugs, 136
trifluoperazine, 252, 253, 267
trifluoroacetic anhydride, 137, 193
trifluoroethanol, 193
trifluperazine, 284
triforine, 196
triglycerides, 333
trihalomethanes, 103, 117, 270
trimethoprim, 59, 160
trimethylamine, 94, 96
trimethylanilinium hydroxide, 194
trimethylbenzenes, 119
trimethylphenylammonium acetate, 194
trimethylprostaglandin, 250
trimetoquinol, 252
trinitrofluoren-9-one, 164
trinitrotoluene, 55, 94
tropospheric pollution, 335

true content, 38
mean value, 31
value, 45, 51, 52
trychotecene mycotoxins, 168
tryptamine, 57
tryptophan, 121, 125, 176, 275, 319
adenosine, 258
tuberculostearic acid, 250
tubocararine, 160
tubular reactor, 188, 189
tunable lasers, 224, 235
tungstate-ammonium sulphate, 125
turimycin, 190
twin-electrode thin-layer detector cell, 327
two-sided confidence limits, 38, 39
type I error, 40
type of distribution, 41
tyramine, 166, 250
tyrosine, 323, 324

U

U-2 aircraft, 334
UDP-glucuronic acid, 319
ultramicro method, 19
ultrasonic vibration, 126
ultraviolet absorption, 161
ultraviolet detection, 156
ultraviolet light, 223
ultraviolet spectrophotometry, 221
ultraviolet spectrum, 74
Umwelt-Chemikaliengesetz, 21
unknown type of distribution, 40
urban air, 119
urea, 117, 182, 318
herbicides, 263
uremia, 338
uric acid, 117, 164, 275, 319
urine, 104

V

valproic acid, 185, 262
vancomycin, 284
vanillin, 288
vanillylmandelic acid, 125, 250, 275
vapour-phase ir spectra, 234
vapour pressure, 97
vegetable oils, 233
venous blood, 102
ventilation system, 69
verapamil, 160
vinyl chloride, 54, 55, 93, 94, 103, 106, 107, 233, 247, 248, 263, 268, 272, 318, 327
visible light, 223
vitamin A, 284
vitamin B, 284, 288, 319
vitamin D, 63, 132, 167, 286
vitamin E, 164
vitamins, 27
volatiles, 107
from human blood, 119
voltammetric detectors, 258
voltammetry, 272
vomitoxin, 164
vortexing, 126

W

wall-paint, 122
Walsh-test, 41
wash-bottles, 120
waste water, 234
water, 20, 92, 98, 116, 238, 247, 317
weighing, 74
WHO – standards in PAHS in water, 21
Wilcoxon-test, 41
wipe tests, 332
working areas, 93, 222
interval, 30
range of concentration, 222
workplaces, 93, 317
work-up procedure, 75

X

XAD resin, 93, 99, 334
XAD-1, 101
XAD-2, 101
XAD-4, 101
XAD-7, 101
XAD-8, 101
XE 60, 173
xylazine, 267
xylene, 270
xylenes, 26, 119
xyloidine, 255
xylose, 166

Y

yield, dependences, 134, 135
yields, 134

Z

zearalanone, 59
zearalenone, 59, 164, 168
zeatin, 164
Zeichen-test, 41
zone melting, 180